Beyond First Order Model Theory

Beyond First Order Model Theory

Edited by

José Iovino

CRC Press
Taylor & Francis Group
Boca Raton London New York

CRC Press is an imprint of the
Taylor & Francis Group, an **informa** business
A CHAPMAN & HALL BOOK

CRC Press
Taylor & Francis Group
6000 Broken Sound Parkway NW, Suite 300
Boca Raton, FL 33487-2742

First issued in paperback 2020

© 2017 by Taylor & Francis Group, LLC
CRC Press is an imprint of Taylor & Francis Group, an Informa business

No claim to original U.S. Government works

ISBN-13: 978-1-4987-5397-5 (hbk)
ISBN-13: 978-0-367-65780-2 (pbk)

Visit the Taylor & Francis Web site at
http://www.taylorandfrancis.com

and the CRC Press Web site at
http://www.crcpress.com

*To Martha and our children,
Abigail and Luca.*

Contents

Foreword

It is natural for me and for this book to consider what the right frameworks are for developing model theory. Being asked to write an introductory note, it seems reasonable to decide to do what nobody else can do better: try to somewhat describe the evolution of my thoughts on the subject.

In the fall of 1968, having started my Ph.D. studies at the Hebrew University, I started to read systematically what was available on model theory. Why model theory? I liked mathematical logic not for an interest in philosophical questions, but for its generality, which appeared to me as the epitome of mathematics; and so model theory seemed right; like algebra, but replacing fields, rings, etc. by the class of models of a first order theory. There was also some randomness — my M.Sc. thesis was on infinitary logics.

The central place of first order theory, equivalently, elementary classes, is not in doubt. The reasons are inherent and also rooted in history — in answering how to axiomatize mathematical theories, in particular set theory, and later Tarski and Malcev starting model theory. Though in the late sixties some thought that first order model theory was done, the news of its demise was premature, as it continued to be the central case in pure model theory, as well as in many applications.

But from quite early on, there were other frames. Mostowski had suggested various infinitary logics and generalized quantifiers.

A syntax-free way to treat model theory is to deal with Abstract Elementary Classes (aec). Why were they introduced ([Sh:88])? Looking at sentences in the logics $\mathbb{L}_{\aleph_1,\aleph_0}$ or $\mathbb{L}(Q)$ or their combinations, with exactly one or just few models in $\aleph_1$ (up to isomorphism, of course, [Sh:48]), there was no real reason to choose exactly those logics. We can expand the logic by various quantifiers stronger than Q (there are unaccountably many), and still have similar results. The answer was to try to axiomatize "an elementary class" using only the most basic properties. This, of course, does not cover some interesting cases (like the $\aleph_1$-saturated models of a first order theory, or classes of complete metric spaces), but still, it covers lots of ground on one hand, and has an interesting model theory on the other.

Later, attending a conference in California in 1986 led me to start to develop a more restricted frame — universal classes (encouraged by [Sh:155]).

Editor's note: The complete list of Shelah's publications can be found online at http://shelah.logic.at.

This Foreword is listed as E:84 in Shelah's list.

This led to introducing orbital types. The point was that an important step in the sixties was moving to saturation, i.e., using elementary mappings adding one element at a time, rather than using elementary submodels; this led to moving from "model homogeneous" to "sequence homogeneous". This move to types of singletons was essential for developing the theory of superstable classes. A major point of [Sh:300] (mainly for aec with amalgamation) was having the best of all possible worlds — dealing with types of elements, but saturation is expressed by embedding of models (where embedding is of the relevant kind, as axiomatized in aec). This is justified, as it is proved that the two definitions of saturated are equivalent (for aec with amalgamation).

Now aec seem close to classes defined as "the class of models of a sentence from $\mathbb{L}_{\lambda^+,\aleph_0}$" when the class has LST-number $\leq \lambda$. However they have closure properties which are better for some aspects; my belief in this direction being worthwhile is witnessed by having written a book. Why not logics like $\mathbb{L}_{\aleph_1,\aleph_1}$ and stronger? For a long time, everybody, including me, took for granted that for such logics we can generalize "elementary model theory" but not classification theory, though we can for more special cases (as mentioned above). Relatively recently, reexamining this has changed my view to some extent ([Sh:1019]).

However, all this does not imply for me that using and looking for logics has passed its day. Moreover, maybe, maybe there is a logic hidden from our view, by which we can get aec's or even a better family of classes. This is given some support by the following (see [Sh:797]).

Assuming for notational simplicity that λ is an inaccessible cardinal, we know that $\mathbb{L}_{\lambda,\aleph_0}$ fails interpolation, and also $\mathbb{L}_{\lambda,\lambda}$ fails it; but the pair satisfies it, so it was asked: is there a "nice" logic which lies between those two logics? Now $\mathbb{L}^1_\lambda$ from [Sh:797] seems a reasonable solution: it lies between them (i.e., $\mathbb{L}_{\lambda,\aleph_0} \leq \mathbb{L}^1_\lambda \leq \mathbb{L}_{\lambda,\lambda}$), has interpolation, has a characterization (as Lindström's characterization of first order logic), well ordering is not definable in it, and it has addition theorems; see more in [Sh:1101]. But this logic, $\mathbb{L}^1_\lambda$, strongly fails the upward Löwenheim-Skolem theorem (quite naturally being maximal under conditions as above). Can we find a similar logic for which this holds, for example, that has EM models? The search for such a logic has failed so far. Still, every aec can be characterized by a suitable sentence in $\mathbb{L}_{\lambda^+,\kappa^+}$ where κ is the (downward) LST number of the class and λ not much larger than κ plus the cardinality of the vocabulary.

Another exciting direction is dealing with complete metric spaces. Though we have learned much, there is much to be desired. For me, the glaring omission is that we do not have a theory generalizing the one for superstable theories.

In conclusion, our horizons have widened, but much remains a mystery. Surely, the interplay of syntax and semantics will continue to puzzle us on one hand and provide us with illuminations on the other.

Saharon Shelah
Jerusalem, December 14, 2016

Preface

Most of mathematics takes place in higher order logics, not just first order. However, the model theory of first order logic is one of the most powerful tools of contemporary logic. In the first order context, our understanding of the manner in which models of mathematical theories can be classified in terms of abstract properties has improved profoundly in the last few decades; this has been the fruit of the groundbreaking work of giants like Chang, Keisler, Morley, Scott, and Vaught, the dissemination of Shelah's formidable output, and beautiful connections with other areas of mathematics found by Hrushovski, Krivine, Zilber and others.

The development of a full model theory for second order logic has been more elusive. Nevertheless, in recent years, along with the successes in the first order realm, model theory has seen a vibrant renaissance of investigations for frameworks that lie beyond first order. This volume aims to present some of the most active areas of research. The book is composed of articles that are mutually independent. Each chapter is intended to serve both as introduction to a current research direction and as presentation of results that are not available elsewhere.

The material is divided into three parts. Part I is on model theory of strong logics (that is, extensions of first-order by infinitary connectives or special quantifiers), Part II, on model theory of special classes of structures (structures of algebra, geometry, analysis, and probability), and Part III, on abstract elementary classes. Such a compartmentalization is necessarily imperfect, since several of the articles fit naturally into more than one of these categories. For instance, Zilber's article on analytic Zariski structures appears in Part II, but its model-theoretic setting is that of abstract elementary classes. Likewise, Shelah's article on 0-1 laws in strong logics appears in Part I, but it would fit in Part II as well.

The volume evolved from a series of lectures given at The University of Texas at San Antonio in January 2015. The editor is thankful to the contributors for their enthusiasm and to Taylor & Francis/CRC Press for its support. A special debt of gratitude is owed to Saf Khan of CRC Press, and to the anonymous referees, without whose selfless work this book would not have been possible.

José Iovino

San Antonio, Winter Solstice, 2016

Contributors

John T. Baldwin
University of Illinois at Chicago
Chicago, Illinois, USA

Will Boney
Harvard University
Cambridge, Massachusetts, USA

Xavier Caicedo
Universidad de los Andes
Bogotá, Colombia

Eduardo Dueñez
The University of Texas at San
    Antonio
San Antonio, Texas, USA

Christopher J. Eagle
University of Victoria
Victoria, British Columbia,
    Canada

José Iovino
The University of Texas at San
    Antonio
San Antonio, Texas, USA

H. Jerome Keisler
University of Wisconsin-Madison
Madison, Wisconsin, USA

Paul B. Larson
Miami University
Oxford, Ohio, USA

Saharon Shelah
The Hebrew University of
    Jerusalem
Jerusalem, Israel

Rutgers, The State University of
    New Jersey
Piscataway, New Jersey, USA

Sebastien Vasey
Carnegie Mellon University
Pittsburgh, Pennsylvania, USA

Boris Zilber
University of Oxford
Oxford, United Kingdom

Part I

Model Theory of Strong Logics

Chapter 1

Expressive power of infinitary [0, 1]-logics

Christopher J. Eagle

University of Victoria
Victoria, British Columbia, Canada

1   Introduction

In the last several years there has been considerable interest in the *continuous first-order logic for metric structures* introduced by Ben Yaacov and Usvyatsov in the mid-2000s and published in [BYU10]. This logic is suitable for studying structures based on metric spaces, including a wide variety of structures encountered in analysis. Continuous first-order logic is a generalization of first-order logic and shares many of its desirable model-theoretic properties, including the compactness theorem. While earlier logics for considering metric structures, such as Henson's logic of positive bounded formulas (see [HI03]), were equivalent to continuous first-order logic, the latter has emerged as the current standard first-order logic for developing the model theory of metric structures. The reader interested in a detailed history of the interactions between model theory and analysis can consult [Iov14].

In classical discrete logic there are many examples of logics that extend first-order logic, yet are still tame enough to allow a useful model theory to be developed; many of the articles in [BF85] describe such logics. The most fruitful extension of first-order logic is the infinitary logic $\mathcal{L}_{\omega_1,\omega}$, which extends the formula creation rules from first-order to also allow countable conjunctions and disjunctions of formulas, subject only to the restriction that the total number of free variables remains finite. While the compactness theorem fails for $\mathcal{L}_{\omega_1,\omega}$, it is nevertheless true that many results from first-order model theory can be translated in some form to $\mathcal{L}_{\omega_1,\omega}$ - see [Kei71] for a thorough development of the model theory of $\mathcal{L}_{\omega_1,\omega}$ for discrete structures.

Many properties considered in analysis have an infinitary character. It is therefore natural to look for a logic that extends continuous first-order logic by allowing infinitary operations. In order to be useful, such a logic should still have desirable model-theoretic properties analogous to those of the discrete infinitary logic $\mathcal{L}_{\omega_1,\omega}$. There have recently been proposals for such a logic by Ben Yaacov and Iovino [BYI09], Sequeira [Seq13], and the author [Eag14]; we call these logics $\mathcal{L}_{\omega_1,\omega}^C$, $\mathcal{L}_{\omega_1,\omega}^C(\rho)$, and $\mathcal{L}_{\omega_1,\omega}$, respectively. The superscript C is intended to emphasize the continuity of the first two of these logics, in a sense to be described below. The goal of Section 2 is to give an overview of some of the model-theoretic properties of each of these logics, particularly with respect to their expressive powers. Both $\mathcal{L}_{\omega_1,\omega}$ and $\mathcal{L}_{\omega_1,\omega}^C$ extend continuous first-order logic by allowing as formulas some expressions of the form $\sup_n \phi_n$, where the ϕ_n's are formulas. The main difference between $\mathcal{L}_{\omega_1,\omega}^C$ and $\mathcal{L}_{\omega_1,\omega}$ is that the former requires infinitary formulas to define *uniformly continuous* functions on all structures, while the latter does not impose any continuity requirements. Allowing discontinuous formulas provides a significant increase in expressive power, including the ability to express classical negation (Proposition 2.8), at the cost of a theory which is far less well-behaved with respect to metric completions (Example 2.7). The logic $\mathcal{L}_{\omega_1,\omega}^C(\rho)$ is obtained by adding an additional operator ρ to $\mathcal{L}^C_{\omega_1,\omega}$, where $\rho(x,\phi)$ is interpreted as the distance from x to the zeroset of ϕ. We show in Theorem 2.6 that ρ can be defined in $\mathcal{L}_{\omega_1,\omega}$.

One of the most notable features of the discrete logic $\mathcal{L}_{\omega_1,\omega}$ (in a countable signature) is that for each countable structure M there is a sentence σ of $\mathcal{L}_{\omega_1,\omega}$ such that a countable structure N satisfies σ if and only if N is isomorphic to M. Such sentences are known as *Scott sentences*, having first appeared in a paper of Scott [Sco65]. In Section 3 we discuss some consequences of the existence of Scott sentences for complete separable metric structures. The existence of Scott sentences for complete separable metric structures was proved by Sequeira [Seq13] in $\mathcal{L}^C_{\omega_1,\omega}(\rho)$ and Ben Yaacov, Nies, and Tsankov [BYNT14] in $\mathcal{L}^C_{\omega_1,\omega}$. Despite having shown in Section 2 that the three logics we are considering have different expressive powers, we use Scott sentences to prove the following in Proposition 3.4:

Theorem. *Let M and N be separable complete metric structures in the same countable signature. The following are equivalent:*

- $M \cong N$,

- $M \equiv N$ *in* $\mathcal{L}^C_{\omega_1,\omega}$,

- $M \equiv N$ *in* $\mathcal{L}^C_{\omega_1,\omega}(\rho)$,

- $M \equiv N$ *in* $\mathcal{L}_{\omega_1,\omega}$.

Scott's first use of his isomorphism theorem was to prove a definability result, namely that a predicate on a countable discrete structure is automorphism invariant if and only if it is definable by an $\mathcal{L}_{\omega_1,\omega}$ formula. The main new result of this note is a metric version of Scott's definability theorem (Theorem 4.1):

Theorem. *Let M be a separable complete metric structure, and $P : M^n \to [0,1]$ be a continuous function. The following are equivalent:*

- P *is invariant under all automorphisms of M,*

- *there is an $\mathcal{L}_{\omega_1,\omega}$ formula $\phi(\vec{x})$ such that for all $\vec{a} \in M^n$, $P(\vec{a}) = \phi^M(\vec{a})$.*

The proof of the above theorem relies heavily on replacing the constant symbols in an $\mathcal{L}_{\omega_1,\omega}$ sentence by variables to form an $\mathcal{L}_{\omega_1,\omega}$ formula; Example 3.5 shows that this technique cannot be used in $\mathcal{L}^C_{\omega_1,\omega}$ or $\mathcal{L}^C_{\omega_1,\omega}(\rho)$, so our method does not produce a version of Scott's definability theorem in $\mathcal{L}^C_{\omega_1,\omega}$ or $\mathcal{L}^C_{\omega_1,\omega}(\rho)$.

Acknowledgments

Some of the material in this chapter appears in the author's Ph. D. thesis [Eag15], written at the University of Toronto under the supervision of José

Iovino and Frank Tall. We thank both supervisors for their suggestions and insights, both on the work specifically represented here, and on infinitary logic for metric structures more generally. We also benefited from discussions with Ilijas Farah and Bradd Hart, which led to Theorem 2.6 and Example 4.5. The final version of this chapter was completed during the Focused Research Group "Topological Methods in Model Theory" at the Banff International Research Station. We thank BIRS for providing an excellent atmosphere for research and collaboration, and we also thank Xavier Caicedo, Eduardo Duéñez, José Iovino, and Frank Tall for their comments during the Focused Research Group.

2   Infinitary logics for metric structures

Our goal is to study infinitary extensions of first-order continuous logic for metric structures. To begin, we briefly recall the definition of metric structures and the syntax of first-order continuous logic. The reader interested in an extensive treatment of continuous logic can consult the survey [BYBHU08].

Definition 2.1. A *metric structure* is a metric space (M, d^M) of diameter at most 1, together with:

- A set $(f_i^M)_{i \in I}$ of uniformly continuous *functions* $f_i : M^{n_i} \to M$,

- A set $(P_j^M)_{j \in J}$ of uniformly continuous *predicates* $P_j : M^{m_j} \to [0, 1]$,

- A set $(c_k^M)_{k \in K}$ of distinguished *elements* of M.

We place no restrictions on the sets I, J, K, and frequently abuse notation by using the same symbol for a metric structure and its underlying metric space.

Metric structures are the semantic objects we will be studying. On the syntactic side, we have *metric signatures*. By a *modulus of continuity* for a uniformly continuous function $f : M^n \to M$ we mean a function $\delta : \mathbb{Q} \cap (0, 1) \to \mathbb{Q} \cap (0, 1)$ such that for all $a_1, \ldots, a_n, b_1, \ldots, b_n \in M$ and all $\epsilon \in \mathbb{Q} \cap (0, 1)$,

$$\sup_{1 \leq i \leq n} d(a_i, b_i) < \delta(\epsilon) \implies d(f(a_i), f(b_i)) \leq \epsilon.$$

Similarly, δ is a modulus of continuity for $P : M^n \to [0, 1]$ means that for all $a_1, \ldots, a_n, b_1, \ldots, b_n \in M$,

$$\sup_{1 \leq i \leq n} d(a_i, b_i) < \delta(\epsilon) \implies |P(a_i) - P(b_i)| \leq \epsilon.$$

Definition 2.2. A *metric signature* consists of the following information:

- A set $(f_i)_{i \in I}$ of *function symbols*, each with an associated arity and modulus of uniform continuity,

- A set $(P_j)_{j \in J}$ of *predicate symbols*, each with an associated arity and modulus of uniform continuity,

- A set $(c_k)_{k \in K}$ of *constant symbols*.

When no ambiguity can arise, we say "signature" instead of "metric signature".

When S is a metric signature and M is a metric structure, we say that M is an *S-structure* if the distinguished functions, predicates, and constants of M match the requirements imposed by S. Given a signature S, the *terms* of S are defined recursively, exactly as in the discrete case.

Definition 2.3. Let S be a metric signature. The *S-formulas* of continuous first-order logic are defined recursively as follows.

1. If t_1 and t_2 are terms, then $d(t_1, t_2)$ is a formula.

2. If $t_1, \ldots, t_n$ are S-terms, and P is an n-ary predicate symbol, then $P(t_1, \ldots, t_n)$ is a formula.

3. If $\phi_1, \ldots, \phi_n$ are formulas, and $f : [0,1]^n \to [0,1]$ is continuous, then $f(\phi_1, \ldots, \phi_n)$ is a formula. We think of each such f as a *connective*.

4. If ϕ is a formula and x is a variable, then $\inf_x \phi$ and $\sup_x \phi$ are formulas. We think of $\sup_x$ and $\inf_x$ as *quantifiers*.

Given a metric structure M, a formula $\phi(\vec{x})$ of the appropriate signature, and a tuple $\vec{a} \in M$, we define the value of ϕ in M at $\vec{a}$, denoted $\phi^M(\vec{a})$, in the obvious recursive manner. We write $M \models \phi(\vec{a})$ to mean $\phi^M(\vec{a}) = 0$. The basic notions of model theory are then defined in the expected way by analogy to discrete first-order logic.

The only difference between our definitions and those of [BYBHU08] is that in [BYBHU08] it is assumed that the underlying metric space of each structure is complete. We do not want to make the restriction to complete metric spaces in general, so our definition of structures allows arbitrary metric spaces, and we speak of *complete metric structures* when we want to insist on completeness of the underlying metric. In first-order continuous logic there is little lost by considering only complete metric structures, since every structure is an elementary substructure of its metric completion. This is also true in $\mathcal{L}^C_{\omega_1, \omega}$ and $\mathcal{L}^C_{\omega_1, \omega}(\rho)$, but not in $\mathcal{L}_{\omega_1, \omega}$, as Example 2.7 below illustrates.

In continuous logic the connectives max and min play the roles of $\wedge$ and $\vee$, respectively, in the sense that for a metric structure M, formulas $\phi(\vec{x})$ and $\psi(\vec{x})$, and a tuple $\vec{a}$, we have $M \models \max\{\phi(\vec{a}), \psi(\vec{a})\}$ if and only if $M \models \phi(\vec{a})$ and $M \models \psi(\vec{a})$, and similarly for min and disjunction. Consequently, the most direct adaptation of $\mathcal{L}_{\omega_1, \omega}$ to metric structures is to allow the formation of formulas $\sup_n \phi_n$ and $\inf_n \phi_n$, at least provided that the total number of free variables remains finite (the restriction on the number of free variables is usually assumed even in the discrete case). However, one of the important features of continuous logic is that it is a *continuous* logic, in the sense that each formula $\phi(x_1, \ldots, x_n)$ defines a continuous function $\phi^M : M^n \to [0,1]$ on each structure M. The pointwise supremum or infimum of a sequence of continuous functions is not generally continuous.

A second issue arises from the fact that one expects the metric version of $\mathcal{L}_{\omega_1,\omega}$ to have the same relationship to separable metric structures as $\mathcal{L}_{\omega_1,\omega}$ has to countable discrete structures. Separable metric structures are generally not countable, so some care is needed in arguments whose discrete version involves taking a conjunction indexed by elements of a fixed structure. For instance, one standard proof of Scott's isomorphism theorem is of this kind (see [Kei71, Theorem 1]). Closely related to the question of whether or not indexing over a dense subset is sufficient is the issue of whether the zeroset of a formula is definable.

With the above issues in mind, we present some of the infinitary logics for metric structures that have appeared in the literature. The first and third of the following logics were both called "$\mathcal{L}_{\omega_1,\omega}$" in the papers where they were introduced, and the second was called "$\mathcal{L}_{\omega_1,\omega}(\rho)$"; we add a superscript "C" to the first and second logics to emphasize that they are continuous logics.

Definition 2.4. The three infinitary logics for metric structures we will be considering are:

- $\mathcal{L}^C_{\omega_1,\omega}$ (Ben Yaacov-Iovino [BYI09]): Allow formulas $\sup_{n<\omega} \phi_n$ and $\inf_{n<\omega} \phi_n$ as long as the total number of free variables remains finite, and the formulas ϕ_n satisfy a common modulus of uniform continuity.

- $\mathcal{L}^C_{\omega_1,\omega}(\rho)$ (Sequeira [Seq13]): Extend $\mathcal{L}^C_{\omega_1,\omega}$ by adding an operator $\rho(x, \phi)$, interpreted as the distance from x to the zeroset of ϕ.

- $\mathcal{L}_{\omega_1,\omega}$ (Eagle [Eag14]): Allow formulas $\sup_{n<\omega} \phi_n$ and $\inf_{n<\omega} \phi_n$ as long as the total number of free variables remains finite, without regard to continuity.

The logic $\mathcal{L}_{\omega_1,\omega}$ was further developed by Grinstead [Gri14], who in particular provided an axiomatization and proof system.

Other infinitary logics for metric structures which are not extensions of continuous first-order logic have also been studied. In a sequence of papers beginning with his thesis [Ort97], Ortiz develops a logic based on Henson's positive bounded formulas and allows infinitary formulas, but also infinite strings of quantifiers. An early version of [CL16] had infinitary formulas in a logic where the quantifiers sup and inf were replaced by category quantifiers.

Remark 2.5. We will often write formulas in any of the above logics in forms intended to make their meaning more transparent, but sometimes this can make it less obvious that the expressions we use are indeed valid formulas.

For example, in the proof of Theorem 2.6 below, we will be given an $\mathcal{L}_{\omega_1,\omega}$ formula $\phi(\vec{x})$, and we will define

$$\rho_\phi(\vec{x}) = \inf_{\vec{y}} \min\left\{\left(d(\vec{x}, \vec{y}) + \sup_{n \in \mathbb{N}} \min\{n\phi(\vec{y}), 1\}\right), 1\right\}.$$

The preceding definition can be seen to be a valid formula of $\mathcal{L}_{\omega_1,\omega}$ as follows. For each $n \in \mathbb{N}$ define $u_n : [0,1] \to [0,1]$ by $u(z) = \min\{nz, 1\}$. Define $v : [0,1]^2 \to [0,1]$ by $v(z, w) = \min\{z + w, 1\}$. Then each u_n is continuous, as is v, and we have

$$\rho_\phi(\vec{x}) = \inf_{\vec{y}} v\left(d(\vec{x}, \vec{y}), \sup_{n \in \mathbb{N}} u_n(\phi(\vec{x}))\right).$$

A similar process may be used throughout the remainder of the chapter to see that expressions we claim are formulas can indeed be expressed in the form of Definitions 2.3 and 2.4.

The remainder of this section explores some of the relationships between $\mathcal{L}^C_{\omega_1,\omega}$, $\mathcal{L}^C_{\omega_1,\omega}(\rho)$, and $\mathcal{L}_{\omega_1,\omega}$. It is clear that each $\mathcal{L}^C_{\omega_1,\omega}$ formula is both an $\mathcal{L}_{\omega_1,\omega}$ formula and an $\mathcal{L}^C_{\omega_1,\omega}(\rho)$ formula. The next result shows that the ρ operation is implemented by a formula of $\mathcal{L}_{\omega_1,\omega}$, so each $\mathcal{L}^C_{\omega_1,\omega}(\rho)$ formula is also equivalent to an $\mathcal{L}_{\omega_1,\omega}$ formula.

Theorem 2.6. *For every $\mathcal{L}_{\omega_1,\omega}$ formula $\phi(\vec{x})$ there is an $\mathcal{L}_{\omega_1,\omega}$ formula $\rho_\phi(\vec{x})$ such that for every metric structure M and every $\vec{a} \in M$,*

$$\rho_\phi^M(\vec{a}) = \inf\{d(\vec{a}, \vec{b}) : \phi^M(\vec{b}) = 0\}.$$

Proof. Let ϕ be an $\mathcal{L}_{\omega_1,\omega}$ formula, and define

$$\rho_\phi(\vec{x}) = \inf_{\vec{y}} \min\left\{\left(d(\vec{x}, \vec{y}) + \sup_{n \in \mathbb{N}} \min\{n\phi(\vec{y}), 1\}\right), 1\right\}.$$

(See Remark 2.5 above for how to express this as an official $\mathcal{L}_{\omega_1,\omega}$ formula). Now consider any metric structure M, and any $\vec{y} \in M$. We have

$$\sup_{n \in \mathbb{N}} \min\{n\phi^M(\vec{y}), 1\} = \begin{cases} 0 & \text{if } \phi^M(\vec{y}) = 0, \\ 1 & \text{otherwise.} \end{cases}$$

Therefore for any $\vec{a}, \vec{y} \in M$,

$$\min\left\{\left(d(\vec{a}, \vec{y}) + \sup_{n \in \mathbb{N}} \min\{n\phi(\vec{y}), 1\}\right), 1\right\} = \begin{cases} d(\vec{a}, \vec{y}) & \text{if } \phi^M(\vec{y}) = 0, \\ 1 & \text{otherwise.} \end{cases}$$

Since all values are in $[0,1]$, it follows that:

$$\rho_\phi^M(\vec{a}) = \inf\left(\{d(\vec{a}, \vec{y}) : \phi^M(\vec{y}) = 0\} \cup \{1\}\right) = \inf\{d(\vec{a}, \vec{y}) : \phi^M(\vec{y}) = 0\}.$$

$\square$

Each formula of $\mathcal{L}^C_{\omega_1,\omega}$ or $\mathcal{L}^C_{\omega_1,\omega}(\rho)$ defines a uniformly continuous function on each structure, and just as in first-order continuous logic, the modulus of continuity of this function depends only on the signature, not the particular structure. By contrast, the functions defined by $\mathcal{L}_{\omega_1,\omega}$ formulas need not be continuous at all. The loss of continuity causes complications for the theory, especially when one is interested in *complete* metric structures, as is often the case in applications. Of particular note is the fact that, while every metric structure is an $\mathcal{L}^C_{\omega_1,\omega}$-elementary substructure of its metric completion, this is very far from being true for the logic $\mathcal{L}_{\omega_1,\omega}$:

Example 2.7. Let S be the signature consisting of countably many constant symbols $(q_n)_{n<\omega}$. Consider the $\mathcal{L}_{\omega_1,\omega}$ formula

$$\phi(x) = \inf_{n<\omega} \sup_{R\in\mathbb{N}} \min\{1, Rd(x, q_n)\}.$$

For any a in a metric structure M we have $M \models \phi(a)$ if and only if $a = q_n$ for some n. In particular, if M is a countable metric space which is not complete, and $(q_n)_{n<\omega}$ is interpreted as an enumeration of M, then

$$M \models \sup_x \phi(x) \qquad \text{and} \qquad \overline{M} \not\models \sup_x \phi(x).$$

In particular, $M \not\equiv_{\mathcal{L}_{\omega_1,\omega}} \overline{M}$.

While discontinuous formulas introduce complications, they also give a significant increase in expressive power. As an example, recall that continuous first-order logic lacks an exact negation connective, in the sense that there is no connective $\neg$ such that $M \models \neg\phi$ if and only if $M \not\models \phi$. Indeed there is no continuous function $\neg : [0,1] \to [0,1]$ such that $\neg(x) = 0$ if and only if $x \neq 0$, so $\mathcal{L}^C_{\omega_1,\omega}$ also lacks an exact negation connective. Similarly, the formula $\inf_n \phi_n$ is not the exact disjunction of the formulas ϕ_n, and $\inf_x$ is not an exact existential quantifier, and neither exact disjunction nor exact existential quantification is present in either continuous infinitary logic. In $\mathcal{L}_{\omega_1,\omega}$, we recover all three of these classical operations on formulas.

Proposition 2.8. *The logic $\mathcal{L}_{\omega_1,\omega}$ has an exact countable disjunction, an exact negation, and an exact existential quantifier.*

Proof. We first show that $\mathcal{L}_{\omega_1,\omega}$ has an exact infinitary disjunction. Suppose that $(\phi_n(\vec{x}))_{n<\omega}$ are formulas of $\mathcal{L}_{\omega_1,\omega}$. Define

$$\psi(\vec{x}) = \inf_{n<\omega} \sup_{R\in\mathbb{N}} \min\{1, R\phi_n(\vec{x})\}.$$

Then in any metric structure M, for any tuple $\vec{a}$, we have

$$M \models \psi(\vec{a}) \qquad \Longleftrightarrow \qquad M \models \phi_n(\vec{a}) \text{ for some } n.$$

Using the exact disjunction we define the exact negation. Given any formula $\phi(\vec{x})$, define

$$\neg\phi(\vec{x}) = \bigvee_{n<\omega}\left(\phi(\vec{x}) \geq \frac{1}{n}\right),$$

where $\bigvee$ is the exact disjunction described above. Then for any metric structure M, and any $\vec{a} \in M$,

$$M \models \neg\phi(\vec{a}) \iff (\exists n < \omega)M \models \phi(\vec{a}) \geq \frac{1}{n}$$
$$\iff (\exists n < \omega)\phi^M(\vec{a}) \geq \frac{1}{n}$$
$$\iff \phi^M(\vec{a}) \neq 0$$
$$\iff M \not\models \phi(\vec{a})$$

Finally, with exact negation and the fact that $M \models \sup_{\vec{x}}\phi(\vec{x})$ if and only if $M \models \phi(\vec{a})$ for every $\vec{a} \in M$, we define $\exists\vec{x}\phi$ to be $\neg\sup_{\vec{x}}\neg\phi$, and have that $M \models \exists\vec{x}\phi(\vec{x})$ if and only if there is $\vec{a} \in M$ such that $M \models \phi(\vec{a})$. $\qquad\square$

Remark 2.9. Some caution is necessary when using the negation operation defined in Proposition 2.8. Consider the following properties a negation connective $\sim$ could have for all metric structures M, all tuples $\vec{a} \in M$, and all formulas $\phi(\vec{x})$. These properties mimic properties of negation in classical discrete logic:

1. $M \models \sim\phi(\vec{a})$ if and only if $M \not\models \phi(\vec{a})$,

2. $M \models \sim\sim\phi(\vec{a})$ if and only if $M \models \phi(\vec{a})$,

3. $(\sim\sim\phi)^M(\vec{a}) = \phi^M(\vec{a})$.

Each of (1) and (3) implies property (2). In classical $\{0,1\}$-valued logics there is no distinction between properties (2) and (3), but these properties do not coincide for $[0,1]$-valued logic. Property (2) is strictly weaker than property (1), since the identity connective $\sim\sigma = \sigma$ satisfies (2) but not (1).

The connective $\neg$ defined in the proof of Proposition 2.8 has properties (1) and (2), but does not have property (3), because if $\phi(\vec{a})^M > 0$, then $(\neg\neg\phi)^M(\vec{a}) = 1$. The approximate negation commonly used in continuous first-order logic, which is defined by $\sim \phi(\vec{x}) = 1 - \phi(\vec{x})$, satisfies properties (2) and (3), but not property (1).

In fact, there is no truth-functional connective in *any* $[0,1]$-valued logic that satisfies both (1) and (3). Suppose that $\sim$ were such a connective. Then $\sim : [0,1] \to [0,1]$ would have the following two properties for all $x \in [0,1]$, as consequences of (1) and (3), respectively:

- $\sim(x) = 0$ if and only if $x \neq 0$,

- $\sim(\sim(x)) = x$.

The first condition implies that $\sim$ is not injective, and hence cannot satisfy the second condition.

The expressive power of $\mathcal{L}_{\omega_1,\omega}$ is sufficient to introduce a wide variety of connectives beyond those of continuous first-order logic and the specific ones described in Proposition 2.8.

Proposition 2.10. *Let* $u : [0,1]^n \to [0,1]$ *be a Borel function, with* $n < \omega$, *and let* $(\phi_l(\vec{x}))_{l<n}$ *be* $\mathcal{L}_{\omega_1,\omega}$-*formulas. There is an* $\mathcal{L}_{\omega_1,\omega}$-*formula* $\psi(\vec{x})$ *such that for any metric structure* M *and any* $\vec{a} \in M$,

$$\psi^M(\vec{a}) = u(\phi_1^M(\vec{a}), \ldots).$$

Proof. Recall that the *Baire hierarchy* of functions $f : [0,1]^n \to [0,1]$ is defined recursively, with f being *Baire class* 0 if it is continuous, and *Baire class* α (for an ordinal $\alpha > 0$) if it is the pointwise limit of a sequence of functions each from some Baire class $< \alpha$. The classical Lebesgue-Hausdorff theorem (see [Sri98, Proposition 3.1.32 and Theorem 3.1.36]) implies that a function $f : [0,1]^n \to [0,1]$ is Borel if and only if it is Baire class α for some $\alpha < \omega_1$. Our proof will therefore be by induction on the Baire class α of our connective $u : [0,1]^n \to [0,1]$. The base case is $\alpha = 0$, in which case u is continuous, and hence is a connective of first-order continuous logic.

Now suppose that $u = \lim_{k \to \infty} u_k$ pointwise, with each u_k of a Baire class $\alpha_k < \alpha$. By induction, for each k let $\psi_k(\vec{x})$ be such that for every metric structure M and every $\vec{a} \in M$ $\psi_k^M(\vec{a}) = u_k(\phi_1^M(\vec{a}), \ldots, \phi_n^M(\vec{a}))$. Then we have

$$u(\phi_1^M(\vec{a}), \ldots, \phi_n^M(\vec{a})) = \lim_{k \to \infty} u_k(\phi_1^M(\vec{a}), \ldots, \phi_n^M(\vec{a}))$$

$$= \limsup_{k \to \infty} \psi_k^M(\vec{a})$$

$$= \inf_{k \geq 0} \sup_{m \geq k} \psi_m^M(\vec{a})$$

The final expression shows that the required $\mathcal{L}_{\omega_1,\omega}$ formula is $\inf_{k \geq 0} \sup_{m \geq k} \psi_m(\vec{x})$. $\qquad\square$

Remark 2.11. The case of Proposition 2.10 for sentences appears, with a different proof, in [Gri14, Theorem 1.25].

The expressive power of continuous first-order logic is essentially unchanged if continuous functions of the form $u : [0,1]^\omega \to [0,1]$ are allowed as connectives in addition to the continuous functions on finite powers of $[0,1]$ (see [BYBHU08, Proposition 9.3]). If such infinitary continuous connectives are permitted in $\mathcal{L}_{\omega_1,\omega}$, then the same proof as above also shows that $\mathcal{L}_{\omega_1,\omega}$ implements all Borel functions $u : [0,1]^\omega \to [0,1]$.

In order to obtain the benefits of both $\mathcal{L}_{\omega_1,\omega}$ and $\mathcal{L}_{\omega_1,\omega}^C$ or $\mathcal{L}_{\omega_1,\omega}^C(\rho)$, it is sometimes helpful to work in $\mathcal{L}_{\omega_1,\omega}$ and then specialize to a more restricted logic when continuity becomes relevant. A *fragment* of an infinitary metric logic $\mathcal{L}$ is a set of $\mathcal{L}$-formulas including the formulas of continuous first-order logic, closed under the connectives and quantifiers of continuous first-order logic, closed under subformulas, and closed under substituting terms for variables. In [Eag14] we defined a fragment L of $\mathcal{L}_{\omega_1,\omega}$ to be *continuous* if it has the property that every L-formula defines a continuous function on all structures. The definition of a continuous fragment ensures that if L is a continuous fragment and M is a metric structure, then $M \preceq_L \overline{M}$.

It follows immediately from the definitions that $\mathcal{L}^C_{\omega_1,\omega}$ is a continuous fragment of both $\mathcal{L}_{\omega_1,\omega}$ and $\mathcal{L}^C_{\omega_1,\omega}(\rho)$. The construction of the ρ operation as a formula of $\mathcal{L}_{\omega_1,\omega}$ in Theorem 2.6 uses discontinuous formulas as subformulas, so $\mathcal{L}^C_{\omega_1,\omega}(\rho)$ is not a continuous fragment of $\mathcal{L}_{\omega_1,\omega}$, although it would be if we viewed the formula ρ_ϕ from Theorem 2.6 as having only ϕ as a subformula. While it is a priori possible that there are continuous fragments of $\mathcal{L}_{\omega_1,\omega}$ that are not subfragments of $\mathcal{L}^C_{\omega_1,\omega}(\rho)$, we are not aware of any examples. It also remains unclear whether or not the ρ operation of $\mathcal{L}^C_{\omega_1,\omega}(\rho)$ can be implemented by an $\mathcal{L}^C_{\omega_1,\omega}$ formula. We therefore ask:

Question 2.12. Suppose that $\phi(\vec{x})$ is an $\mathcal{L}_{\omega_1,\omega}$ formula such that for every subformula ψ of ϕ, $\psi^M : M^n \to [0,1]$ is *uniformly* continuous, with the modulus of uniform continuity not depending on M. Is ϕ equivalent to an $\mathcal{L}^C_{\omega_1,\omega}$ formula? Is $\rho(\vec{y},\phi)$ equivalent to an $\mathcal{L}^C_{\omega_1,\omega}$ formula?

A positive answer to the first part of Question 2.12 would imply that every continuous fragment of $\mathcal{L}_{\omega_1,\omega}$ is a fragment of $\mathcal{L}^C_{\omega_1,\omega}$. In the first part of the question, the answer is negative if we only ask for ϕ to define a uniformly continuous function. For example, consider the sentence $\sigma = \sup_x \phi(x)$ from Example 2.7. For any M we have $\sigma^M : M^0 \to [0,1]$ is constant, yet we saw that this σ can be a witness to $M \not\equiv_{\mathcal{L}_{\omega_1,\omega}} \overline{M}$, and hence is not equivalent to any $\mathcal{L}^C_{\omega_1,\omega}$ sentence. This example can be easily modified to produce examples of $\mathcal{L}_{\omega_1,\omega}$ formulas with free variables that are uniformly continuous but not equivalent to any $\mathcal{L}^C_{\omega_1,\omega}$ sentence (for example, $\max\{\sigma, d(y,y)\}$).

---

3   Consequences of Scott's Isomorphism Theorem

The existence of Scott sentences for complete separable metric structures was first proved by Sequeira [Seq13] in $\mathcal{L}^C_{\omega_1,\omega}(\rho)$. Sequeira's proof of the existence of Scott sentences is a back-and-forth argument, generalizing the standard proof in the discrete setting. An alternative proof of the existence of Scott sentences in $\mathcal{L}^C_{\omega_1,\omega}$ goes by first proving a metric version of the López-Escobar Theorem, which characterizes the isomorphism-invariant bounded Borel functions on a space of codes for structures as exactly those functions of the form $M \mapsto \sigma^M$ for an $\mathcal{L}^C_{\omega_1,\omega}$-sentence σ. Using this method Scott sentences in $\mathcal{L}^C_{\omega_1,\omega}$

were found by Coskey and Lupini [CL16] for structures whose underlying metric space is the Urysohn sphere, and such that all of the distinguished functions and predicates share a common modulus of uniform continuity. Shortly thereafter, Ben Yaacov, Nies, and Tsankov obtained the same result for all complete metric structures.

Theorem 3.1 ([BYNT14, Corollary 2.2]). *For each separable complete metric structure M in a countable signature there is an $\mathcal{L}^C_{\omega_1,\omega}$ sentence σ such that for every other separable complete metric structure N of the same signature,*

$$\sigma^N = \begin{cases} 0 & \text{if } M \cong N \\ 1 & \text{otherwise} \end{cases}$$

We note that a positive answer to Question 2.12 would imply that Sequeira's proof works in $\mathcal{L}^C_{\omega_1,\omega}$, and hence would give a more standard back-and-forth proof of Theorem 3.1.

Remark 3.2. Even with the increased expressive power of $\mathcal{L}_{\omega_1,\omega}$ over $\mathcal{L}^C_{\omega_1,\omega}$, we cannot hope to prove the existence of Scott sentences for arbitrary (i.e., possibly incomplete) separable metric structures, because there are $2^{2^{\aleph_0}}$ pairwise non-isometric separable metric spaces ([KN51, Theorem 2.1]), but only $2^{\aleph_0}$ sentences of $\mathcal{L}_{\omega_1,\omega}$ in the empty signature.

We can easily reformulate Theorem 3.1 to apply to incomplete structures, but little is gained, as we only get uniqueness at the level of the metric completion.

Corollary 3.3. *For each separable metric structure M in a countable signature there is an $\mathcal{L}^C_{\omega_1,\omega}$ sentence σ such that for every other separable metric structure N of the same signature,*

$$\sigma^N = \begin{cases} 0 & \text{if } \overline{M} \cong \overline{N} \\ 1 & \text{otherwise} \end{cases}$$

Proof. Let σ be the Scott sentence for $\overline{M}$, as in Theorem 3.1. Since σ is in $\mathcal{L}^C_{\omega_1,\omega}$, we have

$$\sigma^N = \sigma^{\overline{N}} = \begin{cases} 0 & \text{if } \overline{M} \cong \overline{N} \\ 1 & \text{otherwise} \end{cases}.$$

$\square$

The following observation should be compared with Example 2.7 and Proposition 2.8, which showed that there are $\mathcal{L}_{\omega_1,\omega}$ formulas (and even sentences) that are not $\mathcal{L}^C_{\omega_1,\omega}$ or $\mathcal{L}^C_{\omega_1,\omega}(\rho)$ formulas.

Proposition 3.4. *For any separable complete metric structures M and N in the same countable signature, the following are equivalent:*

1. $M \cong N$,

2. $M \equiv_{\mathcal{L}_{\omega_1,\omega}} N$,

3. $M \equiv_{\mathcal{L}^C_{\omega_1,\omega}(\rho)} N$,

4. $M \equiv_{\mathcal{L}^C_{\omega_1,\omega}} N$.

Proof. It is clear that (1) implies (2). By Theorem 2.6 each $\mathcal{L}^C_{\omega_1,\omega}(\rho)$ formula can be implemented as an $\mathcal{L}_{\omega_1,\omega}$ formula, so (2) implies (3). Similarly, each $\mathcal{L}^C_{\omega_1,\omega}$ formula is an $\mathcal{L}^C_{\omega_1,\omega}(\rho)$ formula, so (3) implies (4). Finally, if $M \equiv_{\mathcal{L}^C_{\omega_1,\omega}}$ N then, in particular, N satisfies M's Scott sentence, and both are complete separable metric structures, so $M \cong N$ by Theorem 3.1. $\square$

The formula creation rules for $\mathcal{L}_{\omega_1,\omega}$ imply that if $\phi(\vec{x})$ is an $\mathcal{L}_{\omega_1,\omega}$-formula in a signature with a constant symbol c, then the expression obtained by re-placing each instance of c by a new variable y is an $\mathcal{L}_{\omega_1,\omega}$-formula $\psi(\vec{x}, y)$. In particular, the usual identification of formulas with sentences in a language with new constant symbols can be used in $\mathcal{L}_{\omega_1,\omega}$. By contrast, when this proce-dure is performed on an $\mathcal{L}^C_{\omega_1,\omega}$ or $\mathcal{L}^C_{\omega_1,\omega}(\rho)$ formula, the result is not necessarily again an $\mathcal{L}^C_{\omega_1,\omega}$ or $\mathcal{L}^C_{\omega_1,\omega}(\rho)$ formula, because it may not have the appropriate continuity property. Scott's isomorphism theorem provides a plentiful supply of examples.

Example 3.5. Let M be a complete separable connected metric structure such that $\mathrm{Aut}(M)$ does not act transitively on M. Pick any $a \in M$, and let $\mathcal{O}_a$ be the $\mathrm{Aut}(M)$-orbit of a. The fact that $\mathrm{Aut}(M)$ does not act transitively implies $\mathcal{O}_a \neq M$. Let $\theta_a(x)$ be the $\mathcal{L}_{\omega_1,\omega}$ formula obtained by replacing a by a new variable x in the $\mathcal{L}^C_{\omega_1,\omega}$ Scott sentence of (M, a). Then for any $b \in X$,

$$
\theta_a^M(b) = \begin{cases} 0 & \text{if } (M,a) \cong (M,b) \\ 1 & \text{otherwise} \end{cases}
$$

$$
= \begin{cases} 0 & \text{if } b \in \mathcal{O}_a, \\ 1 & \text{otherwise}. \end{cases}
$$

Since M is connected and the image of θ_a^M is $\{0, 1\}$, the function θ_a^M is not continuous. Therefore θ_a is not an $\mathcal{L}^C_{\omega_1,\omega}$ or $\mathcal{L}^C_{\omega_1,\omega}(\rho)$ formula.

The fact that the formula θ_a in the above example *is* an $\mathcal{L}_{\omega_1,\omega}$ formula will be relevant in the proof of Theorem 4.1 below.

4   Definability in $\mathcal{L}_{\omega_1,\omega}$

The original use of Scott's isomorphism theorem in [Sco65] was to prove a definability theorem. We obtain an analogous definability theorem for the metric logic $\mathcal{L}_{\omega_1,\omega}$.

Theorem 4.1. *Let M be a separable complete metric structure in a countable signature. For any continuous function $P : M^n \to [0,1]$, the following are equivalent:*

1. *There is an $\mathcal{L}_{\omega_1,\omega}$ formula $\phi(\vec{x})$ such that for all $\vec{a} \in M^n$, $\phi^M(\vec{a}) = P(\vec{a})$,*

2. *P is fixed by all automorphisms of M (in the sense that for all $\Phi \in \mathrm{Aut}(M)$, $P = P \circ \Phi$).*

Proof. The proof that (1) implies (2) is a routine induction on the complexity of formulas, so we only prove that (2) implies (1).

Fix a countable dense subset $D \subseteq M$. For each $\vec{a} \in D$, let $\theta_{\vec{a}}(\vec{x})$ be the formula obtained by replacing each occurrence of $\vec{a}$ in the Scott sentence of $(M, \vec{a})$ by a tuple of new variables $\vec{x}$. The Scott sentence is obtained from Theorem 3.1. Observe that this formula has the following property, for all $\vec{b} \in M^n$:

$$\theta_{\vec{a}}^M(\vec{b}) = \begin{cases} 0 & \text{if there is } \Phi \in \mathrm{Aut}(M) \text{ with } \Phi(\vec{b}) = \vec{a} \\ 1 & \text{otherwise} \end{cases}$$

For each $\epsilon > 0$, define:

$$\sigma_\epsilon(\vec{x}) = \inf_{\vec{y}} \max \left\{ d(\vec{x}, \vec{y}), \inf_{\substack{\vec{a} \in D^n \\ P(\vec{a}) < \epsilon}} \theta_{\vec{a}}(\vec{y}) \right\}.$$

Each $\sigma_\epsilon(\vec{x})$ is a formula of $\mathcal{L}_{\omega_1,\omega}(S)$.

Claim 4.2. Consider any $\epsilon \in \mathbb{Q} \cap (0,1)$ and any $\vec{b} \in M^n$.

(a) If $M \models \sigma_\epsilon(\vec{b})$, then $P(\vec{b}) \le \epsilon$.

(b) If $P(\vec{b}) < \epsilon$, then $M \models \sigma_\epsilon(\vec{b})$.

Proof.   (a) Suppose that $M \models \sigma_\epsilon(\vec{b})$. Fix $\epsilon' > 0$, and pick $0 < \delta < 1$ such that if $d(\vec{b}, \vec{y}) < \delta$, then $\left| P(\vec{b}) - P(\vec{y}) \right| < \epsilon'$. This exists because we assumed that P is continuous. Now from the definition of $M \models \sigma_\epsilon(\vec{b})$ we can find $\vec{y} \in M^n$ such that

$$\max \left\{ d(\vec{b}, \vec{y}), \inf_{\substack{\vec{a} \in D^n \\ P(\vec{a}) < \epsilon}} \theta_{\vec{a}}(\vec{y}) \right\} < \delta.$$

In particular, we have that $d(\vec{b}, \vec{y}) < \delta$, so $\left| P(\vec{b}) - P(\vec{y}) \right| < \epsilon'$. On the other hand, $\inf_{\substack{\vec{a} \in D^n \\ P(\vec{a}) < \epsilon}} \theta_{\vec{a}}(\vec{y}) < \delta$, and $\theta_{\vec{a}}(\vec{y}) \in \{0, 1\}$ for all $\vec{a} \in D^n$, so in fact there is $\vec{a} \in D^n$ with $P(\vec{a}) < \epsilon$ and $\theta_{\vec{a}}(\vec{y}) = 0$. For such an $\vec{a}$ there is an automorphism of M taking $\vec{y}$ to $\vec{a}$, and hence by (2) we have that $P(\vec{y}) < \epsilon$ as well. Combining what we have,

$$
\begin{aligned}
P(\vec{b}) &= \left| P(\vec{b}) \right| \\
&\leq \left| P(\vec{b}) - P(\vec{y}) \right| + |P(\vec{y})| \\
&< \epsilon' + \epsilon
\end{aligned}
$$

Taking $\epsilon' \to 0$ we conclude $P(\vec{b}) \leq \epsilon$.

(b) Suppose that $P(\vec{b}) < \epsilon$, and again fix $\epsilon' > 0$. Using the continuity of P, find δ sufficiently small so that if $d(\vec{b}, \vec{y}) < \delta$ then $P(\vec{y}) < \epsilon$. The set D is dense in M, so we can find $\vec{y} \in D^n$ such that $d(\vec{b}, \vec{y}) < \min\{\delta, \epsilon'\}$. Then $P(\vec{y}) < \epsilon$, so choosing $\vec{a} = \vec{y}$ we have

$$
\inf_{\substack{\vec{a} \in D^n \\ P(\vec{a}) < \epsilon}} \theta_{\vec{a}}(\vec{y}) = 0.
$$

Therefore

$$
\max\left\{ d(\vec{b}, \vec{y}), \inf_{\substack{\vec{a} \in D^n \\ P(\vec{a}) < \epsilon}} \theta_{\vec{a}}(\vec{y}) \right\} = d(\vec{b}, \vec{y}) < \epsilon',
$$

and so taking $\epsilon' \to 0$ shows that $M \models \sigma_\epsilon(\vec{b})$.

$\dashv$ - Claim 4.2

Consider now any $\vec{a} \in M^n$. By (a) of the claim $P(\vec{a})$ is a lower bound for $\{\epsilon \in \mathbb{Q} \cap (0, 1) : M \models \sigma_\epsilon(\vec{a})\}$. If α is another lower bound, and $\alpha > P(\vec{a})$, then there is $\epsilon \in \mathbb{Q} \cap (0, 1)$ such that $P(\vec{a}) < \epsilon < \alpha$. By (b) of the claim we have $M \models \sigma_\epsilon(\vec{a})$ for this ϵ, contradicting the choice of α. Therefore

$$
P(\vec{a}) = \inf\{\epsilon \in \mathbb{Q} \cap (0, 1) : M \models \sigma_\epsilon(\vec{a})\}.
$$

Now for each $\epsilon \in \mathbb{Q} \cap (0, 1)$, define a formula

$$
\psi_\epsilon(\vec{x}) = \max\left\{ \epsilon, \sup_{m \in \mathbb{N}} \min\{m\sigma_\epsilon(\vec{x}), 1\} \right\}.
$$

Then for any $\vec{a} \in M^n$,

$$
\psi_\epsilon^M(\vec{a}) = \begin{cases} \epsilon & \text{if } \sigma_\epsilon^M(\vec{a}) = 0, \\ 1 & \text{otherwise.} \end{cases}
$$

Let $\phi(\vec{x}) = \inf_{\epsilon \in \mathbb{Q} \cap (0,1)} \psi_\epsilon(\vec{x})$. Then

$$\phi^M(\vec{a}) = \inf \left\{ \epsilon : \sigma_\epsilon^M(\vec{a}) = 0 \right\} = P(\vec{a}).$$

$\square$

We also have a version where parameters are allowed in the definitions:

Corollary 4.3. *Let M be a separable complete metric structure in a countable signature, and fix a set $A \subseteq M$. For any continuous function $P : M^n \to [0,1]$, the following are equivalent:*

1. *There is an $\mathcal{L}_{\omega_1,\omega}$ formula $\phi(\vec{x})$ with parameters from A such that for all $\vec{a} \in M^n$,*

$$\phi^M(\vec{a}) = P(\vec{a}),$$

2. *P is fixed by all automorphisms of M that fix A pointwise,*

3. *P is fixed by all automorphisms of M that fix $\overline{A}$ pointwise.*

Proof. Since M is a separable metric space, there is a countable set $D \subseteq A$ such that $\overline{D} = \overline{A}$ in M. An automorphism of M fixes A pointwise if and only if it fixes D pointwise if and only if it fixes $\overline{D} = \overline{A}$ pointwise, which establishes the equivalence of (2) and (3). For the equivalence of (1) and (2), apply Theorem 4.1 to the structure obtained from M by adding a new constant symbol for each element of D. $\square$

Theorem 4.1 does not hold, as stated, with $\mathcal{L}_{\omega_1,\omega}$ replaced by $\mathcal{L}_{\omega_1,\omega}^C$ or $\mathcal{L}_{\omega_1,\omega}^C(\rho)$, because we assumed only continuity for the function P, while formulas in $\mathcal{L}_{\omega_1,\omega}^C$ always define uniformly continuous functions. Even if P is assumed to be uniformly continuous, some intermediate steps in our proof use the formulas discussed in Example 3.5, as well as other possibly discontinuous formulas, and hence our argument does not directly apply to give a version of Scott's definability theorem in the other infinitary logics.

Question 4.4. Let M be a separable complete metric structure, and let $P : M^n \to [0,1]$ be uniformly continuous and automorphism invariant. Is P definable in M by an $\mathcal{L}_{\omega_1,\omega}^C$-formula? Is P definable in M by an $\mathcal{L}_{\omega_1,\omega}^C(\rho)$-formula?

To conclude, we give one quite simple example of definability in $\mathcal{L}_{\omega_1,\omega}$ where first-order definability fails.

Example 4.5. Recall that a (unital) C*-algebra is a unital Banach algebra with an involution $*$ satisfying the C*-identity $\|xx^*\| = \|x\|^2$. A formalization for treating C*-algebras as metric structures is presented in [FHS14], where it is also shown that in an appropriate language the class of C*-algebras is $\forall$-axiomatizable in continuous first-order logic. The model theory of C*-algebras has since become an active area of investigation.

A *trace* on a C*-algebra A is a bounded linear functional $\tau : A \to \mathbb{C}$ such that $\tau(1) = 1$, and for all $a, b \in A$, $\tau(a^*a) \geq 0$ and $\tau(ab) = \tau(ba)$. An appropriate way to consider traces as $[0,1]$-valued predicates on the metric structure associated to a C*-algebra is given in [FHS14]. Traces appear as important tools throughout the C*-algebra literature. In the first-order continuous model theory of C*-algebras, traces play a key role in showing that certain important C*-algebras can be constructed as Fraïssé limits [EFH+16], and traces are also related to the failure of quantifier elimination for most finite-dimensional C*-algebras [EFKV15]. Several other uses of traces in the model theory of C*-algebras can be found in [FHL+16]. Of particular interest is the case where a C*-algebra has a unique trace; such algebras are called *monotracial*.

In general, traces on C*-algebras need not be automorphism invariant. For an example, consider $C(2^\omega)$, the C*-algebra of continuous complex-valued functions on the Cantor space. Pick any $z \in 2^\omega$, and define $\tau : C(2^\omega) \to \mathbb{C}$ by $\tau(f) = f(z)$. It is straightforward to verify that τ is a trace. For any other $z' \in 2^\omega$ there is an autohomeomorphism ϕ of 2^ω sending z to z'. The map $\Phi : f \mapsto f \circ \phi$ is then an automorphism of $C(2^\omega)$, and we have $(\tau \circ \Phi)(f) = \tau(f \circ \phi) = (f \circ \phi)(z) = f(z')$, so $\tau \circ \Phi \neq \tau$.

On the other hand, it is easily seen that if τ is a trace on A and $\Phi \in \mathrm{Aut}(A)$, then $\tau \circ \Phi$ is again a trace on A. Thus for monotracial C*-algebras the unique trace *is* automorphism invariant. The following is therefore a direct consequence of Theorem 4.1:

Corollary 4.6. *If A is a separable C*-algebra with a unique trace τ, then τ is $\mathcal{L}_{\omega_1,\omega}$-definable (without parameters) in A.*

It is natural to ask whether $\mathcal{L}_{\omega_1,\omega}$-definability in Corollary 4.6 can be replaced by definability in a weaker logic. Monotracial C*-algebras satisfying certain additional properties do have their traces definable in first-order continuous logic (see [FHL+16]), but the additional assumptions on the C*-algebras are necessary. In [FHL+16] it is shown that the separable monotracial C*-algebra constructed by Robert in [Rob15, Theorem 1.4] has the property that the trace is not definable in first-order continuous logic.

The situation for definability in $\mathcal{L}_{\omega_1,\omega}^C$ is less clear. Any trace on a C*-algebra is 1-Lipschitz, and so in particular is uniformly continuous. An interesting special case of Question 4.4 is then whether or not the trace on a monotracial separable C*-algebra is always $\mathcal{L}_{\omega_1,\omega}^C$-definable.

Bibliography

[BF85]   J. Barwise and S. Feferman (eds.), *Model-theoretic logics*, Springer-Verlag, 1985.

[BYBHU08] I. Ben Yaacov, A. Berenstein, C. W. Henson, and A. Usvyatsov, *Model theory for metric structures*, Model Theory with Applications to Algebra and Analysis, Vol. II (Z. Chatzidakis, D. Macpherson, A. Pillay, and A. Wilkie, eds.), Lecture Notes series of the London Mathematical Society, no. 350, Cambridge University Press, 2008, pp. 315–427.

[BYI09]   I. Ben Yaacov and J. Iovino, *Model theoretic forcing in analysis*, Annals of Pure and Applied Logic **158** (2009), no. 3, 163–174.

[BYNT14]   I. Ben Yaacov, A. Nies, and T. Tsankov, *Metric Scott analysis*, preprint (2014), arXiv:1407.7102.

[BYU10]   I. Ben Yaacov and A. Usvyatsov, *Continuous first order logic and local stability*, Trans. Amer. Math. Soc. **362** (2010), 5213–5259.

[CL16]   S. Coskey and M. Lupini, *A López-Escobar theorem for metric structures, and the topological Vaught conjecture*, Fund. Math., 3486 (2016), 55-72.

[Eag14]   C. J. Eagle, *Omitting types in infinitary [0, 1]-valued logic*, Annals of Pure and Applied Logic **165** (2014), 913–932.

[Eag15]   ———, *Topological aspects of real-valued logic*, Ph.D. thesis, University of Toronto, 2015.

[EFH⁺16]   C. J. Eagle, I. Farah, B. Hart, B. Kadets, V. Kalashnyk, and M. Lupini, *Fraïssé limits of C*-algebras*, J. Symbolic Logic, **81** (2016), 755–773.

[EFKV15]   C. J. Eagle, I. Farah, E. Kirchberg, and A. Vignati, *Quantifier elimination in C*-algebras*, Int. Math. Res. Notices, 2016

[FHL⁺16]   I. Farah, B. Hart, M. Lupini, L. Robert, A. Tikuisis, A. Vignati, and W. Winter, *Model theory of C*-algebras*, preprint (2016), arXiv:1602.08072.

[FHS14]   I. Farah, B. Hart, and D. Sherman, *Model theory of operator algebras II: Model theory*, Israel J. Math. **201** (2014), 477–505.

[Gri14]   J. Grinstead, *Model existence for* [0, 1]*-valued logic*, Master's thesis, Carnegie Mellon University, 2014.

[HI03]   C. W. Henson and J. Iovino, *Ultraproducts in analysis*, Analysis and Logic, London Mathematical Society Lecture Note Series, no. 262, Camb, 2003.

[Iov14]   J. Iovino, *Applications of model theory to functional analysis*, Dover, Mineola, New York, 2014.

[Kei71]   H.J. Keisler, *Model theory for infinitary logic: Logic with countable conjuctions and finite quantifiers*, North-Holland, 1971.

[KN51]   L.M. Kelly and E.A. Nordhaus, *Distance sets in metric spaces*, Trans. Amer. Math. Soc. **7** (1951), 440–956.

[Ort97]   C. Ortiz, *Truth and approximate truth in metric spaces*, Ph.D. thesis, University of Wisconsin - Madison, 1997.

[Rob15]   L. Robert, *Nuclear dimension and sums of commutators*, Indiana Univ. Math. J. **64** (2015), 559–576.

[Sco65]   D. Scott, *Logic with denumerably long formulas and finite strings of quantifiers*, The theory of models (J. W. Addison, ed.), North-Holland, Amsterdam, 1965, pp. 329–341.

[Seq13]   N. Sequeira, *Infinitary continuous model theory*, Master's thesis, McMaster University, 2013.

[Sri98]   S.M. Srivastava, *A course on Borel sets*, Graduate Texts in Mathematics, vol. 180, Springer, New York, 1998.

Chapter 2

Scott processes

Paul B. Larson
Miami University
Oxford, Ohio, USA

Research supported in part by NSF Grants DMS-0801009 and DMS-1201494.
A significant portion of the work in this chapter was done while the author
was participating in the Thematic Program on Forcing and its Applications
at the Fields Institute in Fall 2012. The author thanks John Baldwin, Su Gao
and Caleb Ziegler for comments on earlier drafts.

1 Introduction

The Scott analysis of a structure (introduced in [Sco65]) is a procedure which assigns a transfinite sequence of infinitary formulas to the finite tuples of the structure. In the case of countable structures, this process culminates in a sentence in $\mathcal{L}_{\aleph_1,\aleph_0}$ which characterizes the structure up to isomorphism. This approach was used by Morley [Mor70] to show that if a sentence in $\mathcal{L}_{\aleph_1,\aleph_0}$ has more than $\aleph_1$ many countable models, then it has continuum many. Vaught's Conjecture [Vau61], which remains open, is the corresponding statement with $\aleph_1$ replaced by $\aleph_0$.

In this chapter we give axioms for the class of sequences of sets of formulas which arise in the Scott analysis of infinite structures in a given relational vocabulary. For the case of countable structures, our axioms characterize this class exactly (see Section 6). For uncountable cardinals, while the sequences of formulas given by the Scott analysis satisfy our axioms, it can happen that a sequence satisfying our axioms does not have a model. Remark 10.8 shows this for a sequence of length ω_2; we do not know in general if this can happen for a sequence of cardinality $\aleph_1$ (the material in Section 7 shows that it cannot if we impose additional conditions on the sequence). We use this approach to give new proofs of several classical results on counterexamples to Vaught's Conjecture, including an unpublished theorem of Leo Harrington saying that the Scott ranks of the models of any counterexample to Vaught's Conjecture include a cofinal subset of ω_2 (in fact they include every limit ordinal of cardinality $\aleph_1$). Although our analysis concentrates on the scattered (Vaught's Conjecture counterexample) case, we expect that our approach will have applications to the general study of infinitary model theory.

We fix for this chapter a relational vocabulary τ, and distinct variable symbols $\{x_n : n < \omega\}$. For notational convenience, we assume that τ contains a 0-ary relation symbol, as well as the binary symbol $=$, which is always interpreted as equality. We refer the reader to [Hod93, Kei71, Mar02] for the definition of the language $\mathcal{L}_{\infty,\aleph_0}(\tau)$ and the languages $\mathcal{L}_{\kappa,\aleph_0}(\tau)$, for κ an infinite cardinal. In this chapter, all formulas will have only finitely many free variables. Formally, we consider conjunctions and disjunctions of formulas as unordered, even when we write them as indexed by an ordered set (in this way, for instance, a formula in $\mathcal{L}_{\aleph_2,\aleph_0}(\tau)$ becomes a member of $\mathcal{L}_{\aleph_1,\aleph_0}(\tau)$ in a forcing extension in which the ω_1 of the ground model is countable). We begin by recalling the standard definition of the Scott process corresponding to an infinite τ-structure M, as introduced in [Sco65] (see also [Hod93, Mar02]), slightly modified to require the sequences $\bar{a}$ to consist of distinct elements.

Definition 1.1. Given an infinite τ-structure M over a relational vocabulary τ, we define for each finite ordered tuple $\bar{a} = \langle a_0, \ldots, a_{|\bar{a}|-1} \rangle$ of distinct elements of M and each ordinal α the $|\bar{a}|$-ary $\mathcal{L}_{\infty,\aleph_0}(\tau)$-formula $\phi^M_{\bar{a},\alpha}$, recursively on α, as follows.

1. Each formula $\phi_{\bar{a},0}^M$ is the conjunction of all expressions of the two following forms:

   - $R(x_{f(0)}, \ldots, x_{f(k-1)})$, for R a k-ary relation symbol from τ and f a function from k to $|a|$, such that $M \models R(a_{f(0)}, \ldots, a_{f(k-1)})$,

   - $\neg R(x_{f(0)}, \ldots, x_{f(k-1)})$, for R a k-ary relation symbol from τ and f a function from k to $|a|$, such that $M \models \neg R(a_{f(0)}, \ldots, a_{f(k-1)})$.

2. Each formula $\phi_{\bar{a},\alpha+1}^M$ is the conjunction of the following three formulas:

   - $\phi_{\bar{a},\alpha}^M$,

   - $\bigwedge_{c \in M \setminus \{a_0, \ldots, a_{|\bar{a}|-1}\}} \exists x_{|\bar{a}|} \phi_{\bar{a}^\frown \langle c \rangle, \alpha}^M$,

   - $\forall x_{|\bar{a}|} \notin \{x_0, \ldots, x_{|a|-1}\} \bigvee_{c \in M \setminus \{a_0, \ldots, a_{|\bar{a}|-1}\}} \phi_{\bar{a}^\frown \langle c \rangle, \alpha}^M$.

3. For limit ordinals β, $\phi_{\bar{a},\beta}^M = \bigwedge_{\alpha < \beta} \phi_{\bar{a},\alpha}^M$.

We call $\phi_{\bar{a},\alpha}^M$ the *Scott formula* of $\bar{a}$ in M at level α.

For each infinite τ-structure M, each finite injective tuple $\bar{a}$ from M and each ordinal α, $\phi_{\bar{a},\alpha}^M \in \mathcal{L}_{|M \cup \tau \cup \alpha|^+, \aleph_0}(\tau)$ and $M \models \phi_{\bar{a},\alpha}^M(\bar{a})$. The following well-known fact can be proved by induction on α (see Theorem 3.5.2 of [Hod93]). Again, we refer the reader to [Hod93, Kei71, Mar02] for the definition of the *quantifier depth* of a formula, and note that each formula $\phi_{\bar{a},\alpha}^M$ as defined above has quantifier depth exactly α.

Theorem 1.2. *Given infinite τ-structures M and N, $n \in \omega$, an ordinal α and injective n-tuples $\bar{a}$ from M and $\bar{b}$ from N, $\phi_{\bar{a},\alpha}^M = \phi_{\bar{b},\alpha}^N$ if and only if, for each n-ary $\mathcal{L}_{\infty,\omega}(\tau)$ formula ψ of quantifier depth at most α, $\bar{a}$ satisfies ψ in M if and only if $\bar{b}$ satisfies ψ in N.*

Definition 1.3. Given an infinite τ-structure M and an ordinal β, we let $\Phi_\beta(M)$ denote the set of all formulas of the form $\phi_{\bar{a},\beta}^M$, for $\bar{a}$ a finite tuple of distinct elements of M. We call the class-length sequence $\langle \Phi_\alpha(M) : \alpha \in \mathrm{Ord} \rangle$ the *Scott process* of M.

This chapter studies a proposed axiomatization of the class set-length initial segments of Scott processes of infinite τ-structures. Section 2 introduces an array of sets of formulas (properly) containing all the formulas appearing in the Scott process of any infinite τ-structure, and vertical and horizontal projection functions acting on this array. Section 3 introduces our general notion of a Scott process (i.e., without regard to a fixed τ-structure). Section 4 develops some of the basic consequences of this definition, showing the equivalence of our definition with a natural variation, and Section 5 defines the rank of a Scott process. The material in these two sections checks that Scott processes in general, as defined here, satisfy various basic properties of Scott processes of τ-structures as defined by Scott. Section 6 shows that a

Scott process of countable length whose last level is countable is an initial segment of the Scott process of some τ-structure. Theorem 6.12 shows that if countable structures M and N have the same Scott process through level δ, for δ a countable ordinal, and N has Scott rank δ, then M is isomorphic to a quantifier-depth-δ-elementary substructure of N. Section 7 shows (following Harrington) how to build models of cardinality $\aleph_1$ for certain Scott processes (roughly, those corresponding to Scott sentences). Section 8 develops more basic material on Scott processes, studying the way they reflect finite blocks of existential quantifiers. Section 9 looks at extending Scott processes of limit length. Section 10 is largely disjoint from the rest of the chapter, and presents an argument showing that in some cases (for instance, counterexamples to Vaught's Conjecture in $\mathcal{L}_{\omega_1,\omega}(\tau)$) a Scott process which exists in a forcing extension can be shown to exist in the ground model. Put together, the material in Sections 7, 9 and 10 gives Harrington's theorem that a counterexample to Vaught's Conjecture has models of cofinally many Scott ranks below ω_2. Section 11 produces a second class of models of a counterexample to Vaught's Conjecture, among other things. In Section 12 we analyze the isomorphism relation on Scott subprocesses, and use it to give a new proof using Scott processes of the fact that if there is a counterexample to Vaught's Conjecture in $\mathcal{L}_{\aleph_1,\aleph_0}$, then there is one of quantifier depth ω. In Section 13 we define a class of structures corresponding to Scott processes (where the infinitary formulas become points) and observe that if there is a counterexample to Vaught's Conjecture, then there is one given by a subclass of this class.

The material in this chapter was inspired by the slides of a talk given by David Marker on Harrington's theorem [Mar11]. Our proof is different in some respects from the proof outlined there. Marker's talk outlines a recursion-theoretic argument, assuming the existence of a counterexample ϕ to Vaught's Conjecture, for finding a sentence in $\mathcal{L}_{\omega_2,\omega}$ which will be the Scott sentence of a model of ϕ (of suitably high Scott rank) in a forcing extension collapsing ω_1. This part of the proof is replaced here by a forcing-absoluteness argument in Section 10 (essentially equivalent versions of these arguments appear in Section 1 of [Hjo96]). The remainder of Harrington's proof builds a model of this Scott sentence. We do this in Section 7, guided by the argument in Marker's slides.

Other, different, proofs of Harrington's theorem appear in [BFKL16] and [KMS16].

---

2   Formulas and projections

For each $n \in \omega$, we let X_n denote the set $\{x_m : m < n\}$ and i_n denote the identity function on X_n. For all $m \leq n$ in ω, we let $\mathfrak{I}_{m,n}$ denote the set of injections from X_m into X_n.

We start by defining a class of formulas which contains every formula appearing in the Scott process of any infinite τ-structure (see Remark 2.5). The sets Ψ_α defined below also contain formulas that do not appear in the Scott process of any τ-structure (i.e., which are not satisfiable). This extra degree of freedom is sometimes useful (for instance, in Definition 5.14); in any case strengthening the definition to rule out such formulas would raise issues that we would rather defer (see Remark 2.16, however). For the moment, the important point is that the sets Ψ_β ($\beta \in \mathrm{Ord}$) are small enough to carry the projection functions $V_{\alpha,\beta}$ and H_α^n defined below.

Definition 2.1. We define, for each ordinal α and each $n \in \omega$, the sets Ψ_α and Ψ_α^n, by recursion on α, as follows.

1. For each $n \in \omega$, Ψ_0^n is the set of all n-ary formulas which are conjunctions consisting of, for each atomic τ-formula using variables from X_n, exactly one of the formula and its negation, including an instance of the formula $x_i = x_i$ for each $x_i \in X_n$, and an instance of $x_i \neq x_j$ for each pair of distinct x_i, x_j from X_n.

2. For each ordinal α and each $n \in \omega$, $\Psi_{\alpha+1}^n$ is the set of formulas ϕ for which there exist a formula $\phi' \in \Psi_\alpha^n$ and a nonempty $E \subseteq \Psi_\alpha^{n+1}$ such that ϕ is the conjunction of ϕ' with the following two formulas.

    (a) $\bigwedge_{\psi \in E} \exists x_n \psi$;
    (b) $\forall x_n (x_n \notin \{x_0, \dots, x_{n-1}\} \rightarrow \bigvee_{\psi \in E} \psi)$.

3. For each limit ordinal α and each $n \in \omega$, Ψ_α^n is the set of conjunctions which consist of exactly one formula ψ_β from each Ψ_β^n, for $\beta < \alpha$, satisfying the following conditions.

    (a) For each $\beta < \alpha$, ψ_β is the formula ϕ' with respect to $\psi_{\beta+1}$, as in condition (2) (i.e., the unique conjunct of quantifier depth β).
    (b) For each limit ordinal $\beta < \alpha$, $\psi_\beta = \bigwedge \{\psi_\gamma : \gamma < \beta\}$.

4. For each ordinal α, $\Psi_\alpha = \bigcup_{n \in \omega} \Psi_\alpha^n$.

We can think of the sets Ψ_α^n as forming an array, with the rows indexed by α and the columns indexed by n. In the rest of this section, we define the functions $V_{\alpha,\beta}$, which map between rows while preserving column rank, and the functions H_α^n which map between columns while preserving row rank.

Remark 2.2. Each Ψ_α is a set of $\mathcal{L}_{\infty,\omega}(\tau)$ formulas of quantifier depth α, so the sets Ψ_α are disjoint for distinct α. Similarly, for each $n \in \omega$ and each ordinal α, X_n is the set of free variables for each formula in each Ψ_α^n.

Remark 2.3. As we require our vocabulary to contain a 0-ary relation as well as the binary relation $=$, Ψ_α^n is nonempty for each ordinal α and each $n \in \omega$.

Definition 2.4. For each ordinal α, and each formula ϕ in $\Psi_{\alpha+1}$, we let $E(\phi)$ denote the set E from condition (2) of Definition 2.1.

Remark 2.5. If M is a τ-structure, α is an ordinal and $\bar{a}$ is a finite tuple of distinct elements of M, then the Scott formula of $\bar{a}$ in M at level α as defined in Definition 1.1 (i.e., $\phi_{\bar{a},\alpha}^M$) is an element of $\Psi_\alpha^{|\bar{a}|}$. It follows that $\Phi_\alpha(M) \subseteq \Psi_\alpha$.

The functions $V_{\alpha,\beta}$, as defined below, are the *vertical projection functions*.

Definition 2.6. The functions $V_{\alpha,\beta} \colon \Psi_\beta \to \Psi_\alpha$, for all pairs of ordinals $\alpha \leq \beta$ are defined as follows.

1. Each function $V_{\alpha,\alpha}$ is the identity function on Ψ_α.

2. For each ordinal α, and each $\phi \in \Psi_{\alpha+1}$, $V_{\alpha,\alpha+1}(\phi)$ is the first conjunct of ϕ, i.e., the formula ϕ' in condition (2) of Definition 2.1.

3. For each limit ordinal β, each formula $\phi \in \Psi_\beta$, and each $\alpha < \beta$, $V_{\alpha,\beta}(\phi)$ is the unique conjunct of ϕ in Ψ_α.

4. For all ordinals $\alpha < \beta$, $V_{\alpha,\beta+1} = V_{\alpha,\beta} \circ V_{\beta,\beta+1}$.

Definition 2.7. For formulas ϕ, ψ in $\bigcup_{\alpha \in \mathrm{Ord}} \Psi_\alpha$, we write $\phi \leq_V \psi$ to mean that $V_{\alpha,\beta}(\psi) = \phi$, where $\phi \in \Psi_\alpha$ and $\psi \in \Psi_\beta$.

Remark 2.8. Conditions (2) and (3) of Definition 2.1 imply the following stronger version of condition (4) of Definition 2.6 : for all ordinals $\alpha \leq \beta \leq \gamma$, $V_{\alpha,\gamma} = V_{\alpha,\beta} \circ V_{\beta,\gamma}$.

Remark 2.9. For all ordinals $\alpha \leq \beta$, each $n \in \omega$, and each $\phi \in \Psi_\beta^n$, $V_{\alpha,\beta}(\phi)$ is in Ψ_α^n, so ϕ and $V_{\alpha,\beta}(\phi)$ have the same free variables.

Remark 2.10. Since the domains of the functions $V_{\alpha,\beta}$ are disjoint for distinct β, one could drop β and simply write V_α (which would then be a definable class-sized function from $\bigcup_{\beta \in (\mathrm{Ord} \backslash \alpha)} \Psi_\beta$ to Ψ_α). We retain both subscripts for clarity.

We define the *horizontal projection functions* as follows.

Definition 2.11. The functions H_α^n, for each ordinal α and each $n \in \omega$, are defined recursively on α, as follows.

1. The domain of each H_α^n consists of all pairs (ϕ, j), where $\phi \in \Psi_\alpha^n$ and, for some $m \leq n$, $j \in \mathfrak{I}_{m,n}$.

2. For all $m \leq n$ in ω, all formulas $\phi \in \Psi_0^n$, and all $j \in \mathfrak{I}_{m,n}$, $H_0^n(\phi, j)$ is the conjunction of all conjuncts from ϕ whose variables are all in the range of j, with these variables replaced by their j-preimages.

3. For each ordinal α, all $m \leq n$ in ω, each $\phi \in \Psi^n_{\alpha+1}$, and each $j \in \mathcal{I}_{m,n}$, $H^n_{\alpha+1}(\phi, j)$ is the formula $\psi \in \Psi^m_{\alpha+1}$ such that

$$V_{\alpha,\alpha+1}(\psi) = H^n_\alpha(V_{\alpha,\alpha+1}(\phi), j)$$

and $E(\psi) = H^{n+1}_\alpha[E(\phi) \times \{j \cup \{(x_m, y)\} \mid y \in (X_{n+1} \setminus \mathrm{range}(j))\}]$.

4. For each limit ordinal α, each $m \leq n$ in ω, each $j \in \mathcal{I}_{m,n}$ and each $\phi \in \Psi^n_\alpha$,

$$H^n_\alpha(\phi, j) = \bigwedge\{H^n_\beta(V_{\beta,\alpha}(\phi), j) : \beta < \alpha\}.$$

Remark 2.12. Since the domains of the functions H^n_α are disjoint for distinct pairs (α, n), one could drop α and n and simply write H. We retain them for clarity. One could further streamline the notation used here by writing $\phi^{(j)}$ for $H^n_\beta(\phi, j)$ and $\phi_{(\alpha)}$ for $V_{\alpha,\beta}(\phi)$, for appropriate α, β, j and n. Again, we will stick to the more explicit notation in this chapter.

Remark 2.13. For all ordinals α, all $m \leq n$ in ω, all $j \in \mathcal{I}_{m,n}$ and all $\phi \in \Psi^n_\alpha$, $H^n_\alpha(\phi, j)$ is an element of Ψ^m_α, and $H^n_\alpha(\phi, i_n) = \phi$.

We leave it to the reader to verify (by induction on α) that if

- M is a τ-structure,

- α is an ordinal,

- $m \leq n$ are elements of ω,

- $\bar{b} = \langle b_0, \ldots, b_{n-1} \rangle$ is a sequence of distinct elements of M,

- $j^*: m \to n$ is an injection,

- $\bar{a}$ is the sequence $\langle b_{j^*(0)}, \ldots, b_{j^*(m-1)} \rangle$ and

- $j \in \mathcal{I}_{m,n}$ is such that $j(x_p) = x_{j^*(p)}$ for each $p < m$,

then $H^n_\alpha(\phi^M_{\bar{b},\alpha}, j) = \phi^M_{\bar{a},\alpha}$.

Remark 2.14. The following facts can be easily verified by induction on α.

1. For each ordinal α, each $n \in \omega$, each $\phi \in \Psi^n_\alpha$ and each $j \in \mathcal{I}_{n,n}$, $H^n_\alpha(\phi, j)$ is the result of replacing each free variable in ϕ (i.e., each member of X_n) with its j-preimage.

2. For each ordinal α, all $m \leq n \leq p$ in ω, all $\phi \in \Psi^p_\alpha$, all $j \in \mathcal{I}_{n,p}$ and all $k \in \mathcal{I}_{m,n}$, $H^n_\alpha(H^p_\alpha(\phi, j), k) = H^p_\alpha(\phi, j \circ k)$.

The following proposition shows that the vertical and horizontal projection functions commute appropriately.

Proposition 2.15. *For all ordinal* $\alpha \leq \beta$, *all* $m \leq n \in \omega$, *all* $j \in \mathcal{I}_{m,n}$, *and all* $\phi \in \Psi_\beta^n$,

$$V_{\alpha,\beta}(H_\beta^n(\phi, j)) = H_\alpha^n(V_{\alpha,\beta}(\phi), j).$$

Proof. When $\alpha = \beta$, both sides are equal to $H_\alpha^n(\phi, j)$. When $\beta = \alpha + 1$, the proposition is part of condition (3) of Definition 2.11. When β is a limit ordinal, it follows from condition (3) of Definition 2.6 and condition (4) of Definition 2.11. The remaining cases can be proved by induction on β, fixing α, using the induction hypotheses for the pairs α, β and $\beta, \beta + 1$ at successor stages of the form $\beta + 1$.

$\square$

Remark 2.16. For all ordinals α, all $n \in \omega$, all $\phi \in \Psi_{\alpha+1}^n$ appearing in the Scott process of a τ-structure, and all $\psi \in E(\phi)$, $H_\alpha^{n+1}(\psi, i_n) = V_{\alpha,\alpha+1}(\phi)$. Having defined the horizontal and vertical projection functions, we could now thin the sets Ψ_α^n by adding this as an additional requirement, but choose not to.

Example 2.17. Suppose that τ contains a single binary relation symbol R, along with $=$ and the 0-ary relation symbol S. The set Ψ_0^0 then consists of the sentences S and $\neg S$. The set Ψ_0^1 contains four formulas, $S \wedge R(x_0, x_0) \wedge x_0 = x_0$, $S \wedge \neg R(x_0, x_0) \wedge x_0 = x_0$, $\neg S \wedge R(x_0, x_0) \wedge x_0 = x_0$ and $\neg S \wedge \neg R(x_0, x_0) \wedge x_0 = x_0$. Call the first two of these formulas ψ_0^1 and ϕ_0^1, respectively. Then

$$H_0^1(\psi_0^1) = H_0^1(\phi_0^1) = S.$$

The set Ψ_0^2 then contains 32 formulas, for instance,

$$S \wedge \neg R(x_0, x_1) \wedge \neg R(x_1, x_0) \wedge R(x_0, x_0) \wedge R(x_1, x_1) \wedge x_0 \neq x_1 \wedge x_0 = x_0 \wedge x_1 = x_1$$

and

$$S \wedge \neg R(x_0, x_1) \wedge \neg R(x_1, x_0) \wedge R(x_0, x_0) \wedge \neg R(x_1, x_1) \wedge x_0 \neq x_1 \wedge x_0 = x_0 \wedge x_1 = x_1.$$

Call these formulas ψ_0^2 and ϕ_0^2, respectively. Then

$$H_0^2(\psi_0^2, i_1) = \psi_0^1$$

and

$$H_0^2(\phi_0^2, \{(x_0, x_1)\}) = \phi_0^1,$$

as defined in Definition 2.11. The set Ψ_0^3 then contains 2^{10} formulas, including the conjunction of S with every instance of $R(y, z)$ for $y, z \in X_3$ (and the requisite formulas involving $=$). In general, Ψ_0^n contains $2^{(n^2+1)}$ formulas.

The set Ψ_1^0 contains the sentence

$$S \wedge (\exists x_0 S \wedge R(x_0, x_0) \wedge x_0 = x_0) \wedge (\forall x_0 S \wedge R(x_0, x_0) \wedge x_0 = x_0)$$

(omitting one instance each of $\wedge$ and $\vee$, corresponding to a conjunction and a disjunction of of size 1, and a subformula asserting that x_0 is not in the emptyset) and the conjunction of the three following formulas

- S

- $(\exists x_0 S \wedge R(x_0, x_0) \wedge x_0 = x_0) \wedge (\exists x_0 S \wedge \neg R(x_0, x_0) \wedge x_0 = x_0)$

- $\forall x_0 (x_0 \notin \emptyset \to ((S \wedge R(x_0, x_0) \wedge x_0 = x_0) \vee (S \wedge \neg R(x_0, x_0) \wedge x_0 = x_0)))$.

Call these sentences ψ_1^0 and ϕ_1^0, respectively. Then $E(\psi_1^0) = \{\psi_0^1\}$ and $E(\phi_1^0) = \{\psi_0^1, \phi_0^1\}$, as defined in Definition 2.4. The set Ψ_1^1 contains the formulas

$$\psi_0^1 \wedge (\exists x_1 \psi_0^2) \wedge (\forall x_1\, x_1 \neq x_0 \to \psi_0^2)$$

and

$$\psi_0^1 \wedge (\exists x_1 \phi_0^2) \wedge (\forall x_1\, x_1 \neq x_0 \to \phi_0^2),$$

with the same omissions as ψ_1^0. Call these formulas ψ_1^1 and ϕ_1^1. Then $E(\psi_1^1) = \{\psi_0^2\}$, $E(\phi_1^1) = \{\phi_0^2\}$,

$$V_{0,1}(\psi_1^1) = V_{0,1}(\phi_1^1) = \psi_0^1,$$

$H_1^1(\psi_1^1, i_0) = \psi_1^0$ and $H_1^1(\phi_1^1, i_0) = \phi_1^0$. Note that the function H_1^1 changes the bound variables (as well as the free variables).

---

3   Scott processes

This section introduces the central topic of the chapter, the class of Scott processes (for a relational vocabulary τ).

Definition 3.1. A *Scott process* is a sequence $\langle \Phi_\alpha : \alpha < \delta \rangle$, for some nonzero ordinal δ (the *length* of the process), satisfying the following conditions, where for each ordinal α and each $n \in \omega$, Φ_α^n denotes the set $\Phi_\alpha \cap \Psi_\alpha^n$.

1. **The Formula Conditions**

   (a) Each Φ_α is a nonempty subset of the corresponding set Ψ_α.

   (b) For each ordinal of the form $\alpha + 1 < \delta$, and each $\phi \in \Phi_{\alpha+1}$, $E(\phi)$ is a subset of Φ_α.

   (c) For all $\alpha < \beta < \delta$, $\Phi_\alpha = V_{\alpha,\beta}[\Phi_\beta]$.

   (d) For all $\alpha < \delta$, all $n \in \omega$, all $j \in \mathcal{I}_{n,n}$ and all $\phi \in \Phi_\alpha^n$, $H_\alpha^n(\phi, j) \in \Phi_\alpha^n$.

   (e) For all $\alpha < \delta$, and all $m < n$ in ω, $\Phi_\alpha^m = H_\alpha^n[\Phi_\alpha^n \times \{i_m\}]$.

2. **The Coherence Conditions**

   (a) For each ordinal of the form $\alpha + 1$ below δ, each $n \in \omega$ and each $\phi \in \Phi_{\alpha+1}^n$,

   $$E(\phi) = V_{\alpha,\alpha+1}[\{\psi \in \Phi_{\alpha+1}^{n+1} \mid H_{\alpha+1}^{n+1}(\psi, i_n) = \phi\}].$$

(b) For all $\alpha < \beta < \delta$, all $n \in \omega$ and all $\phi \in \Phi_\beta^n$,

$$E(V_{\alpha+1,\beta}(\phi)) \subseteq V_{\alpha,\beta}[\{\psi \in \Phi_\beta^{n+1} \mid H_\beta^{n+1}(\psi, i_n) = \phi\}].$$

(c) For all $\alpha < \delta$, n, m in ω, $\phi \in \Phi_\alpha^n$ and $\psi \in \Phi_\alpha^m$, there exist $\theta \in \Phi_\alpha^{n+m}$ and $j \in \mathfrak{I}_{m,n+m}$ such that $\phi = H_\alpha^{n+m}(\theta, i_n)$ and $\psi = H_\alpha^{n+m}(\theta, j)$.

The sets Φ_α are called the *levels* of the Scott process.

Remark 3.2. Condition (2b) of Definition 3.1 includes the left to right inclusion in condition (2a). We prefer the given formulation of condition (2a), as it gives a better sense of the meaning of $E(\phi)$.

Remark 3.3. Proposition 4.4 shows that equality holds in condition (2b) of Definition 3.1, for any Scott process, so that conditions (2a) and (2b) could equivalently be replaced by condition (2b) alone with $=$ in place of $\subseteq$.

Remark 3.4. Conditions (1d) and (1e) of Definition 3.1 combine to give the following: for all $\alpha < \delta$, all $m \leq n$ in ω and all $j \in \mathfrak{I}_{m,n}$, $\Phi_\alpha^m = H_\alpha^n[\Phi_\alpha^n \times \{j\}]$.

Proposition 3.5 shows that each level of a Scott process contains a unique sentence.

Proposition 3.5. *Whenever $\langle \Phi_\alpha : \alpha < \delta \rangle$ is a Scott process, Φ_α^0 has a unique element, for each $\alpha < \delta$.*

Proof. That each Φ_α^0 is nonempty follows from conditions (1a) and (1e) of Definition 3.1. Now, suppose that ϕ and ψ are elements of Φ_α^0. By condition (2c) of Definition 3.1, there exist $\theta \in \Phi_\alpha^0$ and $j \in \mathfrak{I}_{0,0}$ such that $\phi = H_\alpha^0(\theta, i_0)$ and $\psi = H_\alpha^0(\theta, j)$. However, i_0 is the unique element of $\mathfrak{I}_{0,0}$, so $\phi = \psi$. $\square$

It follows from Proposition 3.5 and conditions (1c), (1e) and (2a) of Definition 3.1 that whenever $\langle \Phi_\alpha : \alpha < \delta \rangle$ is a Scott process and α is an ordinal with $\alpha + 1 < \delta$, if ϕ is the unique element of $\Phi_{\alpha+1}^0$, then $E(\phi) = \Phi_\alpha^1$.

---

4   Consequences of coherence

In this section we prove some basic facts about Scott processes, primarily about sets of the form $E(\phi)$ and the projection functions. The main result of the section is Proposition 4.4, which was referred to in Remark 3.3. We fix for this section a Scott process $\langle \Phi_\alpha : \alpha < \delta \rangle$.

Proposition 4.1 follows from Proposition 2.15 (i.e., the commutativity of the horizontal and vertical projections). The failure of the reverse inclusion is witnessed whenever a set of the form $V_{\alpha,\beta}^{-1}[\{\rho\}]$ has distinct members ϕ_1, ϕ_2.

Proposition 4.1. *For all $\alpha \leq \beta < \delta$, all $m \leq n \in \omega$, all $j \in \mathfrak{I}_{m,n}$, and all $\phi \in \Phi_\beta^m$,*

$$V_{\alpha,\beta}[\{\psi \in \Phi_\beta^n \mid H_\beta^n(\psi, j) = \phi\}] \subseteq \{\theta \in \Phi_\alpha^n \mid H_\alpha^n(\theta, j) = V_{\alpha,\beta}(\phi)\}.$$

The right-to-left inclusion in Proposition 4.2 says that every one-point extension of a formula ϕ at level α is a member of $E(\psi)$, for some $\psi \in V_{\alpha,\alpha+1}^{-1}[\phi]$. This proposition is used in Remark 5.15.

Proposition 4.2. *For each ordinal of the form $\alpha + 1$ below δ, each $n \in \omega$ and each $\phi \in \Phi_\alpha^n$,*

$$\bigcup \{E(\psi) \mid \psi \in V_{\alpha,\alpha+1}^{-1}[\{\phi\}]\} = \{\theta \in \Phi_\alpha^{n+1} \mid H_\alpha^{n+1}(\theta, i_n) = \phi\}.$$

Proof. The left-to-right inclusion follows from Proposition 4.1 and condition (2a) of Definition 3.1. The reverse inclusion follows from conditions (1c) and (2a) of Definition 3.1, and Proposition 2.15. $\square$

Proposition 4.3 is the successor case of Proposition 4.4, modulo condition (2a) of Definition 3.1.

Proposition 4.3. *For all $\alpha \leq \beta$ such that $\beta + 1 < \delta$, and all $\phi \in \Phi_{\beta+1}$,*

$$E(V_{\alpha+1,\beta+1}(\phi)) = V_{\alpha,\beta}[E(\phi)].$$

Proof. Fix $n \in \omega$ such that $\phi \in \Phi_{\beta+1}^n$. For the forward direction, condition (2b) of Definition 3.1 gives that

$$E(V_{\alpha+1,\beta+1}(\phi)) \subseteq V_{\alpha,\beta+1}[\{\psi \in \Phi_{\beta+1}^{n+1} \mid H_{\beta+1}^{n+1}(\psi, i_n) = \phi\}],$$

which by condition (4) of Definition 2.6 is equal to

$$V_{\alpha,\beta}[V_{\beta,\beta+1}[\{\psi \in \Phi_{\beta+1}^{n+1} \mid H_{\beta+1}^{n+1}(\psi, i_n) = \phi\}]],$$

which by condition (2a) of Definition 3.1 is equal to $V_{\alpha,\beta}[E(\phi)]$.

For the reverse direction we have from condition (2a) of Definition 3.1 that $V_{\alpha,\beta}[E(\phi)]$ is equal to

$$V_{\alpha,\beta}[V_{\beta,\beta+1}[\{\psi \in \Phi_{\beta+1}^{n+1} \mid H_{\beta+1}^{n+1}(\psi, i_n) = \phi\}]],$$

which by Remark 2.8 is equal to

$$V_{\alpha,\alpha+1}[V_{\alpha+1,\beta+1}[\{\psi \in \Phi_{\beta+1}^{n+1} \mid H_{\beta+1}^{n+1}(\psi, i_n) = \phi\}]],$$

which by Proposition 4.1 is contained in

$$V_{\alpha,\alpha+1}[\{\theta \in \Phi_{\alpha+1}^{n+1} \mid H_{\alpha+1}^{n+1}(\theta, i_n) = V_{\alpha+1,\beta+1}(\phi)\}]],$$

which by condition (2a) of Definition 3.1 is equal to $E(V_{\alpha+1,\beta+1}(\phi))$. $\square$

We now show that the reverse inclusion of condition (2b) of Definition 3.1 holds for any Scott process.

Proposition 4.4. *For all $\alpha < \beta < \delta$, for all $n \in \omega$ and all $\phi \in \Phi_\beta^n$,*

$$E(V_{\alpha+1,\beta}(\phi)) = V_{\alpha,\beta}[\{\psi \in \Phi_\beta^{n+1} \mid H_\beta^{n+1}(\psi, i_n) = \phi\}].$$

Proof. When β is a successor ordinal, this is Proposition 4.3, using condition (2a) of Definition 3.1. For any β, the left-to-right inclusion is condition (2b) of Definition 3.1. For the reverse inclusion,

$$V_{\alpha,\beta}[\{\psi \in \Phi_\beta^{n+1} \mid H_\beta^{n+1}(\psi, i_n) = \phi\}]$$

is equal to

$$V_{\alpha,\alpha+1}[V_{\alpha+1,\beta}[\{\psi \in \Phi_\beta^{n+1} \mid H_\beta^{n+1}(\psi, i_n) = \phi\}]]$$

by Remark 2.8, and this is contained in

$$V_{\alpha,\alpha+1}[\{\psi \in \Phi_{\alpha+1}^{n+1} \mid H_{\alpha+1}^{n+1}(\psi, i_n) = V_{\alpha+1,\beta}(\phi)\}],$$

by Proposition 4.1. Finally, this last term is equal to $E(V_{\alpha+1,\beta}(\phi))$ by condition (2a) of Definition 3.1. □

---

5 Ranks and Scott sentences

The *Scott rank* of a τ-structure M (see [Hod93, Mar02], for instance) is the least ordinal α such that $V_{\alpha,\alpha+1}$ is injective on $\Phi_{\alpha+1}(M)$ (which we defined in the introduction). If α is the Scott rank of M, then $V_{\beta,\beta+1}$ injective on $\Phi_{\beta+1}(M)$ for all $\beta \geq \alpha$ as well. Proposition 5.5 below verifies that Scott processes have the same property. We isolate the successor step of the proof as a separate proposition. The second part of the proposition is used in Remark 9.11.

Proposition 5.1. *Let β be an ordinal, and let $\langle \Phi_\alpha : \alpha \leq \beta + 2 \rangle$ be a Scott process. If ϕ is an element of $\Phi_{\beta+1}$, then each of the following conditions implies that $V_{\beta+1,\beta+2}^{-1}[\{\phi\}] \cap \Phi_{\beta+2}$ is a singleton.*

1. $V_{\beta,\beta+1}^{-1}[\{\psi\}] \cap \Phi_{\beta+1}$ *is a singleton for each* $\psi \in E(\phi)$.

2. *There exists a* $\psi \in E(\phi)$ *such that* $V_{\beta,\beta+2}^{-1}[\{\psi\}] \cap \Phi_{\beta+2}$ *is a singleton.*

Proof. Condition (1c) of Definition 3.1 implies that $V_{\beta+1,\beta+2}^{-1}[\{\phi\}] \cap \Phi_{\beta+2}$ is nonempty. Suppose that $\phi' \in \Phi_{\beta+2}$ is such that $V_{\beta+1,\beta+2}(\phi') = \phi$.

If (1) holds then, by Proposition 4.3, $V_{\beta,\beta+1}[E(\phi')] = E(\phi)$. Since

$$V_{\beta,\beta+1}^{-1}[\{\psi\}] \cap \Phi_{\beta+1}$$

is a singleton for each $\psi \in E(\phi)$, this implies that $E(\phi') = V_{\beta,\beta+1}^{-1}[E(\phi)] \cap \Phi_{\beta+1}$, which uniquely determines ϕ'.

If (2) holds, let ψ' be the unique member of $V_{\beta,\beta+2}^{-1}[\{\psi\}] \cap \Phi_{\beta+2}$. Since $\psi \in E(\phi)$, $V_{\beta+1,\beta+2}(\psi')$ is a member of $E(\phi')$, by Proposition 4.3. Let $n \in \omega$ be such that $\phi \in \Phi_{\beta+1}^n$. Then $\phi' = H_{\beta+2}^{n+1}(\psi', i_n)$, by part (2a) of Definition 3.1. $\square$

Corollary 5.2 is a consequence of part (1) of Proposition 5.1, and Corollary 5.3 is a consequence of part (2) (using Proposition 4.3).

Corollary 5.2. *Let β be an ordinal, and let $\langle \Phi_\alpha : \alpha \leq \beta + 2 \rangle$ be a Scott process. Suppose that $n \in \omega$ is such that $V_{\beta,\beta+1}$ is injective on $\Phi_{\beta+1}^{n+1}$. Then $V_{\beta+1,\beta+2}$ is injective on $\Phi_{\beta+2}^n$.*

Corollary 5.3. *Let β and δ be ordinals with $\delta \geq \beta + 2$, and let*

$$\langle \Phi_\alpha : \alpha \leq \delta \rangle$$

be a Scott process. Suppose that $\phi \in \Phi_{\beta+1}$ and $\psi \in E(\phi)$ are such that $V_{\beta,\delta}^{-1}[\{\psi\}] \cap V_\delta$ is a singleton. Then $V_{\beta+1,\delta}^{-1}[\{\phi\}] \cap \Phi_\delta$ is a singleton.

Remark 5.4. It is natural to ask whether part (1) of Proposition 5.1 has a converse, in the sense that if $\langle \Phi_\alpha : \alpha \leq \beta + 1 \rangle$ is a Scott process and $\phi \in \Phi_{\beta+1}$ and $\psi \in E(\phi)$ are such that $V_{\beta,\beta+1}^{-1}[\{\psi\}]$ has at least two members, then there must exist a set $\Phi_{\beta+2}$ such that $\langle \Phi_\alpha : \alpha \leq \beta + 2 \rangle$ is a Scott process and $V_{\beta+1,\beta+2}^{-1}[\{\phi\}]$ is not a singleton. This is not the case in general, however, as by Proposition 3.5, each function of the form $V_{\alpha,\alpha+1} \restriction \Phi_{\alpha+1}^0$ is always injective.

Proposition 5.5. *If $\langle \Phi_\alpha : \alpha < \delta \rangle$ is a Scott process, $\beta < \gamma$ are ordinals with $\gamma + 1 < \delta$, and $V_{\beta,\beta+1} \restriction \Phi_{\beta+1}$ is injective, then $V_{\gamma,\gamma+1} \restriction \Phi_{\gamma+1}$ is injective.*

Proof. Letting η be such that $\gamma = \beta + \eta$, we prove the proposition by induction on η, for all β and δ simultaneously. Applying the induction hypotheses, the limit case follows from Remark 2.8, and the successor case follows from part (1) of Proposition 5.1 (and also from Corollary 5.2). $\square$

Definition 5.6. The *rank* of a Scott process $\langle \Phi_\alpha : \alpha < \delta \rangle$ is the least β such that $V_{\beta,\beta+1} \restriction \Phi_{\beta+1}$ is injective, if such a β exists, and undefined otherwise. We say that a Scott process is *terminating* (or *terminates*) if its rank is defined, and *nonterminating* otherwise.

The rank of (any sufficiently long set-sized initial segment of) the Scott process of a τ-structure M is the same as the Scott rank of M.

Remark 5.7. Suppose that β and γ are ordinals, and $n \in \omega$ is such that $\gamma > \beta + n + 1$. Suppose that $\langle \Phi_\alpha : \alpha < \gamma \rangle$ is a Scott process, and that $V_{\beta, \beta+1}$ is injective on $\Phi^m_{\beta+1}$, for all $m > n$ in ω. By Corollary 5.2, the rank of $\langle \Phi_\alpha : \alpha < \gamma \rangle$ is at most $\beta + n$ (since each Φ^0_α is a singleton, $V_{\alpha, \alpha+1} {\restriction} \Phi^0_{\alpha+1}$ is injective for all α).

In the following definition, j can equivalently be replaced with i_n, by condition (1d) of Definition 3.1.

Definition 5.8. Let β and γ be ordinals such that $\gamma > \beta + 1$, and let

$$\langle \Phi_\alpha : \alpha < \gamma \rangle$$

be a Scott process. Let n be an element of ω, and let ϕ be an element of Φ^n_β. We say that the Scott process $\langle \Phi_\alpha : \alpha < \gamma \rangle$ is *injective beyond* ϕ if for all $m \in \omega \backslash n$, all $j \in \mathfrak{I}_{n,m}$ and all $\psi \in \Phi^m_\beta$ such that $\phi = H^m_\beta(\psi, j)$, $V^{-1}_{\beta, \beta+1}[\{\psi\}] \cap \Phi_{\beta+1}$ is a singleton.

Remark 5.9. Let $\beta < \gamma$ be ordinals, and let $\langle \Phi_\alpha : \alpha \leq \gamma \rangle$ be a Scott process. Let n be an element of ω, and let $\phi \in \Phi^n_\beta$ be such that $\langle \Phi_\alpha : \alpha < \gamma \rangle$ is injective beyond ϕ. Applying part (1) of Proposition 5.1, one can show by induction that for all $\delta \in [\beta, \gamma]$,

- $V^{-1}_{\beta, \delta}[\{\phi\}] \cap \Phi_\delta$ is a singleton;

- if $\delta < \gamma$ then $\langle \Phi_\alpha : \alpha \leq \delta \rangle$ is injective beyond the unique member of $V^{-1}_{\beta, \delta}[\{\phi\}]$;

- for all $m \in \omega \setminus n$, all $j \in \mathfrak{I}_{n,m}$ and all $\psi \in \Phi^m_\beta$ such that

$$\phi = H^m_\beta(\psi, j),$$

$V^{-1}_{\beta, \delta}[\{\psi\}] \cap \Phi_\delta$ is a singleton.

Remark 5.10. Let β be an ordinal, and n an element of ω. Suppose that

$$\langle \Phi_\alpha : \alpha \leq \beta + 1 \rangle$$

is a Scott process, and that $\phi \in \Phi^n_\beta$ is such that $\langle \Phi_\alpha : \alpha \leq \beta + 1 \rangle$ is injective beyond ϕ. The proof of Scott's Isomorphism Theorem (Theorem 2.4.15 of [Mar02]; using $\bar{a}$ in place of $\emptyset$ at stage 0) shows that for any two countable τ-structures M and N whose Scott processes agree with $\langle \Phi_\alpha : \alpha \leq \beta + 1 \rangle$ through level $\beta + 1$, if $\bar{a}$ is an n-tuple from M and $\bar{b}$ is an n-tuple from N, each satisfying ϕ in their respective models, then there is an isomorphism of M and N sending $\bar{a}$ to $\bar{b}$. Alternately, one can show that for each ordinal $\gamma > \beta + 1$, there is a unique Scott process of length γ extending β, using either Remark 5.9 or Proposition 9.3.

Remark 5.11. In the situation of Definition 5.8, $\langle \Phi_\alpha : \alpha \leq \beta + 1 \rangle$ need not have rank β. To see this, consider the Scott process of a countably infinite undirected graph G consisting of an infinite set of nodes which are not connected to anything, and another infinite set of nodes which are all connected to each other, but not to themselves. The formula in $\Phi_0^2(G)$ corresponding to a connected pair has the property of ϕ in Remark 5.10, but the Scott rank of G is 1, not 0, since the unique member of $\Phi_0^1(G)$ has two successors in $\Phi_1^1(G)$.

The following definition is inspired by Remarks 5.10 and 5.11. There may be some connection between the notion of pre-rank and the subject of [Mon15].

Definition 5.12. The *pre-rank* of a Scott process $\langle \Phi_\alpha : \alpha < \beta \rangle$ is the least $\gamma \leq \beta$ such that for all ordinals $\eta > \gamma$, there exists a unique Scott process of length η extending $\langle \Phi_\alpha : \alpha < \gamma \rangle$ (if such a γ exists). The Scott *pre-rank* of a τ-structure is the pre-rank of any terminating initial segment of its Scott process.

By Proposition 5.5, the pre-rank of a terminating Scott process is at most one more than its rank; Remark 5.11 shows that it can be smaller. By Proposition 9.24, if a Scott process has countable length, and all of its levels are countable, then its rank is at most ω more than its pre-rank. Proposition 5.13 gives a tighter bound in the situation of Definition 5.8.

Proposition 5.13. *Let β be an ordinal, and n an element of ω. Suppose that*

$$\langle \Phi_\alpha : \alpha \leq \beta + n + 1 \rangle$$

is a Scott process, and that $\phi \in \Phi_\beta^n$ is such that $\langle \Phi_\alpha : \alpha \leq \beta + n + 1 \rangle$ is injective beyond ϕ. Then $\langle \Phi_\alpha : \alpha \leq \beta + n + 1 \rangle$ has rank at most $\beta + n$.

Proof. For each $p \leq n$, let Υ_p be the set of $\psi \in \Phi_{\beta+p}$ for which

$$V_{\beta+p,\beta+n+1}^{-1}[\{\psi\}] \cap \Phi_{\beta+n+1}$$

is a singleton. We want to see that $\Upsilon_n = \Phi_{\beta+n}$.

By Remark 5.9, we have the following, for each $p \leq n$:

- $V_{\beta,\beta+p}^{-1}[\{\phi\}]$ has a single element (which we call ϕ_p);

- for each $m \in \omega \setminus n$, Υ_p contains each $\psi \in \Phi_{\beta+p}^m$ for which $\phi_p = H_\beta^m(\psi, j)$ for some $j \in \mathfrak{I}_{n,m}$.

We prove the following statement by induction on $p \leq n$: if $k \in \omega$ is such that $k + p \geq n$ and $\theta \in \Phi_{\beta+p}^k$ is such that $\theta = H_{\beta+p}^{k+p}(\rho, i_k)$ for some $\rho \in \Phi_{\beta+p}^{k+p}$ such that $V_{\beta,\beta+p}(\rho) \in \Upsilon_0$, then θ is in Υ_p. For $p = 0$ this is immediate.

For the induction step from p to $p + 1$, fix

- $k \in \omega$ such that $k + p + 1 \geq n$,

- $\theta \in \Phi_{\beta+p+1}^k$ and

- $\rho \in \Phi_{\beta+p+1}^{k+p+1}$

such that $V_{\beta,\beta+p+1}(\rho)$ is in Υ_0 and $\theta = H_{\beta+p+1}^{k+p+1}(\rho, i_k)$. The induction hypothesis gives that

$$V_{\beta+p,\beta+p+1}(H_{\beta+p+1}^{k+p+1}(\rho, i_{k+1})),$$

is in Υ_p, which (as this formula is in $E(\theta)$) by Corollary 5.3 shows that θ is in Υ_{p+1} as desired.

Finally, the statement for $n = p$ implies the proposition, applying condition (2c) of Definition 3.1 to an arbitrary $\theta \in \Phi_{\beta+n}$ and ϕ_n to obtain the desired formula ρ. $\qquad\square$

Definition 5.14. Given an ordinal δ and a set $\Phi \subseteq \Psi_\delta$, the *maximal completion* of Φ is the set of $\phi \in \Psi_{\delta+1}$ such that for some $n \in \omega$ and some $\phi' \in \Phi \cap \Psi_\delta^n$, $V_{\delta,\delta+1}(\phi) = \phi'$, and

$$E(\phi) = \{\psi \in \Phi \cap \Psi_\delta^{n+1} \mid H_\delta^{n+1}(\psi, i_n) = \phi'\}.$$

The extension of a Scott process of successor length by the maximal completion of its last level may not be a Scott process (see Proposition 5.19 below).

Remark 5.15. By Proposition 4.2, if $\langle \Phi_\alpha : \alpha < \delta \rangle$ is a Scott process, and β is an ordinal such that $\beta + 1 < \delta$, then $V_{\beta,\beta+1} \restriction \Phi_{\beta+1}$ is injective if and only if $\Phi_{\beta+1}$ is the maximal completion of Φ_β.

The following definition describes the situation in which no formula ϕ has incompatible horizontal extensions.

Definition 5.16. Given an ordinal δ, a set $\Phi \subseteq \Psi_\delta$ satisfies the *amalgamation property* (or *amalgamates*) if for all $m < n \in \omega$, $\phi \in \Phi \cap \Psi_\delta^{m+1}$, and $\psi \in \Phi \cap \Psi_\delta^n$ such that $H_\delta^{m+1}(\phi, i_m) = H_\delta^n(\psi, i_m)$, there exist $\theta \in \Phi \cap \Psi_\delta^{n+1}$ and $y \in X_{n+1} \setminus X_m$ such that $H_\delta^{n+1}(\theta, i_m \cup \{(x_m, y)\}) = \phi$ and $H_\delta^{n+1}(\theta, i_n) = \psi$.

Remark 5.17. Given an ordinal δ and a set $\Phi \subseteq \Phi_\delta$ satisfying condition (1d) of Definition 3.1 (i.e., closure under the functions H_δ^n ($n \in \omega$)), the amalgamation property for a set $\Phi \subseteq \Psi_\delta$ is equivalent to the statement that for all $m \leq n \in \omega$, $\phi \in \Phi \cap \Psi_\delta^m$, $j \in \mathcal{F}_{m,n}$ and $\psi \in \Phi \cap \Psi_\delta^n$ such that $\phi = H_\delta^n(\psi, j)$,

$$\{\theta \in \Phi \cap \Psi_\delta^{m+1} \mid H_\delta^{m+1}(\theta, i_m) = \phi\}$$

is the same as

$$H_\delta^{n+1}[\{\rho \in \Phi \cap \Psi_\delta^{n+1} \mid H_\delta^{n+1}(\rho, i_n) = \psi\} \times \{j \cup \{(x_m, y)\} \mid y \in (X_{n+1} \setminus \mathrm{range}(j))\}].$$

This follows immediately from the definitions (using part (3) of Definition 2.11).

Remark 5.18. The set in the second displayed formula in Remark 5.17 is always contained in the set in the first, by part (2) of Remark 2.14.

Proposition 5.19. *The extension of a nonempty Scott process of nonlimit length by the maximal completion of its last level induces a Scott process if and only if its last level amalgamates.*

Proof. Let $\langle \Phi_\alpha : \alpha \leq \delta \rangle$ be a Scott process. Conditions (1a)-(1c) of Definition 3.1 are always satisfied by the extension by the maximal completion. The other conditions depend on whether the functions $H_{\delta+1}^n$ ($n \in \omega$) lift the actions of the functions H_δ^n ($n \in \omega$), i.e., whether whenever $n \in \omega$, $j \in \bigcup_{m \leq n} \mathfrak{I}_m$, $\psi \in \Phi_\delta^n$ and ψ' is the unique member of $V_{\delta,\delta+1}^{-1}[\{\psi\}]$ in the maximal completion of Φ_δ, $H_{\delta+1}^n(\psi', j)$ is the unique member of $V_{\delta,\delta+1}^{-1}[\{H_\delta^n(\psi, j)\}]$ in the maximal completion of Φ_δ. Comparing the condition (3) of Definition 2.11 with Definition 5.14 shows that is exactly the statement that Φ_δ amalgamates as expressed in Remark 5.17. $\qquad\square$

We conclude this section by giving a restatement of the amalgamation property which will be useful in Section 7. A failure of amalgamation gives a counterexample to Proposition 5.20 with $n = m + 1$.

Proposition 5.20. *Suppose that $\langle \Phi_\alpha : \alpha \leq \delta \rangle$ is Scott process whose last level amalgamates, and that $m, n, p \in \omega$ are such that $m \leq \min\{n, p\}$. Suppose now that $j \in \mathfrak{I}_{m,n}$, $k \in \mathfrak{I}_{m,p}$, $\psi \in \Phi_\delta^n$ and $\theta \in \Phi_\delta^p$ are such that*

$$H_\delta^n(\psi, j) = H_\delta^p(\theta, k).$$

Then there exist $q \in \omega \setminus \max\{n, p\}$, a formula $\rho \in \Phi_\delta^q$ and functions $j' \in \mathfrak{I}_{n,q}$ and $k' \in \mathfrak{I}_{p,q}$ such that $X_q = \mathrm{range}(j') \cup \mathrm{range}(k')$, $j' \circ j = k' \circ k$, $\psi = H_\delta^q(\rho, j')$ and $\theta = H_\delta^q(\rho, k')$.

Proof. Fixing m and p, we prove the proposition by induction on n. If $n = m$, then we can let $q = p$, $\rho = \theta$, $k' = i_p$ and $j' = k \circ j^{-1}$, using the second half of Remark 2.14, which we do repeatedly throughout this proof.

Suppose now that the proposition holds for some $n \in \omega \setminus m$. Let $j \in \mathfrak{I}_{m,n+1}$, $k \in \mathfrak{I}_{m,p}$, $\psi \in \Phi_\delta^{n+1}$ and $\theta \in \Phi_\delta^p$ be such that $H_\delta^{n+1}(\psi, j) = H_\delta^p(\theta, k)$. Let $f \in \mathfrak{I}_{n+1,n+1}$ be i_{n+1} if $x_n \notin \mathrm{range}(j)$; otherwise, fix n' such that $x_{n'} \notin \mathrm{range}(j)$ and let f map x_n and $x_{n'}$ to each other and fix the rest of X_{n+1}. Then $f^{-1} = f$ and $x_n \notin \mathrm{range}(f \circ j)$. Let $\psi_0 = H_\delta^{n+1}(\psi, f)$; then $\psi = H_\delta^{n+1}(\psi, f)$. Let $\psi_1 = H_\delta^{n+1}(\psi_0, i_n)$. By the second part of Remark 2.14,

$$
\begin{aligned}
H_\delta^n(\psi_1, f \circ j) &= H_\delta^n(H_\delta^{n+1}(\psi_0, i_n), f \circ j) \\
&= H_\delta^{n+1}(\psi_0, i_n \circ (f \circ j)) \\
&= H_\delta^{n+1}(\psi_0, f \circ j) \\
&= H_\delta^{n+1}(H_\delta^{n+1}(\psi_0, f), j) \\
&= H_\delta^{n+1}(\psi, j) \\
&= H_\delta^p(\theta, k).
\end{aligned}
$$

Applying the induction hypothesis to $f \circ j$, k, ψ_1 and θ, we get $q_0 \in \omega \setminus$

$\max\{n,p\}$, a formula $\rho_0 \in \Phi_\delta^{q_0}$ and functions $j_0 \in \mathfrak{I}_{n,q_0}$ and $k' \in \mathfrak{I}_{p,q_0}$ such that

$$X_{q_0} = \text{range}(j_0) \cup \text{range}(k'),$$

$j_0 \circ (f \circ j) = k' \circ k$, $\psi_1 = H_\delta^{q_0}(\rho_0, j_0)$ and $\theta = H_\delta^{q_0}(\rho_0, k')$.

Suppose first that there exists a $y \in X_{q_0} \setminus \text{range}(j_0)$ such that

$$\psi_0 = H_\delta^{q_0}(\rho_0, j_0 \cup \{(x_n, y)\}).$$

Then q_0, ρ_0 and k' are as desired. If $f = i_{n+1}$, then we can let $j' = j_0 \cup \{(x_n, y)\}$ and we are done. Otherwise, let j' send $x_{n'}$ to y, x_n to $j_0(x_{n'})$ and every other member of X_n to the same place that j_0 does (i.e., let $j' = (j_0 \cup \{(x_n, y)\}) \circ f$). Then $j' \circ j = k' \circ k$, and

$$\begin{aligned}
\psi &= H_\delta^{n+1}(\psi_0, f) \\
&= H_\delta^{n+1}(H_\delta^q(\rho, j_0 \cup \{(x_n, y)\}), f) \\
&= H_\delta^q(\rho, (j_0 \cup \{(x_n, y)\}) \circ f) \\
&= H_\delta^q(\rho, j'),
\end{aligned}$$

as desired.

Finally, suppose that there is no such $y \in X_{q_0} \setminus \text{range}(j_0)$. Putting together the amalgamation property of Φ and the equation $\psi_1 = H_{\delta+1}^{n+1}(\psi_0, i_n)$, we get that there exist a formula $\rho \in \Phi_\delta^{q_0+1}$ such that $H_\delta^{q_0+1}(\rho, i_{q_0}) = \rho_0$ and a $y \in X_{q_0+1} \setminus \text{range}(j_0)$ such that $H_\delta^{q_0+1}(\rho, j_0 \cup \{(x_n, y)\}) = \psi_0$. Then k', ρ, and $q = q_0 + 1$ are as desired. If $f = i_{n+1}$, then we can let $j' = j_0 \cup \{(x_n, y)\}$, and we are done. Otherwise, as above, let $j' = (j_0 \cup \{(x_n, y)\}) \circ f$. Then again $j' \circ j = k' \circ k$ and $\psi = H_\delta^q(\rho, j')$, as desired. $\qquad\square$

6   Building countable models

We say that a τ-structure M is a *model* of a Scott process $\langle \Phi_\alpha : \alpha < \delta \rangle$ if $\Phi_\alpha = \Phi_\alpha(M)$ for all $\alpha < \delta$. In this section, we show that any Scott process of successor length has a countable model if its last level is countable. This in turn implies that such a sequence can be extended to any given ordinal length (although the rank of the Scott process of length ω_1 corresponding to a countable model is countable).

Definition 6.1. Given an ordinal β, and a countable set $\Phi \subseteq \Psi_\beta$, a *thread* through Φ is a set of formulas $\{\phi_n : n \in \omega\} \subseteq \Phi$ such that

1. for all $n \in \omega$, $\phi_n \in \Psi_\beta^n$;

2. for all $m < n$ in ω, $\phi_m = H_\beta^n(\phi_n, i_m)$;

3. for all $m \in \omega$, all $\alpha < \beta$, and all $\psi \in E(V_{\alpha+1,\beta}(\phi_m))$, there exist an $n \in \omega \backslash (m+1)$ and a $y \in X_n \backslash X_m$ such that $\psi = V_{\alpha,\beta}(H_\beta^n(\phi_n, i_m \cup \{(x_m, y)\}))$.

Remark 6.2. If β is a successor ordinal, condition (3) of Definition 6.1 is equivalent to the restriction of the condition to the case where $\alpha = \beta - 1$. This follows from Proposition 4.3. Similarly, condition (3) of Definition 6.1 is equivalent to the restriction of the condition to the set of α in any cofinal subset of β.

Remark 6.3. Suppose that $\langle \Phi_\alpha : \alpha \leq \delta \rangle$ is a Scott process, and $\beta < \delta$ is such that $V_{\beta,\delta} \restriction \Phi_\delta$ is injective. Then the $V_{\beta+1,\delta}$-preimage of a thread through $\Phi_{\beta+1}$ is a thread through Φ_δ. This follows from Remark 2.9, Proposition 2.15 and Proposition 4.3. For conditions (1) and (2) of Definition 6.1 this is almost immediate; for condition (3) it requires tracing through the horizontal and vertical projections.

The proof of Proposition 6.4 shows how to build thread for a Scott process whose last level is countable. The proof of Theorem 6.5 then shows how to use such a thread to build a model of the Scott process.

Proposition 6.4. *If $\langle \Phi_\alpha : \alpha \leq \delta \rangle$ is a Scott process with Φ_δ countable, then there exists a thread through Φ_δ.*

Proof. The case $\delta = 0$ follows easily from condition (1e) of Definition 3.1, so suppose that δ is positive. By Remark 6.3, it suffices to consider the cases where δ is either a successor ordinal or an ordinal of cofinality ω. Let A be $\{\delta - 1\}$ in the case where δ is a successor ordinal, and a countable cofinal subset of δ otherwise. We choose the formulas ϕ_n recursively, meeting instances of condition (3) of Definition 6.1 for $\alpha \in A$ while satisfying condition (2). Note that ϕ_0 is the unique element of Φ_δ^0. To satisfy an instance of condition (3), we need to see that if $m \leq n$ are in ω, $\alpha \in A$, ϕ_n has been chosen, and $\psi \in E(V_{\alpha+1,\beta}(\phi_m))$ is not equal to $V_{\alpha,\beta}(H_\beta^n(\phi_n, i_m \cup \{(x_m, y)\}))$ for any $y \in X_n \setminus X_m$, then ϕ_{n+1} can be chosen so that

$$\psi = V_{\alpha,\beta}(H_\beta^{n+1}(\phi_{n+1}, i_m \cup \{(x_m, x_n)\}))$$

(since Φ_δ is countable, the set of such formulas ψ is also countable). The existence of such a ϕ_{n+1} follows from condition (3) of Definition 2.11 applied to $V_{\alpha+1,\beta}(\phi_n)$ and i_m, giving a $\theta \in E(V_{\alpha+1,\beta}(\phi_n))$ such that

$$H_\alpha^{n+1}(\theta, i_m \cup \{(x_m, x_n)\}) = \psi,$$

followed by condition (2b) of Definition 3.1 applied to ϕ_n, giving ϕ_{n+1} as desired. $\square$

Theorem 6.5. *Given a Scott process $\langle \Phi_\alpha : \alpha \leq \delta \rangle$ with Φ_δ countable, a thread $\langle \phi_n : n \in \omega \rangle$ through Φ_δ and an injective sequence $C = \{c_n : n \in \omega\}$, there is a τ-structure with domain C in which each tuple $\langle c_m : m < n \rangle$ satisfies ϕ_n.*

Proof. Let each tuple $\langle c_m : m < n \rangle$ satisfy all the atomic formulas indicated by $V_{0,\delta}(\phi_n)$. We show by induction on $\alpha \leq \delta$ that each tuple $\langle c_m : m < n \rangle$ satisfies the formula $V_{\alpha,\delta}(\phi_n)$. This follows immediately for limit stages. For the induction step from α to $\alpha + 1$, $\langle c_m : m < n \rangle$ satisfies $V_{\alpha+1,\delta}(\phi_n)$ if and only if

$$E(V_{\alpha+1,\delta}(\phi_n)) = V_{\alpha,\delta}[\{H_\delta^p(\phi_p, i_n \cup \{(x_n, y)\}) : p \in \omega \setminus (n+1), y \in X_p \setminus X_n\}].$$

That is, checking that $\langle c_m : m < n \rangle$ satisfies $V_{\alpha+1,\delta}(\phi_n)$ means showing that the left side of the equality is the set of formulas from Φ_α^{n+1} satisfied by extensions of $\langle c_m : m < n \rangle$ by one point, which by the induction hypothesis is what the right side is. The left-to-right containment follows from condition (3) of Definition 6.1. For the other direction, note first that by Proposition 4.4,

$$E(V_{\alpha+1,\delta}(\phi_n)) = V_{\alpha,\delta}[\{\theta \in \Phi_\delta^{n+1} \mid H_\delta^{n+1}(\theta, i_n) = \phi_n\}].$$

That

$$\{H_\delta^p(\phi_p, i_n \cup \{(x_n, y)\}) : p \in \omega \setminus (n+1), y \in X_p \setminus X_n\}$$

is contained in $\{\theta \in \Phi_\delta^{n+1} \mid H_\delta^{n+1}(\theta, i_n) = \phi_n\}$ follows from the assumption that $\phi_n = H_\delta^p(\phi_p, i_n)$, by part (2) of Remark 2.14. □

Definition 6.6. Given an ordinal β and a countable set $\Phi \subseteq \Psi_\beta$, a thread

$$\{\phi_n : n \in \omega\}$$

through Φ is *complete* if for all $m \in \omega$ and all $\psi \in \Phi \cap \Psi_\beta^m$, there exist $n \in \omega$ and $j \in \mathcal{I}_{m,n}$ such that $\psi = H_\beta^n(\phi_n, j)$.

Remark 6.7. The thread through Φ_δ given by Proposition 6.4 induces (via Theorem 6.5) a model of $\langle \Phi_\alpha : \alpha < \delta \rangle$ for which the δ-th level of the corresponding Scott process is contained in the given Φ_δ. The δ-th level is equal to Φ_δ if and only if the thread is complete. Condition (2c) of Definition 3.1 implies that one can add stages to the construction in Proposition 6.4 to produce a complete thread.

Proposition 6.4, Theorem 6.5 and Remark 6.7 give the following.

Theorem 6.8. *Every Scott process $\langle \Phi_\alpha : \alpha \leq \delta \rangle$ with Φ_δ countable has a countable model.*

Remark 6.9. Theorem 9.9 gives a stronger version of Theorem 6.8, showing that every Scott process with all levels countable (and possibly of limit length) has a model.

Remark 6.10. The proof of Proposition 6.4 shows that each Scott process $\mathcal{P} = \langle \Phi_\alpha : \alpha \leq \delta \rangle$ with Φ_δ countable induces a partial order $\mathbb{Q}_\mathcal{P}$ of finite partial threads, which is essentially the same as Cohen forcing, except that there need not be splitting below each condition. Theorem 6.5 shows that each thread induces a model of $\mathcal{P}$. As we shall see in Section 10, if $\mathcal{P}$ gives rise to a counterexample to Vaught's Conjecture, then all Scott processes extending $\mathcal{P}$ which exist in any forcing extension exist already in the ground model. It follows then that (in this case) densely many conditions in $\mathbb{Q}_{\mathcal{P}}$ decide the Scott process of the structure corresponding to generic thread, and, letting γ be the rank of this Scott process, that there is a $\mathbb{Q}_{\mathcal{P}}$-name for a function associating each $\langle c_m : m < n \rangle$ in the structure associated to the generic thread with the formula at level γ of this Scott process which it is forced to satisfy. By a homogeneity argument, all conditions in $\mathbb{Q}_{\mathcal{P}}$ must force the same Scott process for this generic structure.

Given $n \leq p$ in ω, $\phi \in \Phi_\delta^n$ and ψ in Φ_δ^p such that $H_\delta^p(\psi, i_n) = \phi$, say that ϕ *amalgamates over* ψ if for all $q \in \omega \setminus p$ and all $\theta \in \Phi_\delta^q$

$$\{H_\delta^{q+1}(\rho, i_n \cup \{(x_n, y)\}) : y \in X_{q+1} \setminus X_n, \rho \in \Phi_\delta^{q+1}, H_\delta^{q+1}(\rho, i_p) = \theta\}$$

is equal to

$$\{H_\delta^{p+1}(\rho, i_n \cup \{(x_n, y)\}) : y \in X_{p+1} \setminus X_n, \rho \in \Phi_\delta^{p+1}, H_\delta^{p+1}(\rho, i_p) = \psi\}.$$

The argument of the previous paragraph shows that if $\mathcal{P}$ gives rise to a counterexample to Vaught's Conjecture, then for each $n \in \omega$ and each $\phi \in \Phi_\delta^n$, there exist $p \in \omega \setminus n$ and $\psi \in \Phi_\delta^p$ such that $H_\delta^p(\psi, i_n) = \phi$ and ϕ amalgamates above ψ (considering ϕ and ψ as conditions in $\mathbb{Q}_{\mathcal{P}}$, this happens if ψ decides $\phi_{\langle c_m : m < n \rangle, \delta + 1}^M$, where M denotes the structure induced by the generic thread).

Remark 6.11. If $\langle \Phi_\alpha : \alpha \leq \delta \rangle$ is a Scott process, $\gamma < \delta$ and $\{\phi_n : n \in \omega\}$ is a thread through Φ_δ, then $\{V_{\gamma,\delta}(\phi_n) : n \in \omega\}$ is a thread through Φ_γ (this follows from Proposition 2.15). This thread induces (as in the proof of Theorem 6.5) the same class-length Scott process as $\{\phi_n : n \in \omega\}$.

We insert here two arguments for constructing pairs of models. With respect to Theorem 6.12, note that (by Theorem 6.8 and Scott's Isomorphism Theorem) if $\mathcal{P} = \langle \Phi_\alpha : \alpha \leq \delta \rangle$ is a Scott process such that Φ_δ is countable and amalgamates, there is exactly one model of $\mathcal{P}$ of Scott rank at most δ, up to isomorphism. Whether or not the Scott rank of this model is less than δ depends on whether or not $\mathcal{P}$ terminates.

Theorem 6.12. *Let* $\mathcal{P} = \langle \Phi_\alpha : \alpha \leq \delta \rangle$ *be a Scott process such that* Φ_δ *is countable and amalgamates, and let* M *be a countable* τ-*structure such that every finite tuple from* M *satisfies a member of* Φ_δ. *Then there is a countable* τ-*structure* N, *modeling* $\mathcal{P}$, *such that* M *is a quantifier-depth-δ-elementary substructure of* N, *and* N *has Scott rank at most* δ.

Proof. Let $\langle c_n : n \in \omega \rangle$ be an enumeration of the domain of M. By Theorem 6.5, it suffices to find a thread $\{\phi_n : n \in \omega\}$ through Φ_δ and an infinite set $Y \subseteq \omega$ such that, for each $n \in \omega$,

1. letting

- j_n be the order preserving map from X_n to the first n elements of the set $\{x_m : m \in Y\}$, and

- k_n be the least element of ω such that $|Y \cap k_n| = n$,

$$H_\delta^{k_n}(\phi_{k_n}, j_n) = \phi_{\langle c_0, \ldots, c_{n-1}\rangle, \delta}^M;$$

2. for each $\psi \in \Phi_\delta^{n+1}$ such that $H_\delta^{n+1}(\psi, i_n) = \phi_n$, there exist $m > n$ in ω and $y \in X_m \setminus X_n$ such that $H_\delta^m(\phi_m, i_n \cup \{(x_n, y)\}) = \psi$.

Condition (2), and the assumption that Φ_δ amalgamates, will then give the following:

- $\{\phi_n : n \in \omega\}$ is complete (one can show by induction on m, for instance, that

$$\Phi_\delta \cap \Psi_\delta^m = \{H_\delta^n(\phi_n, j) : n \in \omega \setminus m, j \in \mathfrak{I}_{m,n}\}$$

for all $m \in \omega$);

- the model given by Theorem 6.8 will have Scott rank δ, as the induced $(\delta+1)$-st level of the Scott process of N will be the maximal completion of Φ_δ.

Then $\{\phi_n : n \in \omega\}$ induces the desired τ-structure N, and the set Y induces the desired copy of M inside N.

We start (as we must) with ϕ_0 as the unique element of Φ_δ^0. In our construction, we alternate stages for putting new elements in Y (while preserving condition (1)) and meeting condition (2). At each stage we will have chosen ϕ_p and decided $Y \cap p$ for some $p \in \omega$. As we construct, we maintain the following condition (*) : for each $p \in \omega$, if ϕ_p and $Y \cap p$ have been chosen, and $|Y \cap p| = n$, then there do not exist $y \in X_p \setminus \{x_m : m \in Y \cap p\}$ and $c \in M \setminus \{c_0, \ldots, c_{n-1}\}$ such that $H_\delta^p(\phi_n, j_n \cup \{(x_n, y)\}) = \phi_{\langle c_0, \ldots, c_{n-1}, c\rangle, \delta}^M$. As long as we do this, our assumption that Φ_δ amalgamates implies that we can choose ϕ_{n+1} in such a way that we can put $n \in Y$ and maintain condition (1). To meet condition (2), suppose that $p \in \omega$ is maximal such that ϕ_p and $Y \cap p$ have been chosen, and let ψ be given as in condition (2). Again, since Φ_δ amalgamates, we may assume that $H_\delta^{p+1}(\psi, i_p) = \phi_p$ (that is, to meet the condition for some $n \leq p$ we can meet it for p). If possible (while maintaining condition (1)), we satisfy this instance of condition (2) with a formula $\phi_q \in \Phi_\delta^q$ (for some $q > p$) while putting all of $q \setminus p$ in Y. If this is not possible, then we can let ϕ_{p+1} be ψ, and condition (*) is preserved. $\qquad \square$

The proof of the following theorem is similar, but we assume a weaker amalgamation property. Given an ordinal γ and sets $\Upsilon \subseteq \Phi \subseteq \Psi_\gamma$, say that Υ *weakly amalgamates with respect to* Φ if whenever

- m, n and p are in ω, with $m \leq n$,

- ϕ is in $\Phi \cap \Psi_\gamma^n$,

- $k \in \mathfrak{I}_{m,n}$ is such that $H^n_\gamma(\phi, k) \in \Upsilon$ and

- ψ is in $\Upsilon \cap \Psi^p_\gamma$,

there exist $\theta \in \Phi \cap \Psi^{n+p}_\gamma$, $q \le m+p$, $j \in \mathfrak{I}_{p,n+p}$ and $k' \in \mathfrak{I}_{q,n+p}$ extending k such that

- $H^{n+p}_\gamma(\theta, i_n) = \phi$,

- $H^{n+p}_\gamma(\theta, k') \in \Upsilon$ and

- $H^{n+p}_\gamma(\theta, j) = \psi$.

If Φ amalgamates, then it weakly amalgamates with respect to itself. The issue of extending Theorem 6.13 (or the weak version of it where Φ_γ is assumed to amalgamate) to uncountable models is discussed in Remark 7.15.

Theorem 6.13. *Let γ be a countable ordinal, and suppose that $\langle \Phi_\beta : \beta \le \gamma \rangle$ is a Scott process with Φ_γ countable. Let Φ^* be a subset of Φ_γ such that the extension of $\langle \Phi_\beta : \beta < \gamma \rangle$ by Φ^* is also a Scott process, and Φ^* weakly amalgamates with respect to Φ_γ. Then there exists τ-structures M and N such that M is a substructure of N, N is a model of $\langle \Phi_\beta : \beta \le \gamma \rangle$ and M is a model of the extension of $\langle \Phi_\beta :< \gamma \rangle$ by Φ^*.*

Proof. By Theorem 6.5, it suffices to find a complete thread $\bar\phi = \langle \phi_n : n \in \omega \rangle$ through Φ_γ and an infinite set $Y \subseteq \omega$ such that, letting, for each $n \in \omega$,

- j_n be the order preserving map from X_n to the first n elements of the set $\{x_m : m \in Y\}$,

- k_n be the least element of ω such that $|Y \cap k_n| = n$,

$\langle H^{k_n}_\gamma(\phi_{k_n}, j_n) : n \in \omega \rangle$ is a complete thread through Φ^*.

A construction of such a pair $\langle \phi_n : n \in \omega \rangle$, Y can be carried out in essentially the same manner as the proof of Theorem 6.4, recursively putting n into Y whenever $H^n_\gamma(\phi_n, j_{|Y\cap n|} \cup \{(x_{|Y\cap n|}, x_n)\})$ is a member of Φ^*. Again, we let ϕ_0 be the unique member of Φ_0. The only new issue is the completeness of the two threads being constructed. For $\bar\phi$, completeness can be achieved using condition (2c) of Definition 3.1. That is, having chosen ϕ_n (and thus $Y \cap (n+1)$), and given some $\psi \in \Phi^m_\gamma$, we let ϕ_{n+m} be an element of Φ_γ (as given by Condition 2c of Definition 3.1) such that $H^{n+m}_\gamma(\phi_{n+m}, i_n) = \phi_n$ and $H^{n+m}_\gamma(\phi_{n+m}, j) = \psi$, for some $j \in \mathfrak{I}_{m,n+m}$. We then use our recursive rule to decide $Y \cap (n+1, n+m)$.

The notion of weak amalgamation of Φ^* with respect to Φ_γ was defined to make the same argument work for the sequence $\langle H^{k_n}_\gamma(\phi_{k_n}, j_n) : n \in \omega \rangle$. Here we again have ϕ_n and $Y \cap (n+1)$, we are given a $\psi \in \Phi^* \cap \Psi^p_\gamma$, for some $p \in \omega$, and we want to find a formula $\phi_{n+p} \in \Phi^{n+p}_\gamma$ and an extension of $Y \cap (n+1)$ to $Y \cap (n+p+1)$ of size at least p such that $\psi = H^{n+m}_\gamma(\phi_{n+m}, k_p)$. Our recursive condition on Y guarantees that for the extension k' (the desired k_p) given by weak amalgamation, the range of $k' \setminus k$ (where k is $k_{|Y\cap(n+1)|}$) is contained in $X_{n+p} \setminus X_n$, so that we can extend Y as desired. $\square$

7 Models of cardinality $\aleph_1$

In this section, we show how to build models for Scott processes of length a successor ordinal, under the assumption that the last level of the process amalgamates and has cardinality at most $\aleph_1$.

Given two finite sets of ordinals $a \subseteq b$ with $a = \{\alpha_0, \ldots, \alpha_{n-1}\}$ (listed in increasing order), let $j_{a,b}$ be the function j in $\mathfrak{I}_{n,|b|}$ such that $j(x_m) = x_{|b \cap \alpha_m|}$ for all $m < n$.

In the case $\gamma = \omega$, the following definition is essentially the same as Definition 6.1, as the formulas $\{\phi_n : n \in \omega\}$ of the weaving then satisfy Definition 6.1.

Definition 7.1. Suppose that δ is an ordinal and Φ is a subset of Ψ_δ. A *weaving* through Φ is a set of formulas $\{\phi_a : a \in [\gamma]^{<\omega}\} \subseteq \Phi$, for some infinite ordinal γ, such that the following hold.

1. Each $\phi_a \in \Psi_\delta^{|a|}$.

2. For all $a \subseteq b \in [\gamma]^{<\omega}$, $\phi_a = H_\delta^{|b|}(\phi_b, j_{a,b})$.

3. For all $a \in [\gamma]^{<\omega}$, all $\alpha < \delta$, and all $\psi \in E(V_{\alpha+1,\delta}(\phi_a))$, there exist a $b \in [\gamma]^{|a|+1}$ containing a and a $y \in X_{|b|} \setminus \text{range}(j_{a,b})$ such that

$$\psi = V_{\alpha,\delta}(H_\delta^{|a|+1}(\phi_b, j_{a,b} \cup \{(x_{|a|}, y)\})).$$

The proof of Theorem 7.2 is an adaptation of the proof of Theorem 6.5.

Theorem 7.2. *Given a Scott process $\langle \Phi_\alpha : \alpha \leq \delta \rangle$, an infinite ordinal γ, a weaving $\langle \phi_a : a \in [\gamma]^{<\omega} \rangle$ through Φ_δ and an injective sequence $C = \langle c_\alpha : \alpha < \gamma \rangle$, there is a τ-structure with domain C in which, for each $a \in [\gamma]^{<\omega}$, the tuple $\langle c_\alpha : \alpha \in a \rangle$ satisfies ϕ_a.*

Proof. For each $a \in [\gamma]^{<\omega}$, let the tuple $\langle c_\alpha : \alpha \in a \rangle$ satisfy all the atomic formulas indicated by $V_{0,\delta}(\phi_a)$. We show by induction on $\beta < \delta$ that each tuple $\langle c_\alpha : \alpha \in a \rangle$ satisfies the formula $V_{\beta,\delta}(\phi_a)$. This follows immediately for limit stages. For the induction step from β to $\beta + 1$, $\langle c_\alpha : \alpha \in a \rangle$ satisfies $V_{\beta+1,\delta}(\phi_a)$ if and only if $E(V_{\beta+1,\delta}(\phi_a))$ is equal to

$$V_{\beta,\delta}[\{H_\delta^{|b|}(\phi_b, j_{a,b} \cup \{(x_{|a|}, y)\}) : a \subseteq b \in [\gamma]^{<\omega}, y \in X_{|b|} \setminus \text{range}(j_{a,b})\}].$$

The left-to-right containment follows from condition (3) of Definition 7.1. For the other direction, note first that by Proposition 4.4,

$$E(V_{\beta+1,\delta}(\phi_a)) = V_{\beta,\delta}[\{\theta \in \Phi_\delta^{|a|+1} \mid H_\delta^{|a|+1}(\theta, i_{|a|}) = \phi_a\}].$$

That

$$\{H_\delta^{|b|}(\phi_b, j_{a,b} \cup \{(x_{|a|}, y)\}) : a \subseteq b \in [\gamma]^{<\omega}, y \in X_{|b|} \setminus \text{range}(j_{a,b})\}$$

is contained in $\{\theta \in \Phi_\delta^{|a|+1} \mid H_\delta^{|a|+1}(\theta, i_{|a|}) = \phi_a\}$ follows from condition (2) of Definition 7.1 and part (2) of Remark 2.14. □

Definition 7.3. Suppose that δ is an ordinal, γ is an infinite cardinal and Φ is a subset of Ψ_δ. A weaving $\{\phi_a : a \in [\gamma]^{<\omega}\}$ through Φ is *complete* if for all $n \in \omega$ and all $\psi \in \Phi \cap \Psi_\delta^n$, there exist $a \in [\gamma]^n$ and $j \in \mathfrak{I}_{n,n}$ such that $\psi = H_\delta^n(\phi_a, j)$.

Remark 7.4. As in Remark 6.7, given a Scott process $\langle \Phi_\alpha : \alpha \leq \delta \rangle$ and a weaving through Φ_δ, the proof of Theorem 7.2 gives a model of $\langle \Phi_\alpha : \alpha < \delta \rangle$, for which the δ-th level of its Scott process is contained in the given Φ_δ. The δ-th level is equal to Φ_δ if and only if the weaving is complete.

Proposition 7.10 below show that if $\mathcal{P} = \langle \Phi_\alpha : \alpha \leq \delta \rangle$ is a Scott process such that Φ_δ amalgamates and has cardinality $\aleph_1$, then there exists a complete weaving through $\mathcal{P}$. To simplify the proof, we introduce a notion of strong weaving (at the cost of limiting the set of models of $\mathcal{P}$ we can construct, see Remark 7.8).

Definition 7.5. Suppose that δ is an ordinal and Φ is a subset of Ψ_δ. A *strong weaving* through Φ is a set

$$\{\phi_a : a \in [\gamma]^{<\omega}\} \subseteq \Phi,$$

for some infinite ordinal γ, satisfying conditions (1) and (2) of Definition 7.1 plus the following condition: for all $a \in [\gamma]^{<\omega}$, and all $\psi \in \Phi \cap \Psi_\delta^{|a|+1}$ such that $H_\delta^{|a|+1}(\psi, i_{|a|}) = \phi_a$, there exist a $b \in [\gamma]^{|a|+1}$ containing a and a $y \in X_{|b|} \setminus \text{range}(j_{a,b})$ such that

$$\psi = H_\delta^{|a|+1}(\phi_b, j_{a,b} \cup \{(x_{|a|}, y)\})).$$

Remark 7.6. In condition (3) of Definition 7.1 and in Definition 7.5, the variable y is in fact the unique member of $X_{|b|} \setminus \text{range}(j_{a,b})$.

Proposition 7.7. *Suppose that δ is an ordinal and Φ is a subset of Ψ_δ. Every strong weaving through Φ is both a weaving and complete.*

Proof. That a strong weaving satisfies condition (3) of Definition 7.1 follows from condition (2b) of Definition 3.1. Completeness for formulas in $\Phi \cap \Psi_\delta^n$ follows by induction on n. □

Remark 7.8. Let $\mathcal{P} = \langle \Phi_\alpha : \alpha \leq \delta \rangle$ be a Scott process. If there is a strong weaving through Φ_δ, then Φ_δ amalgamates. Moreover, the τ-structures induced by strong weavings through Φ_δ (as in the proof of Theorem 7.2) are, up to isomorphism, the models of $\mathcal{P}$ of Scott rank at most δ. This is in contrast to threads and weavings: every model of a Scott process $\mathcal{P}$ of successor length is induced (up to isomorphism) by weaving through the last level of $\mathcal{P}$.

A subset S of a collection C of sets is $\subseteq$-*cofinal* in C if every member of C is contained in a member of S.

Proposition 7.9. *Suppose that $\langle \Phi_\alpha : \alpha \leq \delta \rangle$ is a Scott process such that Φ_δ amalgamates. Let $\mathcal{W} = \{\phi_a : a \in [\gamma]^{<\omega}\}$ be a subset of Φ satisfying conditions (1) and (2) of Definition 7.1, such that the set of $a \in [\gamma]^{<\omega}$ for which the condition in Definition 7.5 is satisfied is $\subseteq$-cofinal in $[\gamma]^{<\omega}$. Then $\mathcal{W}$ is a strong weaving.*

Proof. Suppose that we have $a \subseteq b \in [\gamma]^{<\omega}$, and that the condition in Definition 7.5 holds for b. Suppose that $\psi \in \Phi_\delta^{|a|+1}$ is such that $H_\delta^{|a|+1}(\psi, i_{|a|}) = \phi_a$. By Proposition 5.20, there is formula $\theta \in \Phi_\delta^{|b|+1}$ such that $H_\delta^{|b|+1}(\theta, i_{|b|}) = \phi_b$ and $H_\delta^{|b|+1}(\theta, j_{a,b} \cup \{(x_{|a|}, x_{|b|})\}) = \psi$. Then there exist $\beta \in \gamma \setminus b$ and

$$y \in X_{|b|+1} \setminus \mathrm{range}(j_{b,b\cup\{\beta\}})$$

such that

$$\theta = H_\delta^{|b|+1}(\phi_{b\cup\{\beta\}}, j_{b,b\cup\{\beta\}} \cup \{(x_{|b|}, y)\}),$$

which implies that

$$\psi = H_\delta^{|a|+1}(\phi_{a\cup\{\beta\}}, j_{a,a\cup\{\beta\}} \cup \{(x_{|a|}, y)\}),$$

for y the unique element of $X_{|a|+1} \setminus \mathrm{range}(j_{a,a\cup\{\beta\}})$. $\qquad\square$

The proof of Proposition 7.10 implements the final part of Harrington's argument as it appears in Marker's slides [Mar11].

Proposition 7.10. *If $\langle \Phi_\alpha : \alpha \leq \delta \rangle$ is a Scott process such that Φ_δ amalgamates and has cardinality $\aleph_1$, then there is a strong weaving through Φ_δ.*

Proof. We recursively pick suitable formulas ϕ_a, for $a \in [\omega_1]^{<\omega}$. To begin with, let ϕ_n ($n \in \omega$) be any elements of Φ_δ with the property that $H_\delta^n(\phi_n, i_m) = \phi_m$, for all $m \leq n < \omega$. Suppose now that we have $\alpha < \omega_1$ and that ϕ_a has been chosen for each finite subset of α (note that a choice of ϕ_a determines a choice of ϕ_b for each subset of b, where a is a finite subset of ω_1) and no other subsets of ω_1. Using some wellordering of $[\omega_1]^{<\omega} \times \Phi_\delta$ in ordertype ω_1, we fix the least pair $a \in [\alpha]^{<\omega}$, $\psi \in \Phi_\delta^{|a|+1}$ as in Definition 7.5 for which the corresponding condition has not been met (which must exist since Φ_δ is uncountable), and let $\phi_{a\cup\{\alpha\}}$ be this ψ. Fixing a bijection $\pi \colon \omega \to (\alpha \setminus a)$, we now successively choose the formulas $\phi_{a\cup\{\alpha\}\cup\pi[n]}$. For each positive n, the choice of $\phi_{a\cup\{\alpha\}\cup\pi[n]}$ requires amalgamating $\phi_{a\cup\{\alpha\}\cup\pi[n-1]}$ with $\phi_{a\cup\pi[n]}$, which have already been chosen. The fact that Φ_δ amalgamates (via Proposition 5.20) implies that there exists a suitable choice for $\phi_{a\cup\{\alpha\}\cup\pi[n]}$. Since $\phi_{a\cup\{\pi(n-1)\}}$ did not satisfy the third condition of Definition 7.5 with respect to a and ψ, this choice of $\phi_{a\cup\{\alpha\}\cup\pi[n]}$ does not require identifying $\pi(n-1)$ and α. Proceeding in this fashion completes the construction of the desired strong weaving. $\qquad\square$

Putting together Theorems 6.8 (for the case where Φ_δ is countable) and 7.2 with Propositions 7.7 and 7.10, we have the following.

Theorem 7.11. *If $\langle \Phi_\alpha : \alpha \leq \delta \rangle$ is a Scott process, Φ_δ amalgamates and $|\Phi_\delta| \leq \aleph_1$, then $\langle \Phi_\alpha : \alpha \leq \delta \rangle$ has a model of Scott rank at most δ.*

Remark 7.12. One might naturally try to adapt the proof of Theorem 7.2 to build a model of size $\aleph_2$ by assigning a formula from Φ_δ to each finite tuple from ω_2. Doing this in the manner of the proof of Theorem 7.2, one finds oneself with an uncountable $\alpha < \omega_2$ such that formulas have been assigned for all finite subsets of α, but not for $\{\alpha\}$. Choosing formulas for all finite subsets of $\alpha + 1$, one comes to a point where, for some countably infinite $B \subseteq \alpha$, formulas have been chosen for all sets of the form $\{\alpha\} \cup b$, for b a finite subset of B. Then, for some $\beta \in \alpha \setminus B$, one would like to choose a formula for some finite superset c of $\{\alpha, \beta\}$ intersecting B. Finally, consider $\gamma \in B \setminus c$. We have at this point that formulas have been chosen for $\{\alpha, \gamma\}$, $\{\beta, \gamma\}$ and c, but not for $\{\alpha, \beta, \gamma\}$, and our assumptions do not give us suitable choice for $\{\alpha, \beta, \gamma\}$ that extends the choices already made. One can naturally define a notion of 3-amalgamation such that this construction could succeed under the assumption that this property holds.

Remark 7.13. Given an ordinal δ and $n \in \omega \setminus 2$, say that a set $\Phi \subseteq \Psi_\delta$ *n-amalgamates* if for all $m \in \omega$ and all

$$\{\phi_a : a \in [(m+n) \setminus m]^{n-1}\} \subseteq \Phi \cap \Psi_\delta^{m+n-1},$$

if (using the notation $j_{a,b}$ from the beginning of this section, restricted to finite subsets of ω),

$$H_\delta^{m+n-1}(\phi_a, j_{m\cup(a\cap b),m\cup a}) = H_\delta^{m+n-1}(\phi_b, j_{m\cup(a\cap b),m\cup b})$$

for all $a, b \in [(m+n) \setminus m]^{n-1}$ then there exists $\theta \in \Phi \cap \Psi_\delta^{m+n}$ such that

$$H_\delta^{m+n}(\theta, j_{m\cup a,m+n}) = \phi_a$$

for all a. The proof of Proposition 7.10 then gives that for all $n \in \omega \setminus 2$ and all Scott processes

$$P = \langle \Phi_\alpha : \alpha \leq \delta \rangle$$

such that $|\Phi_\delta| \leq \aleph_{n-1}$, if Φ_δ n-amalgamates then $\mathcal{P}$ has a model of cardinality at most $\aleph_{n-1}$ with Scott rank at most δ. We leave the details to the interested reader, as well as the verification that 2-amalgamation is equivalent to amalgamation.

Theorem 7.14. *If $P = \langle \Phi_\alpha : \alpha \leq \delta \rangle$ is a Scott process such that, for all $n \in \omega \setminus 2$, Φ_δ n-amalgamates, then P has a model of cardinality $|\Phi_\delta|$ and Scott rank at most δ.*

Proof. Let $\kappa = |\Phi_\delta|$. We build a strong weaving $\{\phi_a : a \in [\gamma]^{<\omega}\}$ through Φ_δ, for some ordinal $\gamma \in [\kappa, \kappa^+)$, where the value of γ is determined by the construction. At each stage of our construction, we will have chosen formulas $\phi_a \in \Phi_\delta^{|a|}$ for all a in some subset of $[\eta]^{<\omega}$ closed under subsets, for some $\eta < \kappa^+$, satisfying condition (2) of Definition 7.1. As always, we let $\phi_{\langle\rangle}$ be the unique member of Φ_δ^0. We have tasks of two types:

- Choosing a formula for each nonempty $a \in [\eta]^{<\omega}$, once we have chosen formula for each element of $[a]^{|a|-1}$. This we can do by $|a|$-amalgamation (with $m = 0$).

- Satisfying Definition 7.5. When doing this for a given $a \in [\eta]^{<\omega}$ (for which ϕ_a has been chosen) and $\psi \in \Phi_\delta^{|a|+1}$ (for which $H_\delta^{|a|+1}(\psi, i_{|a|}) = \phi_a$), we choose the corresponding b from $[\eta]^{|a|+1}$ if possible (given the choices already made). If this is not possible, we let $\phi_{a\cup\{\eta\}} = \psi$ (and increase η by one).

The ordinal γ will then be the supremum of the values of η during the construction. □

Say that a set $\Phi \subseteq \Psi_\delta$ is *rigid* if there is no $\phi \in \Phi \cap \Psi_\delta^2$ such that $H_\delta^2(\phi, i_1) = H_\delta^2(\phi, \{(x_0, x_1)\})$. Intuitively, this says that no two distinct points satisfy the same formula at level δ. A countable τ-structure M of Scott rank δ is rigid in the usual sense if and only if $\Phi_\delta(M)$ is rigid in this sense. If Φ is rigid then Φ satisfies n-amalgamation for all $n \in \omega$. A question due to Arnie Miller asks: Can there exist a ϕ in $\mathcal{L}_{\aleph_1,\aleph_0}$ with uncountably many rigid models but not perfectly many? Remark 7.13 shows that such a ϕ would have models of all infinite cardinalities. Part (4) of Theorem 11.1 shows that it would have nonrigid models as well. One could ask similar questions for n-amalgamation.

Remark 7.15. The natural attempt to combine the proofs of Theorem 6.13 (in the simplified case where Φ^* satisfies amalgamation) and Proposition 7.10 to produce a version of Theorem 6.13 for models of size $\aleph_1$ runs into a problem similar to the one in Remark 7.12. In this case, we have a Scott process $\langle \Phi_\alpha : \alpha \le \beta \rangle$, for some $\beta \in [\omega_1, \omega_2)$ such that, letting Φ^* be the set of isolated threads in Φ_β,

- Φ^* is a proper subset of Φ_β,

- the extension of $\langle \Phi_\alpha : \alpha < \beta \rangle$ by Φ^* gives a Scott process.

We could then try to build a strong weaving $\{\phi_a : a \in [\omega_1]^{<\omega}\}$ through Φ_β, and an uncountable set $Y \subseteq \omega_1$ such that $\{\phi_a : a \in [Y]^{<\omega}\}$ is a strong weaving through Φ^* (or, more precisely, induces one via some bijection between Y and ω_1). Carrying out this construction, we come to a point where, for some infinite $\gamma < \omega_1$, ϕ_a has been chosen for every finite subset of γ, and for $\{\gamma\} \cup a$, for some finite $a \subseteq \gamma$ intersecting Y as so far constructed, but not contained in it. At some stages it will also be that this γ has been put into Y. Now suppose that δ is in $Y \cap \gamma$, as constructed so far, but that no formula for $\{\delta, \gamma\}$ has been chosen. Then we need to choose a formula for $a \cup \{\delta, \gamma\}$ such that the induced formula for $(a \cap Y) \cup \{\delta, \gamma\}$ is in Φ^*. Since Φ_δ amalgamates, we can choose a formula for $a \cup \{\delta, \gamma\}$, but we cannot guarantee that the induced formula for $(a \cap Y) \cup \{\delta, \gamma\}$ will be in Φ^*. Similarly, since Φ^* amalgamates we can choose a formula for $(a \cap Y) \cup \{\delta, \gamma\}$ in Φ^*. Then we have the same 3-amalgamation issue as in Remark 7.12, as we would then need to amalgamate the chosen formulas for $(a \cap Y) \cup \{\delta, \gamma\}$, $a \cup \{\delta\}$ and $a \cup \{\gamma\}$ in Φ_β.

---

8   Finite existential blocks

The function E defined in Definition 2.4 corresponds to a single existential quantifier. In this section, we extend E to the function F which corresponds to finite blocks of existential quantifiers. The analysis of F in this section is used in the following section. Most of this section consists of consequences of Proposition 8.5, which gives an alternate characterization of F.

Definition 8.1. For each ordinal β, each $m \in \omega$ and each $\phi \in \Psi_\beta^m$, $F(\phi)$ is the set of ψ such that for some $n \in \omega$ and some ordinal α with $\alpha + n \leq \beta$, $\psi \in \Psi_\alpha^{m+n}$ and there exist $\psi_0, \ldots, \psi_n$ such that

- $\psi_0 = \psi$;

- for all $p \in \{0, \ldots, n-1\}$, $\psi_p \in E(\psi_{p+1})$;

- $\psi_n = V_{\alpha+n,\beta}(\phi)$.

Remark 8.2. Suppose that α, β, m, ϕ and $\psi_0, \ldots, \psi_n$ are as in Definition 8.1. Then by condition (1b) of Definition 3.1, each ψ_i is in $\Psi_{\alpha+i}^{m+n-i}$.

Remark 8.3. Given α, β, n, ϕ and ψ as in Definition 8.1, the issue of whether or not ψ is in $F(\phi)$ depends only on $V_{\alpha+n,\beta}(\phi)$ (as opposed to ϕ). It follows that if $\psi \in F(\phi)$, then $\psi \in F(\theta)$ for any formula $\theta \in \Psi_\gamma^m$ (for some ordinal $\gamma \geq \alpha + n$) such that $V_{\alpha+n,\gamma}(\theta) = V_{\alpha+n,\beta}(\phi)$.

Remark 8.4. An iterated application of condition (1b) of Definition 3.1 gives that if $\langle \Phi_\alpha : \alpha < \delta \rangle$ is a Scott process, $\alpha < \beta$ are in δ, $\phi \in \Phi_\beta$ and $\psi \in F(\phi)$, then $\psi \in \Phi_\beta$.

Fix for the rest of this section a Scott process $\langle \Phi_\alpha : \alpha < \delta \rangle$.

Proposition 8.5. *Suppose that $m, n \in \omega$ and $\alpha, \beta < \delta$ are such that $\alpha + n \leq \beta$. Let ϕ and ψ be elements of Φ_β^m and Φ_α^{m+n}, respectively. Then $\psi \in F(\phi)$ if and only if there is a formula $\theta \in \Phi_\beta^{m+n}$ such that $H_\beta^{m+n}(\theta, i_m) = \phi$ and $V_{\alpha,\beta}(\theta) = \psi$.*

Proof. By induction on n. In the case $n = 1$, $\psi \in F(\phi)$ if and only if $\psi \in E(V_{\alpha+1,\beta}(\phi))$. In this case, the proposition is Proposition 4.4. The induction step from $n = p$ to $n = p + 1$ follows from the induction hypothesis in the cases $n = p$ and $n = 1$ (applied twice). $\qquad\square$

Remarks 8.6 and 8.7 and Propositions 8.8, 8.9 and 8.10 list several useful properties of the function F.

Remark 8.6. Applying Propositions 2.15 and 8.5, and condition (1e) of Definition 3.1, we get that if $m, n, p \in \omega$ and $\alpha, \beta < \delta$ are such that $\alpha + n + p \leq \beta$, and if $\phi \in \Phi_\beta^m$, then for each $\psi \in \Phi_\alpha^{m+n} \cap F(\phi)$ there exists a $\rho \in \Phi_\alpha^{m+n+p} \cap F(\phi)$ such that $H_\alpha^{m+n+p}(\rho, i_{m+n}) = \psi$.

Remark 8.7. Fix $m, n \in \omega$ and suppose $\alpha, \beta < \delta$ are such that $\alpha + n \leq \beta$. Let ϕ be an element of Φ_β^{m+n}, let f be an element of $\mathcal{I}_{m,m+n}$ and let g be any element of $\mathcal{I}_{m+n,m+n}$ extending f. Then, by Proposition 8.5 and part (2) of Remark 2.14, $V_{\alpha,\beta}(H_\beta^{m+n}(\phi, g))$ is in $F(H_\beta^{m+n}(\phi, f))$.

Proposition 8.8 follows from Proposition 8.5 and Remark 2.8. It shows, for suitable ϕ and α, that $F(\phi)$ is closed under $V_{\gamma,\alpha}$ for all $\gamma \leq \alpha$.

Proposition 8.8. *Fix $\phi \in \Phi_\beta^m$, for some $\beta < \delta$ and $m \in \omega$. Let $\psi \in \Phi_\alpha^{m+n}$ be an element of $F(\phi)$, for some $n \in \omega$ and some ordinal α with $\alpha + n \leq \beta$. Then for all $\gamma < \alpha$, $V_{\gamma,\alpha}(\psi) \in F(\phi)$.*

Propositions 2.15 and 8.5 imply that members of $F(\phi)$ project horizontally to vertical projections of ϕ.

Proposition 8.9. *Suppose that $\alpha < \beta < \delta$, $m \leq n \in \omega$, $\phi \in \Phi_\beta^m$ and $\psi \in \Phi_\alpha^n \cap F(\phi)$. Then $H_\alpha^n(\psi, i_m) = V_{\alpha,\beta}(\phi)$.*

Proposition 8.10 is used in the proof of Theorem 9.9.

Proposition 8.10. *Suppose that $m, n \in \omega$, $\alpha < \beta$ are such that $\beta + n < \delta$, $\phi \in \Phi_{\beta+n}^m$ and $\psi \in \Phi_\alpha^{m+n} \cap F(\phi)$. Then there exists a $\psi' \in V_{\alpha,\beta}^{-1}[\{\psi\}] \cap F(\phi)$.*

Proof. By Proposition 8.5, there is a $\theta \in \Phi_{\beta+n}^{m+n}$ such that $V_{\alpha,\beta+n}(\theta) = \psi$ and $H_{\beta+n}^{m+n}(\theta, i_m) = \phi$. By Proposition 8.5 again, $V_{\beta,\beta+n}(\theta) \in F(\phi)$. $\qquad\square$

Proposition 8.11 is a version of condition (2c) of Definition 3.1 for $F(\phi)$.

Proposition 8.11. *For all $m, n, p \in \omega$, all $\alpha, \beta < \delta$ such that $\beta \geq \alpha + n + p$, and all $\phi \in \Phi_\beta^m$, $\psi \in \Phi_\alpha^{m+n} \cap F(\phi)$ and $\theta \in \Phi_\alpha^{m+p} \cap F(\phi)$, there exist $j \in \mathcal{I}_{m+p,m+n+p}$ and $\rho \in \Phi_\alpha^{m+n+p} \cap F(\phi)$ such that*

- $j \circ i_m = i_m$;

- $H_\alpha(\rho, i_{m+n}) = \psi$;

- $H_\alpha(\rho, j) = \theta$.

Proof. This can be proved by induction on p, for all m and n simultaneously. In the case where $p = 0$ there is nothing to show (since then $\theta = V_{\alpha,\beta}(\phi)$), so suppose that p is positive. Since $\theta \in F(\phi)$, there is a $\theta' \in \Phi_{\alpha+1}^{m+p-1} \cap F(\phi)$ such that $\theta \in E(\theta')$. By Proposition 8.10, there is a $\psi' \in \Phi_{\alpha+1}^{m+n} \cap F(\phi)$ such that $V_{\alpha,\alpha+1}(\psi') = \psi$. Let $\rho' \in \Phi_{\alpha+1}^{m+p+p-1}$ be the result of applying the induction hypothesis to ψ' and θ'. Since $\theta \in E(\theta')$, the desired ρ can be found in $E(\rho')$ by applying condition (3) of Definition 2.11. □

Proposition 8.12 shows that the set $F(\phi)$ is closed under suitable (horizontal) restrictions. The proposition follows immediately from Propositions 2.15 and 8.5.

Proposition 8.12. *Suppose that $\alpha < \beta < \delta$, $m, n \in \omega$, $\phi \in \Phi_\beta^m$ and $\psi \in \Phi_\alpha^{m+n}$ are such that $\alpha + n \leq \beta$ and $\psi \in F(\phi)$. Fix $p \in [m, m+n]$ and let $j \in \mathcal{I}_{p,m+n}$ be such that $j \upharpoonright X_m = i_m$. Then $H_\alpha^{m+n}(\psi, j) \in F(\phi)$.*

---

9   Extending a process of limit length

Suppose that δ is a limit ordinal and $\mathcal{P}$ is a Scott process. Must there be a Scott process properly extending $\mathcal{P}$? We show in this section that there exists such a proper extension if δ has countable cofinality and each level of $\mathcal{P}$ is countable (Theorem 9.9). We also derive a positive answer from various scatteredness conditions on $\mathcal{P}$ (see Proposition 9.13, for instance). In general the question remains open, as far as we know.

Definition 9.1. Given a limit ordinal β and a sequence $\langle \Phi_\alpha : \alpha < \beta \rangle$ such that each Φ_α is a subset of Ψ_α, a *path through* $\langle \Phi_\alpha : \alpha < \beta \rangle$ is a formula ϕ in Ψ_β such that $V_{\alpha,\beta}(\phi) \in \Phi_\alpha$ for each $\alpha < \beta$.

Remark 9.2. For each limit ordinal α, Ψ_α is the set of paths through the sequence $\langle \Psi_\beta : \beta < \alpha \rangle$. If $\langle \Phi_\alpha : \alpha < \delta \rangle$ is a Scott process, and $\beta < \delta$ is a limit ordinal, then each member of Φ_β is a path through $\langle \Phi_\alpha : \alpha < \beta \rangle$.

The issue of extending a given Scott process $\mathcal{P}$ of limit length then is whether there exists a set of paths through $\mathcal{P}$ large enough to satisfy conditions (1c) and (2b) of Definition 3.1 while also satisfying condition and (2c) (conditions (1d) and (1e) can then be achieved by closing under horizontal projections).

Proposition 9.3 implies in particular that every path through a Scott process of limit length determines the entire process (recall from Definition 8.1 that the set $F(\phi)$ depends only on ϕ, and not on a particular Scott processes containing ϕ).

Proposition 9.3. *Suppose that $\langle \Phi_\alpha : \alpha < \beta \rangle$ is a Scott process. Fix $\gamma < \beta$, $n \in \omega$ and $\phi \in \Phi_\gamma^n$. Then for each $\alpha < \beta$ and $m \in \omega$ such that $\alpha + m \leq \gamma$, the set Φ_α^m is equal to $\{H_\alpha^n(\psi, j) : \psi \in F(\phi) \cap \Phi_\alpha^{n+m}, j \in \mathcal{I}_{m,n+m}\}$.*

Proof. Let ψ be a member of Φ_α^m. By condition (1c) of Definition 3.1, there is a $\psi' \in \Phi_\gamma^m$ such that $V_{\alpha,\gamma}(\psi') = \psi$. By condition (2c) of Definition 3.1, there exist $j \in \mathcal{I}_{m,n+m}$ and $\theta \in \Phi_\gamma^{n+m}$ such that $H_\gamma^{n+m}(\theta, i_n) = \phi$ and $H_\gamma^{n+m}(\theta, j) = \psi'$. Then $V_{\alpha,\gamma}(\theta)$ is in $F(\phi)$ by Proposition 8.5, and is as desired by Proposition 2.15. $\square$

Proposition 9.3 implies that "according to $\langle \Phi_\alpha : \alpha < \beta \rangle$" is unnecessary in the following definition, if $\langle \Phi_\alpha : \alpha < \beta \rangle$ is a Scott process.

Definition 9.4. Let β be a limit ordinal β and let $\langle \Phi_\alpha : \alpha < \beta \rangle$ be such that each Φ_α is a subset of Ψ_α. Let ϕ be a path through $\langle \Phi_\alpha : \alpha < \beta \rangle$, and let $n \in \omega$ be such that $\phi \in \Psi_\beta^n$. The *minimal set* of ϕ according to $\langle \Phi_\alpha : \alpha < \beta \rangle$ is the set of paths θ through $\langle \Phi_\alpha : \alpha < \beta \rangle$ for which there exist

- $m \in \omega \setminus n$;

- $p \in m + 1$;

- $\alpha_0 < \beta$;

- $\psi_0 \in \Phi_{\alpha_0}^m \cap F(\phi)$;

- $j \in \mathcal{I}_{p,m}$;

such that for all $\alpha \in [\alpha_0, \beta)$ and all $\psi \in \Phi_\alpha^m \cap F(\phi)$ such that $V_{\alpha_0,\alpha}(\psi) = \psi_0$, $H_\alpha^m(\psi, j) = V_{\alpha,\beta}(\theta)$.

We write $\mathrm{ms}(\phi)$ for the minimal set of ϕ.

Remark 9.5. In the case where $\mathcal{P} = \langle \Phi_\alpha : \alpha < \beta \rangle$ is a Scott process and each element of $\bigcup\{\Phi_\alpha : \alpha < \beta\}$ is extended by a path through $\mathcal{P}$ (for instance, if β has countable cofinality, or $\mathcal{P}$ is scattered (see Definition 9.15)), the part of Definition 9.4 after the itemized list can equivalently be replaced by "such that for all $\psi \in \Psi_\beta$ such that $V_{\alpha_0,\beta}(\psi) = \psi_0$ and $H_\beta^m(\psi, i_n) = \phi$, $H_\beta^m(\psi, j) = \theta$."

Remark 9.6. Let $p \leq n$ be elements of ω, let j be an element of $\mathcal{I}_{p,n}$, let β be a limit ordinal, and let ϕ and ψ be paths through a Scott process $\langle \Phi_\alpha : \alpha < \beta \rangle$, with $\psi \in \mathrm{ms}(\phi)$ and $\psi \in \Psi_\beta^n$. Then $H_\beta^n(\psi, j)$ is an element of $\mathrm{ms}(\phi)$.

Remark 9.7. Let the *weakly minimal set* of a formula ϕ (in the context of Definition 9.4) be the set of formulas $\upsilon \in \Psi_\beta^p$ for which membership in $\mathrm{ms}(\phi)$ is witnessed with $j = i_p$. One obtains an equivalent definition of the minimal set of ϕ by taking the closure of the weakly minimal set under permutations of free variables (i.e., including all formulas of the form $H_\beta^p(\theta, j)$, where $\theta \in \Psi_\beta^p$ is in the weakly minimal set of ϕ and j is in $\mathcal{I}_{p,p}$. This follows from the second part of Remark 2.2, and condition (1d) of Definition 3.1.

Remark 9.8. Suppose that β is a limit ordinal, $\langle \Phi_\alpha : \alpha \leq \beta \rangle$ is a Scott process and ρ is an element of Φ_β. Then $\mathrm{ms}(\rho) \subseteq \Phi_\beta$. This follows from Proposition 8.5.

Theorem 9.9. *Suppose that δ is a limit ordinal of countable cofinality and $\langle \Phi_\alpha : \alpha < \delta \rangle$ is a Scott process such that each Φ_α is countable. Let ρ be a path through $\langle \Phi_\alpha : \alpha < \delta \rangle$. Then there exists a countable $\Phi_\delta \subseteq \Psi_\delta$ such that $\rho \in \Phi_\delta$ and $\langle \Phi_\alpha : \alpha \leq \delta \rangle$ is a Scott process.*

Furthermore, if Υ is a countable subset of Ψ_δ disjoint from $\mathrm{ms}(\rho)$, Φ_δ can be chosen to be disjoint from Υ.

Proof. In order to make $\langle \Phi_\alpha : \alpha \leq \delta \rangle$ a Scott process, we need to pick Φ_δ so that conditions (1c), (1d), (1e), (2b) and (2c) of Definition 3.1 are satisfied. Let $\langle \gamma_p^0 : p < \omega \rangle$ be an increasing sequence cofinal in δ. We will recursively pick formulas θ_p $(p < \omega)$, a nondecreasing sequence of ordinals γ_p $(p < \omega)$ below δ and a nondecreasing unbounded sequence of integers n_p $(p < \omega)$ such that $\rho \in \Psi_\delta^{n_0}$ and such that, for each $p \in \omega$,

- $\gamma_p \geq \gamma_p^0$;

- $\theta_p \in \Phi_{\gamma_p}^{n_p} \cap F(\rho)$;

- $H_{\gamma_p}^{n_{p+1}}(V_{\gamma_p, \gamma_{p+1}}(\theta_{p+1}), i_{n_p}) = \theta_p.$

These conditions imply that $\theta_0 = V_{\gamma_0, \delta}(\rho)$.

Having chosen the θ_p's, for each $n \in \omega$ we let ϕ_n be the path through $\langle \Phi_\alpha : \alpha < \delta \rangle$ determined by $\{H_{\gamma_p}^{n_p}(\theta_p, i_n) \mid p \in \omega, n_p \geq n\}$. Then for all $m \leq n \in \omega$ we will have that $\phi_m = H_\delta(\phi_n, i_m)$, and we will let

$$\Phi_\delta = \bigcup_{n < \omega} \{H_\delta^n(\phi_n, j) : m \leq n, \, j \in \mathcal{I}_{m,n}\}.$$

This is enough to ensure that conditions (1d), (1e) and (2c) from Definition 3.1 are met. For condition (1d) this is immediate. For condition (1e), the right-to-left containment follow from condition (1d). For the other direction, fix $m \leq n$ in ω. An arbitrary formula $\psi \in \Phi_\delta^m$ has the form $H_\delta^q(\phi_q, j)$, for some $q \in \omega \setminus m$ and some $j \in \mathcal{I}_{m,q}$. Since $\phi_n = H_\delta^p(\phi_p, i_n)$ for all $p \geq n$ in ω, we may assume that $q \geq n$. Letting $j' \in \mathcal{I}_{n,q}$ be such that $j \restriction X_m = j' \restriction X_m$, we have that $H_\delta^q(\phi_q, j') \in \Phi_\delta^n$, and that $\psi = H_\delta^n(H_\delta^q(\phi_q, j'), i_m)$, by part (2) of Remark 2.14.

To see that condition (2c) holds, fix $n, m \in \omega$, $\phi \in \Phi_\delta^n$ and $\psi \in \Phi_\delta^m$. Then there exist $p, q \in \omega$, $j \in \mathcal{I}_{n,p}$ and $k \in \mathcal{I}_{m,q}$ such that $\phi = H_\delta^p(\phi_p, j)$ and $\psi = H_\delta^q(\phi_q, k)$. Since

$$\phi_p = H_\delta^{\max\{p,q\}}(\phi_{\max\{p,q\}}, i_p)$$

and

$$\phi_q = H_\delta^{\max\{p,q\}}(\phi_{\max\{p,q\}}, i_q),$$

we may assume by part 2 of Remark 2.14 that $p = q$. Similarly, we may assume that $p \geq m + n$. Let A be a subset of X_p of size $m + n$ which contains the ranges of both j and k. Let $j' \colon X_{m+n} \to A$ be a bijection such that $j = j' \circ i_n$. Then

$$\phi = H_\delta^p(\phi_p, j' \circ i_n) = H_\delta^{m+n}(H_\delta^p(\phi_p, j'), i_n),$$

by part 2 of Remark 2.14, and $H_\delta^p(\phi_p, j') \in \Phi_\delta$. Finally, let $k' \in \mathfrak{I}_{m,m+n}$ be such that $k = j' \circ k'$. Then $\psi = H_\delta^{m+n}(H_\delta^p(\phi_p, j'), k')$, as desired.

To complete the proof, we show how to choose the formulas θ_p so that conditions (1c) and (2b) of Definition 3.1 are satisfied, and also so that no member of Υ is in Φ_δ. We let $\theta_0 = V_{\gamma_0, \delta}(\rho)$, as above. Suppose that $p \in \omega$ is such that θ_p has been chosen, but θ_{p+1} has not.

To satisfy condition (1c), let γ_{p+1} be the least member of $\{\gamma_q^0 : q \in \omega\}$ which is at least as big as both γ_p and γ_{p+1}^0, and suppose that ψ is an element of Φ_α^m, for some $\alpha \leq \gamma_{p+1}$ and some $m \in \omega$. By Proposition 8.5, we can find a formula $\theta_p' \in \Phi_{\gamma_{p+1}+m+n_p}^{n_p}$ such that $V_{\gamma_p, \gamma_{p+1}+m+n_p}(\theta_p') = \theta_p$ and

$$H_{\gamma_{p+1}+m+n_p}^{n_p}(\theta_p', i_{n_0}) = V_{\gamma_{p+1}+m+n_p, \delta}(\rho).$$

By condition (1c), there is a $\psi' \in \Phi_{\gamma_{p+1}+m+n_p}^m$ such that $V_{\alpha, \gamma_{p+1}+m+n_p}(\psi') = \psi$. Applying condition (2c) of Definition 3.1, we can choose $\theta_p'' \in \Phi_{\gamma_{p+1}+m+n_p}^{m+n_p}$ and $j \in \mathfrak{I}_{m,n_p+m}$ such that

$$H_{\gamma_{p+1}+m+n_p}^{m+n_p}(\theta_p'', i_{n_p}) = \theta_p'$$

and

$$H_{\gamma_{p+1}+m+n_p}^{m+n_p}(\theta_p'', j) = \psi'.$$

Then $\theta_{p+1} = V_{\gamma_{p+1}, \gamma_{p+1}+m+n_p}(\theta_p'')$ is as desired, by Propositions 2.15 and 8.5.

To satisfy condition (2b), fix $m \in \omega$. Each element of Φ_δ^m will have the form $H_\delta^{n_p}(\phi_p, j)$ for some $n_p \geq m$ and $j \in \mathfrak{I}_{m,n_p}$, in which case $V_{\alpha+1,\delta}(\phi)$ will be $V_{\alpha+1,\gamma_p}(H_{\gamma_p}^{n_p}(\theta_p, j))$. Suppose then that for some $p \in \omega$ we have chosen $n_p \geq m$ and θ_p but not θ_{p+1}, and that $j \in \mathfrak{I}_{m,n_p}$ and ψ in $E(V_{\alpha+1,\gamma_p}(H_{\gamma_p}^{n_p}(\theta_p, j)))$ are given. By Proposition 8.10, it suffices to find a $\theta_p' \in \Phi_{\gamma_p}^{n_p+1} \cap F(\rho)$ such that $H_{\gamma_p}^{n_p+1}(\theta_p', i_{n_p}) = \theta_p$, and such that

$$H_\alpha^{n_p+1}(V_{\alpha,\gamma_p}(\theta_p'), j \cup \{(x_m, y)\}) = \psi$$

for some $y \in X_{n_p+1} \setminus \mathrm{range}(j)$. By Proposition 2.15,

$$V_{\alpha+1,\gamma_p}(H_{\gamma_p}^{n_p}(\theta_p, j)) = H_{\alpha+1}^{n_p}(V_{\alpha+1,\gamma_p}(\theta_p), j).$$

By condition (3) of Definition 2.11, there is a $\psi' \in E(V_{\alpha+1,\gamma_p}(\theta_p))$ such that

$$\psi = H_\alpha^{n_p+1}(\psi', j \cup \{(x_m, y)\})$$

for some $y \in X_{n_p+1} \setminus \mathrm{range}(j)$. By Proposition 8.10, there is a $\theta_p^* \in \Phi_{\gamma_p+1}^{n_p} \cap F(\rho)$

such that $\theta_p = V_{\gamma_p, \gamma_p+1}(\theta_p^*)$. By Proposition 4.3, there is a $\theta_p' \in E(\theta_p^*)$ such that $V_{\alpha, \gamma_p}(\theta_p') = \psi'$. Then θ_p' is as desired.

Finally let us see how to avoid the members of Υ. Fix $m \leq n_p$, $j \in \mathcal{I}_{m,n_p}$ and $\upsilon \in \Upsilon \cap \Psi_\delta^m$. It suffices to show that we can find γ_{p+1} in the interval $(\max\{\gamma_p, \gamma_{p+1}^0\}, \delta)$ and a $\theta_{p+1} \in \Phi_{\gamma_{p+1}}^{n_p} \cap F(\rho)$ such that $H_\delta^{n_p}(\theta_{p+1}, j) \neq \upsilon$. Since Υ is disjoint from $\mathrm{ms}(\rho)$, there exists such a θ_{p+1} as desired. $\qquad\square$

We now turn our attention to extending Scott processes of limit length in the scattered case, which includes the case of counterexamples to Vaught's Conjecture.

Definition 9.10. Given a limit ordinal β and sets Φ_β ($\alpha < \beta$) such that each Φ_α is a subset of Ψ_α, a path $\bigwedge\{\psi_\alpha : \alpha < \beta\}$ through $\langle \Phi_\alpha : \alpha < \beta \rangle$ is *isolated* (with respect to $\langle \Phi_\alpha : \alpha < \beta \rangle$) if for some $\alpha_0 < \beta$, $|V_{\alpha_0, \alpha}^{-1}[\{\phi_{\alpha_0}\}] \cap \Phi_\alpha| = 1$ for all $\alpha \in (\alpha_0, \beta)$.

Proposition 9.3 shows that the term "with respect to $\langle \Phi_\alpha : \alpha < \beta \rangle$" is unnecessary in Definition 9.10, if $\langle \Phi_\alpha : \alpha < \beta \rangle$ is a Scott process.

Remark 9.11. Suppose that β is a limit ordinal, and $\mathcal{P} = \langle \Phi_\alpha : \alpha < \beta \rangle$ is a Scott process. Suppose that $m \leq n$ are elements of ω, $j \in \mathcal{I}_{m,n}$ and $\phi \in \Psi_\beta^n$ is an isolated path through $\mathcal{P}$. Then $H_\beta^n(\phi, j)$ is isolated. To see this, note first that the case $m = n$ follows from part (1) of Remark 2.14. This fact allows us to reduce to the case where $j = i_m$. Then a proof by induction reduces to the case where $n = m + 1$. This case follows part (2) of Proposition 5.1.

Remark 9.12. Given a limit ordinal β and sets Φ_β ($\alpha < \beta$) such that each Φ_α is a subset of Ψ_α, the isolated paths through $\langle \Phi_\alpha : \alpha < \beta \rangle$ are exactly the minimal set of the sentence formed by taking the conjunction of the unique members of each set Φ_α^0. This follows from Remark 9.11 and Proposition 8.5.

In Proposition 9.13, we do not require δ to have countable cofinality (whereas we did for Theorem 9.9).

Proposition 9.13. *Suppose that δ is a limit ordinal, and that $\mathcal{P} = \langle \Phi_\alpha : \alpha < \delta \rangle$ is a Scott process such that each element of $\bigcup\{\Phi_\alpha : \alpha < \delta\}$ is extended by an isolated path through $\mathcal{P}$. Letting Φ_δ be the set of isolated paths through $\mathcal{P}$, $\langle \Phi_\alpha : \alpha \leq \delta \rangle$ is a Scott process. Furthermore, Φ_δ then satisfies amalgamation, and every Scott process properly extending $\langle \Phi_\alpha : \alpha \leq \delta \rangle$ has rank at most δ.*

Proof. Checking that Φ_δ induces a Scott process involves checking conditions (1e), (2b) and (2c) of Definition 3.1. Remark 9.11 gives one direction of (1e). The other conditions can be shown by applying the corresponding fact at levels above the ordinal α_0 witnessing that the formulas in question are isolated.

That Φ_δ amalgamates also follows from the definition of the functions $H_{\alpha+1}^n$ ($n \in \omega$) for any ordinal α above the ordinal α_0 witnessing that the formulas in question are isolated. By Proposition 5.19, it also follows from the fact that some Scott properly extending $\langle \Phi_\alpha : \alpha \leq \delta \rangle$ has rank δ, which follows from the next paragraph.

To see that every Scott process $\langle \Phi_\alpha : \alpha \leq \delta + 1 \rangle$ extending $\langle \Phi_\alpha : \alpha \leq \delta \rangle$ has rank δ, suppose that we have $n \in \omega$, $\phi \in \Phi_{\delta+1}^n$ and $\psi \in \Phi_\delta^{n+1}$ such that $H_\delta^{n+1}(\psi, i_n) = V_{\delta,\delta+1}(\phi)$. Let $\beta < \delta$ be such that $V_{\delta,\delta+1}(\phi)$ and ψ are the unique members of $V_{\beta,\delta}^{-1}[\{V_{\beta,\delta+1}(\phi)\}]$ and $V_{\beta,\delta}^{-1}[\{V_{\beta,\delta}(\psi)\}]$, respectively. Then

$$H_{\beta+1}^{n+1}(V_{\beta+1,\delta}(\psi), i_n) = V_{\beta+1,\delta+1}(\phi)$$

by Proposition 2.15, so $V_{\beta,\delta}(\psi) \in E(V_{\beta+1,\delta+1}(\phi))$ by condition (2a) of Definition 3.1. Then conditions (2a) and (2b) of Definition 3.1 imply that $\psi \in E(\phi)$. $\qquad\square$

Definition 9.14. A *Scott subprocess* is a set of the form $\langle \Phi_\alpha : \alpha \in I \rangle$, for some Scott process $\{\Phi_\alpha : \alpha < \beta\}$ and $I \subseteq \beta$.

Definition 9.15. Given partial orders $(P, \leq_P)$ and $(Q, \leq_Q)$, let us say that $(P, \leq_P)$ *contains a copy of* $(Q, \leq_Q)$ if there is a function π from Q to P such that for all q_1, q_2 in Q, $q_1 \leq_Q q_2$ if and only if $\pi(q_1) \leq \pi(q_2)$. We say that a partial order is *scattered* if it does not contain a copy of $2^{<\omega}$, ordered by extension. We say that a Scott subprocess $\langle \Phi_\alpha : \alpha \in I \rangle$ is *scattered* if $(\bigcup_{\alpha \in I} \Phi_\alpha, \leq_V)$ is scattered in this sense (recall Definition 2.7). We say that a Scott process $\langle \Phi_\alpha : \alpha < \beta \rangle$ is *eventually scattered* if $\langle \Phi_\alpha : \alpha \in I \rangle$ is scattered for some cofinal $I \subseteq \beta$.

Remark 9.16. If a Scott process $\langle \Phi_\alpha : \alpha < \beta \rangle$ of limit length is eventually scattered, then there is a $\gamma < \beta$ such that $\langle \Phi_\alpha : \alpha \in (\gamma, \beta) \rangle$ is scattered.

Remark 9.17. Whether or not a given partial order is scattered is absolute between wellfounded models of ZFC containing the partial order. Suppose that $(T, \leq_T)$ is a tree ordering (a partial order such that the predecessors of any point are wellordered by $\leq_T$) with $T \subseteq L[T]$. The Cantor-Bendixon analysis (iteratively removing nodes without incompatible extensions) shows that if $(T, \leq_T)$ is scattered, then every maximal linearly ordered subset of T is a member of $L[T]$. The Cantor-Bendixon *rank* of $(T, \leq_T)$ is the (possibly transfinite) number of steps it takes for this analysis to terminate. If a Scott subprocess $\mathcal{P} = \langle \Phi_\alpha : \alpha \in I \rangle$ is scattered, we call the Cantor-Bendixon rank of the partial order $(\bigcup\{\Phi_\alpha : \alpha \in I\}, \leq_V)$ the *Cantor-Bendixon rank* of $\mathcal{P}$. If an ordinal γ is greater than the Cantor-Bendixon rank of $\mathcal{P}$, then every path through $\mathcal{P}$ is an element of $L_{\beta+\gamma}[\mathcal{P}]$.

Definition 9.18. We say that a τ-structure N is *Scott rank atomic* if, letting δ be the Scott rank of N, δ is a limit ordinal, and every element of $\Phi_\delta(N)$ is isolated in $\langle \Phi_\alpha(N) : \alpha < \delta \rangle$.

Combining Proposition 9.13 with Theorems 1.2, 6.8 and 7.11, we get the following model-existence result. Recall that for any ordinal γ, $\mathrm{Col}(\omega, \gamma)$ is the partial order which adds a function (generically, a surjection) from ω to γ by finite pieces, ordered by inclusion.

Theorem 9.19. *Let ϕ be a sentence of $\mathcal{L}_{\omega_1,\omega}(\tau)$ and let η be the quantifier depth of ϕ. Let $\beta \in (\eta, \omega_2)$ be an ordinal such that ϕ has a model of Scott rank β, but only countably many models of Scott rank γ for each countable ordinal γ in the interval (η, β). Then for every limit ordinal $\delta \in (\eta, \beta)$, ϕ has a Scott rank atomic model of Scott rank δ.*

Proof. Let M be a model (which by taking the transitive collapse of a suitable elementary substructure if necessary we may assume to be of cardinality at most $\aleph_1$) of ϕ of Scott rank β, and fix a limit ordinal $\delta < \beta$. Let

$$\mathcal{P} = \langle \Phi_\alpha(M) : \alpha \in (\eta, \delta) \rangle.$$

We claim first that $\mathcal{P}$ is scattered. To see this, suppose to the contrary that π embeds $2^{<\omega}$ into $(\bigcup\{\Phi_\alpha(M) : \alpha \in (\eta, \delta)\}, \leq_V)$. Let X be a countable elementary submodel of $L_{\omega_3}[\mathcal{P}]$ with $\eta \subseteq X$ and $\pi \in X$. Let γ be the ordertype of $X \cap \delta$. Let Q be the transitive collapse of X, and let $\mathcal{P}' = \langle \Phi_\alpha : \alpha \in (\eta, \gamma) \rangle$ and π' be the images of $\mathcal{P}$ and π under this collapse. For each $\alpha \leq \eta$, let Φ_α be $\Phi_\alpha(M)$, and let $\mathcal{P}^*$ be $\langle \Phi_\alpha : \alpha < \gamma \rangle$. Let g be Q-generic for $\mathrm{Col}(\omega, \omega_1^Q)$, the partial order adding a surjection from ω to ω_1^Q by finite pieces.

For each $x \in 2^\omega$, $\langle \pi'(x{\upharpoonright}n) : n \in \omega \rangle$ gives a path through an initial segment of $\mathcal{P}^*$ properly extending $\langle \Phi_\alpha(M) : \alpha \leq \eta \rangle$. Continuum many $x \in 2^\omega$ are generic over $Q[g]$ for Cohen forcing. For each such x, the corresponding path is a formula ϕ_x which by Theorem 9.9 is part of a Scott process in $Q[g][x]$ of successor length properly extending $\langle \Phi_\alpha(M) : \alpha \leq \eta \rangle$ and having a countable top level. By Theorem 6.8, each of these formulas has a countable model N_x (of Scott rank less than ω_2^Q) in the corresponding $Q[g][x]$, and by Theorem 1.2 they are all models of ϕ. Each ϕ_x then has the form $\phi_{\bar{a},\zeta}^{N_x}$, for some ordinal ζ and some finite tuple $\bar{a}$ from N_x. Since the formulas ϕ_x are pairwise $\leq_V$-incompatible, no τ-structure can satisfy more than one of them with the same finite tuple, so no countable τ-structure can satisfy more than countably many of them. It follows then that there exist continuum many models of ϕ of Scott rank less than ω_2^Q, which is countable, giving a contradiction.

Now Proposition 9.13 and Theorem 7.11 give a model of ϕ of Scott rank δ, as desired (the model cannot have Scott rank less than δ, since $\mathcal{P}$ is non-terminating). $\square$

Remark 9.20. Suppose that $\mathcal{P} = \langle \Phi_\alpha : \alpha < \beta \rangle$ is a Scott process of countable length, with all levels countable, having only countably many models of Scott rank γ for each countable ordinal γ. Let T be the (class-sized) tree of Scott processes extending $\mathcal{P}$, ordered by extension. A minor variation of the first paragraph of the proof of Theorem 9.19 shows that T is scattered.

Definition 9.21. Let $\mathcal{P} = \langle \Phi_\alpha : \alpha < \beta \rangle$ be a Scott process of limit length, let $m \leq n$ be elements of ω and let $\phi \in \Psi_\beta^m$ and $\psi \in \Psi_\beta^n$ be paths through $\mathcal{P}$. Let f be an element of $\mathcal{I}_{m,n}$. We say that ψ is (f, ϕ)-*isolated* if there exists a $\gamma < \beta$ such that, for all $\delta \in (\gamma, \beta)$, $V_{\delta,\beta}(\psi)$ is the unique $\theta \in \Phi_\delta^n$ such that $V_{\gamma,\delta}(\theta) = V_{\gamma,\beta}(\psi)$ and $H_\delta^n(\theta, f) = V_{\delta,\beta}(\phi)$.

Remark 9.22. If $\mathcal{P} = \langle \Phi_\alpha : \alpha < \beta \rangle$ is an eventually scattered Scott process of limit length, $k \in \omega$ and $\phi \in \Psi_\beta^k$ is a path through $\mathcal{P}$, then $\mathrm{ms}(\phi)$ is the set of formulas of the form $H_\beta^n(\psi, g)$, where for some $m \le n$ in ω (with $n \ge k$), $\psi \in \Phi_\beta^n$ is (i_k, ϕ)-isolated and g is in $\mathcal{I}_{m,n}$. That this set is contained in $\mathrm{ms}(\phi)$ follows from Remarks 9.5 and 9.6. The other direction follows from the usual argument that in an eventually scattered partial order every node is extended by an isolated path, applied to $F(\phi)$.

Remark 9.23. Theorem 9.9 shows that if δ is a limit ordinal and $\langle \Phi_\alpha : \alpha < \delta \rangle$ is a Scott process with just countably many paths, then for each such path ρ, letting Φ_δ be $\mathrm{ms}(\rho)$ we get a Scott process $\langle \Phi_\alpha : \alpha \le \delta \rangle$. Since $\mathrm{ms}(\phi)$ and being scattered are absolute to forcing extensions, we get the same conclusion from the assumption that $\langle \Phi_\alpha : \alpha < \delta \rangle$ is eventually scattered, without any countability assumption. In this context, then, since $\mathrm{ms}(\rho)$ is the smallest set one can add to $\langle \Phi_\alpha : \alpha < \delta \rangle$ to get a Scott process with ρ in its last level, it follows (again, in the case where $\langle \Phi_\alpha : \alpha < \delta \rangle$ is eventually scattered) that if ϕ and ψ are paths through $\langle \Phi_\alpha : \alpha < \delta \rangle$ with $\phi \in \mathrm{ms}(\psi)$, then $\mathrm{ms}(\phi)$ is a subset of $\mathrm{ms}(\psi)$.

In the following proposition, the countability assumption on the sets Φ_α can be replaced by the assumption that $\langle \Phi_\alpha : \alpha < \gamma \rangle$ is eventually scattered, using Remark 9.23 (recall that pre-rank was defined in Definition 5.12).

Proposition 9.24. *Let β be an ordinal, and let γ be the least limit ordinal greater than or equal to β. Suppose that $\langle \Phi_\alpha : \alpha \le \gamma + 1 \rangle$ is a Scott process of pre-rank β, and that Φ_α is countable for all $\alpha < \gamma$. Then the rank of $\langle \Phi_\alpha : \alpha \le \gamma + 1 \rangle$ is at most γ.*

Proof. Since Φ_α is countable for all $\alpha < \gamma$, β is countable. By the definition of pre-rank, $\langle \Phi_\alpha : \alpha \le \gamma \rangle$ is the unique Scott process of length $\gamma + 1$ extending $\langle \Phi_\alpha : \alpha < \gamma \rangle$. By Theorem 9.9, $\langle \Phi_\alpha : \alpha < \gamma \rangle$ has only countably many paths. By Proposition 9.13, all of them are isolated, and $\langle \Phi_\alpha : \alpha \le \gamma + 1 \rangle$ has rank at most γ. $\qquad\square$

10   Absoluteness

In this section, we record various standard absoluteness results concerning counterexamples to Vaught's Conjecture. We assume here that our relational vocabulary τ is countable. The set of τ-structures with domain ω is then naturally seen as a Polish space X_τ, where a basic open set is given by the set of structures in which $R(i_0, \ldots, i_{n-1})$ holds, for R an n-ary relation symbol from τ and $i_0, \ldots, i_{n-1} \in \omega$ (see Section 11.3 of [Gao09], for instance). Given a sentence $\phi \in \mathcal{L}_{\omega_1,\omega}(\tau)$, the set of models of ϕ with domain ω is a Borel subset of X_τ. By a theorem of Lopez-Escobar [LE65], every Borel subset of X_τ which is closed under isomorphism is also the set of models (with domain ω) of some $\mathcal{L}_{\omega_1,\omega}(\tau)$ sentence.

Let us call the following (false) statement the *analytic Vaught Conjecture*: for every countable relational vocabulary τ, every analytic subset of X_τ (closed under isomorphism) which contains uncountably many nonisomorphic structures contains a perfect set of nonisomorphic structures. Steel [Ste78] presents two examples of analytic counterexamples to Vaught's Conjecture (for certain relational vocabularies), one due to H. Friedman and the other to K. Kunen. In this section we use a forcing-absoluteness argument to prove the following, which was presumably well known. As mentioned in the introduction, the forcing-absoluteness arguments in this section appear in essentially identical form in Section 1 of [Hjo96].

Theorem 10.1. *Suppose that $\mathcal{A}$ is a counterexample to the analytic Vaught Conjecture, and let $x \subseteq \omega$ be such that $\mathcal{A}$ is Σ_1^1 in x. Fix $M \in \mathcal{A}$, and let β be an ordinal. Then $\langle \Phi_\alpha(M) : \alpha < \beta \rangle \in L[x]$.*

Applying this theorem in forcing extensions of V, we get the following ostensibly stronger fact.

Corollary 10.2. *Suppose that $\mathcal{A}$ is a counterexample to the analytic Vaught Conjecture, and let $x \subseteq \omega$ be such that $\mathcal{A}$ is Σ_1^1 in x. Let M be a member of the reinterpreted version of $\mathcal{A}$ in a forcing extension of V, and let β be an ordinal. Then $\langle \Phi_\alpha(M) : \alpha < \beta \rangle \in L[x]$.*

Before beginning the proof of Theorem 10.1 (which is short), we make a few observations.

Remark 10.3. Remarks 9.17 and 9.20, along with Theorem 1.2, show that if ϕ is a counterexample to (the usual) Vaught's Conjecture, γ is a limit ordinal greater than the quantifier depth of ϕ and M is an inner model of ZFC containing

$$\langle \Phi_\alpha(N) : \alpha < \beta \rangle$$

for each τ-structure $N \models \phi$ and each $\beta < \gamma$, then M contains $\langle \Phi_\alpha(N) : \alpha < \gamma \rangle$ for each τ-structure $N \models \phi$.

In what follows we will talk of sufficient fragments of ZFC. The theory ZFC$^\circ$ from [BL16] is one such fragment.

Remark 10.4. As we are assuming that τ is countable, we can associate each atomic or negated atomic formula from τ to an element of ω, and each Scott process over τ of length 1 to a subset of $\mathcal{P}(\omega)$. For a countable τ-structure, this subset of $\mathcal{P}(\omega)$ will be countable. Let $=^+$ be the equivalence relation on functions from $\omega \times \omega$ to 2 defined by setting $f =^+ g$ if

$$\{\{m : f(n, m) = 1\} : n \in \omega\} = \{\{m : g(n, m) = 1\} : n \in \omega\}.$$

Then $=^+$ is easily seen to be a Borel equivalence relation, and it follows for instance from Silver's theorem on coanalytic equivalence relations (Theorem 5.3.5 of [Gao09]) that every analytic set $A \subseteq {}^{\omega \times \omega}2$ containing representatives of uncountably many equivalence classes contains a perfect $=^+$-inequivalent set (for $=^+$ this can be proved more easily, considering separately the cases where

$$\{\{m : f(n, m) = 1\} : n \in \omega, f \in A\}$$

is countable or uncountable, and in the former case arguing as in the proof of Theorem 4.6 of [BL16]).

Let $\mathcal{A}$ be an analytic family of τ-structures on ω. It follows from the previous paragraph that if the set of Scott processes of length 1 corresponding to structures in $\mathcal{A}$ is uncountable, then there exists a perfect subset of ${}^{\omega \times \omega}2$ coding distinct elements of this set, and, via the proofs of Theorems 6.4 and 6.5, a perfect subset of ${}^{\omega}\omega$ coding distinct structures in $\mathcal{A}$. Working by induction, essentially the same analysis (breaking into successor and limit cases) shows that if $\beta > 0$ is a countable ordinal and the set of Scott processes of length of less than β corresponding to structures in $\mathcal{A}$ is countable, then if there are uncountably many Scott processes of length of β corresponding to structures in $\mathcal{A}$, then there is a perfect subset of ${}^{\omega}\omega$ coding distinct elements of $\mathcal{A}$. If $\mathcal{A}$ is a counterexample to the analytic Vaught Conjecture, then, the set of Scott processes length β corresponding to structures in $\mathcal{A}$ is countable for each $\beta < \omega_1$.

For any analytic family $\mathcal{A}$ of τ-structures, and any countable (possibly empty) set of Scott processes of length $\beta < \omega_1$, the assertion that there exists a member of the family whose Scott process up to length β is not in this countable set is Σ_1^1 in codes for β, the family and the countable set, and thus absolute to any model of (a sufficient fragment of) ZFC that contains them. Furthermore, if such a model thought that there were uncountably many Scott processes of length β corresponding to structures in $\mathcal{A}$, it could find a perfect subset of ${}^{\omega}\omega$ coding distinct Scott processes in this family. It follows that if $\mathcal{A}$ is a counterexample to the analytic Vaught Conjecture, then any inner model N of (a sufficient fragment of) ZFC containing a real parameter code for $\mathcal{A}$ contains all sequences of the form $\langle \Phi_\alpha(M) : \alpha < \beta \rangle$, for $M \in \mathcal{A}$ and $\beta < \omega_1^N$. This gives Theorem 10.1 for initial segments of Scott processes of length less than $\omega_1^{L[x]}$.

Proof of Theorem 10.1. Let $\theta > \beta$ be a regular cardinal of $L[x]$ such that $L_\theta[x]$ satisfies a sufficient fragment of ZFC as in Remark 10.4 (for instance, let θ be a regular cardinal of V greater than $2^{2^{(|\beta| + \omega_1)}}$). Let X be a countable (in V) elementary submodel of $L_\theta[x]$ containing $\{x, \langle \Phi_\alpha : \alpha < \beta \rangle\} \cup \beta$. Let γ be such that the transitive collapse of X is $L_\gamma[x]$. By the last paragraph of Remark 10.4, whenever g is an $L_\gamma[x]$-generic filter for $\mathrm{Col}(\omega, \beta)$, $\langle \Phi_\alpha : \alpha < \beta \rangle$ is in $L_\gamma[x][g]$. This means that $\langle \Phi_\alpha : \alpha < \beta \rangle$ is in $L_\gamma[x]$ (this is a classical forcing fact due to Solovay; the point is that otherwise one could choose a generic filter while ensuring that each name in $L_\gamma[x]$ realizes to some value other than $\langle \Phi_\alpha : \alpha < \beta \rangle$). $\qquad\square$

Remark 10.5. Let $\mathcal{A}$ be an analytic family of τ-structures on ω. The assertion that $\mathcal{A}$ is a counterexample to the analytic Vaught Conjecture is Π_2^1 in a real parameter x for $\mathcal{A}$, and therefore absolute to $L[x]$.[1] It follows (assuming that $\mathcal{A}$ is a counterexample to the analytic Vaught Conjecture) that for every ordinal γ, there are cofinally many ordinals below $(|\gamma|^+)^{L[x]}$ which are the Scott rank of a countable structure in $\mathcal{A}$, in any forcing extension of $L[x]$ via the partial order $\mathrm{Col}(\omega, \gamma)$ (all levels of the Scott processes of these structures are then countable in the corresponding forcing extensions). Applying Theorem 9.19, this gives (in the case where $\mathcal{A}$ is Borel) that this set of ordinals (in such a forcing extension) includes coboundedly many limit ordinals below $(\kappa^+)^{L[x]}$.

Theorems 7.11 and 9.19, along with Corollary 10.2 and Remark 10.5, give the following unpublished theorem of Leo Harrington from the 1970s. The arguments we have given here give a slightly stronger version of Harrington's theorem than the one in [Mar11]. A similar result (for countable models) appears in [Sac07]. Theorem 11.2 gives non-Scott-rank-atomic models (as does [Sac07]).

Theorem 10.6 (Harrington). *Suppose that τ is a countable relational vocabulary and that $\phi \in \mathcal{L}_{\omega_1,\omega}(\tau)$ gives a counterexample to Vaught's Conjecture. Let α be the quantifier depth of ϕ. Then for every limit ordinal δ in the interval $[\alpha, \omega_2)$, ϕ has a Scott rank atomic model of Scott rank δ.*

Proof. By Theorem 9.19, it suffices to show that for cofinally many $\beta < \omega_2$, ϕ has a model of Scott rank at least β. Fix such a β. By Remark 10.5, in some forcing extension by the partial order $\mathrm{Col}(\omega, \beta)$, ϕ has a countable model with Scott rank in the interval (β, ω_2). Let γ be the Scott rank of this model. By Corollary 10.2, the Scott process of this model of length $\gamma + 1$ exists already in V, and since the levels of this Scott process are countable in the $\mathrm{Col}(\omega, \beta)$ extension, they have cardinality at most $\aleph_1$ in V. By Proposition 5.19, the top level of this Scott process amalgamates. By Theorem 7.11, a model of this Scott process exists. By Theorem 1.2, this model is a model of ϕ. $\qquad\square$

Standard arguments show that if there is a counterexample to Vaught's Conjecture, then there is one of quantifier depth at most ω, in an expanded language. We give a new proof of this fact in Section 12.

Remark 10.7. The proof of Theorem 1 of [HM77] can be rephrased in terms of the arguments given here, showing that any counterexample to Vaught's

---

[1]There exist perfectly many nonisomorphic structures in $\mathcal{A}$ if and only if some well-founded countable model of a sufficient fragment of ZFC thinks there exist perfectly many nonisomorphic structures in $\mathcal{A}$ (see the proof of Theorem 4.6 of [BL16], for instance), and this later statement is easily seen to be Σ_2^1. The statement that there are countable models in $\mathcal{A}$ of unboundedly many Scott ranks below ω_1 is easily seen to be Π_2^1.

Conjecture can be strengthened to a minimal counterexample. The point again is that if $\sigma \in \mathcal{L}_{\omega_1,\omega}(\tau)$ is a counterexample to Vaught's Conjecture, and α is the quantifier depth of σ, then there is a sentence $\sigma' \in \mathcal{L}_{\omega_1,\omega}(\tau)$ which is the unique member of $\Phi^0_\alpha(M)$ for uncountably many countable models M satisfying σ. Then all models of σ' are models of σ, by Theorem 1.2, and σ' is also a counterexample to Vaught's Conjecture. Let S be the set of all countable length Scott processes which have σ' as their unique sentence at level α and are initial segments of the Scott process of some model of uncountable Scott rank. Since σ' is a counterexample to Vaught's Conjecture, S is not empty, by Theorem 10.6. On the other hand, since σ' does not have perfectly many countable models, there will be a $\mathcal{P}$ in S without incompatible extensions in S. Since any extension of $\mathcal{P}$ in S will have the same property, there is such a member of S with successor length. Let ϕ be the unique sentence in the last level of this process. Then ϕ is a counterexample to Vaught's Conjecture, and all uncountable models of ϕ satisfy the same $\mathcal{L}_{\omega_1,\omega}(\tau)$-theory.

Remark 10.8. Hjorth [Hjo07] showed that if there exists a counterexample to Vaught's Conjecture, then there is one with no model of cardinality $\aleph_2$. Recently, this has been extended by Baldwin, Friedman, Koerwien and Laskowski [BFKL16], who showed (among other things) that if there exists a counterexample to Vaught's Conjecture, then there is one with the property that for some countable $\mathcal{L}_{\aleph_1,\aleph_0}$-fragment T, no model of cardinality $\aleph_1$ has a T-elementary extension. Roughly speaking, Hjorth's argument (as reformulated by [BFKL16]), finds an absolutely definable method for taking any countable structure M in a relational language and building a structure $H(M)$ in such a way that (1) the Scott sentence of $H(M)$ cannot have a model of cardinality $\aleph_2$; (2) $H(M)$ contains a copy of M; (3) if M and N are isomorphic then so are $H(M)$ and $H(N)$. It follows from (3) that for any structure M in a relational vocabulary (countable or not), the structure $H(M)$ has the same Scott process in each forcing extension in which M is countable. From the classical fact due to Solovay mentioned in the proof of Theorem 10.1, it follows that every initial segment of the Scott process of $H(M)$ (as constructed in any such forcing extension) exists in V. It follows from (1) and (2) that if M has Scott rank $\gamma \geq \omega_2$, then the initial segment of the Scott process of $H(M)$ of length γ does not have a model in V (so the condition $|\Phi_\delta| \leq \aleph_1$ in the statement of Theorem 7.11 is necessary).

11   A local condition for amalgamation

In this section we give a sufficient condition for showing that the set of all paths through a Scott process of limit length amalgamates, and use it to produce models of a counterexample to Vaught's Conjecture which are not

Scott rank atomic. As in Remark 9.17 above, if γ in the theorem below is greater than the Cantor-Bendixon rank of $\mathcal{P}$, then $\Phi_\beta \subseteq M$. Parts (1) and (5) of the theorem can be phrased more generally as theorems about scattered trees.

Theorem 11.1. *Suppose that $\mathcal{P} = \langle \Phi_\alpha : \alpha < \beta \rangle$ is an eventually scattered Scott process, where β is a countable limit ordinal. Let $\gamma > \beta$ be an ordinal, and let $M = L_\gamma[\langle \Phi_\alpha : \alpha < \beta \rangle]$. Suppose that, in M, the cofinality of β is greater than $|\Phi_\alpha|$, for each $\alpha < \beta$. Let Φ_β be the set of all paths through $\mathcal{P}$.*

1. *Let A be a subset of $\bigcup\{\Phi_\alpha : \alpha < \beta\}$, in M, such that A contains a member of Φ_α for cofinally many $\alpha < \beta$. Then there is a $\phi \in \Phi_\beta$ such that, for each $\alpha < \beta$, there exist $\delta \in \beta \setminus \alpha$ and $\psi \in A \cap \Phi_\delta$ such that $V_{\alpha,\beta}(\phi) = V_{\alpha,\delta}(\psi)$.*

2. *If $\Phi_\beta \subseteq M$, then $\langle \Phi_\alpha : \alpha \leq \beta \rangle$ is a Scott process.*

3. *If $\Phi_\beta \subseteq M$, then Φ_β amalgamates.*

4. *If $\Phi_\beta \subseteq M$ and $\mathcal{P}$ is nonterminating, then no model of $\langle \Phi_\alpha : \alpha \leq \beta \rangle$ of Scott rank β is rigid.*

5. *Suppose that $\gamma \geq (\beta^+)^M$, $n \in \omega$, A is a stationary subset of β in M, and $\{\psi_{\alpha,i} : \alpha \in A, i < n\}$ is a set in M such that, for each $\alpha \in A$ and $i < n$, $\psi_{\alpha,i} \in \Phi_\alpha$. Then there exist $\phi_i \in \Phi_\beta$ $(i < n)$ such that, in M, for stationarily many $\alpha \in A$, for all $i < n$, $V_{\alpha,\beta}(\phi_i) = \psi_{\alpha,i}$.*

6. *If $\gamma \geq (\beta^+)^M$, and $\mathcal{P}$ is nonterminating, then Φ_β has a nonisolated path.*

7. *If $\gamma \geq (\beta^+)^M$, then, for club many $\alpha < \beta$, Φ_α amalgamates.*

Proof. Since $\mathcal{P}$ is eventually scattered, we may work in a forcing extension in which β is countable, as the set of paths through $\mathcal{P}$ is the same in any such extension.

For part (1), let $\langle \gamma_i : i \in \omega \rangle$ be a cofinal increasing sequence in β. Let Θ be the set of θ such that, for some $i \in \omega$, $\theta \in \Phi_{\gamma_i}$, and, for cofinally many $\delta < \beta$, there exists a $\psi \in \Phi_\delta \cap A$ with $V_{\gamma_i,\delta}(\psi) = \theta$. Since in M there is no cofinal function from any Φ_α to β, $\Theta \cap \Phi_{\gamma_0}$ is nonempty, and, for each $i \in \omega$ and each θ in $\Theta \cap \Phi_{\gamma_i}$, there is $\rho \in \Theta \cap \Phi_{\gamma_{i+1}}$ with $V_{\gamma_i,\gamma_{i+1}}(\rho) = \theta$. The existence of a ϕ as desired follows.

For part (2), all parts of Definition 3.1 are immediate (from the assumption that $\mathcal{P}$ is eventually scattered, which implies that every member of each Φ_α $(\alpha < \beta)$ is part of an element of Φ_β), aside from conditions (2b) and (2c). Each of these follows easily from part (1). For condition (2b), fix $\phi \in \Phi_\beta^n$ (for some $n \in \omega$), $\alpha < \beta$ and $\theta \in E(V_{\alpha+1,\beta}(\phi))$. We wish to find a $\psi \in \Phi_\beta^{n+1}$ such that $H_\beta^{n+1}(\psi, i_n) = \phi$ and $V_{\alpha,\beta}(\psi) = \theta$. Applying part (1), it suffices to show that for each $\delta \in (\alpha, \beta)$, there is a $\rho \in \Phi_\delta^{n+1}$ such that $H_\delta^{n+1}(\rho, i_n) = V_{\delta,\beta}(\phi)$ and $V_{\alpha,\delta}(\rho) = \theta$. The existence of such a ρ follows from condition (2b) applied to $V_{\delta,\beta}(\phi)$. The argument for condition (2c) is similar, but we use the fact that the union of the sets $\mathcal{I}_{m,n}$ is countable.

For part (3), fix $m < n \in \omega$, $\phi \in \Phi_\beta^{m+1}$ and $\psi \in \Phi_\beta^n$ such that $H_\beta^{m+1}(\phi, i_m) = H_\beta^n(\psi, i_m)$. Applying the first conclusion, it suffices to show that for each $\alpha < \beta$, there exist $\theta \in \Phi_\alpha^{n+1}$ and $y \in X_{n+1} \setminus X_m$ such that

$$H_\alpha^{n+1}(\theta, i_m \cup \{(x_m, y)\}) = V_{\alpha,\beta}(\phi)$$

and $H_\alpha^{n+1}(\theta, i_n) = V_{\alpha,\beta}(\psi)$. Fix α. Since $V_{\alpha,\beta}(\phi)$ is in $E(V_{\alpha+1,\beta}(H_\beta^{m+1}(\phi, i_m)))$ and

$$V_{\alpha+1,\beta}(H_\beta^{m+1}(\phi, i_m)) = V_{\alpha+1,\beta}(H_\beta^n(\psi, i_m)) = H_{\alpha+1}^n(V_{\alpha+1,\beta}(\psi), i_m),$$

(by Proposition 2.15) there exist $\theta \in E(V_{\alpha+1,\beta}(\psi), i_m)$ and $y \in X_{n+1} \setminus X_m$ such that

$$H_\alpha^{n+1}(\theta, i_m \cup \{(x_m, y)\}) = V_{\alpha,\beta}(\phi).$$

Then θ is as desired.

For part (4), let f be $\{(x_0, x_1)\}$. We need to find a $\phi \in \Phi_\beta^2$ such that $H_\beta^2(\phi, i_1) = H_\beta^2(\phi, f)$. Applying the first conclusion, it suffices to show that for each $\alpha < \beta$, there is a $\theta \in \Phi_\alpha^2$ such that $H_\alpha^2(\phi, i_1) = H_\alpha^2(\phi, f)$. Fix α. Since $\mathcal{P}$ is nonterminating, there exist $n \in \omega$ and distinct $\rho_1, \rho_2 \in \Phi_{\alpha+1}^n$ such that $V_{\alpha,\alpha+1}(\rho_1) = V_{\alpha,\alpha+1}(\rho_2)$. By condition (2c) of Definition 3.1, there exists an $\upsilon \in \Phi_{\alpha+1}^{2n}$, and $f, g \in \mathcal{I}_{n,2n}$ such that $H_{\alpha+1}^{2n}(\upsilon, f) = \rho_1$ and $H_{\alpha+1}^{2n}(\upsilon, g) = \rho_2$. There must be some $x_i \in X_n$ then such that $f(x_i) \neq g(x_i)$. Let $h = \{(x_0, f(x_i)), (x_1, g(x_i))\}$. Then $H_\alpha^{2n}(V_{\alpha,\alpha+1}(\upsilon), h)$ is as desired.

Part (5) follows from repeated application of the result for the case $n = 1$. We prove this case. Fix $\xi < \beta$ such that $\langle \Phi_\alpha : \alpha \in (\xi, \beta) \rangle$ is scattered, and let η be the Cantor-Bendixon rank of $\langle \Phi_\alpha : \alpha \in (\xi, \beta) \rangle$. Assuming that there is no ϕ as desired there is a $\zeta \in (\beta + \eta, \beta^+)$ such that, letting $M' = L_\zeta[\mathcal{P}]$, there exist, in M' an enumeration $\langle \phi_\alpha : \alpha < \beta \rangle$ of Φ_β and a club $C \subseteq \beta$ such that for all $\delta \in C \cap A$ and all $\alpha < \delta$, $V_{\delta,\beta}(\phi_\alpha) \neq \psi_\delta$. Working in M, we can find an elementary submodel X of M' such that $\xi \subseteq X$, $C \in X$ and $X \cap \beta \in C \cap A$. Let $\delta = X \cap \beta$. Then the Cantor-Bendixon rank of $\langle \Phi_\alpha : \alpha \in (\xi, \delta) \rangle$ is less than the ordertype of $X \cap \zeta$, which means that every path through $\langle \Phi_\alpha : \alpha \in (\xi, \delta) \rangle$, in particular ψ_δ, is in the transitive collapse of X. This means that $\psi_\delta = V_{\delta,\beta}(\phi_\alpha)$ for some $\alpha < \delta$, giving a contradiction.

For part (6), the assumption that $\mathcal{P}$ is nonterminating, plus Proposition 9.13, implies that for every limit ordinal $\alpha < \beta$, Φ_α has a nonisolated path. Applying part (5) gives a nonisolated ϕ.

For part (7), let A be the set of $\alpha < \beta$ for which Φ_α does not amalgamate. Working in M, for each $\alpha \in A$, pick $m_\alpha < n_\alpha$ in ω, $\phi_\alpha \in \Phi_\alpha^{m_\alpha+1}$ and $\psi_\alpha \in \Phi_\alpha^{n_\alpha}$ such that $H_\alpha^{m_\alpha+1}(\phi_\alpha, i_{m_\alpha}) = H_\alpha^{n_\alpha}(\psi_\alpha, i_{m_\alpha})$, and for which there is no amalgamating formula in Φ_α. Applying part (5), there exist ϕ and ψ in Φ_β and a stationary set $B \subseteq A$ such that, for all $\alpha \in B$, $V_{\alpha,\beta}(\phi) = \phi_\alpha$ and $V_{\alpha,\beta}(\psi) = \psi_\alpha$. Letting p and n be such that $\phi \in \Phi_\beta^p$ and $\psi \in \Phi_\beta^n$, it follows that $0 < p \leq n$, and $H_\beta^p(\phi, i_{p-1}) = H_\beta^n(\psi, i_{p-1})$. Applying part (3), there exist $\theta \in \Phi_\beta^{n+1}$ and $y \in X_{n+1} \setminus X_{p-1}$ such that $H_\beta^{n+1}(\theta, i_n) = \psi$ and $H_\beta^{n+1}(\theta, i_{p-1} \cup \{(x_{p-1}, y)\}) = \phi$. Then, for any $\alpha \in B$, $V_{\alpha,\beta}(\theta)$ contradicts our assumption that $\alpha \in A$. $\qquad\square$

By Theorem 7.11, there is exactly one model of a Scott process

$$\langle \Phi_\alpha : \alpha \leq \beta \rangle$$

of Scott rank β, if $|\Phi_\beta| \leq \aleph_1$, $\langle \Phi_\alpha : \alpha < \beta \rangle$ is nonterminating and Φ_β amalgamates.

Putting Theorem 11.1 together with the results of Section 10, we get Theorem 11.2 below. In conjunction with Theorem 10.6, we have the following theorem of Sacks (see [Sac83, Sac07]): if ϕ is a counterexample to Vaught's Conjecture, then for club many ordinals α below each of ω_1 and ω_2, ϕ has two nonisomorphic models of Scott rank α.

Theorem 11.2. *Let τ be a countable relational vocabulary, and suppose that $\phi \in L_{\omega_1,\omega}(\tau)$ is a counterexample to Vaught's Conjecture. Then for club many ordinals α below each of ω_1 and ω_2, ϕ has a model Scott rank α which is not Scott rank atomic.*

Proof. Let κ be either ω or ω_1 (of V). Work in $L[\phi]$. By Corollary 10.2 (and the fact that ϕ remains a counterexample to Vaught's Conjecture in any forcing extension, as discussed in Remark 10.5), there exists a nonterminating Scott process $\mathcal{P} = \langle \Phi_\alpha : \alpha < \kappa^+ \rangle$ such that

- in any forcing extension in which κ^+ is countable, $\mathcal{P}$ is satisfied by a model of ϕ;

- each Φ_α has cardinality less than κ^+.

Letting η be the quantifier depth of ϕ, we have by Theorem 1.2 that any model of $\langle \Phi_\alpha : \alpha \leq \eta \rangle$ is a model of ϕ. By part (7) of Theorem 11.1, Φ_β amalgamates, for club many $\beta < \kappa^+$. Fix such a β. Since $\mathcal{P}$ is a nonterminating Scott process of length greater than β, Φ_β contains a nonisolated path, by Proposition 9.13. By Theorem 7.11, there is a model of $\langle \Phi_\alpha : \alpha \leq \beta \rangle$ of Scott rank β (it cannot be of Scott rank less than β, since $\mathcal{P}$ is nonterminating). $\qquad\square$

The following question, a natural follow-up to Theorems 10.6 and 11.2, appears to be open.

Question 11.3. Let τ be a countable relational vocabulary, and suppose that $\phi \in L_{\omega_1,\omega}(\tau)$ is a counterexample to Vaught's Conjecture. Must there be an ordinal α such that ϕ has three nonisomorphic models of Scott rank α?

A positive answer to the previous question would follow from a positive answer to both parts of the following question. The question has several natural variations (for instance, one could strengthen the assumption on γ, as in parts (5) and (6) of Theorem 11.1).

Question 11.4. Suppose that $\mathcal{P} = \langle \Phi_\alpha : \alpha < \beta \rangle$ is a nonterminating scattered Scott process, where β is a limit ordinal. Let $\gamma > \beta$ be an ordinal, and let $M = L_\gamma[\langle \Phi_\alpha : \alpha < \beta \rangle]$. Suppose that, in M, the cofinality of β is greater than $|\Phi_\alpha|$, for each $\alpha < \beta$. Let Φ_β be the set of paths through $\mathcal{P}$, and suppose that $\Phi_\beta \subseteq M$.

1. Must $\mathrm{ms}(\phi)$ amalgamate, for each $\phi \in \Phi_\beta$?

2. Must there be a non-isolated $\phi \in \Phi_\beta$ such that $\mathrm{ms}(\phi) \neq \Phi_\beta$?

---

12   Isomorphic subprocesses

Recall that a *Scott subprocess* is a set of the form $\langle \Phi_\alpha : \alpha \in I \rangle$, for some Scott process $\{\Phi_\alpha : \alpha < \beta\}$ and $I \subseteq \beta$. An *isomorphism* between Scott subprocesses $\langle \Phi_\alpha : \alpha \in I \rangle$ and $\langle \Upsilon_\alpha : \alpha \in J \rangle$ is a bijection

$$\pi \colon \bigcup \{\Phi_\alpha : \alpha \in I\} \to \bigcup \{\Upsilon_\alpha : \alpha \in J\}$$

which commutes with the vertical and horizontal projection functions, i.e., such that there is an order preserving bijection σ from I to J and,

- for all $\alpha \leq \gamma$ in I, and all $\phi \in \Phi_\gamma$, $V_{\sigma(\alpha),\sigma(\gamma)}(\pi(\phi)) = \pi(V_{\alpha,\gamma}(\phi))$;

- for all $\alpha \in I$, $m \leq n$ in ω, $\phi \in \Phi_\alpha^n$ and $j \in \mathcal{I}_{m,n}$,

$$H_{\sigma(\alpha)}^n(\pi(\phi), j) = \pi(H_\alpha^n(\phi, j)).$$

The following theorem shows, among other things, that a Scott process of successor length is essentially determined by how the projection functions act on its first and last levels.

Theorem 12.1. *Let $\langle \Phi_\alpha : \alpha \leq \beta \rangle$ and $\langle \Upsilon_\alpha : \alpha \leq \gamma \rangle$ be nonterminating Scott processes. Let $\delta < \beta$ and $\epsilon < \gamma$ such that the Scott subprocesses $\{\Phi_\delta, \Phi_\beta\}$ and $\{\Upsilon_\epsilon, \Upsilon_\gamma\}$ are isomorphic. Then the intervals $[\delta, \beta]$ and $[\epsilon, \gamma]$ have the same ordertype, and $\langle \Phi_\alpha : \alpha \in [\delta, \beta] \rangle$ and $\langle \Upsilon_\alpha : \alpha \in [\epsilon, \gamma] \rangle$ are isomorphic.*

Proof. Let $\pi \colon \Phi_\delta \cup \Phi_\beta \to \Upsilon_\epsilon \cup \Upsilon_\gamma$ be an isomorphism. Let ζ be such that $\delta + \zeta = \beta$. Without loss of generality, we may assume that $\epsilon + \zeta \leq \gamma$. We define recursively, for $\eta < \zeta$ a $\subseteq$-increasing sequence of isomorphisms

$$\pi_\eta \colon \bigcup \{\Phi_\alpha : \alpha \in [\delta, \delta + \eta] \cup \{\beta\}\} \to \bigcup \{\Upsilon_\alpha : \alpha \in [\epsilon, \epsilon + \eta] \cup \{\gamma\}\},$$

with π_0 as π. When η is a limit ordinal, the existence of a unique extension as desired follows from condition (1c) of Definition 3.1, applied to $\Phi_{\delta+\eta}$, Φ_β,

$\Upsilon_{\epsilon+\eta}$. The successor case is similar, but slightly more involved: for each $n \in \omega$ and each $\phi \in \Phi_{\delta+\eta+1}^n$, the formula $V_{\delta+\eta,\beta}(\psi)$ and the set

$$V_{\delta+\eta,\beta}[\{\theta \in \Phi_\beta^{n+1} : H_\beta^{n+1}(\theta, i_n) = \psi\}]$$

are the same for all $\psi \in \Phi_\beta \cap V_{\delta+\eta+1,\beta}^{-1}[\{\phi\}]$. It follows that the formula $V_{\epsilon+\eta,\gamma}(\pi_\eta(\psi))$ and the set

$$V_{\epsilon+\eta,\gamma}[\{\theta \in \Upsilon_\gamma^{n+1} : H_\gamma^{n+1}(\theta, i_n) = \pi_\eta(\psi)\}],$$

and therefore, $V_{\epsilon+\eta+1,\gamma}(\pi_\eta(\psi))$ are the same for all such ψ, by Proposition 4.4. This common value of $V_{\epsilon+\eta+1,\gamma}(\pi_\eta(\psi))$ is the appropriate value for $\pi_{\eta+1}(\phi)$.

Finally, having defined π_η for each $\eta < \zeta$, let π^* be the union of these functions. Then π^* is an isomorphism from $\bigcup\{\Phi_\alpha : \alpha \in [\delta, \beta]\}$ to $\bigcup\{\Upsilon_\alpha : \alpha \in [\epsilon, \epsilon + \zeta), \gamma\}$. It follows then that $V_{\epsilon+\zeta,\gamma}$ is injective on Υ_γ (essentially by the argument just given for constructing the maps π_η). Since $\langle \Upsilon_\alpha : \alpha \leq \gamma \rangle$ was assumed to be nonterminating, we have that $\epsilon + \zeta = \gamma$. $\square$

Remark 12.2. The assumption that the Scott subprocesses in Theorem 12.1 are nonterminating is mostly for notational convenience. Without this assumption, and adding the assumption that the ordertype of the interval $[\delta, \beta]$ is at most that of the interval $[\xi, \gamma]$, the proof gives directly that $\langle \Phi_\alpha : \alpha \in [\delta, \beta] \rangle$ is isomorphic to $\langle \Upsilon_\alpha : \alpha \in [\epsilon, \epsilon + \zeta) \cup \{\gamma\} \rangle$, where ζ is such that $\delta + \zeta = \beta$. In general, if $\langle \Phi_\alpha : \alpha \leq \delta \rangle$ is a Scott process, γ is an element of δ, I is a subset of γ and $V_{\gamma,\delta} \upharpoonright \Phi_\delta$ is injective, then $\langle \Phi_\alpha : \alpha \in I \cup \{\gamma\} \rangle$ is isomorphic to $\langle \Phi_\alpha : \alpha \in I \cup \{\delta\} \rangle$.

The following theorem shows that, up to isomorphism, the tree of Scott processes extending a given Scott process of successor length is determined by the last two levels of the Scott processes, up to isomorphism.

Theorem 12.3. *Let $\mathcal{P} = \langle \Phi_\alpha : \alpha \leq \beta+1 \rangle$ and $\mathcal{Q} = \langle \Upsilon_\alpha : \alpha \leq \gamma+1 \rangle$ be Scott processes such that the Scott subprocesses $\{\Phi_\beta \cup \Phi_{\beta+1}\}$ and $\{\Upsilon_\gamma \cup \Upsilon_{\gamma+1}\}$ are isomorphic. Then for each ordinal $\delta > 0$ and each Scott process*

$$\langle \Phi_\alpha : \alpha \leq \beta+1+\delta \rangle$$

extending $\mathcal{P}$ there is a unique Scott process $\langle \Upsilon_\alpha : \alpha < \gamma+1+\delta \rangle$ extending $\mathcal{Q}$ such that the Scott subprocesses $\langle \Phi_\alpha : \alpha \in [\beta, \beta+1+\delta) \rangle$ and $\langle \Upsilon_\alpha : \alpha \in [\gamma, \gamma+1+\delta) \rangle$ are isomorphic.

Proof. By induction on δ. The limit case is immediate. For the case $\delta + 1$, the desired set $\Upsilon_{\gamma+1+\delta+1}$ is induced by the fact that each $\phi \in \Phi_{\beta+1+\delta+1}$ is uniquely determined by $V_{\beta+1+\delta,\beta+1+\delta+1}(\phi)$ and $E(\phi)$. Checking that this induced set $\Upsilon_{\gamma+1+\delta+1}$ gives a Scott process is routine for essentially all the conditions of Definition 3.1. For condition (2b), it follows from the fact that the domain of our given isomorphism contains cofinally many levels below $\gamma + 1 + \delta$. $\square$

Remark 12.4. We needed to start with an isomorphism on a pair of levels in Theorem 12.3, as opposed to just one level, in order to ensure that the successor levels of $\mathfrak{Q}$ would satisfy condition (2b) of Definition 3.1. In Theorem 12.5 below this issue does not arise, since we start our construction of $\mathfrak{Q}$ at level 0, so there are no instances of condition (2b) to consider.

In combination with the previous theorem, Theorem 12.5 shows that if there is a counterexample to Vaught's Conjecture, then there is one given by a Scott process of length 2. This gives another proof of the well-known fact that if there is a counterexample to Vaught's Conjecture, then there is one given by a $\mathcal{L}_{\aleph_1, \aleph_0}$ sentence of quantifier depth ω.

Theorem 12.5. *Let β, γ be ordinals, and let $\mathcal{P} = \langle \Phi_\alpha : \alpha \in [\beta, \beta + \gamma) \rangle$ be a Scott subprocess. Then there is a Scott process $\mathfrak{Q} = \langle \Upsilon_\alpha : \alpha < \gamma \rangle$ isomorphic to $\mathcal{P}$, over a distinct vocabulary.*

Proof. By Theorem 12.3 and Remark 12.4, it suffices to produce a Scott process of length 1 whose unique level is isomorphic to Φ_β. Let τ be the vocabulary corresponding to $\mathcal{P}$. Let μ be the vocabulary consisting of, in addition to $=$, a relation symbol R_ϕ for each $\phi \in \Phi_\beta$, where R_ϕ and ϕ have the same arity. We construct Υ_0 and the desired isomorphism π by defining the formula $\pi(\phi)$ for each $\phi \in \Phi_\beta$.

Fix $n \in \omega$ and Φ_β^n. The formula $\pi(\phi)$ will be a conjunction consisting of one instance each of the formula $x_i \neq x_j$, for distinct pair x_i, x_j from X_n, and for each formula of the form $R_\psi(y_0, \ldots, y_{m-1})$ (for $m \in \omega$, $\psi \in \Phi_\beta^m$ and $\{y_0, \ldots, y_{m-1}\} \subseteq X_n$) either this formula or its negation. If there exist $i < j < m$ such that $y_i = y_j$ (in particular, if $m > n$) we choose the negation. Otherwise, letting $f \in \mathcal{I}_{m,n}$ be such that $y_i = x_{f(i)}$ for each $i < m$, we choose $R_\psi(y_0, \ldots, y_{m-1})$ if and only if $H_\beta^n(\phi, f) = \psi$. This determines $\pi(\phi)$.

To check that this works, consider $n \in \omega$, $\phi \in \Phi_\beta^n$, $k \leq n$ and $g \in \mathcal{I}_{k,n}$. Let $\theta = H_\beta^n(\phi, g)$. We want to see that $\pi(\theta)$ is the set of conjuncts from $\pi(\phi)$ whose variables are contained in the range of g, with each variable replaced by its g-preimage. For the conjuncts of the form $x_i \neq x_j$ this is clear. Now suppose that we have a formula of the form $R_\psi(y_0, \ldots, y_{m-1})$, for some $m \in \omega$, $\psi \in \Phi_\beta^m$ and $\{y_0, \ldots, y_{m-1}\} \subseteq X_k$. Exactly one of $R_\psi(y_0, \ldots, y_{m-1})$ and its negation is a conjunct of $\pi(\theta)$, and we want to see that $R_\psi(y_0, \ldots, y_{m-1})$ is a conjunct of $\pi(\theta)$ if and only if $R_\psi(g(y_0), \ldots, g(y_{m-1}))$ is a conjunct of ϕ. The case where there exists an $i < j < m$ such that $y_i = y_j$ works out (in each direction), since g is an injection. In the other case, let $f \in \mathcal{I}_{k,m}$ be such that $y_i = x_{f(i)}$ for each $i < m$. Then

$$H_\beta^n(\phi, g \circ f) = H_\beta^k(\theta, f),$$

by part (2) of Remark 2.14, so $H_\beta^k(\theta, f) = \phi$ if and only if $H_\beta^n(\phi, g \circ f) = \psi$, as desired. For the reverse direction, suppose that we have a formula of the form $R_\psi(y_0, \ldots, y_{m-1})$, for some $m \in \omega$ and $\psi \in \Phi_\beta^m$, with $\{y_0, \ldots, y_{m-1}\}$ contained in the range of g. Then the formula $R_\psi(g^{-1}(y_0), \ldots, g^{-1}(y_{m-1}))$ is of the type just considered (i.e., $\{g^{-1}(y_0), \ldots, g^{-1}(y_{m-1})\} \subseteq X_k$), and we are done. $\qquad \square$

---

13 Projection structures

Two structures with isomorphic Scott processes need not be isomorphic: consider for instance a structure M with a single unary predicate P, where P^M and $|M| \setminus P^M$ have different cardinalities; the Scott process of M is isomorphic to the Scott process of the structure on $|M|$ with the interpretation of P reversed. Structures with isomorphic Scott processes must have the same Scott rank, however. The results of Section 12 show then that if there is a counterexample to Vaught's Conjecture, then there is one whose models are essentially the Scott processes of the structures from the given counterexample. We make this explicit in this section. However, we leave most of the verification to the reader, as the details are essentially the same as the arguments of the previous section.

We consider in this section structures whose relations satisfy the properties of the projection functions on Scott subprocesses, and whose points play the role of the formulas in a Scott process. These structures could be defined more generally, but we concentrate on a case (i.e., subprocesses with four levels, corresponding to levels 0, 1, γ and $\gamma+1$ of the Scott process of a structure of Scott rank γ) that seems more relevant to Vaught's Conjecture.

We let μ^* be the vocabulary consisting of $=$, unary predicate symbols $\mathrm{P}_{n,i}$ ($n \in \omega$, $i \in 4$) and unary function symbols v_i for $i \in 4$ and h_f for f in

$$\bigcup \{ \mathfrak{I}_{m,n} : m \leq n < \omega \}.$$

If M is a μ^*-structure, we let L_i^M and $L_{\geq i}^M$ denote the sets $\bigcup \{ \mathrm{P}_{n,i}^M : n \in \omega \}$ and $\bigcup \{ \mathrm{P}_{n,j}^M : n \in \omega, j \in 4 \setminus i \}$ respectively, for each $i \in 4$. Similarly, we let R_m^M and $R_{\geq m}^M$ denote the sets $\bigcup \{ \mathrm{P}_{m,i}^M : i \in 4 \}$ and $\bigcup \{ \mathrm{P}_{n,i}^M : n \in \omega \setminus m, i \in 4 \}$ respectively, for each $m \in \omega$. We say that a *projection structure* is a μ^*-structure M such that the following hold (the following lists established properties of the projection functions, plus parts of the definition of Scott process corresponding to a Scott subprocess whose first two levels are the first two levels of the corresponding process, and such that the vertical projection function from the last level to the second-to-last level is injective).

1. The sets $\mathrm{P}_{n,i}^M$ ($n \in \omega$, $i \in 4$) are nonempty and partition the domain of M. (The $\mathrm{P}_{n,i}^M$'s correspond to Φ_α^n's.)

2. Each v_i^M has domain $L_{\geq i}^M$ and range L_i^M, and is the identity function

on L_i^M. (Each $\mathbf{v}_i^M$ corresponds to a function of the form $\bigcup_{\beta \in I} V_{\alpha,\beta}$, for I the set of levels at or above α in the corresponding subprocess.)

3. For all $i \leq j < 4$, $\mathbf{v}_i^M \circ \mathbf{v}_j^M = \mathbf{v}_i^M \upharpoonright L_{\geq j}^M$. (This corresponds to Remark 2.8.)

4. The function $\mathbf{v}_2^M \upharpoonright L_3^M$ is injective. (Level L_2^M of M corresponds to the γ-th level of a structure of Scott rank γ, and level L_3^M corresponds to the $(\gamma + 1)$-st level.)

5. For all $i \leq j < 4$, and $n \in \omega$, $\mathbf{P}_{n,i}^M = \mathbf{v}_i^M[\mathbf{P}_{n,j}^M]$. (This corresponds to part (1c) of Definition 3.1.)

6. For all $m \leq n$ in ω and all $f \in \mathfrak{I}_{m,n} \setminus \bigcup\{\mathfrak{I}_{m,p} : p \in [m,n)\}$, $\mathbf{h}_f^M$ has domain $R_{\geq n}^M$ and range R_m^M. (Each $\mathbf{h}_f^M$ corresponds to a function of the form $\phi \mapsto \overline{H}(\phi, f)$, omitting the subscripts and superscripts on H.)

7. For all $i < j < 4$, $m \leq n$ in ω, $f \in \mathfrak{I}_{m,n}$ and $p \in \mathbf{P}_{n,j}^M$,

$$\mathbf{v}_i[\{q \in \mathbf{P}_{m+1,j}^M : \mathbf{h}_{i_m}^M(q) = \mathbf{h}_f^M(p)\}]$$

is equal to

$$\{\mathbf{h}_{i_m \cup \{(x_m, y)\}}^M(q) : y \in X_{n+1} \setminus X_m, q \in \mathbf{v}_i[\{r \in \mathbf{P}_{n+1,j}^M : \mathbf{h}_{i_n}^M(r) = p\}]\}$$

(This is a combination of part (3) of Definition 2.11 with Proposition 4.4.)

8. For all $i < 4$, all $n \in \omega$, all $f \in \mathfrak{I}_{n,n}$ and all $p \in \mathbf{P}_{n,i}^M$, $\mathbf{h}_f^M(p) \in \mathbf{P}_{n,i}^M$. (This corresponds to part (1d) of Definition 3.1.)

9. For all $i < 4$, and all $m \leq n$ in ω, $\mathbf{P}_{m,i}^M = \mathbf{h}_{i_m}^M[\mathbf{P}_{n,i}^M]$. (This corresponds to part (1e) of Definition 3.1.)

10. For all $i < 4$, all $m \leq n \leq p$ in ω, all $p \in \mathbf{P}_{p,i}^M$, all $f \in \mathfrak{I}_{n,p}$ and all $g \in \mathfrak{I}_{m,n}$, $\mathbf{h}_g^M(\mathbf{h}_f^M(p)) = \mathbf{h}_{f \circ g}^M(p)$. (This corresponds to part (2) of Remark 2.14.)

11. For all $i \leq j < 4$, all $m \leq n$ in ω, all $f \in \mathfrak{I}_{m,n}$, and all $p \in \mathbf{P}_{n,\beta}^M$,

$$\mathbf{v}_i^M(\mathbf{h}_f^M(p)) = \mathbf{h}_f^M(\mathbf{v}_i^M(p)).$$

(This corresponds to Proposition 2.15.)

12. For all $i < j < 4$, all $n \in \omega$ and all distinct $p, q \in \mathbf{P}_{n,j}^M$,

$$\mathbf{v}_i^M[\{r \in \mathbf{P}_{n+1,j}^M : \mathbf{h}_{i_n}^M(r) = p\}] \neq \mathbf{v}_i^M[\{r \in \mathbf{P}_{n+1,j}^M : \mathbf{h}_{i_n}^M(r) = q\}].$$

(This corresponds to the fact that formulas at successor levels of a Scott process are determined by their E-sets, which in turn are definable from the projection functions, by condition (2a) of Definition 3.1.)

13. For all $n \in \omega$, $p \in \mathrm{P}^M_{n,1}$ and $q \in \mathrm{P}^M_{n,2}$ with $\mathrm{v}^M_1(q) = p$,

$$\mathrm{v}^M_0[\{r \in \mathrm{P}^M_{n+1,2} : \mathrm{h}^M_{i_n}(r) = q\}] = \mathrm{v}^M_0[\{r \in \mathrm{P}^M_{n+1,1} : \mathrm{h}^M_{i_n}(r) = p\}].$$

(This corresponds to Proposition 4.4.)

14. For all $i < 4$, n, m in ω, $p \in \mathrm{P}^M_{n,i}$ and $q \in \mathrm{P}^M_{m,i}$, there exist $r \in \mathrm{P}^M_{n+m,i}$ and $f \in \mathfrak{I}_{m,n+m}$ such that $p = \mathrm{h}^M_{i_n}(r)$ and $q = \mathrm{h}^M_f(r)$. (This corresponds to part (2c) of Definition 3.1.)

If $\mathcal{P}$ is a Scott process of length 2 (over a relational vocabulary τ as assumed in this chapter), then there is a surjective correspondence between the Scott processes extending $\mathcal{P}$ whose length is two more than their rank and the projection structures (as defined in this section) whose first two levels are isomorphic to $\mathcal{P}$ (with the horizontal and vertical projection functions corresponding to the functions h_f and v_i). This is essentially Theorem 12.3. It follows that if Vaught's Conjecture is false, then there is a counterexample consisting of all projection structures whose first two levels are isomorphic to some fixed Scott process of length 2 (equivalently, an equivalence class of the class of projection structures under isomorphism of the first two levels).

Bibliography

[BFKL16] John T. Baldwin, Sy D. Friedman, Martin Koerwien, and Michael C. Laskowski, *Three red herrings around Vaught's conjecture*, Trans. Amer. Math. Soc. **368** (2016), no. 5, 3673–3694.

[BL16] John T. Baldwin and Paul B. Larson, *Iterated elementary embeddings and the model theory of infinitary logic*, Ann. Pure Appl. Logic **167** (2016), no. 3, 309–334.

[Gao09] Su Gao, *Invariant descriptive set theory*, Pure and Applied Mathematics (Boca Raton), vol. 293, CRC Press, Boca Raton, FL, 2009.

[Hjo96] Greg Hjorth, *On $\aleph_1$ many minimal models*, J. Symbolic Logic **61** (1996), no. 3, 906–919.

[Hjo07] ――――, *A note on counterexamples to the Vaught conjecture*, Notre Dame J. Formal Logic **48** (2007), no. 1, 49–51.

[HM77] V. Harnik and M. Makkai, *A tree argument in infinitary model theory*, Proc. Amer. Math. Soc. **67** (1977), no. 2, 309–314.

[Hod93] Wilfrid Hodges, *Model theory*, Encyclopedia of Mathematics and its Applications, vol. 42, Cambridge University Press, Cambridge, 1993.

[Kei71] H. Jerome Keisler, *Model theory for infinitary logic. Logic with countable conjunctions and finite quantifiers*, North-Holland Publishing Co., Amsterdam-London, 1971, Studies in Logic and the Foundations of Mathematics, Vol. 62.

[KMS16] Julia Knight, Antonio Montalbán, and Noah Schweber, *Computable structures in generic extensions*, J. Symb. Log. **81** (2016), no. 3, 814–832.

[LE65] E. G. K. Lopez-Escobar, *An interpolation theorem for denumerably long formulas*, Fund. Math. **57** (1965), 253–272.

[Mar02] David Marker, *Model theory*, Graduate Texts in Mathematics, vol. 217, Springer-Verlag, New York, 2002, An introduction.

[Mar11]   ———, *Scott ranks of counterexamples to Vaught's conjecture*, slides from a talk, `http://homepages.math.uic.edu/~marker/harrington-vaught.pdf`, 2011.

[Mon15]   Antonio Montalbán, *A robuster Scott rank*, Proc. Amer. Math. Soc. **143** (2015), no. 12, 5427–5436.

[Mor70]   Michael Morley, *The number of countable models*, J. Symbolic Logic **35** (1970), 14–18.

[Sac83]   Gerald E. Sacks, *On the number of countable models*, Southeast Asian conference on logic (Singapore, 1981), Stud. Logic Found. Math., vol. 111, North-Holland, Amsterdam, 1983, pp. 185–195.

[Sac07]   ———, *Bounds on weak scattering*, Notre Dame J. Formal Logic **48** (2007), no. 1, 5–31.

[Sco65]   Dana Scott, *Logic with denumerably long formulas and finite strings of quantifiers*, Theory of Models (Proc. 1963 Internat. Sympos. Berkeley), North-Holland, Amsterdam, 1965, pp. 329–341.

[Ste78]   John R. Steel, *On Vaught's conjecture*, Cabal Seminar 76–77 (Proc. Caltech-UCLA Logic Sem., 1976–77), Lecture Notes in Math., vol. 689, Springer, Berlin, 1978, pp. 193–208.

[Vau61]   R. L. Vaught, *Denumerable models of complete theories*, Infinitistic Methods (Proc. Sympos. Foundations of Math., Warsaw, 1959), Pergamon, Oxford; Państwowe Wydawnictwo Naukowe, Warsaw, 1961, pp. 303–321.

Chapter 3

Failure of 0-1 law for sparse random graph in strong logics (Sh1062)

Saharon Shelah
The Hebrew University of Jerusalem
Jerusalem, Israel
Rutgers, The State University of New Jersey
Piscataway, New Jersey, USA

Dedicated to Yuri Gurevich on the Occasion of his 75th Birthday

This work was partially supported by European Research Council grant 338821. Publication 1062 on Shelah's list. The author thanks Alice Leonhardt for the beautiful typing.

0   Introduction

3.0(A) The Question

Let $G_{n,p}$ be the random graph with set of nodes $[n] = \{1, \ldots, n\}$, each edge of probability $p \in [0,1]_{\mathbb{R}}$, the edges being drawn independently, see $\boxplus_2$ below. On 0-1 laws (and random graphs) see Spencer [Spe01] or Alon and Spencer [AS08], in particular on the behaviour of the random graph $G_{n,1/n^\alpha}$ for $\alpha \in (0,1)_{\mathbb{R}}$ irrational. On finite model theory see Flum and Ebbinghaus [EF06], e.g., on the logic $\mathbb{L}_{\infty,\mathbf{k}}$ (see §1) and on LFP (least fixed point[1]) logic. A characteristic example of what can be expressed by it is "in the graph G there is a path from the node x to node y"; this is close to what we use. We know that $G_{n,p}$, i.e., p constant satisfies the 0-1 law for first order logic (proved independently by Fagin [Fag76] and Glebskii et al. [GKLT69]). This holds also for many stronger logics like $\mathbb{L}_{\infty,\mathbf{k}}$ and LFP logic. If $\alpha \in (0,1)_{\mathbb{R}}$ is irrational, the 0-1 law holds for $G_{n,(1/n^\alpha)}$ and first order logic, see, e.g., [AS08].

The question we address is whether this holds also for stronger logics as above. Though our main aim is to address the problem for the case of graphs, the proof seems more transparent when we have two random graph relations (we make them directed graphs just for extra transparency). So here we shall deal with two cases A and B. In Case A, the usual graph, we have to show that there are (just first order) formulas $\varphi_\ell(x,y)$ for $\ell = 1, 2$ with some special properties (actually also φ_0), see Claim 1.2. For Case B, those formulas are $R_\ell(x,y), \ell = 1, 2$, the two directed graph relations. Note that (for Case B), the satisfaction of the cases of the R_ℓ are decided directly by the drawing and so are independent, whereas for Case A there are (small) dependencies for different pairs, so the probability estimates are more complicated.

In the case of constant probability $p \in (0,1)_{\mathbb{R}}$, the 0-1 law is strong: it is obtained by proving elimination of quantifier and it works also for stronger logics: $\mathbb{L}_{\infty,\mathbf{k}}$ (see §2) and so also for the LFP logic $\mathbb{L}_{\mathrm{LFP}}$. Another worthwhile case is:

$\boxplus_1$ $G_{n,1/n^\alpha}$ where $\alpha \in (0,1)_{\mathbb{R}}$; so $p_n = 1/n^\alpha$.

Again the edges are drawn independently but the probability depends on n.

The 0-1 law holds if α is irrational, but we have elimination of quantifiers only up to (Boolean combinations of) existential formulas. Do we have 0-1 law also for those stronger logics? We shall show that by proving that for some so-called scheme of interpretation $\bar\varphi$, for random enough $G_n, \bar\varphi$ interpret number theory up to m_n where m_n is not too small, e.g., $m_n \geq \log_2 \log_2(n)$.

---

[1]There are some variants, but those are immaterial for our perspective.

A somewhat related problem asks whether for some logic the 0-1 law holds for $G_{n,p}$ (e.g., $p = \frac{1}{2}$) but does not have the elimination of quantifiers, see [Sh:1077].

We now try to informally describe the proof, naturally concentrating on case B.

Fix reals $\alpha_1 < \alpha_2$ from $(0, \frac{1}{4})_{\mathbb{R}}$ for transparency, so $\bar{\alpha} = (\alpha_1, \alpha_2)$ letting $\alpha(\ell) = \alpha_\ell$;

$\boxplus_2$ let the random digraph $G_{n,\bar{\alpha}} = ([n], R_1, R_2) = ([n], R_1^{G_{n,\bar{\alpha}}}, R_2^{G_{n,\bar{\alpha}}})$ with R_1, R_2 irreflexive 2-place relations drawn as follows:

    (a) for each $a \neq b$, we draw a truth value for $R_2(a, b)$ with probability $\frac{1}{n^{1-\alpha_2}}$ for yes

    (b) for each $a \neq b$, we draw a truth value for $R_1(a, b)$ with probability $\frac{1}{n^{1+\alpha_1}}$ for yes

    (c) those drawings are independent.

Now for random enough digraph $G = G_n = G_{n,\bar{\alpha}} = ([n], R_1, R_2)$ and node $a \in G$; we try to define the set $S_k = S_{G,a,k}$ of nodes of G not from $\cup\{S_m : m < k\}$ by induction on k as follows:

For $k = 0$ let $S_k = \{a\}$. Assume $S_0, \ldots, S_k$ has been chosen, and we shall choose S_{k+1}.

$\boxplus_3$ For $\iota = 1, 2$ we ask: is there an R_ι-edge (a, b) with $a \in S_k$ and b not from $\cup\{S_m : m \leq k\}$?

If the answer is no for both $\iota = 1, 2$, we stop and let $\text{height}(a, G) = k$. If the answer is yes for $\iota = 1$, we let S_{k+1} be the set of b such that for some a the pair (a, b) is as above for $\iota = 1$. If the answer is no for $\iota = 1$ but yes for $\iota = 2$, we define S_{k+1} similarly using $\iota = 2$.

Let the height of G be $\max\{\text{height}(a, G) : a \in G\}$.

Now we can prove that for every random enough G_n, for $a \in G_n$ or easier - for most $a \in G_n$, for every not too large k we have:

$\boxplus_4$ $S_{G_n,a,k}$ is on one hand not empty and on the other hand with $\leq n^{2\alpha_2}$ members.

This is proved by drawing the edges not all at once but in k stages. In stage $m \leq k$ we already can compute $S_{G_n,a,0}, \ldots, S_{G_n,a,m}$ and we have already drawn all the R_1-edges and R_2-edges having first node in $S_{G_n,a,0} \cup \cdots \cup S_{G_n,a,m-1}$; that is for every such pair (a, b) we draw the truth values of $R_1(a, b), R_2(a, b)$. For $m = 0$ this is clear. So arriving to m we can draw the edges having the first node in S_m and not dealt with earlier, and hence can compute S_{m+1}.

The point is that in the question $\boxplus_3$ above, if the answer is yes for $\iota = 1$, then the number of nodes in S_{m+1} will be small, almost surely smaller than in S_m because its expected value is $|S_m| \cdot |[n] - \bigcup_{\ell \leq m} S_\ell| \cdot \frac{1}{n^{1+\alpha_1}} \leq n^{1+2\alpha_2-(1+\alpha_1)} = n^{2\alpha_2-\alpha_1}$ and the drawings are independent so except for an event of very small probability this is what will occur. Further, if for $\iota = 1$ the answer is no but for $\iota = 2$ the answer is yes, then almost surely S_m is smaller than a number near n^{α_1} but it is known that the R_2-valency of any node of G_n is near to n^{α_2}. Of course, the "almost surely" is such that the probability that at least one undesirable event mentioned above occurs is negligible.

So the desired inequality holds.

By a similar argument, if we stop at k then there is no R_2-edge from S_k into $[n]\backslash(S_0 \cup \ldots S_k)$ so the expected value is $\geq |S_k| \cdot (n - \sum_{\ell \leq k}(S_k)) \cdot \frac{1}{n^{1-\alpha_2}}$ hence in $S_0 \cup \cdots \cup S_k$ there are many nodes, e.g., at least near $n/2$ by a crude argument. As each S_m is not too large necessarily the height of G_n is large.

The next step is to express in our logic the relation $\{(a_1, b_1, a_2, b_2)$: for some k_1, k_2 we have $b_1 \in S_{G_n,a_1,k_1}, b_2 \in S_{G_n,a_2,k_2}$ and $k_1 \leq k_2\}$.

By this we can interpret a linear order with height(G_n) members. Again using the relevant logic, this suffices to interpret number theory up to this height. Working more we can define a linear order with n elements, so we can essentially find a formula "saying" n is even (or odd).

For random graphs we have to work harder: instead of having two relations we have two formulas; one of the complications is that their satisfaction for the relevant pairs is not fully independent.

In [Sh:1096] we shall deal with the strong failure of the 0-1 for Case A, i.e., G_{n,p^α}, (e.g., can "express" n is even) and also intend to deal with the α rational case. The irrationality can be replaced by discarding few exceptions.

We thank the referee for helping to improve the presentation.

3.0(B) History

The history is nontrivial having nontrivial opaque points. I have a clear memory of the events but vague on the exact statements and more so on the proof (and a concise entry in my (private F-list, [Sh:F159])) that in January 1996, in a Conference in DIMACS, Monica McArthur gave a lecture claiming that the graph $G_{n,\alpha}$ satisfies the 0-1 law not only for first order logic (by Shelah-Spencer [ShSp:304]) but also for a stronger logic. Joel Spencer said this coud be contradicted in a way he outlined. I thought about this and saw further things, and wrote them in a letter to Monica and Joel. I understood that it was agreed that Monica would write a paper with us saying more, but eventually she left academia.

As the referee found out, MacArthur's claim in [McA97] (DIMACS) failure of the law in $\mathbb{L}^\omega_{\infty,\omega}$, but refers the proof to a paper in preparation with Spencer that never appeared. She claims also that there is 0-1 law for $\mathbb{L}^k_{\infty,\omega}$ if $k = [1/\alpha]$, referring again to the paper in preparation. The later claim is not contradicted by the results of this paper. Lynch [Lyn97] refers also to a joint paper with McArthur and Spencer that never appeared proving that for the TC (transitive closure) logic satisfies the 0-1 law.

Having sent Joel (in 2011) an earlier version of this paper, his recollection of talking to Monica was that "we hadn't really gotten a handle on the situation".

Discussing with Simi Haber (December 2011), this question arose again. Trying to recollect it was not clear to me what the logic was: inductive logic? $\mathbb{L}_{\infty,k}$? Looking at it again, I saw a proof for the logic $\mathbb{L}_{\infty,k}$. No trace of the letter or the notes mentioned above were found. The only tangible evidence is in an entry [Sh:F159] from my F-list. Joel declined a suggestion that Haber, he and I deal with it, and eventually also Haber left.

The notes on §1 are from January 2012; for §2 from Sept. 4, 2012; revised in Nov./Dec. 2014 and expanded March 2015, June 2015.

The intention was that it would appear in the Yurifest, commemorating Yuri Gurevich's 75th birthday, but it was not in a final version in time, so only a short version (with the abstract and §(0A)) appears in the Yurifest volume, [Sh:1061].

3.0(C) Preliminaries

Notation 0.1. 1) $n \in \mathbb{N}\backslash\{0\}$ will be used for "$G_n \in K_n$ random enough".
2) G, H denote graphs and M, N denote more general structures = models.
3) a, b, c, d, e denote nodes of graphs or elements of structures.
4) m, k, ℓ denote natural numbers.
5) τ denotes a vocabulary, M a model with vocabulary $\tau = \tau_M$ (see 0.1(9),(10) below).
6) $\mathscr{L}$ denotes a logic, $\mathbb{L}$ is first order logic, so $\mathbb{L}(\tau)$ is first order language (set of formulas) for the vocabulary τ. $\mathscr{L}(\tau)$ is the language for the logic $\mathscr{L}$ and the vocabulary τ.
7) $\mathbb{L}_{\mathrm{LFP}}$ is the least fixed point logic, abbreviated LFP.
8)

(a) Let ${}^k A$ be the set of sequences η of length k of members of A, i.e., $\eta = \langle a_0, \ldots, a_{k-1} \rangle$ where $\bigwedge_{\ell < k} a_\ell \in A$, so $a_\ell = \eta(\ell)$.

(b) For a set u, e.g., of natural number let $\bar{x}_{[u]} = \langle x_s : s \in u \rangle$.

(c) If $\varphi(\bar{x}_m, \bar{y}) \in \mathscr{L}(\tau)$ and M is a τ-model and $\bar{b} \in {}^{\ell g(\bar{y})} M$ and $\bar{x}_m = \langle x_i : i < m \rangle$, then $\varphi(M, \bar{b}) = \{\bar{a} \in {}^m M : M \models \varphi[\bar{a}, \bar{b}]\}$.

9) Let τ_{gr} denote the vocabulary of graphs, but we may write $\mathbb{L}(\mathrm{graph})$ or $\mathscr{L}(\mathrm{graph})$ instead of $\mathbb{L}(\tau_{\mathrm{gr}}), \mathscr{L}(\tau_{\mathrm{gr}})$. So τ_{gr} consists of one two-place predicate R, (below always interpreted as a symmetric irreflexive relation).
10) Let τ_{dg} consist of two two-place predicates, below always interpreted as irreflexive relations. Let $\tau_{\mathbb{N}}$ be the vocabulary from 0.2(1).

11) We define the function $\log_*$ from $\mathbb{R}_{\geq 0}$ to $\mathbb{N}$ by:
  $\log_*(x)$ is 0 if $x < 2$
  $\log_*(x)$ is $\log_*(\log_2(x)) + 1$ if $x \geq 2$
12) $|u|$ is the cardinality = the number of elements of a set u.

Explanation 0.2. 1) Recall above that the vocabulary of the structure $\mathbb{N}$ is the set of symbols $\{0, 1, +, \times, <\}$ where $0,1$ are individual constants (interpreted in $\mathbb{N}$ as the corresponding elements) and $+, \times$ are two-place function symbols interpreted as $+^{\mathbb{N}}$, $\times^{\mathbb{N}}$ the two-place functions of addition and multiplication, and $<$ is a two-place predicate (relation symbol) interpreted as $<^{\mathbb{N}}$, the usual order on $\mathbb{N}$.
2) In general

(A) a vocabulary is a set of predicates, individual constants and function symbols each with a given arity (number of places); individual constants (like 0,1 above) are considered as 0-place function symbols

(B) a τ-model or a τ-structure M consists of:

  (a) its universe, $|M|$, a nonempty set of elements so $\|M\|$ is their number

  (b) if $P \in \tau$ is an n-place predicate, P^M is a set of n-tuples of members of M

  (c) if $F \in T$ is an n-place function symbol, then F^M is an n-place function from $|M|$ to $|M|$.

Definition 0.3. Let τ be a finite vocabulary, for simplicity with predicates only or we just consider a function as a relation; here we use $\tau_{\mathrm{gr}}, \tau_{\mathrm{dg}}$ only except when we interpret.
1) We say $\bar{\varphi}$ is in a (τ_*, τ)-scheme of interpretaion <u>when</u>: (if τ is clear from the context we may write τ_*-scheme)

(a) $\bar{\varphi} = \langle \varphi_R(\bar{x}_{n_\tau(R)}) : R \in \tau_* \cup \{=\} \rangle$ where $n_\tau(R)$ is the arity (number of places) of R

(b) $\varphi_R \in \mathbb{L}(\tau)$

(c) $\varphi_=(x_0, x_1)$ is always an equivalence relation on $\{y : (\varphi(y,y))\}$; if $\varphi_=$ is $(x_0 = x_1)$, then we may omit it.

2) For a τ-model M (here a graph or diagram) and $\bar{\varphi}$ as above, let $N = N_{M,\bar{\varphi}}$ be the following structure:

(a) $|N|$ the set of elements of N, is $\{a/\varphi_=(M) : a \in M$ and $M \models \varphi_=(a, a)\}$; note that $\varphi(M)$ is an equivalence relation on $\{a : M \models \varphi_=(a, a)\}$

(b) if $R \in \tau$ has arity m, then R^N, the interpretation of r is $\{\langle a_\ell / \varphi_=(M) : \ell < m : M \models \bigwedge_{\ell < m} \varphi_=(a_\ell, a_\ell) \wedge \varphi_R(a_0, \ldots, a_{m-1}) \text{ so } a_0, \ldots, a_{m-1} \in M\}$.

Recall that here "for every random enough G_n" is a central notion.

Definition 0.4. 1) A 0-1 context consists of:

(a) a vocabulary τ, here just the one of graphs or double directed graphs, see $0.1(5),(9),(10)$

(b) for each n, K_n is a set of τ-models with set of elements = nodes $[n]$, in our case graphs or double directed graphs

(c) a distribution μ_n on K_n, i.e., $\mu_n : K_n \to [0,1]_{\mathbb{R}}$ satisfying $\Sigma\{\mu_n(G) : G \in K_n\} = 1$

(d) the random structure is called $G_n = G_{\mu_n}$ and we tend to speak on G_{μ_n} rather than on the context.

2) For a given 0-1 context, let "for every random enough G_n we have $G_n \models \psi$, i.e., G satisfies ψ" and "if G_n is random enough, then ψ", etc. means that the sequence $\langle \text{Prob}(G_n \models \psi) : n \in \mathbb{N} \rangle$ converge to 1; of course, $\text{Prob}(G_n \models \psi) = \Sigma\{\mu_n(G) : G \in K_n \text{ and } G \models \psi\}$.
3) For $\bar{p} = \langle p_n = p(n) : n \rangle$ a sequence of probabilities, $G_{n,\bar{p}}$ is the case $K_n =$ graphs on $[n]$ and we draw the edges independently

(a) with probability p when $\bar{p}$ is constantly p, e.g., $\frac{1}{2}$, and

(b) with probability $p(n)$ or p_n when p is a function from $\mathbb{N}$ to $[0,1]_{\mathbb{R}}$.

Below, we add the second context because for it the proof is more transparent.

Context 0.5. 1) Case A:

(a) $a \in (0,1)$ is irrational

(b) $p_n = 1/n^\alpha$.

2) Case B:
$\bar{\alpha}^* = (\alpha_1^*, \alpha_2^*)$ where $\alpha_1^*, \alpha_2^* \in (0, 1/4)$ are irrational numbers, (natural to add linearly independent over $\mathbb{Q}$) such that $0 < \alpha_1^* < \alpha_2^* < \alpha_2^* + \alpha_2^* < 1/2$ and let $\alpha_0^* = \alpha_1^*$.

Definition 0.6. For Case A:

1) Let $K^1 := \bigcup_n K_n^1$ where we let K_n^1 be the set of graphs G on $[n] = \{1, \ldots, n\}$ so $R^G \subseteq \{\{i, j\} : i \neq j \in [n]\}$.

2) For $\alpha \in (0, 1)_\mathbb{R}$ let $G_n = G_{n;\alpha}$ be the random graph on $[n]$ with the probability of an edge being $1/n^\alpha$ and the drawing of the edges being independent.

3) Let $\mu_n = \mu_{n;\alpha}$ be the corresponding distribution on K_n^1; so $\mu_n : K_n^1 \to [0, 1]_\mathbb{R}$ and $1 = \Sigma\{\mu_n(M) : M \in K_n\}$, in fact, $\mu_n(G) = (1/n^\alpha)^{|R^G|} \times (1 - 1/n^\alpha)^{\binom{n}{2} - |R^G|}$.

Convention 0.7. Writing K_n means we intend K_n^1 or K_n^2 (see below), similarly G_n is $G_{n,\alpha}$ if Case A and $G_{n,\bar{\alpha}}$ if Case B and similarly K is K^1 or K^2.

The more transparent related case is the following:

Definition 0.8. On Case B, for $G_{n;\bar{\alpha}}$:

1) Recall τ_{dg} is the vocabulary $\{R_1, R_2\}$ intended to be two directed graph relations.

2) Let $K^2 = \bigcup_n K_n^2$ where we let $K_n^2 = \{G : G = ([n], R_1^G, R_2^G)$ satisfying $([n], R_\ell^G)$ is a directed graph for $\ell = 1, 2$; we may write R_ℓ instead of R_ℓ^G when G is clear from the context$\}$. We assume[2] irreflexivity, i.e., $(a, a) \notin R_\ell^G$ but allow $(a, b), (b, a) \in R_\ell^G$.

3) For reals $\alpha_1 < \alpha_2$ from $(0, \frac{1}{4})_\mathbb{R}$, say from $0.5(2)$ so $\bar{\alpha} = (\alpha_1, \alpha_2)$ let $\alpha(\ell) = \alpha_\ell$; let the random model $G_{n;\bar{\alpha}} = ([n], R_1, R_2) = ([n], R_1^{G_{n;\bar{\alpha}}}, R_2^{G_{n;\bar{\alpha}}})$ with R_1, R_2 irreflexive relations be drawn as follows:

(a) for each $a \neq b$, we draw a truth value for $R_2(a, b)$ with probability $\frac{1}{n^{1-\alpha_2}}$ for yes

(b) for each $a \neq b$, we draw a truth value for $R_1(a, b)$ with probability $\frac{1}{n^{1+\alpha_1}}$ for yes

(c) those drawings are independent.

4) We define the distribution $\mu_{n;\bar{\alpha}}$ as follows:

(a) $\mu_n = \mu_{n;\bar{\alpha}} = \mu_{n;\alpha_1,\alpha_2}$ is the following distributions on K_n^2:

  • $\mu_n(G) = \mu_{n;\alpha_2}^2([n], R_1^G)) \cdot \mu_{n;-\alpha_1}^2([n], R_2^G))$ where

  • $\mu_{n;\alpha}^2([n], R) = (\frac{1}{n^{1-\alpha}})^{|R^G|} \cdot (1 - \frac{1}{n^{1-\alpha}})^{n(n-1)-|R^G|}$

---

[2]We may change the definition of K_n^2 by requiring $R_1^G \cap R_2^G = \emptyset$, this makes little difference. We could further demand R_ℓ is asymmetric, i.e., $(a, b) \in R_\ell^G \Rightarrow (b, a) \notin R_0^G$, again this makes little difference.

(b) $G_n = G_{n;\bar{\alpha}} = G_{n;\alpha_1,\alpha_2}$ denote a random enough $G \in K_n^2$ for $\mu_{n;\bar{\alpha}}$ so n is large enough.

Observation 0.9. *For random enough (recalling 0.4(2))* $G_n = G_{n;\bar{\alpha}} = G_{n;\alpha_1,\alpha_2}$:

(a) *For* $a \in [n]$, *the expected value of the* R_2-*valency of* a, *that is,* $|\{b : aR_2^G b\}|$ *is* $(n-1) \cdot \frac{1}{n^{1-\alpha(2)}} \sim n^{\alpha(2)}$;

(b) *for every random enough* $G_{n;\bar{\alpha}}$ *for every* $a \in [n]$ *this number is close enough to* $n^{\alpha(1)}$, *e.g.,*

     $\bullet_2$ *for some* $\varepsilon \in (0, \alpha_1)_{\mathbb{R}}$, *the probability of the difference being* $\geq$ $n^{\alpha(1)(1-\varepsilon)}$ *for at least one* $a \in [n]$, *goes to zero with* n;

(c) *the expected number of* R_1-*edges is* $n(n-1)/n^{+(1+\alpha_1)} \sim n^{1-\alpha_1}$ *hence the expected value of* $|\{a : aR_1b$ *for some* $b\}|$ *is close to it;*

(d) *for every random enough* $G_{n,\bar{\alpha}}$ *the two numbers in (c) are close enough to* $n^{1-\alpha_1}$ *(similarly to (b)).*

Remark 0.10. 1) For K^2, this is a parallel of Claim 1.2 for K^1.
2) Note that the Clause (a) does not imply clause (b) in 0.9 because the a priori variance may be too large.

1   On the logic $\mathbb{L}_{\infty,\mathbf{k}}$

As the proof for $\mathbb{L}_{\infty,\mathbf{k}}$ is simpler and more transparent than for LFP, we shall explain it.

First, we try to define and then explain the logic $\mathbb{L}_{\infty,\mathbf{k}}$ for $\mathbf{k}$ a finite number.

For a vocabulary τ, we define the set $\mathbb{L}_{\infty,\mathbf{k}}(\tau)$ of formulas as the closure of the set of atomic formulas under some operation similarly to first order logic, but:

- we restrict ourselves to formulas having $< \mathbf{k}$ free variables

- we allow arbitrary conjunctions and disjunctions (that is even infinite[3] ones)

- as in first order logic we allow negation $\neg\varphi$ and existential quantifier (on one variable) $\exists x\varphi(x,\bar{y})$.

So not only does any formula in $\mathbb{L}_{\infty,\mathbf{k}}$ have $< \mathbf{k}$ free variables, but also every subformula does.

It may be helpful to recall the standard game which express equivalence. Recall $(0.3(1))$ that for transparency we assume the vocabulary below has only predicates and is finite.

⊞ we say $\mathscr{F}$ is an $(M_1, M_2) - \mathbb{L}_{\infty,\mathbf{k}}$-equivalence witness <u>when</u> for some vocabulary τ with predicates only

   (a) M_1, M_2 are τ-models

   (b) $\mathscr{F}$ is a nonempty set of partial isomorphisms from M_1 to M_2

   (c) if $f \in \mathscr{F}$, then $|\text{dom}(f)| < \mathbf{k}$

   (d) if $f \in \mathscr{F}, A \subseteq \text{dom}(f), |A| + 1 < \mathbf{k}, \iota \in \{1,2\}$ and $a_\iota \in M_\iota$, <u>then</u> there is g such that $g \in \mathscr{F}, f{\restriction}A \subseteq g$ and $\iota = 1 \Rightarrow a_\iota \in \text{dom}(g)$ and $\iota = 2 \Rightarrow a_\iota \in \text{rang}(g)$.

Now

⊕$_1$ for M_1, M_2 as in (a) of ⊞ above, the following are equivalent:

   (a) M_1, M_2 are $\mathbb{L}_{\infty,\mathbf{k}}$-equivalent, i.e., for every sentence $\psi \in \mathbb{L}_{\infty,\mathbf{k}}(\tau)$, i.e., a formula with no free variables, $M_1 \models \psi \Leftrightarrow M_2 \models \psi$

---

[3]As we consider only finite models, countable conjunctions and injunctions are enough.

(b) there is an $(M_1, M_2) - \mathbb{L}_{\infty,\mathbf{k}}$-equivalence witness $\mathscr{F}$, i.e., as in $\boxplus$.

Also

$\oplus_2$ for $M_1, M_2, \mathscr{F}$ as in $\boxplus$ above we have

(c) if $k < \mathbf{k}, a_0, \ldots, a_{k-1} \in M_1$ and $g \in \mathscr{F}$ and $\{a_\ell : \ell < k\} \subseteq \mathrm{dom}(g)$, <u>then</u> for every formula $\varphi(x_0, \ldots, x_{k-1}) \in \mathbb{L}_{\infty,\mathbf{k}}(\tau)$ we have

$$M_1 \models \varphi[a_0, \ldots, a_{k-1}] \Leftrightarrow M_2 \models \varphi[g(a_0), \ldots, g(a_{k-1})].$$

*   *   *

Having explained the logic, how can we prove for it the failure of the 0-1 law? Consider Case B where we have two kinds of edges, R_1 and R_2. Consider η a sequence from $^k\{1, 2\}$, see 0.1(11) and $a \neq b$. There may be $(\eta, 0, k)$-pre-paths from a to b in G, see Definition 1.6, i.e., $a = a_0, a_1, \ldots, a_k = b$ such that $(a_\ell, a_{\ell+1})$ is an $R_{\eta(\ell)}$-edge for $\ell < k$.

Now depending on η there may be many such pre-paths or few. If η is constantly 2 and $k > 1/\alpha_2^*$, then there are many such pre-paths - as fixing a in $G_{n,\bar{\alpha}^*}$ the expected number of b's for which there is pre-$(\eta, 0, k)$-paths from a to b is 1 for $k = 0$, is $\approx n^{\alpha_2^*}$ for $k = 1$ is $\approx n^{2\alpha_2^*}$ for $k = 2$, etc., so for $k > 1/\alpha_2^*$ it is every $b \in G_n$; not helpful. If η is constantly 1, there are few such pre-paths and they are all short, even $\leq k$ for any random enough G_n, when $1 < \alpha_1^* k$; not helpful.

But we may choose a "Goldilock's" η, that is, such that for every initial segment of η the expected number is not too large and not too small. This means that for some a for every $k' \leq k$ for some b there is such a pre-path but not too many. We need more so that we can define by a formula from $\mathbb{L}_{\infty,\mathbf{k}}$ the set $S_{G_n,a,k'} := \{b : \text{there is such pre-path from } a \text{ to } b \text{ of length } k' \text{ but not a shorter pre-path}\}$ and it is $\neq \emptyset$; moreover, we can define the natural order on the set $\{S_{G_n,a,k} : k \leq n\}$. Fact 1.4 below indicates what kind of η's we need, and we use it proving 1.8; however, in later sections, because we have to estimate the probabilities, we shall use only a closely related definition.

Hypothesis 1.1. 1) Case A of 0.5 holds or Case B there holds.
2)

(a) for case A: $\alpha_\ell^*, \varphi_\ell(x, y), n_\ell^*$ for $\ell = 0, 1, 2$ will be as in Claim 1.2 below

(b) for case B: α_1^*, α_2^* are as in 0.5 and $\varphi_\ell(x, y) = R_\ell(x, y)$ for $\ell = 1, 2$ and let $\alpha_0^* = \alpha_1^*, \varphi_0(x, y) = \varphi_1(x, y)$

(c) let $\bar{\varphi} = \langle \varphi_\ell(x, y) : \ell = 0, 1, 2 \rangle$.

Claim 1.2. Assume we are in Case A. There are $\alpha_\ell^*, \varphi_\ell(x, y)$ and γ_ℓ^* for $\ell = 0, 1, 2$ such that:

(a) $0 < \alpha_1^* < \alpha_0^* < \alpha_2^*$ are reals $\in (0, 1/4)_\mathbb{R}$ and $\gamma_\ell^* \in \mathbb{R}_{>0}$

(b) $\varphi_\ell(x, y)$ are first order formulas (in the vocabulary of graphs) even existential positive formulas such that $\varphi_\ell(x, y) \vdash x \neq y$ for random enough $G_{n;\bar{\alpha}}$

(c) if $G_{n;\bar{\alpha}}$ is random enough, then for every $a \in G_{n;\bar{\alpha}}$ the set $\varphi_2(G_{n;\alpha}, a)$ has $\approx \gamma_\ell^* n^{\alpha_2^*}$ elements, i.e., for some $\varepsilon \in (0, 1)_\mathbb{R}$, if $G_{n;\bar{\alpha}}$ is random enough, then for every $a \in [n]$, the number of members of $\varphi_1(G_{n;\bar{\alpha}}, a)$ belongs to the interval $(\gamma_\ell^* n^{\alpha_2^*} - n^{\alpha_2^*(1-\varepsilon)}, \gamma_2^* n^{\alpha_2^*} + n^{\alpha_2^*(1-\varepsilon)})$

(d) if $\ell = 0, 1$ and $G_{n;\bar{\alpha}}$ is random enough, then $\{a \in [n] : \varphi_\ell(G_{n;\bar{\alpha}}, a) \neq \emptyset\}$ has $\approx \gamma_\ell^* n/n^{\alpha_\ell}$ members.

Remark 1.3. We shall use not just the statements, but also the proof of 1.2 and 1.4.

Proof. Also here we shall use freely the analysis of $G_{n,\alpha}$ for $\alpha \in (0, 1)_\mathbb{R}$ irrational (see, e.g., [AS08]).

Let m_2^*, n_2^* be such that:

(a) n_2^* is large enough

(b) $m_2^* \leq \binom{n_2^*}{2}$

(c) $\alpha_2^* := (n_2^* - 1) - \alpha m_2^*$ is positive but, e.g., $< \frac{1}{12}$.

As $\alpha \in (0, 1)_\mathbb{R}$ is irrational we can find such m_2^*, n_2^*. Let H_2^* be a random enough graph on $[n_2^*]$ with m_2^* edges such that $(1, 2) \notin R^{H_2^*}$. (Note that this "random enough" is just used for the existence proof).

We choose n_1^*, m_1^*, H_1^* similarly except that $-\alpha_1^* := n_1^* - 1 - \alpha m_1^*$ is negative with value close enough, e.g., to $-\alpha_2^*/3$. Lastly, we choose n_0^*, m_0^*, H_0^* similarly except that $-\alpha_0^* = n_0^* - 1 - \alpha m_6^*$ is negative and $\alpha_0^* \in (\alpha_1^*, \alpha_2^*)$.

For $\ell = 1, 2$ let $\varphi_\ell(x, y) = (\exists \ldots x_i \ldots)_{i \in [n_\ell^*]} (x = x_1 \wedge y = x_2 \wedge \bigwedge \{x_i R x_j : i, j \in [m_\ell^*] \text{ satisfies } i R^{H_\ell^*} j\}) \wedge \bigwedge \{x_i \neq x_j : i \neq j \in [n_\ell^*]\}$.

Now check clauses (a)-(d). Clearly α_1^*, α_2^* satisfies clause (a) and φ_1, φ_2 are as in clause (b).

For $\ell = 1, 2$, let $\gamma_\ell^* = 1$. So for any n large enough compared to n_1^*, n_2^* and $a_1 \neq a_2 \in [n]$, the set $\mathscr{F} := \{f : f \text{ is a one-to-one function from } [n_2^*] \text{ to } [n]$ such that $f(1) = a_1, f(2) = a_2\}$ has $\prod_{i < n_2^* - 2} (n - 2 - i) \sim n^{n_2^* - 2}$ members.

For each $f \in \mathscr{F}$ the probability of the event $\mathscr{E}_f = $ "f maps every edge of H_2^* to an edge of $G_{n,\alpha}$" is $(\frac{1}{n^\alpha})^{m_2^*}$ so the expected value of $\{f \in \mathscr{F} : \mathscr{E}_f \text{ occurs}\}$

is $\approx n^{n_2^* - a m_2^* - 2} = \frac{1}{n^{1-a_2^*}}$. Clearly as in 0.9 the expected value is as required in clause (c) and by the well known analysis of $G_{n\alpha}$ (see, e.g., [AS08]), clause (c) holds and see more in §4.

Clause (d) is proved similarly. $\qquad\qquad\qquad\qquad\qquad\qquad\qquad\qquad\square_{1.2}$

Fact 1.4. There is a sequence $\eta \in {}^{\mathbb{N}}\{1,2\}$ such that: for every $n > 0$, $\gamma_n = |(\eta{\restriction}n)^{-1}\{2\}|\alpha_2^* - |(\eta{\restriction}n)^{-1}\{1\}|\alpha_1^*$ belongs[4] to $[\alpha_2^* - \alpha_1^*, \alpha_2^* + \alpha_2^*]_{\mathbb{R}}$.

Proof. We choose $\eta(n)$ by induction on n. Let $\eta(n)$ be 2 if $\gamma_n \leq \alpha_2^*$, e.g., $n = 0$ and $\eta(n)$ be 1 if $\gamma_n > \alpha_2^*$.

Easily η is as required. $\qquad\qquad\qquad\qquad\qquad\qquad\qquad\qquad\qquad\square_{1.4}$

Claim 1.5. 1) If η is as in 1.4, <u>then</u> for any m and every random enough G_n, there is an (η, m)-path in G_n, see below.
2) Moreover, also there is an $(\eta, \varepsilon\lfloor \log(n)\rfloor)$-path and even an $(\eta, \lfloor n^{\varepsilon}\rfloor)$-path for appropriate $\varepsilon \in \mathbb{R}_{>0}$.

Proof. As in [AS08] on $G_{n,1/n^\alpha}$ and see more in §3. $\qquad\qquad\qquad\square_{1.5}$

Definition 1.6. 1) A sequence $\bar{a} = \langle a_\ell : \ell \in [m_1, m_2]\rangle$ of nodes, that is, of members of $G \in K_n$ is called a pre-(ν, m_1, m_2)-path, if $m_1 = 0$ we may omit it, <u>when</u>:

(a) ν is a sequence of length $\geq m_2$ and $i < \ell g(\nu) \Rightarrow \nu(i) \in \{1,2\}$

(b) if $\ell \in \{m_1, m_1 + 1, \ldots, m_2 - 1\}$, then $G \models \varphi_{\nu(\ell)}(a_\ell, a_{\ell+1})$.

2) Above we say "(ν, m_1, m_2)-path" <u>when</u> in addition:

(c) if $m_1 \leq \ell_1 < \ell_2 \leq m_2$ and $\langle a_\ell' : \ell \in [m_1, \ell_2)\rangle$ is a pre-(ν, m_1, ℓ_2)-path, <u>then</u> $a_{m_1}' = a_{m_1} \wedge a_{\ell_2}' = a_{\ell_2} \Rightarrow a_{\ell_1}' = a_{\ell_1}$

(d) if $m_1 \leq \ell_1 < \ell_2 \leq m_2$, <u>then</u> $a_{\ell_1} \neq a_{\ell_2}$.

3) We say "$\bar{a}$ is a (pre)-(ν, m_1, m_2)-path from a to b" <u>when</u> in addition $a_{m_1} = a \wedge a_{m_2} = b$.

Remark 1.7. 1) Note that if $\langle a_\ell : \ell \leq m\rangle$ is a pre-(ν, m)-path, it is possible that $\ell_1 + 1 < \ell_2 \leq m$ and $a_{\ell_1} = a_{\ell_2}$. For a (ν, m)-path this is impossible.
2) In 1.6(2)(c), really the case $\ell_2 = m_2$ suffices.
3) We use the "$\log(n)$" in case 1.5(2), but having $\log(\log(n))$ and even much less has no real effect on the proof.

Conclusion 1.8. Let $\mathbf{k} \geq \max\{n_0^*, n_1^*, n_2^*\}$; <u>then</u> G_n fails the 0-1 law for $\mathbb{L}_{\infty, \mathbf{k}}$.

---

[4]We will also use other intervals and similar sequences.

Remark 1.9. 1) Note that if $\langle a_\ell : \ell \leq m \rangle$ is a pre-(ν, m)-path, it is possible that $\ell_1 + 1 < \ell_2 \leq m$ and $a_{\ell_1} = a_{\ell_2}$. For a (ν, m)-path this is impossible.

2) In 1.6(2)(c), really the case $\ell_2 = m_2$ suffices.

3) We use the "$\log(n)$" in 1.5(2), but having $\log\log(n)$ and even much less has no real effect on the proof.

Note that we rely on 1.5(2) but we prove more in §3.

Proof. For a finite graph G and η as in 1.4 or any $\eta \in {}^{\mathbb{N}}\{1, 2\}$ let $\text{length}_\eta(G)$ be the maximal m such that there is an (η, m)-path in G.

Now consider the statement

$\oplus$ there is a sentence $\psi_m = \psi_{\eta, m} \in \mathbb{L}_{\infty, \mathbf{k}}(\tau_{\text{gr}})$ such that for a finite graph G, $G \models \psi_m$ iff there is an (η, m)-path in G.

Why $\oplus$ is enough? Because then we let

$$\psi = \bigvee \{(\psi_m \wedge \neg\psi_{m+1}) : m \geq 10 \text{ and } (\log_*(m) \text{ is even})\}$$

where $\log_*(m)$ is essentially the inverse of the tower function, see 0.1(3). Note that using 1.4, 1.5(2), of course, we should be able to say much more.

Why $\oplus$ is true? First, we define the formula $\psi_{m_1, m_2}(x, y)$ for $m_1 \leq m_2$ by induction on $m_2 - m_1$ as follows:

$(*)_1$ if $m_1 = m_2$ it is $x = y$

$(*)_2$ if $m_1 < m_2$ it is $(\exists x_1)[\varphi_{\eta(m_1)}(x, x_1) \wedge \psi_{m_1+1, m_2}(x_1, y)]$.

So clearly

$(*)_3$ if $G \in K_n$ and $a, b \in [n]$, then $G \models \psi_{m_1, m_2}(a, b)$ iff there is a pre-(η, m_1, m_2)-path from a to b.

Second, we define $\psi'_{m_2}(x, y)$ as $\psi_{0, m_2}(x, y) \wedge \bigwedge_{\ell_1 < \ell_2 \leq m_2} \neg(\exists z_1', z_1'', z_2)[z_1' \neq z_2'' \wedge$
$\psi_{0, \ell_1}(x, z_1') \wedge \psi_{0, \ell_1}(x, z_1'') \wedge \psi_{\ell_1, \ell_2}(z_1', z_2) \wedge \psi_{\ell_1, \ell_2}(z_1'', z_2) \wedge \psi_{\ell_2, m_2}(z_2, y)]$.

This just formalizes 1.6(2)(c) so

$(*)_4$ $G \models \psi'_{m_2}(a, b)$ iff there is an (η, m_2)-path from a to b.

As said above this is enough. Note that complicating the sentence we may weaken the demand on G_n. $\square_{1.9}$

2 The LFP Logic

In this section we try to interpret an initial segment of number theory in a random enough $G \in K_n$, i.e., with set of nodes $[n]$. In Definition 2.2 for $G \in K_n$ and $a \in G$ we define a model $M_{G,a}$. Now in $M \in \mathbf{M}_{G,a_*}$, the equivalence classes of E^M represent natural numbers. Concentrating on Case B, starting with $\{a_*\}$ as the first equivalence class, its set of R_2-neighbors will usually be the second equivalence class. Generally, if for an equivalence class a/E^M we let the next one be the set $\mathrm{suc}(a/E^M) = \{b \in G : R_2(a',b)$ for some $a' \in a/E^M\}$, then we expect that $\mathrm{suc}(A/E^M)$ has $\approx |a/E^M| \cdot n^{\alpha_2}$ members. So if we continue in this way, shortly we get the equivalence classes cover essentially all the nodes of G. Hence we try to sometimes use the R_1-neighbors instead of the R_2-neighbors, but when? For $\mathbb{L}_{\infty,\mathbf{k}}(\tau_*)$ we can decide a priori, e.g., use η as in 1.4 and the proof of 1.8 so that the expected number will be small. But for LFP logic this is not clear, so we just say: use the R_1-neighbors if there is at least one and the R_2-neighbors otherwise, so this is close to what is done in 1.4, 1.5, 1.8 but not the same.

For case A we use φ_ℓ-neighbors instead of R_ℓ-neighbors for $\ell = 1,2$ except that the question on existence is for φ_0-neighbors.

How do we from equivalence relations and the successor relation reconstruct the initial segment of number theory? This is exactly the power of definition by induction.

Naturally we need just

⊞ letting height(G) be the maximal number of E^M-equivalence classes for $M \in \mathbf{M}_{G,a}, a \in G$, we have:

(*) for every m, for every random enough $G_n, m \leq \mathrm{height}(G)$; moreover, letting $f : \mathbb{N} \to \mathbb{N}$ be $f(n) = \log_*(n)$ for every random enough $G_n, f(n) \leq \mathrm{height}(G_n)$.

For failure of 0-1 laws, ⊞ is enough, but we may wish to prove a stronger version, say finding a sentence ψ which for every random enough G_n, G_n satisfies ψ iff n is even.

We intend to return to it elsewhere; but for now note that for a set $A \subseteq G$ we can define (R is R_2 for Case B, φ_2 for Case A) $cl_{G_n}(A) = \{b : b \in A$ or $b \in G\backslash A$ but for no $c \in G\backslash A\backslash\{a\}$ do we have $(\forall x \in A)[R(x,c) \equiv R(x,b)]\}$. Now from a definition of a linear order on A we can derive one on $cl_{G_n}(A)$. We can replace R by any formula $\varphi(x,y)$. Now in our context, if we know that, with parameters, we define such A of size $\approx n^\varepsilon$ for appropriate ε, then we can define a linear order on $[n]$; why is there such A? because if $M \in \mathbf{M}_{G,a_*}$ and k is not too large, <u>then</u> there is $b \in M, \mathrm{lev}(b,M) = k$ such that there is a unique maximal $<_2^M$-path from a_* to b.

For $\mathbb{L}_{\infty,\mathbf{k}}$ this is much easier.

Context 2.1. (A) or (B):

(A) (case A of 0.5) the vocabulary τ_* is τ_{gr}, the one for the class of graphs, $\varphi_\ell(x,y), \ell = 0,1,2$ are as in 1.2 so $\in \mathbb{L}(\tau_*)$ and $\bar{\alpha}^* = (\alpha_0^*, \alpha_1^*, \alpha_2^*)$, is as there, $G = G_n = G_{n;\alpha}, K_n = K_n^1$ as in Definition 1.4,

(B) (case B of 0.5) $\tau_* = \tau_{dg} = \{R_1, R_2\}, K_n = K_n^2$ and $\bar{\alpha}^* = (\alpha_1^*, \alpha_2^*)$ are as in 0.5(2) and $G_n = G_{n;\bar{\alpha}_*}$ and $\varphi_\ell(x,y) = xR_\ell y$ for $\ell = 1, 2$, with G_n, K_n^2 as in Definition 0.8 and let $\alpha_0^* = \alpha_1^*, \varphi_0(x,y) = \varphi_1(x,y)$.

Definition 2.2. For $G \in K_n$ and $a_* \in G$ we define $\mathbf{M} = \mathbf{M}(G, a_*) = \mathbf{M}_{G,a_*}$ as the set of τ_1-structures of M such that (the vocabulary τ_1 is defined implicitly):

(A) (a) the universe of M is $P^M \subseteq [n]$

   (b) $c_*^M = a_*$, so c_* is an individual constant from τ_1

   (c) E^M is an equivalence relation on M

   (d) $<_1$ is a linear order on P^M / E^M, i.e.,

   (α) $a_1 E^M a_2 \wedge b_1 E^M b_2 \wedge a_1 <_1^M b_1 \Rightarrow a_2 <_1^M b_2$

   (β) for every $a, b \in P^M$ exactly one of the following holds: $a <_1^M b, b <_1^M a$ and $aE^M b$

   (e) $<_2^M$ is a partial order included in $<_1^M$

   (f) (α) a_*/E^M is a singleton and a_* is $<_2^M$-minimal, i.e., $b \in M \backslash \{a_*\} \Rightarrow a <_2^M b$

   (β) if $a <_1^M b <_1^M c$ and $a <_2^M c$, then for some $b' \in b/E^M$ we have $a <_2^M b' <_2^M c$

   (γ) if $a, b \in M$ and b/E^M is the immediate successor of a/E^M, then for some $a' \in a/E^M$, we have $a' <_2^M b$

   (g) if $b \in M$ is a $<_2^M$-immediate successor of $a \in M$ (i.e., $a <_2^M b$ and $\neg(\exists y)(a <_2 y <_2 b)$, equivalently, $\neg(\exists y)(a <_1 y <_1 b))$, then for some $\iota \in \{1, 2\}$ we have $G \models \varphi_\iota[a, b]$

   (h) $P_0, P_1, P_2 = P_+, P_3 = P_\times, P_4 = P_<$ are predicates (of τ_1) with 1,1,3,3,2 places respectively such that using the definitions in clauses (B)(a),(b),(c) below, P_ℓ^M are defined in clauses (B)(d) below

(B) (a) for $a \in M, \mathrm{lev}(a, M)$ is the maximal k such that there are $a_0 <_1^M a_1^M <\ldots<_1^M a_k = a$; so necessarily $a_0 = a_*$

   (b) $\mathrm{height}(M) = \max\{\mathrm{lev}(a, M) : a \in M\}$

(c) for $k < \text{height}(M)$ let $\iota = \iota(k, M) \in \{1, 2\}$ be such that if b, is a $<_2^M$-immediate successor of a and $k = \text{lev}(a, M)$ <u>then</u> $G \models \varphi_\iota[a, b]$, in the unlikely case both $\iota = 1$ and $\iota = 2$ are as required we use $\iota = 1$

(d) (α) $P_0^M = \{a_*\}$

(β) $P_1^M = \{a \in M : \text{lev}(a, M) = 1\}$

(γ) $P_2^M = \{(a, b, c) : a, b, c \in M \text{ and } \mathbb{N} \models \text{``lev}(a, M) + \text{lev}(b, M) = \text{lev}(c, M)\text{''}\}$

(δ) $P_3^M = \{(a, b, c) : a, b, c \in M \text{ and } \mathbb{N} \models \text{``lev}(a, M) \times \text{lev}(b, M) = \text{lev}(c, M)\text{''}\}$

(ε) $P_4^M = \{(a, b) : N \models \text{``lev}(a, M) < \text{lev}(a, N)\text{''}\}$.

Definition 2.3. 1) Let $\iota \in \{1, 2\}$.

We say N in the ι-successor of M in $\mathbf{M}_{G,a}$ <u>when</u> for some k

(a) $M, N \in \mathbf{M}_{G,a}$ so $G \in K_n$ for some n

(b) $M \subseteq N$ as models so $M = N \restriction |M|$, recalling $|M|$ is the set of elements of M

(c) $k = \text{height}(M)$ and $\text{height}(N) = k + 1$

(d) $b \in N \backslash M$ <u>iff</u> $\text{lev}(b, N) = k + 1$ <u>iff</u> $b \in G \backslash M$ and for some[5] $a \in M$ we have $\text{lev}(a, M) = k$ and $G \models \varphi_\iota[a, b]$.

2) We may omit ι above <u>when</u> : $\iota = 1$ <u>iff</u> ($*$) holds where:

($*$) there is $c \in M$ such that $\text{lev}(c, M) = k = \text{height}(G)$ and for some $b \in G \backslash M$ we have $G \models \varphi_0[b, c]$; yes not φ_1! but for case B there is no difference.

3) For a sentence ψ in the vocabulary $\tau_1 \cup \tau_*$, for $M, N \in \mathbf{M}_{G,a}$, we say N is the ψ-successor of M <u>when</u> for some $\iota \in \{1, 2\}$, N is the ι-successor of M and $(G, M) \models \psi \Leftrightarrow (\iota = 1)$.

4) For $G \in K_n, a_* \in G$ and $M \in \mathbf{M}_{G,a_*}$ we define $\mathbb{N}_M$ as the following structure N with the vocabulary of number theory:

(a) set of elements $\{a / E^M : a \in M\}$

(b) $0^N = a_* / E^M = P_0^N$

---

[5] No real harm to demand here (and in 2.2) "unique"

(c) $1^N = P_1^M$

(d) if $\mathbf{a}_\ell = a_\ell/E^M \in N, a_\ell \in M$ for $\ell = 1, 2, 3$, then

  (α) $N \models$ "$\mathbf{a}_1 + \mathbf{a}_2 = \mathbf{a}_3$" iff $(a_1, a_2, a_3) \in P_2^M$

  (β) $N \models$ "$\mathbf{a}_1 \times \mathbf{a}_2 = \mathbf{a}_3$" iff $(a_1, a_2, a_3) \in P_3^M$

  (γ) $N \models$ "$\mathbf{a}_1 < \mathbf{a}_2$" iff $(a_1, a_2) \in P_4^M$.

Claim 2.4. 1) If ι, G, M, a are as in 2.3(1), <u>then</u> there is at most one ι-successor N of M in $\mathbf{M}_{G,a}$.
1A) For some $\psi_* \in \mathbb{L}(\tau_1 \cup \tau_*)$, being a ψ_*-successor is equivalent to being a successor.
2) For a given $G \in K_n, a \in G, M \in \mathbf{M}_{G,a}$ and $\psi \in \mathbb{L}(\tau_1 \cup \tau_*)$ there is at most one ψ-successor N of M in $\mathbf{M}_{G,a}$.
3) For $G \in K_n$ and $a \in G$ there is one and only one sequence $\langle M_k : k \leq k_{G,a} \rangle$ such that:

(a) $M_k \in \mathbf{M}_{G,a}$

(b) M_0 has universe $\{a\}$

(c) M_{k+1} is the successor of M_k in $\mathbf{M}_{G,a}$, recall 2.3(2)

(d) if $k = k_{G,a}$, <u>then</u> there is no $N \in \mathbf{M}_{G,a}$ which is the successor of M_k in $\mathbf{M}_{G,a}$

3A) Above $\mathbb{N}_{M_k,a}$ is isomorphic to $\mathbb{N}\!\restriction\!\{0,\ldots,k\}$. Also, for every sentence ψ in $\underset{\Delta}{\mathbb{L}}$ or even in $\underset{\Delta}{\mathbb{L}_{\mathrm{LFP}}}$ in the vocabulary of number theory there is a sentence $\varphi \in \mathbb{L}_{\mathrm{LFP}}(\tau_*)$ such that $\mathbb{N}_{M_k,a} \models \psi \Rightarrow M_{k,a} \models \varphi$; of course, φ depends on ψ but not on G, a (and k).
4) In the LFP logic for τ_*, we can find a sequence $\bar{\varphi}$ of formulas with[6] variable $x_0, \ldots, y$ such that: for any $G \in K_n$ and $a \in G$, the sequence $\bar{\varphi}$ substituting y by a defines $N = N_{G,a}$ which is M_k for $k = k_{G,a,\psi}$ from part (3).
4A) For ψ as in 2.3(3), i.e., $\psi \in \mathbb{L}(\tau_1 \cup \tau_*)$, a sentence, the parallel of 2.4(3),(4),(5) holds for "ψ-successor", (so we should write $N_{G,a,\psi}$ instead $M_{G,a}$).
5) Letting height$(G) = \max\{$height$(N_{G,a}) : a \in G\}$, in LFP logic there is $\varphi_*(x)$ such that $G \models \varphi_*(a)$ <u>iff</u> $a \in G$ and height$(N_{G,a}) = $ height(G) <u>iff</u> for every $a_1 \in G$, height$(N_{G,a_1}) \leq$ height$(N_{G,a})$.
6) For any sentence ψ in the vocabulary of number theory (in first order or LFP logic) there is a sentence φ in induction logic for τ_* (recalling 2.1) such that for any $G, G \models \varphi$ iff $\mathbb{N}\!\restriction\!\{0,\ldots,$ height$(G)\} \models \psi$.

---

[6]The variable y stands for the parameters a; instead we may define in 2.3 one model M_k coding all $M_{a,\ell} \in \mathbf{M}_{G,a}$ for $\ell \leq k, a \in G$.

Proof. 1) Read Definition 2.3(1).

1A) Read 2.3(2).

2) Read 2.3(3) and recall part (1).

3) We choose M_k and prove its uniqueness by induction on k till we are stuck. Recalling 2.4(3)(d) we are done.

3A) Easy.

4),4A) Should be clear.

5) We can express by induction when "$\operatorname{lev}(b_1, \mathbf{M}_{G,a_1}) \leq \operatorname{lev}(b_2, \mathbf{M}_{G,a_2})$".

6) Should be clear but we elaborate.

Recall the formula $\varphi_*(x) \in \mathbb{L}_{\mathrm{LFP}}(\tau_*)$ from 2.4(5). By the choice of φ_* necessarily for some $a_*, G_n \models \varphi_*[a_*]$ (as in a finite nonempty set there is a maximal member) so $\operatorname{height}(a_*, G_n) = \operatorname{height}(G_n)$.

Now for a given ψ, let $\varphi \in \mathbb{L}_{\mathrm{LFP}}(\tau_*)$ say: for some (equivalently every) $a \in G_n$ such that $G_n \models \varphi_*(a)$, the model $\mathbb{N}_{G_n,a_n}$ defined in 2.3(4), which is isomorphic to $\mathbb{N}{\restriction}\{0, \dots, \operatorname{height}(a, N_{G_n,a_n})\}$, see 2.4(3A), satisfies ψ. $\square_{2.4}$

Conclusion 2.5. *We have "G_n fail the 0-1 law for the LFP logic; moreover; for some $\varphi \in \mathbb{L}_{\mathrm{ind}}(\tau_*)$ we have $\operatorname{Prob}(G_n \models \varphi)$ has $\lim - \sup = 1$ and $\liminf = 0$".*

Proof. Should be clear by the above, in particular 2.4(6), see 3.3, 3.4(2) for details on the probabilistic estimate needed for 1.5(2) on which we rely. But we elaborate.

Note that just the following is not sufficient:

$(*)_1$ some $\bar\varphi, \mathbf{m}$ satisfies

  (a) $\bar\varphi$ an interpretation scheme, see 0.3

  (b) $\mathbf{m}$ is a function from finite graphs to $\mathbb{N}$, depending on the isomorphism type only

  (c) for every m for every random enough $G_n, \mathbf{m}(G_n) \geq m$

  (d) for random enough $G_n, \bar\varphi$ defines an isomorphic copy of $\mathbb{N}{\restriction}\{0, \dots, \mathbf{m}(G_n)\}$.

However, it is enough if we add, e.g.,

$(*)_2$ $\mathbf{m}(G_n) \geq \log_2(\log_2(n))$.

Why it is enough? Let ψ be a first order sentence in the vocabulary such that $\mathbb{N}{\restriction}\{0, \dots, k\} \models \varphi$ iff $\log_*(k)$ belong to $\{10n + \ell : \ell = \{0, 1, 2, 3, 4\}$ and $n \in \mathbb{N}\}$.

Now use the interpretation $\bar\varphi$, i.e., we use 2.4(6) in our case.

Why $(*)_1 + (*)_2$ holds: By 3.3, 3.4(1) and 2.4(3A). $\square_{2.5}$

3 Revisiting induction

As discussed in §2, we need to prove that for random enough G_n, height(G_n) is large enough, equivalently, for some $a \in G_n$ (we shall prove that even for most), height(a, G_n) is large enough. For this a more detailed specific statement is proved - see $(*)$ in the proof of 3.3. That is, we prove that for most $a \in G_n$ (for random enough G_n): on one hand $M_{G,a,=k}$ is not too large, and, on the other hand, is not empty; and for Case A, even not too small. The computation naturally depends on what $\eta_{G,a}$ is, see 3.2(3). This is a delicate point.

For Case B, things are simpler. For each k we ask if there is an R_1-edge out of $M_{G,a,=k}$ to $G \backslash M_{G,a,k}$. If there is, clearly $M_{G,a,=k+1}$ will be quite small but not empty. If not, then necessarily $M_{G,a,=k}$ has $\leq n^{\alpha_2^* - \varsigma}$ members hence the number of R_2-neighbors of members of $M_{G,a,=k}$ cannot be too large (well $< n^{\alpha_2^*} n^{\alpha_2^*}$) so we are done.

Case A seems harder, so we simplify considering only small enough k, see 3.4, hence we can consider all possible η's of length k, that is, summing the probability of the "undesirable" events on all of them; so if each has small enough probability, even the unions of all those events has small enough probability. Now we divide the η's to those which are "reasonable candidates to be $\eta_{G_n,a}$" and those which are not. For the former η's, for almost all $a \in G_n$ there is a pre-$(\eta, 0, k_*)$-path starting with a. So it is enough to prove that for almost all $a \in G_n, \eta_{G,a} \lceil k^*$ is one of those former η's where k_* is the relevant large enough height, e.g., $\geq \lfloor \log_2(\log_2(n)) \rfloor$. For this it is enough to prove that the other η's cannot occur and this is what we do.

In this section we fulfill promises from §2 (and §1).

Context 3.1. As in 2.1.

Below we shall use

Definition 3.2. 1) For $G \in K_n$ we define $M_k(a, G) = M_{G,a,k}$ by induction as in 2.4(3) for $\psi = \psi_*$ from 2.4(1A) and also $k = k_{G,a}$ as there and height(G) as in 2.4(6).
2) Let $M_{G,a,=k} = M_{G,a,k} \backslash \cup \{M_{G,a,m} : m < k\}$ and similarly $M_{G,a,<k}$.
3) Let $\eta = \eta_{G,a}$ be the following sequence of length $k_{G,a}$: if $\ell < k_{G,\ell}$, then $\eta(\ell) = \iota(\ell, M_{G,a}) \in \{1, 2\}$ from Definition 2.2(B)(c).

Claim 3.3. For small enough $\varepsilon \in (0, 1)_\mathbb{R}$, for random enough G_n, for some $a \in [n], k_{G_n,a} \geq k^* = \lfloor n^\varepsilon \rfloor$ in case B and $k_{G_n,a} \geq \lfloor \log(\log(n)) \rfloor$ in case A.

Remark 3.4. It would be nice to use an $\eta \in {}^\mathbb{N}\{1, 2\}$ defined similarly to 1.4, say such that $\gamma_n \in [\alpha_0^*, \alpha_2^* + \alpha_2^*]$, but this is not clear. In case B, in the proof the problem is that the γ_n-s from 1.4 may be very near to α_0^*. Also the parallel problem for case A is that the answer to the question asked there is near the critical stage, so we are not almost sure about the answer.

Proof. For case A, we presently prove it, e.g., for $k^* = \lfloor \log_2(\log_2 n) \rfloor$ and for case B $k^* = \lceil n^\varepsilon \rceil$, an overkill, but this suffices for the failure of the 0-1 law. We intend to fill the general case elsewhere. Actually for any $\varepsilon \in (0,1)_{\mathbb{R}}$ we can get $k^* = \lfloor n^{1-\varepsilon} \rfloor$.

Let $\zeta \in (0,1)_{\mathbb{R}}$ be small enough and k^* be as above.

Clearly it is enough to prove:

($*$) if $a \in [n]$ and $k < k^*$, <u>then</u> the probability that at least one of the following $(i)_{a,k}, (ii)_{a,k}, (iii)_{a,k}$ fails (assuming $\ell < k \Rightarrow (i)_{a,\ell} \wedge (ii)_{a,\ell}$) is small enough; $< \frac{1}{k \log(n)}$ suffices, being $< \frac{1}{kn^i}$ for each i for large enough n is natural)

$(i)_{a,k}$ $k \leq k_{G_n,a}$

$(ii)_{a,k}$ $M_{G_n,a=k}$ has $\leq n^{\alpha_2^* + \alpha_2^*}$ elements

$(iii)_{a,k}$ $M_{G_n,a,k}$, (noting that $(i)_{a,k}$ implies $M_{G_n,\ell,k} \neq \emptyset$) has $\geq n^{\alpha_0^* - \zeta}$ elements[7] except when $k = 0$, <u>not need</u> for Case B.

Why does ($*$) hold?

Case 1 Case B of the Context 2.1 and Definition 0.8

We are given $n \geq 1$ and $a \in [n]$; we draw the edges in k stages so by induction on k. For $k = 0$ draw the edges starting with a (of both kinds, an overkill), i.e., for $\iota \in \{1,2\}$ the truth value of $R_\iota(a,b)$ for every $b \in [n]\setminus\{a\}$, hence we can compute $M_1(a,G)$.

The induction hypothesis on stage k is that $\langle M_{G,a,i} : i \leq k \rangle$ have been computed and we have drawn the truth value of $R_\iota(c,b)$ for $b \in \cup\{M_{G,a,i} : i < k\}$ and $c \in [n]\setminus\{b\}$. If $k < k_*$ we now draw the edges $R_\iota(b,c)$ for $b \in M_{G,a,=k}$ and any $c \neq b$; actually the $c \in M_{G,i,k}$ are irrelevant and so we can compute $M_{G,a,k+1}$. Now we ask: if $(i)_{a,m} + (ii)_{a,m}$ holds for $m \leq k$ what is the probability that $(i)_{a,k+1} + (ii)_{a,k+1}$? (recalling $(iii)_{a,k+1}$ is irrelevant), i.e., is it small enough? This is easy and as required.

In details, we ask

<u>Question:</u> Are there $c \in [n]\setminus M_{G,a,k}$ and $b \in M_{G,a,=k}$ such that $(b,c) \in R_1^G$?
First note

($*$) if $|M_{G,a,=k}| \geq n^{\alpha_2^* - \zeta}$, then the probability that the answer is no is $< 1/2^n$.

---

[7]We can use $\geq n^{\alpha_2^* - \alpha_0^* - \zeta}$.

[Why? We have $M_{G,a,=k} \times ([n]\backslash M_{G,a,k})$ independent drawings so their number is $\geq n^{\alpha_2^* - \varsigma} n/2$, each with probability $\frac{1}{n^{1+\alpha_1^*}}$ of success and $(1 + \alpha_2^* - \varsigma) - (1 + \alpha_1^*) = \alpha_2^* - \varsigma - \alpha_1^* > 0$ so the probability of the no answer is $(1 - \frac{1}{n^{1+\alpha_1^*}})^{n^{(1+\alpha_2^* - \varsigma)}} \sim 1/e^{(n^{\alpha_2^* - \varsigma - \alpha_1^*})/2}$; clearly more than enough.]

By $(*)$ it suffices to deal with the following two possibilities.

Possibility 1: The answer is yes.

In this case $M_{G,a,k+1}$ is well defined and $\iota(k, M_{G,a,k}) = 1, M_{G,a,=k+1} = \{c : c \in G\backslash M_{G,a,k}$ and $(b,c) \in R_1^G$ for some $b \in M_{G,a=k}\}$. Now for each $c \in [n]\backslash M_{G,a,k}$ and $b \in M_{G,a,=k}$ the probability of $(c,b) \in R_1^G$ is $\frac{1}{n^{1+\alpha_1^*}}$ hence by the independence of the drawing, recalling $|M_{G,a,=k}| \leq n^{\alpha_2^* + \alpha_2^*}$ the probability of $|M_{G,a,=k+1}| \geq n^{\alpha_2^* + \alpha_2^*}$ is negligible, e.g., $< 2^n$ so can be ignored. Also by the possibility we are in, $M_{G,a,=k+1} \neq \emptyset$.

Possibility 2: The answer is no and $|M_{G,a,=k}| \leq n^{\alpha_2^* - \varsigma}$.

This is easy, too, recalling that almost surely for every $a' \in G$ the number of R_2-neighbors in the interval $[n^{\alpha_2^*} - n^{\alpha_2^*(1-\varsigma)}, n^{\alpha_2^*} + n^{\alpha_2^*(1-\varsigma)}]$ and so the probability that $M_{G,a,=k+1}$ is too large is negligible.

Case 2: Case A of the context 2.1

Here it helps to use "φ_0, φ_1 are distinct".

Now it suffices to prove:

$(*)_1$ for random enough G_n, for[8] $a \in M$, the following has negligible probability of failure: $(i)_{a,k}, (ii)_{a,k}, (iii)_{a,k}$.

Note that for this it seems more transparent to[9] assume $k < \log_2(\log_2(n))$ and to translate $(*)_1$ to statement on paths.

$(*)_2$ For $k \leq k_*$ let

(a) $\Omega_{=k}$ be the set of $\eta \in {}^k\{1,2\}$ such that $\alpha(\eta) := |\eta^{-1}\{2\}| \cdot \alpha_2^* - |\eta^{-1}\{1\}| \cdot \alpha_1^*$ belongs to the interval $[0, \alpha_2^* + \alpha_2^* + \varsigma]$

(b) Ω_k be the set of $\eta \in {}^k\{1,2\}$ such that for every $\ell < n$ the sequence $\eta{\upharpoonright}\ell = \langle \eta(0), \ldots, \eta(\ell-1) \rangle$ belongs to $\Omega_{=\ell}$

(c) Δ_k be the set of $\eta \in {}^k\{1,2\}$ such that $\eta \in \Omega_{=k}$ and even $\eta \in \Omega_k$ but $\alpha(\eta) \leq \alpha_0^* - \varsigma$

$(*)_3$ recalling that $\varsigma \in (0,1)_{\mathbb{N}}$ is small enough, for any random enough G_n, for every $a \in M$ the following has probability $\leq 1/n^\varsigma$:

---

[8]We allow few a's for which this fails. It suffices to have "for some a", this helps for larger k.

[9]As then, we can consider all the relevant sequences η, (and more).

- for some $k \leq \lfloor \log(\log(n)) \rfloor$ and $\eta \in {}^k\{1, 2\}$ at least one of the following holds

$(a)_\eta$ $\eta \in \Omega_k$ but there is no pre-$(\eta, 0, k)$-path in G_n starting with a

$(b)_\eta$ $\eta \in \Delta_k$ but there is a pre-$(\eta, 0, k+1)$-path in G_n from a to some $b \in G_n \backslash \{a\}$ such that $G_n \models (\exists x)\varphi_0(b, x)$.

Otherwise the proof is as in the earlier case. $\qquad\qquad\qquad\qquad \square_{3.3}$

Claim 3.5. Let $G_n = G_{n;\bar{a}^*}$, i.e., we are in Case B. For some $(\tau_{\mathbb{N}}, \tau_{\mathrm{dn}})$-scheme $\bar{\varphi}$, for every random enough G_n, $\bar{\varphi}$ defines in G_n a structure isomorphic to $\mathbb{N}_{<n}$.

Proof. We make a minor change in Definition 2.3(1),(d).

Clause $(d)^+$: We require the $a \in M$ is unique.

This makes no real difference above because the probability of the occurance if even one "b with two predecessors" is small and we just need that there is one; this makes $M_{G,a,k}$ smaller but not empty.

We start as in the proof of 3.3, for $k_* = \lfloor n^\varsigma \rfloor$, we use Case 1 but for stage k we draw the truth values of $\mathbb{R}_\iota(b, c)$ only when $b \in M_{G,a,=k}$ and $c \in [n]$ but $c \notin M_{G,a,0} \cup \ldots M_{G,a,k-1}$.

So there is $b \in M_{G,a,=k_*}$ and there is a unique sequence $\langle a_\ell : \ell \leq k_* \rangle$, $a_0 = a$, $a_{k_*} = b$ and $(a_\ell, a_{\ell+1}) \in R^{G_n}_\iota$ where ι is such that $M_{G,a,\ell+1}$ is the ι-successor of $M_{G,a,\ell}$ and so there are formulas $\psi_2 \in \mathbb{L}_{\mathrm{LFP}}(\tau_{\mathrm{dg}})$ such that not depending on the pair (a, n), $\psi_2(G, a, b) = \{(a_\ell, a_i) : \ell \leq i \leq k_*\}$.

Now

- the probability of the following event is negligible ($\varsigma < 2$): for some $d_1 \neq d_2 \in [n] \backslash M_{n,k_*}$ for every c: if $\psi_1(c, c, a, b)$, then $(d_1, c) \in R^G_2 \Leftrightarrow (d_2, c) \in R^G_2$.

Ignoring this event, the following formulas define a linear order on $[n] \backslash M_{G,a,k_*}$:

- $\psi_3(d_1, d_2, a, b)$ say: for some c we have $\psi_2(c, c, a, b) \wedge R_2(d_2, c) \wedge \neg R_2(d_1, c)$ and for any c_1 if $\psi_2(c_1, c, a, b)$, then $R_2(d_1, c') \leftrightarrow R_2(d_2, c')$.

So $\psi_3(x, y, a, b)$ defines a linear order on $[n] \backslash M_{G,a,k_*}$ which has $\geq n - \lfloor n^\varsigma \rfloor$ elements. Using the same trick we get $\psi_4 \in \mathbb{L}_{\mathrm{LFP}}(\tau_{\mathrm{dn}})$ and $\psi_4(x, y, a, b)$ defines a linear order on $[n]$. Now the formulas $\psi_2, \ldots, \psi_4$ do not depend on n. Also for some $\psi_5 \in \mathbb{L}_{\mathrm{LFP}}(\tau_{\mathrm{dg}})$

- $G \models \psi_5(a, b)$ iff $\psi_4(-, -, a, b)$ defines a linear order (on G) and for some $\psi_0 \in \mathbb{L}_{\mathrm{LFP}}(\tau_{\mathrm{dg}})$

⊙ $G \models \psi_0[c_1, a_1, b_1, c_2, a_2, b_2]$ iff:

(a)　$(a_1, b_1), (a_2, b_2) \in \psi_5(G)$

(b)　$|\{c : G \models \psi_4[c, c_1, a_2, b_1)\}| = |\{c : G \models \psi_4[c, c_2, a_2, b_2]\}|$.

So the interpretation should be clear.　　　　　　　　　　　　□$_{3.5}$

Conclusion 3.6. *[Case B] For some* $\psi \in \mathbb{L}_{\mathrm{LFP}}(\tau_{\mathrm{dg}})$ *for every random enough* $G_n = G_{n,\bar{\alpha}^*}$ *we have:*

- $G_n \models \psi$ *iff* n *is even.*

Bibliography

[AS08] Noga Alon and Joel H. Spencer, *The probabilistic method*, third ed., Wiley-Interscience Series in Discrete Mathematics and Optimization, John Wiley & Sons, Inc., Hoboken, NJ, 2008, With an appendix on the life and work of Paul Erdős.

[EF06] Heinz-Dieter Ebbinghaus and Jörg Flum, *Finite model theory*, enlarged ed., Springer Monographs in Mathematics, Springer-Verlag, Berlin, 2006.

[Fag76] Ronald Fagin, *Probabilities in finite models*, Journal of Symbolic Logic **45** (1976), 129–141.

[GKLT69] Y.V. Glebskii, D.I. Kogan, M.I. Liagonkii, and V.A. Talanov, *Range and degree of reliability of formulas in restricted predicate calculus*, Kibernetica **5** (1969), 17–27, translation of *Cybernetics* vol 5 pp 142-154.

[Lyn97] J.F. Lynch, *Infinitary logics and very sparse random graphs*, J. Symbolic Logic **62** (1997), 609–623.

[McA97] M. McArthur, *The asymptotic behavior of $L^k_{\infty,\omega}$ on sparse random graphs*, Logic and random structures (New Brunswick, NJ, 1995), DIMACS Ser. Discrete Math. Theoret. Comput. Sci., vol. 33, Amer. Math. Soc., Providence, RI, 1997, pp. 53–63.

[Spe01] Joel Spencer, *The strange logic of random graphs*, Algorithms and Combinatorics, vol. 22, Springer-Verlag, Berlin, 2001.

[Sh:F159] Saharon Shelah, *A letter to Joel Spencer and Monica McArthur*.

[ShSp:304] Saharon Shelah and Joel Spencer, *Zero-one laws for sparse random graphs*, Journal of the American Mathematical Society **1** (1988), 97–115.

[Sh:1061] Saharon Shelah, *On failure of 0-1 laws*, 293–296.

[Sh:1077] _____, *Random graph: stronger logic but with the zero one law*, preprint, arxiv:math.LO/1511.05383.

[Sh:1096] _____, *Strong failure of 0-1 law for LFP and the path logics*, preprint.

Part II

Model Theory of Special Classes of Structures

Chapter 4

Maximality of continuous logic

Xavier Caicedo

Universidad de los Andes
Bogotá, Colombia

1   Introduction

Continuous logic has its prehistory in Chang and Keisler's monograph on logic with values in compact Hausdorff spaces [13]; see also [12]. It had an independent revival in Krivine's successful use of model theoretic methods in Banach spaces [22, 23], work continued by Henson and Iovino [17, 18], and recently generalized to metric spaces by Ben Yaacov and Usvyatsov [5, 6, 3]. Appropriate versions of most techniques and properties of classical model theory generalize to continuous logic making it well suited for exploiting model theoretic methods in analysis: ultraproducts, compactness, Löwenheim-Skolem theorems, omitting types, large portions of stability theory. Noteworthy are the Keisler-Shelah theorem on ultraproducts and Morely's categoricity theorem.

The models of this logic are bounded complete metric structures equipped with uniformly continuous maps and [0,1]-valued predicates. The language has a connective for each continuous truth table $c : [0, 1]^n \to [0, 1]$ and the quantifiers are interpreted as infima and suprema. The metric plays the role of an identity predicate. [1] We will depart of the usual presentations of continuous logic in that we take 1 as the distinguished truth value and, more significant, we allow infinitary continuous connectives $c : [0, 1]^\omega \to [0, 1]$. The aim of this chapter is to prove the following Lindström style result which shows this logic has maximal expressive strength if we wish to maintain the two mentioned model theoretic properties.

(Theorem 6.5) *Let L be an extension of continuous logic closed under Łukasiewicz connectives $\to$, $\neg$ and satisfying compactness and the separable downward Löwenheim-Skolem property. Then any sentence of L is equivalent to a sentence of continuous logic.*

The compactness and Löwenheim-Skolem properties here are natural versions of the analogous classical properties, compactness being asked only for equicontinuous classes of structures, those having a common uniform continuity modulus for each basic function and predicate, and separability being the adequate countable power condition for complete metric structures. Two sentences are said to be equivalent if they attain the same values in all structures. Our main tools are a natural notion of approximation which behaves well in compact extensions, and a characterization of equivalence of models in continuous logic in the style of Fraissé by means of partial approximations. Some topological ideas and the close relation of continuous logic to Łukasiewicz logic will be useful also. We obtain, in fact, a weaker maximality result for Łukasiewicz logic:

---

[1]A more general setting allows unbounded pointed metric spaces, but restricts quantification to closed balls around the distinguished point, see Chapter 5 by Dueñez and Iovino. Our maximality result holds in this setting.

(Theorem 6.3) *Let L be an extension of Łukasiewicz logic closed under Łukasiewicz connectives* →, ¬ *and satisfying compactness and the separable downward Löwenheim-Skolem property. Then any sentence of L has the same models as a countable theory of Łukasiewicz logic in complete structures.*

It follows that continuous and Łukasiewicz logic have the same axiomatizability strength.

This chapter may be seen as a sequel of Iovino's maximality result for the approximation logic of Banach spaces with respect to compactness and the elementary chain property [19], and the characterization by Iovino and the author in terms of uncountable omitting types in [9].

Versions of our results have been announced in [9] and in several talks as early as 2008. The chapter intends to be self-contained but we refer the reader to [6], [3] and [20] for continuous logic, and [15] for Łukasiewicz logic.

2 Preliminaries

We start considering the language of continuous logic interpreted in non-metric $[0, 1]$-valued structures. A (*first order*) *many-sorted signature* is a sequence

$$\tau = \{\mathcal{S}, (R, \bar{s}_R), ..; (f, \bar{s}_f), .., c_{s_c}, ..\},$$

were $\mathcal{S}$ is a nonempty set of sorts, R, f, and c are predicate, function, and constant symbols, respectively, and $\bar{s}_R$, $\bar{s}_f$, s_c, are finite sequences in $\mathcal{S}$ associated to each symbol[2]. In the single-sorted case, these sort sequences may be identified with natural numbers.

A $[0, 1]$-*valued* τ-*structure* is a sequence

$$\mathfrak{A} = (\{A_s : s \in \mathcal{S}\}, R^{\mathfrak{A}}, ..; f^{\mathfrak{A}}..; c^{\mathfrak{A}}, ..),$$

where A_s is a nonempty universe for each sort and

$$R^{\mathfrak{A}} : \Pi_{i=1}^{n} A_{s_{R,i}} \to [0, 1], \quad f^{\mathfrak{A}} : \Pi_{j=1}^{m} A_{s_{f,j}} \to A_{s_{f,m+1}}, \quad c^{\mathfrak{A}} \in A_{s_c}.$$

St_τ will denote the class of all $[0, 1]$-valued τ-structures. This class has natural notions of isomorphism and substructure. An *isomorphism* $h : \mathfrak{A} \to \mathfrak{B}$ is an ordinary isomorphism of the many-sorted algebraic reducts of $\mathfrak{A}$ and $\mathfrak{B}$ satisfying, moreover, $R^{\mathfrak{B}}[h(a_1), .., h(a_n)] = R^{\mathfrak{A}}[a_1, .., a_n]$ for any predicate symbol R in τ and a_i in $A_{s_{R,i}}$. Its existence will be denoted $\mathfrak{A} \simeq \mathfrak{B}$. Similarly, $\mathfrak{A}$ is a *substructure* of $\mathfrak{B}$, denoted $\mathfrak{A} \leq \mathfrak{B}$, if it is a substructure of $\mathfrak{B}$ in the classical sense with respect to the operations in τ, and $R^{\mathfrak{A}}[\bar{a}] = R[\bar{a}]^{\mathfrak{B}}$ for the predicate symbols.

---

[2]Throughout the chapter, $\bar{x}$, $\bar{a}$, ... will denote finite sequences and their components will be denoted, respectively, x_i, a_i, ...

The language of *Łukasiewicz predicate logic,* denoted Ł∀, is built in the same way as the classical one based on the primitive connective symbols →, ¬, and quantifier symbols $\exists^t$, $\forall^t$ for each sort t. The set of formulas built from a signature τ will be denoted Ł∀$_\tau$. Terms, evaluated as in classical logic, give rise to maps $t^A : A_{\bar{s}} \to A_t$ in the usual way, while formulas $\varphi(x_1, ..., x_n)$ give rise to maps $\varphi^{\mathfrak{A}} : A_{\bar{s}} \to [0,1]$ and sentences determine truth values $\varphi^{\mathfrak{A}} \in [0,1]$, defined inductively by composing with Łukasiewicz's functional interpretation of the connectives in [0,1]:

$$p \to q = \min(1 - p + q, 1)$$
$$\neg p = 1 - p$$

(we utilize the same name for the symbol and its interpretation), interpreting quantifiers as suprema and infima. That is, for $\bar{a}$ in $A_{\bar{s}}$:

$$\exists^t x \varphi(x)^{\mathfrak{A}}[\bar{a}] := \sup_{b \in A_t} \varphi^{\mathfrak{A}}[b, \bar{a}]$$

and $\forall x \varphi(x)^{\mathfrak{A}}[\bar{a}] = \neg \exists^t \neg x \varphi(x)^{\mathfrak{A}}[\bar{a}] = \inf_{b \in A_t} \varphi^{\mathfrak{A}}[b, \bar{a}]$.

The language of *restricted continuous logic,* denoted CL°, is obtained by adding a connective symbol c for each continuous map $c : [0,1]^n \to [0,1]$ and closing under these operators: $c(\varphi_1[\bar{x}], .., \varphi_n[\bar{x}])$. The language of *continuous logic,* CL, is obtained similarly by adding all infinitary continuous connectives $c : [0,1]^\omega \to [0,1]$ for the Thychonov topology and closing the language under their application to sequences of formulas in the same finite set of variables: $c(\varphi_1[\bar{x}], \varphi_2[\bar{x}], \varphi_3[\bar{x}]...)$. Notice that a formula of CL may carry a countable infinite vocabulary but have a finite number of free variables. Since Łukasiewicz connectives are clearly continuous, ŁV becomes a sublogic of CL°, in turn a sublogic of CL. The additional strength provided by infinitary continuous connectives amounts to allowing "definable predicates" (see Lemma 2.3). For technical reasons and because the literature is ambiguous about allowing the latter in the syntax of continuous logic, we distinguish CL° and CL.

Clearly, if $h : \mathfrak{A} \to \mathfrak{B}$ is an isomorphism, then $\varphi^{\mathfrak{B}}[h(a_1), .., h(a_n)] = \varphi^{\mathfrak{A}}[a_1, .., a_n]$ for any $\bar{a}$ in A^n and formula φ of CL and a_i in $\mathfrak{A}$. If $\mathfrak{A} \leq \mathfrak{B}$, then $\varphi^{\mathfrak{B}}[a_1, .., a_n] = \varphi^{\mathfrak{A}}[a_1, .., a_n]$ for any quantifier free φ.

Satisfaction of formulas of CL in [0,1]-valued structures is defined in terms of the designed truth value 1:

$$\mathfrak{A} \models \varphi[\bar{a}] \quad (\mathfrak{A} \; satisfies \; \varphi \; at \; \bar{a}) \quad \text{if and only if} \quad \varphi^{\mathfrak{A}}[\bar{a}] = 1.$$

In the literature of continuous logic, 0 is utilized instead and the meaning of the quantifiers is switched, an inessential difference.

If $\mathfrak{A} \models \varphi$ for a sentence $\varphi \in$ CL, we say that $\mathfrak{A}$ is *a model of* φ. This nomenclature extends to theories, that is sets of sentences $T \subseteq$ CL and $Mod_\tau(\varphi)$, $Mod_\tau(T)$ will denote the class of $[0,1]$-valued τ-models of φ or T, respectively. It should be clear that $Mod_\tau(\varphi)$ does not describe the full meaning of φ since it says nothing of the structures where $\varphi^{\mathfrak{A}} = \frac{1}{2}$, for example.

A structure $\mathfrak{A} \in St_\tau$ is *crisp* (two-valued, classical) if its basic relations take values in $\{0, 1\}$. A simple induction in formulas yields that in crisp structures any formula $\varphi \in Ł\forall_\tau$ is two-valued and $\mathfrak{A} \models \varphi[\bar{a}]$ coincides with classical satisfaction. This is clearly not the case for CL.

For simplicity, we will usually refer to the single-sorted case through the rest of the chapter, resorting to the many-sorted case when strictly necessary only.

2.1 Language reductions

Łukasiewicz connectives $\rightarrow$ and $\neg$ have a high expressive power which permits to recover the lattice connectives:

$$p \vee q := \max\{p, q\} = (p \rightarrow q) \rightarrow q$$
$$p \wedge q := \min\{p, q\} = \neg(\neg p \vee \neg q)$$

and a host of arithmetical connectives, as:

$$
\begin{array}{lll}
p \oplus q & := \min\{p + q, 1\} = \neg p \rightarrow q & \text{truncated addition} \\
p \smallsetminus q & := \max\{p - q, 0\} = \neg(p \rightarrow q) & \text{truncated subtraction} \\
|p - q| := (p \smallsetminus q) \oplus (q \smallsetminus p) = (p \smallsetminus q) \vee (q \smallsetminus p) & \text{truth value distance.}
\end{array}
$$

Equivalence becomes the negation of truth distance:

$$p \leftrightarrow q := (p \rightarrow q) \wedge (q \rightarrow p) \equiv (p \rightarrow q) \odot (q \rightarrow p) \equiv \neg|p - q|.$$

Any of $\{\oplus, \neg\}$ or $\{\smallsetminus, \neg\}$ may be taken as complete set of connectives for $Ł\forall$ (this is not the case of $\{\wedge, \neg\}$ or $\{\vee, \neg\}$). The first pair is utilized in Chang's algebraization of Lukasiewicz propositional logic in terms of MV-algebras [10], while the latter is preferred in the literature of continuous logic. McNaughton's theorem [25] characterizes the Łukasiewicz definable connectives as the continuous piecewise linear maps $c : [0, 1]^n \rightarrow [0, 1]$ with integer coefficients. It follows that reflexive rational bounds on truth values of formulas are expressible in Łukasiewicz logic. Irreflexive bounds as $\varphi^{\mathfrak{A}} > r$ are not similarly expressible due to continuity of the connectives.

Lemma 2.1. (Chang [11], Belluce [1], Mundici [26]) *For each rational* $r = (0, 1)$ *there are Łukasiewicz connectives* β_r *and* γ_r *such that*

$$\mathfrak{A} \models \beta_r(\varphi) \text{ iff } \varphi^{\mathfrak{A}} \geq r, \qquad \mathfrak{A} \models \gamma_r(\varphi) \text{ iff } \varphi^{\mathfrak{A}} \leq r.$$

Moreover, $\beta_r(\varphi^{\mathfrak{A}}) \geq 1 - \varepsilon$ *implies* $\varphi^{\mathfrak{A}} \geq r - \varepsilon$, *and* $\gamma_{\frac{n}{m}}(\varphi^{\mathfrak{A}}) \geq 1 - \varepsilon$ *implies* $\varphi^{\mathfrak{A}} \leq r + \varepsilon$, *for any* $\varepsilon \in [0, 1)$.

Proof. The map $\beta_{\frac{n}{m}} : [0, 1] \to [0, 1]$ which is 0 for $x \leq \frac{n-1}{m}$, $mx - n + 1$ in $[\frac{n-1}{m}, \frac{n}{m}]$, and 1 for $x \geq \frac{n}{m}$, satisfies McNaughton's conditions and the claimed properties. And so does $\gamma_{\frac{n}{m}}(x) := 1 - \beta_{\frac{n+1}{m}}(x)$. $\qquad\square$

Notation. $\varphi_{\geq r}$ and $\varphi_{\leq r}$ will abbreviate $\beta_r(\varphi)$ and $\gamma_r(\varphi)$, respectively. As these connectives depend on the actual fraction $\frac{n}{m}$ representing r, the generic notation $\varphi_{\geq r}$, $\varphi_{\leq r}$ will refer to the reduced fraction of r.

It follows from Corollary 1.5 in [6] that any continuous map $c : [0, 1]^n \to [0, 1]$ may be approximated uniformly by combinations of Łukasiewicz connectives and rational constants, thanks to the Stone-Weirstrass theorem and compactness of $[0, 1]^n$ (see also Proposition 1.18 in [9]). This fact extends readily to infinitary continuous connectives if we add projections $[0, 1]^\omega \overset{\pi_{\bar{n}}}{\to} [0, 1]^n$, and it implies that CL may be approximated by the sublanguage ŁɅ$(\mathbb{Q})$ which results of adding to ŁɅ a constant connective of value r for each rational $r \in (0, 1)$.[3]

Lemma 2.2. *For any formula* $\varphi(\overline{x})$ *of CL there is a sequence of formulas* $\varphi_n(\overline{x})$ *in* ŁɅ$(\mathbb{Q})_\tau$ *such that* $\varphi^{\mathfrak{A}}[\overline{a}]$ *converges to* $\varphi_n^{\mathfrak{A}}[\overline{a}]$, *uniformly on* $(\mathfrak{A}, \overline{a}) \in St_{\tau \cup \{\overline{c}\}}$.

Proof. By Induction on complexity of $\varphi(\overline{x})$ and utilizing the fact that the connectives are uniformly continuous, one proves the existence for all n of $\varphi_n(\overline{x})$ such that $|\varphi_n^{\mathfrak{A}}[\overline{a}] - F(\mathfrak{A}, \overline{a})| < \frac{1}{n}$ for all $(\mathfrak{A}, \overline{a})$. $\qquad\square$

The approximation in the lemma cannot be achieved with formulas taken from ŁɅ, since in crisp structures these formulas take values in $\{0, 1\}$ and so must do their limits. As consequence of the lemma the countable theory $T_\varphi\{(\varphi_n)_{\geq 1 - \frac{1}{n}} : n \in \omega\}$ of ŁɅ$(\mathbb{Q})_\tau$ has the same models as φ. This will be improved for continuous structures in Corollary 6.4.

Given a sequence of formulas $\varphi_n(\overline{x})$ of CL converging uniformly on $St_{\tau \cup \{\overline{c}\}}$ to a map $F : St_{\tau \cup \{\overline{c}\}} \to [0, 1]$, then F is a (global) *definable predicate* in the sense of [6, 3]. The next lemma shows that CL is closed under these predicates.

Lemma 2.3. *If* $\varphi_n(\overline{x}) \to F$ *uniformly in* $St_{\tau \cup \{\overline{c}\}}$, *there is a formula* $\varphi(\overline{x})$ *of CL such that* $\varphi^{\mathfrak{A}}[\overline{a}] = F(\mathfrak{A}, \overline{a})$ *for any* $(\mathfrak{A}, \overline{a})$.

---

[3]Called rational Pavelka-Łukasiewicz logic in [15].

Proof. Find a subsequence of φ_{i_n} of $\{\varphi_n\}$ such that $|\varphi_{i_n}^{\mathfrak{A}}[\bar{a}] - \varphi_{i_{n+1}}^{\mathfrak{A}}[\bar{a}]| <$ 2^{-n} for all $(\mathfrak{A}, \bar{a})$. Then Lemma 3.7 in [6] grants that there is a continuous connective $c : [0,1]^\omega \to [0,1]$ such that $c(\varphi_{i_1}^{\mathfrak{A}}[\bar{a}], \varphi_{i_2}^{\mathfrak{A}}[\bar{a}], ...) = \lim \varphi_{i_n}^{\mathfrak{A}}[\bar{a}] =$ $\lim \varphi_n^{\mathfrak{A}}[\bar{a}]$ for all $(\mathfrak{A}, \bar{a})$. $\square$

Lemmas 2.2 and 2.3 put together say that, functionally speaking, CL is the uniform closure of Ł∀(ℚ) in $[0,1]$-valued structures.

2.2   The Löwenheim-Skolem property

$\mathfrak{A}$ is a CL-*substructure* of $\mathfrak{B}$, denoted $\mathfrak{A} \prec_{CL} \mathfrak{B}$, if and only if $\mathfrak{A} \leq \mathfrak{B}$ and $\varphi^{\mathfrak{A}}[\bar{a}] = \varphi^{\mathfrak{B}}[\bar{a}]$ for all $\varphi \in CL_\tau$ and $\bar{a}$ in $\mathfrak{A}$. The following properties of classical elementary embedding hold. In fact, they hold for any language L between Ł∀ and CL generated by the quantifiers and any family of continuous connectives containing $\to$, $\neg$.

Lemma 2.4. a) *Given $\mathfrak{A} \leq \mathfrak{B}$, then $\mathfrak{A} \prec_{CL} \mathfrak{B}$ if and only if for any formula $\varphi(x, \bar{y})$ in CL, $\bar{a}$ in A, b in B, and $r < 1$, $\mathfrak{B} \models \varphi[b, \bar{a}]$ implies there is a $a \in A$ such that $\mathfrak{B} \models \varphi_{\geq r}[a, \bar{a}]$.*

b) *Given $\mathfrak{B}$ and $C \subseteq B$ there is $\mathfrak{A} \prec_{CL} \mathfrak{B}$ with $C \subseteq A$ and $|A| \leq |C| + |\tau| + \omega$.*

Proof. a) One direction is trivial. For the other, show $\varphi^{\mathfrak{A}}[\bar{a}] = \varphi^{\mathfrak{B}}[\bar{a}]$ by induction in the complexity of formulas, the atomic case and the step for connectives are automatic. Assume $\exists x \varphi(x, \bar{a}]^{\mathfrak{A}} > \beta$. Then there is a rational $r > \beta$ and a in A such that $\varphi^{\mathfrak{A}}[a, \bar{a}] \geq r$, equivalent by induction hypothesis to $\varphi^{\mathfrak{B}}[a, \bar{a}] \geq r$, which implies $\exists x \varphi(x, \bar{a}]^{\mathfrak{B}} > \beta$. Reciprocally, $\exists x \varphi(x, \bar{a}]^{\mathfrak{B}} > \beta$ implies $\varphi^{\mathfrak{B}}[b, \bar{a}] \geq r$ for some $r > \beta$ which by the hypothesis of the lemma implies $\mathfrak{B} \models (\varphi_{\geq r})_{1-t}[a, \bar{a}]$ for some a in A and t arbitrarily small. Hence, By Lemma 2.1 $\mathfrak{B} \models (\varphi_{\geq r})_{1-t}[a, \bar{a}] \geq r - t > \alpha$ if t is small enough.

b) A τ-theory of power κ in CL is equivalent to a theory in Ł∀(ℚ)$_\tau$ of power $\kappa + \omega$ by the remark after Lemma 2.2. For each φ in Ł∀(ℚ)$_\tau$ and rational $r \in (0,1)$ choose a Skolem function $f_{\varphi,r}(\bar{y})$ such that $\mathfrak{B} \models \exists x \varphi(x, \bar{a}]$ implies $\varphi^{\mathfrak{B}}[f_{\varphi,r}(\bar{a}), \bar{a}] \geq r$. Let A be the closure of C under these maps and apply (a) to the induced substructure. As the number of Skolem functions is bounded by $\kappa + \omega$, then the bound of $|A|$ is attained. $\square$

Taking countable τ and $C = \emptyset$ in part (b) of the Lemma we obtain:

Proposition 2.5. (Löwenheim-Skolem property) *Any countable theory of CL having $[0,1]$-valued models has a finite or countable model.*

2.3   Continuous structures

The natural notion of identity in a $[0,1]$-valued structure $\mathfrak{A}$ is a distinguished binary predicate $\approx^{\mathfrak{A}}: A^2 \to [0,1]$ such that $\approx^{\mathfrak{A}}(a, b) = 1$ implies

$a = b$ and satisfying the usual congruence axioms. As noticed first by Katz [21] and independently in [15] and [5], these axioms become the axioms of a metric for the predicate $d(x, y) := \neg x \approx y$ under which the basic relations and functions result 1-Lipschitz continuous. Continuous logic takes d as the primitive notion and generalizes the Lipschitz conditions to arbitrary uniform continuity.

A $[0,1]$-valued structures is a *continuous structure* if it has a distinguished relation symbol $d_s \in \tau$ for each sort s, interpreted by a metric $d_s^{\mathfrak{A}} : A_s^2 \to [0, 1]$, under which the relations $R^{\mathfrak{A}}$ and maps $f^{\mathfrak{A}}$ ($R, f \in \tau$) are uniformly continuous with respect to the sup metric induced in $\Pi_i A_{s_{R,i}}$ and $\Pi_i A_{s_{f,i}}$, respectively.

This means that there is a system $S = \{m_\alpha\}_{\alpha \in \tau}$ of uniform continuity moduli $m_\alpha : (0, 1) \to (0, 1)$ such that for any $\varepsilon > 0$:

$$d(\overline{a}, \overline{b}) \quad < \quad m_R(\varepsilon) \text{ implies } |R(\overline{a}) - R(\overline{b})| \leq \varepsilon$$
$$d(\overline{a}, \overline{b}) \quad < \quad m_f(\varepsilon) \text{ implies } d(f(\overline{a}), f(\overline{b})) \leq \varepsilon,$$

where $d(\overline{a}, \overline{a}') = \max_i d(a_i, a_i')$.

It is not necessary to specify m for the metrics d_s or the constant symbols since a metric is always 1-Lipschitz and the constants satisfy any moduli. Without loss of generality, we may assume that all moduli satisfy $m(\varepsilon) \leq \varepsilon$, and they are given for $\varepsilon = \frac{1}{n}$ only. Therefore, thanks to Lemma 2.1, the above conditions are expressible in $L\forall$ by a theory U_S which has the following countable schema for each relation and function symbol in τ:

$$U_S(R) : \forall \overline{x}\overline{y}(d_s(\overline{x}, \overline{y})_{\geq r} \vee |R(\overline{x}) - R(\overline{y})|_{\leq \frac{1}{n}}), \quad n \in \omega^+, r < m_R(\tfrac{1}{n}), r \in \mathbb{Q}^+$$
$$U_S(f) : \forall \overline{x}\overline{y}(d_s(\overline{x}, \overline{y})_{\geq r} \vee d(f(\overline{x}), f(\overline{y}))_{\leq \frac{1}{n}}), \quad n \in \omega^+, r < m_f(\tfrac{1}{n}), r \in \mathbb{Q}^+.$$

$St_{\tau, S}$ will denote the class of metric τ-structures satisfying the moduli in S, that we will call *S-models*. These classes overlap; for example, crisp τ-structures with the discrete metric belong to all $St_{\tau, S}$. We will write $Mod_{\tau, S}(T)$ for the S-models of T.

Call a class $K \subseteq St_\tau$ *equicontinuous* if all structures in K share a common moduli system, that is, $K \subseteq St_{\tau, S}$ for some S. Perhaps the most important property of CL in continuous structures is the validity of a Łoś theorem for metric ultraproducts of equicontinuous families which yields compactness for satisfaction in each class $St_{\tau, S}$ (see [3]).

Proposition 2.6. (Compactness) *Given a moduli system S for τ, if any finite part of a theory $T \subseteq CL$ has S-models, then the theory has S-models.*

2.4 Complete structures

Our maximality results will depend strongly on having complete continuous models.

A continuous structure is *complete* if each metric space $(A_s, d_s^{\mathfrak{A}})$ is complete. $St_{\tau,S}^c$ will denote the class of complete S-structures and $Mod_{\tau,S}^c(T)$ will denote the class of complete S-models of T.

Any continuous structure $\mathfrak{A} \in St_{\tau,S}$ has a *(metric) completion* $\widehat{\mathfrak{A}}$ in $St_{\tau,S}^c$ which is well defined because uniformly continuous functions between metric spaces extend univocally to their completions with the same uniform convergence modulus. Moreover, the uniqueness of the extensions of the maps $t^A : A^n \to A$, $\varphi^{\mathfrak{A}} : A^n \to [0,1]$ to $\widehat{A}$ grants

$$\mathfrak{A} \prec_{CL} \widehat{\mathfrak{A}}.$$

Therefore, taking the completion of the model provided by Proposition 2.6, it may be seen that the compactness property holds in the class of complete structures. This may be obtained also from the fact that metric ultra-products may be chosen ω_1-saturated. Similarly, taking the completion of the model provided by Proposition 2.5, the Löwenheim-Skolem property becomes in complete structures:

Proposition 2.7. (Separable Löwenheim-Skolem property) *Any countable theory of CL having S-models has a separable S-model.*

3 Characterizing equivalence

In this section, we characterize Ł∀-equivalence of (not necessarily complete) continuous structures by means of ranked families of partial approximations. Equivalence of structures is defined in the obvious way for any language L between Ł∀ and CL: $\mathfrak{A} \equiv_L \mathfrak{B}$ if and only if $\varphi^{\mathfrak{A}} = \varphi^{\mathfrak{B}}$ for all suitable sentences $\varphi \in L$. Whenever L is closed under Łukasiewicz connectives this reduces by Lemma 2.1 to

$$\mathfrak{A} \equiv_L \mathfrak{B} \text{ if and only if } Th_L(\mathfrak{A}) = Th_L(\mathfrak{B})$$

Our characterization of Ł∀-equivalence below will imply that CL and Ł∀ share the same equivalence in continuous structures (Corollary 3.4):

$$\mathfrak{A} \equiv_{CL} \mathfrak{B} \text{ if and only if } \mathfrak{A} \equiv_{Ł∀} \mathfrak{B}.$$

For the proof of the next lemma it will be convenient to consider the Łukasiewicz connectives $\wedge$, $(\ )_{\geq \frac{k}{n}}$, and $(\ )_{\leq \frac{k}{n}}$, as primitive and define a complexity rank on formulas of Ł∀ as follows:

For terms, $\rho(t)$ is the usual syntactic depth (0 for variables and constants) $\rho(\alpha(t_1, .., t_k)) = \Sigma_i \rho(t_i)$, for a n-ary predicate symbol $\alpha \in \tau$

$$\rho(\varphi \to \psi) = \max\{\rho(\varphi), \rho(\psi)\} + 1$$
$$\rho(\neg\varphi) = \rho(\varphi) + 1$$
$$\rho(\varphi \wedge \psi) = \max\{\rho(\varphi), \rho(\psi)\}$$
$$\rho(\varphi_{\geq \frac{k}{n}}) = \rho(\varphi_{\leq \frac{k}{n}}) = \rho(\varphi) + n$$
$$\rho(\exists v\varphi) = \rho(\varphi) + 1.$$

Denote by $Ł\forall^\ell_{\mu,\bar{v}}$ the set of Lukasiewicz μ-formulas of rank $\leq \ell$ with free variables in $\bar{v}$. It is easy to verify by induction on ℓ that for any finite signature μ and finite list of variables $\bar{v}$, there are finitely many terms and finitely many nonequivalent formulas in $Ł\forall^\ell_{\mu,\bar{v}}$. For terms this is clear, and a fortiori for atomic formulas. Moreover, any nonatomic formula of rank $\leq \ell + 1$ has one of the forms: $\varphi \to \psi$, $\neg\varphi$, $\varphi_{\geq \frac{k}{n}}$, $\varphi_{\leq \frac{k}{n}}$, $\exists v\varphi$, where φ and ψ, have rank $\leq \ell$ and $n \leq \ell + 1$, or it is equivalent to a $\wedge$-combination of formulas of these forms. By induction hypothesis there are finitely many representatives for the displayed forms and the result follows because a finitely generated $\wedge$-semi-lattice is finite.

Proviso. For a signature τ, τ^* will denote in the rest of this chapter the set of atomic formulas of rank at most 1; that is, formulas $\alpha(t_1, .., t_k)$ where $\alpha \in \tau \cup \{d\}$ and at most one term t_i has the form $f(\bar{x})$ for a function symbol $f \in \tau$, the other being variables or constant symbols.

Definition 3.1. *Given a finite $\mu \subseteq \tau$, and $\varepsilon > 0$, a partial $\mu\varepsilon$-approximation between two τ-structures $\mathfrak{A}$ and $\mathfrak{B}$ is a relation $R \subseteq A \times B$ such that $|\varphi^{\mathfrak{A}}[\bar{a}] - \varphi^{\mathfrak{B}}[\bar{b}]| \leq \varepsilon$ for all $\varphi \in \mu^*$ and all suitable $\bar{a}, \bar{b}$ such that $(a_i, b_i) \in R$.*

Notice that the empty relation is a partial $\mu\varepsilon$-approximation if and only if $|\varphi^{\mathfrak{A}} - \varphi^{\mathfrak{B}}| \leq \varepsilon$ for any sentence $\varphi \in \mu^*$. We may identify a finite partial approximation $R = \{(a_i, b_i)\}_{i=1,..t} \subseteq A \times B$ with the ordered pair $(\bar{a}, \bar{b}) = ((a_1, ...a_t), (b_1, ...b_t)) \in A^t \times B^t$, the empty relation being identified with the pair $(\Lambda, \Lambda) \in A^0 \times B^0$, where $\Lambda = \emptyset$ stands for the empty sequence.

Lemma 3.2. *If $\mathfrak{A} \equiv_{Ł\forall(\tau)} \mathfrak{B}$, then for each finite $\mu \subseteq \tau$, $n \in \omega^+$, and $\varepsilon > 0$, there is a sequence $I_0, ..., I_n$ of nonempty sets of (finite) partial $\mu\varepsilon$-approximations between A and B with the following extension property: $R \in I_{j+1}$ and $a \in A$ $(b \in B)$ imply there is $R' \in I_j$ such that $R' \supseteq R$ and $a \in domR'$ $(b \in ranR')$.*

Proof. By the previous remarks, to prove the claim it is enough (given μ, ε and $n \in \omega$) to define nonempty sets $I_j \subseteq A^{n-j} \times B^{n-j}$, $j = 0, .., n$, satisfying:

(1) If $(\bar{a}, \bar{b}) \in I_j$, then $|\varphi^{\mathfrak{A}}[\bar{a}] - \varphi^{\mathfrak{B}}[\bar{b}]| \leq \varepsilon$ for all atomic $\varphi(\bar{x}) \in \mu^*$ with length$(\bar{x}) = n - j$.
(2) If $(\bar{a}, \bar{b}) \in I_{j+1}$ and $a \in A$ $(b \in B)$, then there is $b \in B$ $(a \in A)$ such that $(\bar{a}a, \bar{b}b) \in I_j$.

In the next definition let $\bar{v}_j$ denote a sequence of j many variables and $\bar{v}_0$ the empty sequence of variables. Fix N such that $\frac{1}{3^N} \leq \varepsilon$ and define for $j \leq n$

$$I_j = \{(\bar{a}, \bar{b}) \in A^{n-j} \times B^{n-j} : |\varphi^{\mathfrak{A}}[\bar{a}] - \varphi^{\mathfrak{B}}[\bar{b}]| \leq \frac{1}{3^{N+j}} \text{ for all } \varphi \in Ł\forall^{3^{N+j+1}}_{\mu, \bar{v}_{n-j}}\}.$$

Then (1) holds by definition. To prove the extension property (2), assume $(\bar{a},\bar{b}) \in I_{j+1}$, $j < n$, and given $a \in A$ consider the formula

$$Th_j(\bar{x}, v) := \bigwedge_i (\varphi_i(\bar{x}v)_{\geq \frac{k_i}{3^{N+j+1}}} \wedge \varphi_i(\bar{x}v)_{\leq \frac{k_i+1}{3^{N+j+1}}})$$

where $\{\varphi_i(\bar{x}v)\}_i$ is a finite list of equivalence representatives of the formulas in $\mathsf{L}\forall^{3^{N+j+1}}_{\mu^*,\bar{v}_{n-j}}$, and k_i is chosen so that $\varphi_i^{\mathfrak{A}}(\bar{a}a) \in [\frac{k_i}{3^{N+j+1}}, \frac{k_i+1}{3^{N+j+1}}]$. Clearly, $\exists v Th_j(\bar{x}v)$ has rank $2 \cdot 3^{N+j+1} + 1 \leq 3^{N+j+2}$ (belongs to $\mathsf{L}\forall^{3^{N+j+2}}_{\mu^*,\bar{v}_{n-(j+1)}}$), and since $(\bar{a},\bar{b}) \in I_{j+1}$ then

$$|\exists v Th_j(\bar{a}v)^{\mathfrak{A}} - \exists v Th_j(\bar{b}v)^{\mathfrak{B}}| \leq \frac{1}{3^{N+j+1}}.$$

As $A \models \exists v Th_j(\bar{a}v)$ by construction, then $\exists v Th_j(\bar{b}v)^B \geq 1 - \frac{1}{3^{N+j+1}}$ and we may find $b \in B$ such that $Th_j^{\mathfrak{B}}[\bar{b}b] \geq 1 - \frac{2}{3^{N+j+1}}$, thus for each φ_i

$$[\varphi_i[\bar{b}b]_{\geq \frac{k_i}{3^{N+j+1}}}], [\varphi_i[\bar{b}b]_{\leq \frac{k_i+1}{3^{N+j+1}}}]^{\mathfrak{B}} \geq Th_j^{\mathfrak{B}}[\bar{b}b] \geq 1 - \frac{2}{3^{N+j+1}},$$

and by Lemma 2.1: $\frac{k_i}{3^{N+j+1}} - \frac{2}{3^{N+j+1}} \leq \varphi_i^{\mathfrak{B}}[\bar{b}b] \leq \frac{k_i+1}{3^{N+j+1}} + \frac{2}{3^{N+j+1}}$. As $\varphi_i^{\mathfrak{A}}[\bar{a}a] \in [\frac{k_i}{3^{N+j+1}}, \frac{k_i+1}{3^{N+j+1}}]$ by construction, this means $|\varphi_i^{\mathfrak{A}}[\bar{a}a] - \varphi_i^{\mathfrak{B}}[\bar{b}b]| \leq 3\frac{1}{3^{N+j+1}} = \frac{1}{3^{N+j}}$. Since this is true for each $\varphi \in \mathsf{L}\forall^{3^{N+j+1}}_{\mu,\bar{v}}$ we conclude that $(\bar{a}a, \bar{b}b) \in I_j$. The other direction is similar. To finish, notice that $(\Lambda, \Lambda) \in I_n = A^0 \times B^0$ because $\mathfrak{A} \equiv^{3^{N+n}}_{\mathsf{L}\forall(\mu)} \mathfrak{B}$. Hence, by the extension property all I_j are nonempty. $\square$

We prove the reciprocal for CL° utilizing the simpler rank notion $r(c(\theta_1, .., \theta_k)) = 1 + \max_i r(\theta_i)$ for all the continuous connectives. $\mathfrak{A} \equiv^n_{\mathrm{CL}^\circ(\mu)} \mathfrak{B}$ will denote equivalence of A and B with respect to μ-sentences of CL° of rank r at most n.

Lemma 3.3. *Let $n \in \omega$ and $\mu \subseteq \tau$ be finite. If there is for each $\varepsilon > 0$ a sequence $I_0^\varepsilon, ..., I_n^\varepsilon$ of nonempty sets of partial $\mu\varepsilon$-approximations with the extension property between continuous structures $\mathfrak{A}$ and $\mathfrak{B}$ as in Lemma 3.2, then $\mathfrak{A} \equiv^n_{\mathrm{CL}^\circ(\mu)} \mathfrak{B}$.*

Proof. Fix μ, n and let $I_0^\varepsilon, ..., I_n^\varepsilon$ be given for each $\varepsilon > 0$ with the described properties. Write $(\bar{a}, \bar{b}) \in^* I_j^\varepsilon$ to mean that there is $R \in I_j^\varepsilon$ such that $(a_i, b_i) \in R$ for $i = 1, .., s$. Since $I_j^\varepsilon \neq \emptyset$, we have always $(\Lambda, \Lambda) \in^* I_j^\varepsilon$. We will show by induction in the rank $r(\varphi)$ of $\varphi(\bar{x}) \in \mathrm{CL}^{\circ n}_\mu$ that for any $\varepsilon > 0$ there is $\delta \leq \varepsilon$ such that for all $\bar{a}, \bar{b}$ of the same length $\geq \mathrm{length}(\bar{x})$:

$$r(\varphi) \leq j, \ \rho \leq \delta \ \text{and} \ (\bar{a}, \bar{b}) \in^* I_j^\rho \ \text{imply} \ |\varphi^{\mathfrak{A}}[\bar{a}] - \varphi^{\mathfrak{B}}[\bar{b}]| \leq \varepsilon.$$

- For φ atomic of rank ≤ 1 take $\delta = \varepsilon$, the result follows by definition of partial $\mu\rho$-approximation because $\varphi \in \mu^*$.

- For φ atomic of rank ≥ 2, $\varphi := \psi(t_1(\overline{x}), ..t_k(\overline{x}), \overline{x})$ where ψ is atomic of rank 1 and at least one t_i has positive rank; thus $r(\varphi) = \Sigma_i r(t_i) + 1 \geq k + 1$. As the atomic formula $\gamma_i := d(t_i(x), y)$ has rank $r(t_i) < r(\varphi)$, there is δ_i satisfying the induction hypothesis for γ_i and $m_\psi(\frac{\varepsilon}{2})$ instead of ε, where m_ψ is the continuity modulus of ψ. Define $\delta = \min_i \delta_i$ and assume: $r(\varphi) \leq j$, $\rho \leq \delta$, and $(\overline{a}, \overline{b}) \in^* I_j^\rho$. Then $k < r(\varphi) \leq j$ and applying the extension property k times we may find $b_1, ...b_k$ such that $(\overline{a}t_1[\overline{a}]...t_k[\overline{a}], \overline{b}b_1...b_k) \in^* I_{j-k}^\rho$. Since $\psi \in \mu^*$, this implies

$$|\psi^{\mathfrak{A}}(t_1[\overline{a}], .., t_k[\overline{a}]) - \psi^{\mathfrak{B}}(b_1, .., b_k)| \leq \rho \leq \delta \leq \delta_{\gamma_1} \leq m_\psi(\frac{\varepsilon}{2}) \leq \frac{\varepsilon}{2}.$$

On the other hand, as $r(\gamma_i) = r(t_i) \leq \Sigma_i(r(t_i) - 1) + 1 \leq j - k$, $\rho \leq \delta \leq \delta_i$ and $(\overline{a}...t_i[\overline{a}]..,\overline{b}...b_i..) \in^* I_{j-k}^\rho$, we may conclude by the properties of δ_i that $|d(t_i[\overline{a}], t_i[\overline{a}]) - d(t_i[\overline{b}], b_i)| \leq m_\psi(\frac{\varepsilon}{2})$; hence, $d(t_i[\overline{b}], b_i) \leq m_\psi(\frac{\varepsilon}{2})$ for all i and thus:

$$|\psi^{\mathfrak{B}}(t_1[\overline{b}], .., t_k[\overline{a}]) - \psi^{\mathfrak{B}}(b_1, .., b_k)| \leq \frac{\varepsilon}{2}.$$

Putting all together, $|\psi^{\mathfrak{A}}(t_1[\overline{a}], ..) - \psi^{\mathfrak{B}}(t_1[\overline{b}], .., t_k[\overline{a}])| \leq \varepsilon$.

- If φ has the form $\exists v\theta$ define $\delta_\varphi(\varepsilon) = \delta_\theta(\varepsilon)$. If $(\overline{a}, \overline{b}) \in^* I_j^\rho$ with $j \geq r(\varphi) \geq 1$, $\rho \leq \delta_\varphi(\varepsilon)$ then for each $a \in A$ there is $b \in B$ such that $(\overline{a}a, \overline{b}b) \in I_{j-1}^{n\rho}$; hence $|\theta^{\mathfrak{A}}[\overline{a}a] - \theta^{\mathfrak{B}}[\overline{b}b]| \leq \varepsilon$ by induction hypothesis. Taking suprema, first over a and then over b, $|(\exists v\rho)^{\mathfrak{A}}[\overline{a}] - (\exists v\rho)^{\mathfrak{B}}[\overline{b}]| \leq \varepsilon$.

- If φ has the form $c(\theta_1, .., \theta_k)$ where c is a continuous connective, each θ_i has rank $< r(\varphi)$ and thus $\delta_{\theta_i}(\varepsilon)$ exists. Let m_c be the uniform continuity modulus of c and define $\delta_\varphi(\varepsilon) = \min_i \delta_{\theta_i}(m_c(\varepsilon))$. If $(\overline{a}, \overline{b}) \in^* I_j^{n\rho}$ with $j \geq r(\varphi)$, $\rho \leq \delta_\varphi(\varepsilon)$ then $r(\theta_i) \leq j$ and $\rho \leq \delta_{\theta_i}(m_c(\varepsilon))$; hence, $|\theta_i^{\mathfrak{A}}[\overline{a}] - \theta_i^{\mathfrak{B}}[\overline{b}]| \leq m_c(\varepsilon)$ by induction hypothesis, and thus $|c(\theta_1, .., \theta_n)^{\mathfrak{A}}[\overline{a}] - c(\theta_1, .., \theta_n)^{\mathfrak{B}}[\overline{b}| \leq \varepsilon$.

- Finally, given a sentence $\varphi \in \mathrm{CL}_\mu^{on}$, pick any $(\overline{a}, \overline{b}) \in^* I_n^{\delta_\varphi(\varepsilon)}$, then $|\varphi^{\mathfrak{A}} - \varphi^{\mathfrak{B}}| = |\varphi^{\mathfrak{A}}[\overline{a}] - \varphi^{\mathfrak{B}}[\overline{b}]| \leq \varepsilon$. As ε is arbitrary, $|\varphi^{\mathfrak{A}} - \varphi^{\mathfrak{B}}| = 0$. □

Remark. The complexity of the proof of Lemma 3.3 in the atomic case is due to the presence of function symbols. For purely relational signatures, the proof would be simpler.

Corollary 3.4. *If $\mathfrak{A}$ and $\mathfrak{B}$ are continuous structures, then $\mathfrak{A} \equiv_{L\forall} \mathfrak{B}$ implies $\mathfrak{A} \equiv_{CL} \mathfrak{B}$.*

Proof. From lemmas 3.2 and 3.3 we have that $\mathfrak{A} \equiv_{L\forall} \mathfrak{B}$ implies $\mathfrak{A} \equiv_{CL^\circ} \mathfrak{B}$. Since any φ sentence of CL may be approximated uniformly by sentences φ_n of $L\forall(\mathbb{Q}) \subseteq CL^\circ$ we then have $|\varphi^{\mathfrak{A}} - \varphi^{\mathfrak{B}}| = |\lim \varphi_n^{\mathfrak{A}} - \lim \varphi_n^{\mathfrak{B}}| = \lim |\varphi_n^{\mathfrak{A}} - \varphi_n^{\mathfrak{B}}| = 0$. □

From this corollary and Proposition 2.6 it follows by a topological argument (Proposition 2.4 in [9], Proposition 4.4 in the next section) that $L\forall$ and CL have the same axiomatizability strength in any equicontinuous class of structures.

4 General $[0, 1]$-valued logics

A $[0,1]$-*valued logic* is a triple $L = (L, \overline{St}, V)$ where L and $\overline{St}$ are applications assigning to each signature τ a set of sentences L_τ, and a subclass $\overline{St}_\tau \subseteq St_\tau$, respectively, and $V : L_\tau \times \overline{St}_\tau \to [0, 1]$ is a semantical map subject to the following axioms where we write $\varphi^{\mathfrak{A}}$ for $V(\varphi, \mathfrak{A})$:

- *Isomorphism.* $\mathfrak{A} \simeq \mathfrak{B}$ implies $\varphi^{\mathfrak{A}} = \varphi^{\mathfrak{B}}$
- *Reducts.* If $\tau \subseteq \mu$ then $L_\tau \subseteq L_\mu$ and $\varphi^{\mathfrak{A}} = \varphi^{\mathfrak{A}}$ for any $\varphi \in L_\tau$, $\mathfrak{A} \in \overline{St}_\mu$.
- *Renaming.* A bijection $\alpha : \tau \to \mu$ preserving type of symbols induces a translation $\widehat{\alpha} : L_\tau \to L_\mu$ such that $(\widehat{\alpha}\varphi)^{\mathfrak{A}} = \varphi^{\mathfrak{A}^\alpha}$ for any $\mathfrak{A} \in \overline{St}(L)_\mu$, and $\varphi \in L_\tau$, where $\mathfrak{A}^\alpha \in \overline{St}(L)_\tau$ is the renaming of $\mathfrak{A}$ via α.

This is a slight reformulation of Definition 1.10 of a $[0,1]$-valued logic in [9] which paraphrases Lindström definition of model theoretic logic in [24]. In the translation axiom α may permute sorts, which means sentences may be duplicated in disjoint sort systems.

For our purposes $\overline{St}_\tau = \cup_S St_{\tau, S}$ (or $\overline{St}_\tau = \cup_S St_{\tau, S}^c$) the class of continuous (complete) structures, if τ has distinguished distance predicates, and $\overline{St}_\tau$ is empty otherwise.

All the model theoretic concepts and properties we have considered so far for CL on these classes may be applied to L, particularly compactness in the classes $St_{\tau, S}$ and the (separable) Löwenheim-Skolem property. In addition, we will utilize the following closure property:

Definition 4.1. L will be L *regular* if the collection of maps $F_\varphi : \overline{St} \to [0, 1]$, $\mathfrak{A} \mapsto \varphi^{\mathfrak{A}}$, for all $\varphi \in L_\tau$, is closed under composition with Łukasiewicz connectives $\to, \neg$.

There are at least two ways to compare $[0,1]$-valued logics:

- *Full strength*: $L \leq L'$ if for each sentence $\varphi \in L_\tau$ there is a sentence $\psi \in L'_\tau$ such that $\varphi^{\mathfrak{A}} = \psi^{\mathfrak{A}}$ for any $\mathfrak{A} \in \overline{St}_\tau$.

- *Axiomatic strength*: $L \leq_{ax} L'$ if for each sentence $\varphi \in L_\tau$ there is a theory $T \subseteq L'_\tau$ of power such that $Mod_\tau(\varphi) = Mod_\tau(T)$.

The second relation is weaker than the first even if T is a sentence. Natural (full) extensions of CL are:

By connectives. These must be discontinuous which necessarily destroy compactness. Indeed, if $c : [0, 1] \to [0, 1]$ and there is $x_n \to x$ in $[0, 1]$ with $\liminf_n c(x_n) < c(x)$, there are rationals r, s such that $c(x_n) \leq r < s \leq c(x)$ for infinitely many $n \in \omega$,, then the theory $T = \{c(p)_{\leq r}, c(q)_{\geq s}, (p \leftrightarrow q)_{\geq n/n+1} : n \in \omega\}$ is finitely satisfiable but unsatisfiable.

Infinitary continuous logic, $CL_{\omega_1\omega}$, introduced by Ben-Yaacov and Iovino in [4], it allows countable conjunctions $\bigwedge \varphi_n(\overline{x})$ when the φ_n share the same uniform convergence modulus. Compactness is lost but formulas stay continuous with respect to the metric and it shares the good behavior of classical $L_{\omega_1 \omega}$, see [7]. Infinitary logic on continuous structures without restriction on the countable conjunctions has been considered by Eagle [14].

Second order continuous logic, CL^{II}. Allow formulas $\exists X_m \varphi(X)$, $\forall X_m \varphi(X)$ where X_m is an n-ary predicate variable X with a specified moduli of continuity m, with the interpretation:

$$\exists X_m \varphi(X)^{\mathfrak{A}} = \sup_{f \in U\mathcal{C}_m(A^n, [0,1])} \varphi(X)^{(\mathfrak{A}, f)},$$

where $U\mathcal{C}_m(A^n, [0,1])$ is the set of functions $f : A^n \to [0,1]$ with uniform continuity modulus m.

Second order existential continuous logic $\Sigma_1^1(CL)$. The existential fragment of CL^{II}. It is not difficult to show that this logic inherits from CL compactness, the (separable) downward Löwenheim-Skolem property, the witnessed model property, and the complete model property.

By quantifiers. Further extensions may be obtained adding generalized quantifiers. For example, if κ is a cardinal:

$W_\kappa x \varphi(x)^{\mathfrak{A}} = \sup\{\varepsilon : (\varphi^{\mathfrak{A}})^{-1}([0, \varepsilon])$ has weight $\geq \kappa\}$.
$I x \varphi(x)^{\mathfrak{A}} = \sup\{\varepsilon : \varphi^{\mathfrak{A}}(A) \supseteq [0, \varepsilon]\}$.

4.1   A topological view

The *L-topology* of a [0,1]-valued logic L is the one generated in each $\overline{St}_\tau$ by the classes $Mod(\theta)$, $\theta \in L_\tau$, as a subbasis of closed classes. We denote this topology $\Gamma_\tau(L)$. The spaces $(\overline{St}_\tau, \Gamma_\tau(L))$ are proper classes but $\Gamma_\tau(L)$ is parametrized by sets (of sets of sentences) which permits to use safely all topological concepts (cf. [9], [8]).

Lemma 4.2. *(i)* $\Gamma_\tau(L)$ *is invariant under isomorphisms.*

(ii) If the logic is closed under disjunctions, then the closed subbasis is a basis and the closed classes are exactly the axiomatizable classes: $Mod(T)$, $T \subseteq L_\tau$.

(iii) A L-regular logic is topologically regular; that is, points and closed classes are separable by open classes in $(St_\tau, \Gamma_\tau(L))$. *These spaces are in fact completely regular, and thus uniformizable, having for uniformity basis the relations*

$$U_{\varepsilon, \varphi}(A, B) \text{ iff } |\varphi^{\mathfrak{A}} - \varphi^{\mathfrak{B}}| \leq \varepsilon, \quad \varphi \in L, \varepsilon > 0.$$

(iv) Model theoretic compactness of L is equivalent to topological compactness of the spaces $(St_\tau, \Gamma_\tau(L))$.

(v) $L \leq_{ax} L'$ *if and only if* $\Gamma_\tau(L')$ *is finer than* $\Gamma_\tau(L)$ *for any* τ.

Proof. (i), (ii), and (v) are immediate, (iv) follows by Alexander subbasis lemma. For (iii) notice that a Ł-regular logic must be closed under disjunctions and the connectives $(\ )_{\leq r}$, $(\ )_{\geq r}$. Moreover, $\mathfrak{A} \notin C = Mod(T)$ implies $\varphi^{\mathfrak{A}} < r < s < 1$ for some rationals r, s. and thus the complements in St_τ of the class $Mod(\varphi_{\geq r})$ and $Mod(\varphi_{\leq s})$ separate $\mathfrak{A}$ and C. □

Define in any topological space X the equivalence relation $x \equiv y$ if and only if $cl\{x\} = cl\{y\}$, where cl denotes topological adherence. Thus, $x \equiv y$ if and only if x and y belong to the same closed (open) subsets of X. It is a topology exercise to show that if X is a regular space, then the quotient space $X_{/\equiv}$ is Hausdorff. The next lemma will be useful in the proof of the maximality theorems.

Lemma 4.3. *If K_1 and K_2 are disjoint compact subclasses of a regular topological space X, which cannot be separated by a finite intersection of basic closed sets of a given basis, then there exist $x_i \in K_i$, $i = 1, 2$, such that $x_1 \equiv x_2$.*

Proof. Let $\eta : X \to X_{/\equiv}$ be the quotient map. By continuity, the images ηK_1 and ηK_2 are compact in $X_{/\equiv}$ and thus closed. They cannot be disjoint; otherwise, their inverse images would be disjoint closed sets separating K_1 and K_2, and by a compactness argument again the separation could be achieved by a finite intersection of basic sets. Pick $x_i \in K_i$, $i = 1, 2$, with $\eta x_1 = \eta x_2 \in \eta K_1 \cap \eta K_2$, then $x_1 \equiv x_2$. □

It should be clear that the topological relation $\equiv$ in $(St_\tau, \Gamma_\tau(L))$ coincides with the logical relation $\equiv_L$. We have then,

Proposition 4.4. *Let L and L' be two Ł-regular [0,1]-valued logics on continuous structures with $L \leq L'$ and L' compact in each $St_{\tau,S}$. If $\mathfrak{A} \equiv_L \mathfrak{B}$ implies $\mathfrak{A} \equiv_{L'} \mathfrak{B}$ for any pair of models, then $L \sim_{ax} L'$ in any $St_{\tau,S}$. More precisely, any $\varphi \in L'_\tau$ has the same models in $St_{\tau,S}$ as a countable theory of L.*

Proof. Let $\varphi \in L'$ and consider $\varphi_{\leq r}$ for rational $r \in (0, 1)$, then $Mod(\varphi)$ and $Mod(\varphi_{\leq r})$ are disjoint compact subclasses of $St_{\tau,S}$ in the L'-topology, a fortiori in the L-topology. Then they must be separated by an intersection of basic closed sets of the latter topology, otherwise by Lemma 4.3 there should exist S-models $\mathfrak{A} \models \varphi$, $\mathfrak{B} \models \varphi_{\leq r}$ with $\mathfrak{A} \equiv_L \mathfrak{B}$ and thus $\mathfrak{A} \equiv_{L'} \mathfrak{B}$, but the last equivalence is impossible because $\varphi, \varphi_{\leq r} \in L'$. As L is closed under disjunctions, the separating class has the form $C_r = Mod(F_r)$ for a finite theory F_r. As $Mod(\varphi) \subseteq C_r$ and $Mod(\varphi_{\leq r}) \cap C_r = \emptyset$ for each $r < 1$, then $Mod(\varphi) = \cap_r C_r = Mod(\cup_r F_r)$. □

5   Compact extensions and approximations

We prove in this section several properties of compact extensions of Ł∀ on continuous complete structures utilizing a simple and robust notion of approximation. We prove in particular that arbitrarily approximable model-classes of the logic are jointly satisfiable, approximable complete structures are equivalent in the logic, and each sentence of the logic has countable dependence number in any equicontinuous class of complete structures.

Definition 5.1. Given τ-structures $\mathfrak{A}$ and $\mathfrak{B}$, a subsignature $\mu \subseteq \tau$, and $\varepsilon \in (0, 1)$, a relation $R \subseteq A \times B$ is a $\mu\varepsilon$-*approximation* between $\mathfrak{A}$ and $\mathfrak{B}$ if

i) $|\varphi^{\mathfrak{A}}[\bar{a}] - \varphi^{\mathfrak{B}}[\bar{b}]| \leq \varepsilon$ for all $\varphi \in \mu^*$ and all suitable $\bar{a}, \bar{b}$ such that $(a_i, b_i) \in R$.

ii) $d^{\mathfrak{A}}(a, dom R) \leq \varepsilon$ and $d^{\mathfrak{B}}(b, rang R) \leq \varepsilon$ for all $a \in A$, and $b \in B$.

We will write $\mathfrak{A} \simeq_{\mu\varepsilon} \mathfrak{B}$ ($R : \mathfrak{A} \simeq_{\mu\varepsilon} \mathfrak{B}$) to indicate that there is (R is) such an approximation. Notice that a $\mu\varepsilon$-approximation between crisp structures is an ordinary μ-isomorphism. The next lemma shows that if we have $\mu\varepsilon$-approximations for arbitrarily small ε, then we may assume that these are total functions having an inverse "up to ε".

Lemma 5.2. *Given τ-structures $\mathfrak{A}$, $\mathfrak{B}$, a finite $\mu \subseteq \tau$, and $\varepsilon > 0$, there is $\rho = \rho(\mu, \varepsilon) > 0$ such that if $\mathfrak{A} \approx_{\mu, \rho} \mathfrak{B}$,, then there are maps $h : \mathfrak{A} \to \mathfrak{B}$ and $g : \mathfrak{B} \to \mathfrak{A}$ such that*

(1) $|\varphi^{\mathfrak{A}}[\bar{a}] - \varphi^{\mathfrak{B}}[h(a_1), .., h(a_k)]| \leq \varepsilon$ *for all $\varphi \in \mu$ and $\bar{a}$ in A*

(2) $d^{\mathfrak{B}}[h(f^{\mathfrak{A}}(\bar{a})), f^{\mathfrak{B}}(h(a_1), .., h(a_k)))] \leq \varepsilon$ *for all $f \in \mu$ and $\bar{a}$ in A, and similarly for g. Moreover,*

(3) $d^{\mathfrak{A}}(a, gh(a)) \leq \varepsilon$ *and* $d^{\mathfrak{B}}(hg(b), b) \leq \varepsilon$ *for any $a \in A$, $b \in B$.*

Proof. Let $\rho = \min\{m_{\varphi^{\mathfrak{A}}}(\varepsilon/3), m_{\varphi^{\mathfrak{B}}}(\varepsilon/3) : \varphi \in \mu^*\}$ and assume $R : A \approx_{\mu, \rho} B$. We show first that there is $R' : A \simeq_{\mu, \varepsilon} B$ with $dom R' = A$ and $rang R' = B$. Define $(a', b') \in R'$ if and only if there is $(a, b) \in R$ such that $d(a', a)$, $d(b', b) \leq \rho(\mu, \varepsilon)$. Then $(a_i', b_i') \in R'$ implies for any $\varphi \in \mu^* : |\varphi^{\mathfrak{A}}\bar{a}' - \varphi^{\mathfrak{B}}\bar{b}'| \leq |\varphi^{\mathfrak{A}}\bar{a}' - \varphi^{\mathfrak{A}}\bar{a}| + |\varphi^{\mathfrak{A}}\bar{a} - \varphi^{\mathfrak{B}}\bar{b}| + |\varphi^{\mathfrak{B}}\bar{b} - \varphi^{\mathfrak{B}}\bar{b}'| \leq \varepsilon/3 + \rho + \varepsilon/3 \leq \varepsilon$; the first and last bounds follow by uniform continuity of φ in A and B, the second the defining properties of R. Moreover, for any $a' \in A$ there is $(a, b) \in R_{\mu, \rho}$ such that $d(a', a) \leq \rho$ by condition (ii), then $(a', b) \in R'$, showing that $dom R'$ is all of A, similarly, its image is all of B. Now let $h : A \to B$, $g : B \to A$ be choice functions for R' and R'^{-1}, respectively. Then (1) follows by construction, and (2) follows applying (1) to the formula $d(f(\bar{x}), y) \in \mu^*$ to obtain $|d^{\mathfrak{A}}(f(\bar{a}), a) - d^{\mathfrak{B}}(f(h(a_1), .., h(a_k)), h(a))| \leq \varepsilon$, and then making $a := f(\bar{a})$. For (3) observe that $(a, h(a)), (g(b), b) \in R'$, then $|d^{\mathfrak{A}}(a, g(b)) - d^{\mathfrak{B}}(h(a), b)| \leq \varepsilon$. Making $b := h(a)$ we obtain $d^{\mathfrak{A}}(a, g(h(a))) \leq \varepsilon$ and making $a := g(b)$ yields $d^{\mathfrak{B}}(h(g(b)), b) \leq \varepsilon$. $\square$

Proposition 5.3. *(joint consistency)* Let T_1, T_2 be τ-theories of a compact extension of $L\forall$ on complete continuous structures. If for each finite $\mu \subseteq \tau$ and $\varepsilon > 0$ there are equicontinuous families $\{\mathfrak{A}_{\mu,\varepsilon}\}_{\mu,\varepsilon}$ and $\{\mathfrak{B}_{\mu,\varepsilon}\}_{\mu,\varepsilon}$ of complete structures such that $\mathfrak{A}_{\mu,\varepsilon} \simeq_{\mu,\varepsilon} \mathfrak{B}_{\mu,\varepsilon}$ and $\mathfrak{A}_{\mu,\varepsilon} \models T_1$, $\mathfrak{B}_{\mu,\varepsilon} \models T_2$, then $T_1 \cup T_2$ is satisfiable.

Proof. By Lemma 5.2, we may assume the approximations are of the form $h_{\varepsilon,\mu} : A_{\mu,\varepsilon} \to B_{\mu,\varepsilon}$ with a right ε-inverse $g_{\varepsilon,\mu} : B_{\mu,\varepsilon} \to A_{\mu,\varepsilon}$ The obvious idea of using compactness to transform them in an isomorphism $h : \mathfrak{A} \to \mathfrak{B}$ with $\mathfrak{A} \models T_1$, $\mathfrak{B} \models T_2$ does not work directly because the $h_{\varepsilon,\mu}$, $g_{\varepsilon,\mu}$ are not necessarily uniformly continuous. To circumvent this problem, we consider diagrams of continuous structures of the form

$$(A^\circ, \Delta_{A^\circ}, D_{A^\circ}, ...) \overset{h}{\underset{g}{\rightleftarrows}} (B^\circ, \Delta_{B^\circ}, D_{B^\circ}, ...)$$
$$\downarrow \pi_1 \qquad\qquad\qquad \downarrow \pi_2$$
$$(A, d_A, ..) \qquad\qquad\qquad (B, d_B, ..),$$

where
- A and B have signature τ and distinguished metrics d_A, d_B; moreover, and $A \models T_1$, $B \models T_2$.
- A° and B° have type $\tau \cup \{D\}$, where D_{A°, D_{B° are pseudometrics, and the discrete metrics Δ_{A°, Δ_{B° as distinguished metrics.
- π_1 is a map with dense image in A, preserving the truth value of the predicates in τ (that is, $\varphi^A(\pi_1 a_1, .., \pi_k a_k) = \varphi^{A^\circ}(\pi_1, .., \pi_k)$) and the operations in τ (that is, $\pi_1 f^{A^\circ}(a_1, .., a_k) = f^A(\pi_1 a_1, .., \pi_k a_k)$), but not preserving the distinguished metrics and sending instead the pseudometric D_{A° to the metric d_A (that is, $d_A(\pi_1 a, \pi_2 b) = D_{A^\circ}(a, b)$). Similarly for π_2.
- h preserves the value of the predicates in $\tau \cup \{D\}$ but it is not asked to preserve Δ (that is, it is not asked to be one to one) and it preserves the operations in τ only up to D; that is, $D(h f^{A^\circ}(a_1, .., a_k), f^{B^\circ}(ha_1, .., ha_k)) = 0$.
- g is a left inverse of h up to D; that is, it preserves the value of D and $D(gh(a)), a) = 0$.

These diagrams may be taken as many-sorted structures or coded as single-sorted structures $([A^\circ, B^\circ, A, B], ...)$. In any case, the properties described above may be axiomatized by the following theory, where the predicates P_1, P_2, P_1°, P_2° specify the sorts of the variables and formulas in the many-sorted version, or denote relativizations in the single-sorted version. For convenience, the properties of h are not stated sharply but by a sequence of approximations. For the sake of clearness, we use sometimes the identity $\approx$ associated to the distinguished metric in the sorts P_i:

For $i = 1, 2$:
A1 γ^{P_i} for all $\gamma \in T_i$
A2 $\forall x^{P_i^\circ} y^{P_i^\circ} z^{P_i^\circ} (\neg D(x, x) \wedge (D(x, y) \to D(y, x)) \wedge (D(x, y) \odot D(y, z) \to D(x, z)))$
A3 $\forall y^{P_i} \exists x^{P_i^\circ} (\pi_i(x) \approx y)$

A4 $\forall \overline{x}^{P_i^\circ}(\varphi^{P_i}(\pi_i(x_1)...\pi_i(x_k))) \leftrightarrow \varphi^{P_i^\circ}(\overline{x}))$ for each predicate symbol $\varphi \in \tau$

A5 $\forall \overline{x}^{P_i^\circ}(f^{P_i}(\pi_i(x_1)...\pi_i(x_k))) \approx \pi_i(f^{P_i^\circ}(\overline{x})))$ for each function symbol $f \in \tau$

A6 $\forall x^{P_i^\circ} y^{P_i^\circ}(d^{P_i}(\pi_i(x), \pi_i(y))) \leftrightarrow D^{P_i^\circ}(x, y))$

For each n :

A7 $\forall \overline{x}^{P_1^\circ}[\varphi^{P_2^\circ}(h(x_1), .., h(x_k))) \leftrightarrow \varphi^{P_1^\circ}(\overline{x})]_{\geq 1-\frac{1}{n}}$ for each predicate symbol $\varphi \in \tau \cup \{D\}$

A8 $\forall \overline{x}^{P_1^\circ}[D^{P_2^\circ}(h(f^{P_1^\circ}(\overline{x})), f^{P_2^\circ}(h(\overline{x})))]_{\leq \frac{1}{n}}$ for each function symbol $f \in \tau$

A9 $\forall x^{P_2^\circ} y^{P_2^\circ}[D^{P_1^\circ}(g(x), g(y)) \leftrightarrow D^{P_2^\circ}(x, y)]_{\geq 1-\frac{1}{n}}$

A10 $\forall y^{P_2^\circ}[D(h(g(y)), y)]_{\leq \frac{1}{n}}$.

Given a finite part Σ of this theory, where μ is the subsignature of τ occurring in Σ and ε the minimum $\frac{1}{n}$ occurring, then the structure $C_{\mu,\varepsilon}$

$$
\begin{array}{ccc}
A^\circ_{\mu,\varepsilon} & \overset{h_{\mu,\varepsilon}}{\underset{g_{\mu,\varepsilon}}{\rightleftarrows}} & B^\circ_{\mu,\varepsilon} \\
\downarrow \pi_1 & & \downarrow \pi_2 \\
A_{\mu,\varepsilon} & & B_{\mu,\varepsilon}
\end{array}
$$

where $A^\circ_{\mu,\varepsilon}$ ($B^\circ_{\mu,\varepsilon}$) is a copy of $A_{\mu,\varepsilon}$ ($B_{\mu,\varepsilon}$) with the discrete metric Δ as distinguished metric, D interpreted by d, and π_1, π_2 the identity functions, is a model of Σ. Axioms A1-A6 are satisfied trivially, axioms A7, A8 are satisfied by properties of functional $\mu\varepsilon$-approximations, and A9, A10 by Lemma 5.2.

The predicates and operations in $A^\circ_{\mu,\varepsilon}$, $B^\circ_{\mu,\varepsilon}$, and the maps $h_{\varepsilon,\mu}$, $g_{\mu,\varepsilon}$, π_1, π_2 are trivially uniformly continuous with respect to Δ, say with Lipschitz constant $\frac{1}{2}$, and the structures $A_{\mu,\varepsilon}$, $B_{\mu,\varepsilon}$ are equicontinuous by hypothesis. Hence, the $C_{\mu,\varepsilon}$ are equicontinuous and thus we may apply compactness to obtain a model C of T. Given such a model, and referring to the first diagram, $\Delta_{A^\circ}, \Delta_{B^\circ}, d_A$ and d_B are metrics, D_{A° and D_{B° are pseudometrics by A2 but might not be metrics, the images of π_1 and π_2 are dense substructures of A and B, respectively, by axiom A3.

Let $x \sim_D y$ iff $D(x, y) = 0$ be the similarity induced by the pseudometric D in A° and B°, then we have $d(\pi_i(x), \pi_i(y)) = D(x, y)$ by A6. In particular, since d is a metric,

$$\pi_i(x) = \pi_i(y) \text{ if and only if } x \sim_D y.$$

Moreover, we have $D^{A^\circ}(x, y) = D^{B^\circ}(h(x), h(y))$ by A7, and thus

$$x \sim_D y \text{ if and only if } h(x) \sim_D h(y).$$

This means that the map $\widehat{h} : A \to B$ defined as $\widehat{h}(\pi_1(x)) := \pi_2 h(x)$ is a well-defined bijection between the images of π_1 and π_2, since

$$\pi_1(x) = \pi_1(y) \Leftrightarrow x \sim_D y \Leftrightarrow h(x) \sim_D h(y) \Leftrightarrow \pi_2(x)) = \pi_2 h(y).$$

It is actually an isomorphism because by A6 and A7 applied to D:

$d^A(\pi_1(x), \pi_1(y)) \overset{(A6)}{=} D^{A^\circ}(x, y) \overset{(A7)}{=} D^{B^\circ}(h(x), h(y)) \overset{(A6)}{=} d^B(\pi_2 h(x), \pi_2 h(y)) \overset{def}{=}$
$d_B(\widehat{h}(\pi_1(x)), \widehat{h}(\pi_1(y)))$. In particular, h is uniformly continuous. Moreover, by A4 and A7, for any relation symbol $\varphi \in \tau : \varphi^{\mathfrak{A}}(p_1(x), ..) \overset{(A4)}{=}$
$\varphi^{A^\circ}(x, ..) \overset{(A7)}{=} \varphi^{B^\circ}(h(x), ..) \overset{(A4)}{=} \varphi^{\mathfrak{B}}(p_2(h(x), ..) \overset{def}{=} \varphi^{\mathfrak{B}}(\widehat{h}(p_1(x)), ..)$; and by A8, $h(f^{A^\circ}(x, ..))) \sim_D f^{B^\circ}(h(x), ..)$ for any function symbol $f \in \tau$; thus $\widehat{h}(f^A(\pi_1(x), ..)) \overset{(A5)}{=} \widehat{h}(\pi_1(f^{A^\circ}(x, ..)) \overset{def}{=} \pi_2 h(f^{A^\circ}(x, ..)) = \pi_2(f^{B^\circ}(h(x), ..))$
$\overset{(A5)}{=} f^B(\pi_2 h(x)), ..) \overset{def}{=} f^B(\widehat{h}(\pi_1(x)), ..)$.

Moreover, $\widehat{h} : \pi_1(A^\circ) \to \pi_2(B^\circ)$ is surjective because g may be lifted similarly to $\widehat{g} : \pi_2(B^\circ) \to \pi_1(A^\circ)$ as $\widehat{g}(\pi_2(x)) = \pi_1 g(x)$ by A9, and A10 plus A6 imply $\pi_2(y) = \pi_2(h(g(y))) = \widehat{h}(\pi_1(g(x))) = \widehat{h}(\widehat{g}(\pi_2(x)))$. As A and B are complete, the uniformly continuous maps and $\widehat{h}, \widehat{g}$ may be extended to an isomorphism between A and B. $\square$

The previous proposition does not require closure of the logic L under Łukasiewicz connectives, but the following corollaries do.

Corollary 5.4. *For any L-regular compact extension $L \geq L\forall$ and complete τ-structures $\mathfrak{A}, \mathfrak{B} : \mathfrak{A} \simeq_{\mu\varepsilon} \mathfrak{B}$ for all finite $\mu \subseteq \tau$ and $\varepsilon > 0$ implies $\mathfrak{A} \equiv_L \mathfrak{B}$.*

Proof. Assume $\varphi^{\mathfrak{A}} \leq r < s \leq \varphi^{\mathfrak{B}}$ for some φ and rationals $r < s$. Then $\mathfrak{A} \models \varphi_{\leq r}, \mathfrak{B} \models \varphi_{\geq s}$ and by Proposition 5.3 $\{\varphi_{\leq r}\} \cup \{\varphi_{\geq s}\}$ would be satisfiable, a contradiction. $\square$

We may refine the previous result. It is easy to verify that the relations $\approx_{\mu,\varepsilon}$ for $\mu \subseteq_{fin} \tau$, $\varepsilon > 0$, form a uniformity basis in $St_{\tau,S}$ which generates what we call the *approximation uniformity*. The following corollary says that the approximation uniformity is finer than the canonical uniformity of the L-topology (see Lemma 4.2 (iii)).

Corollary 5.5. *Let $L \geq L\forall$ be a compact L-regular logic, then for each sentence $\varphi \in L_\tau$ and $\varepsilon > 0$ there are a finite $\mu \subseteq \tau$ and $\delta > 0$ such that $\mathfrak{A} \approx_{\mu,\delta} \mathfrak{B}$ implies $|\varphi^{\mathfrak{A}} - \varphi^{\mathfrak{B}}| \leq \varepsilon$ for all complete τ-structures $\mathfrak{A}, \mathfrak{B}$.*

Proof. Suppose no, then for some $\varepsilon > 0$ and each μ, δ, there are $A_{\mu,\delta} \approx_{\mu,\delta} B_{\mu,\delta}$ such that $|\varphi^{A_{\mu,\delta}} - \varphi^{B_{\mu,\delta}}| \geq \varepsilon$. Then $[A_{\mu,\delta}, B_{\mu,\delta}] \models (\varphi^P \leftrightarrow \varphi^{P'})_{\leq 1-\varepsilon}$ while $[A_{\mu,\delta}, A_{\mu,\delta}] \models \varphi^P \leftrightarrow \varphi^{P'}$ (trivially). But $[A_{\mu,\delta}, A_{\mu,\delta}] \approx_{\mu,\delta} [A_{\mu,\delta}, B_{\mu,\delta}]$ for all μ, δ by construction which contradicts Proposition 5.3. $\square$

A sentence $\varphi \in L_\tau$ is said to have *dependence number* κ if there is a $\mu \subseteq \tau$ of power less or equal than κ such that for any pair of models $\mathfrak{A}, \mathfrak{B} \in St_\tau$, $\mathfrak{A} \upharpoonright \mu \simeq \mathfrak{B} \upharpoonright \mu$ implies $\varphi^{\mathfrak{A}} = \varphi^{\mathfrak{B}}$. Any sentence of a compact extension closed under Boolean connectives of classical logic has finite dependence (Lindström [24]). This is not the case of CL. However,

Corollary 5.6. (Countable dependence) *Any sentence of a L-regular compact extension L of $L\forall$ has countable dependence on each equicontinuous class of complete structures.*

Proof. Given φ, and $n \in \omega^+$ find finite μ_n, $\delta_n > 0$ such that $\mathfrak{A} \approx_{\mu_n, \delta_n} \mathfrak{B}$ implies $|\varphi^{\mathfrak{A}} - \varphi^{\mathfrak{B}}| \leq \frac{1}{n}$, for any complete $\mathfrak{A}$, $\mathfrak{B}$. Let $\mu = \cup_n \mu_n$ then, trivially, $\mathfrak{A} \upharpoonright \mu \simeq \mathfrak{B} \upharpoonright \mu$ implies $\mathfrak{A} \approx_{\mu_n, \delta_n} \mathfrak{B}$ and thus $|\varphi^{\mathfrak{A}} - \varphi^{\mathfrak{B}}| \leq \frac{1}{n}$ for all n, and thus, $\varphi^{\mathfrak{A}} = \varphi^{\mathfrak{B}}$. □

6   Lindström's theorem

The following property generalizes the separable Löwenheim-Skolem property and will be enough to obtain our maximality results.

Definition 6.1. A logic L has the *approximate Löwenheim-Skolem property* if any satisfiable countable theory has for each $\varepsilon > 0$ a model A which may be covered with countably many ε-balls. We call this a ε-separable model, and the set of balls centers we will call a ε-dense subset.

Our main result hinges in the next separation lemma which does not assume L-regularity of L.

Proposition 6.2. (Separation) *Let $L \geq L\forall$ be a compact logic satisfying the approximate Löwenheim-Skolem property, S a moduli system S for τ, and $T_1 \cup T_2 \subseteq L_\tau$ unsatisfiable in $St^c_{\tau,S}$. Then T_1 and T_2 are separable in $St^c_{\tau,S}$ by a sentence $\varphi \in L\forall_\tau$. That is,*

$$Mod^c_{\tau,S}(T_1) \subseteq Mod^c_{\tau,S}(\varphi), \quad Mod\psi_{\tau,S}(T_2) \cap Mod^c_{\tau,S}(\varphi) = \emptyset.$$

Proof. If $T_1 \cup T_2$ is unsatisfiable by S-models, then by compactness of L there are finite subsets $\Delta_i \subseteq T_i$ such that $\Delta_1 \cup \Delta_2$ is similarly unsatisfiable and it is enough to separate Δ_1 and Δ_2. The closed classes $Mod^c_{\tau,S}(\Delta_i)$ are compact in the L-topology of $St^c_{\tau,S}$, a fortiori in the weaker $L\forall$-topology. Assume the claimed separation is not possible, since the closed basics of the $L\forall$-topology are closed under intersections, then by regularity of the $L\forall$-topology and Lemma 4.3 there exist $\mathfrak{A}, \mathfrak{B} \in St_{\tau,S}$ such that $\mathfrak{A} \models \Delta_1$, $\mathfrak{B} \models \Delta_2$ and $\mathfrak{A} \equiv_{L\forall} \mathfrak{B}$. By Lemma 3.2 there is for each finite $\mu \subseteq \tau$, $\varepsilon > 0$ and each $n \in \omega$ a family of relations $I^n_0, ..., I^n_n$ coding sets of partial $\mu\frac{\varepsilon}{6}$-approximations with the extension property between $\mathfrak{A}$ and $\mathfrak{B}$ (the choice $\varepsilon/6$ will be useful later). For the rest of the proof we fix μ, ε, and show how to transform, using compactness of L, the various families $I^n_0, ..., I^n_n$ (depending on n) in a single $\mu\varepsilon$-approximation $R_{\mu,\varepsilon} : \mathfrak{A}' \approx \mathfrak{B}'$, where $\mathfrak{A}' \models \Delta_1$, $\mathfrak{B}' \models \Delta_2$ and $\mathfrak{A}', \mathfrak{B}'$ share the convergence moduli S (this is extremely important for the proof). Since μ, ε are arbitrary, this will imply by Proposition 5.3 that $\Delta_1 \cup \Delta_2$ is satisfiable, a contradiction.

To obtain the claimed μ, ε-approximation, consider the following countable theory in the signature $\{P, P'\} \cup \mu \cup \mu' \cup \{I_n\}_{n \in \omega}$, where P, P' are two sorts, u

is considered of sort P, μ' is a copy of μ in sort P', and for each μ-sentence φ the corresponding sentences in sorts P, P' are denoted φ^P, $\varphi^{P'}$, respectively.

A1. Δ_1^P, $\Delta_2^{P'}$
A2. $\forall \overline{xy}(I_j \overline{xy} \to \wedge_i(P(x_i) \wedge P'(y_i)))$, for each $j \in \omega$
A3. $\exists xy I_1 xy$
A4. $\forall \overline{xy}yx(I_j \overline{xy} \to \exists w I_{j+1} \overline{x}x\overline{y}w \wedge \exists w I_{j+1} \overline{x}w\overline{y}y)$, for each $j \geq 1$
A5. $\forall \overline{xy}(I_j \overline{xy} \to (|\varphi^P[\overline{x}] - \varphi^{P'}[\overline{x}]|_{\leq \varepsilon/2}))$, for each atomic $\varphi \in \mu^*$ and $j \geq 1$.

Recall that μ and ε will remain fixed through the argument. Each finite part Σ of this theory where $N = n_\Sigma$ is the largest subindex of the I_j occurring in Σ has the following non-continuous model

$$M_N = ([\mathfrak{A}, \mathfrak{B}], I_1^N, ..., I_N^N, \emptyset, \emptyset, ...).$$

Since the I_j^N are not necessarily equicontinuous in $[A, B]$ for varying N, we cannot apply compactness to obtain a model of the full theory. To overcome this problem we make them equicontinuous taking convenient "distance" predicates (cf. [3]), the finiteness of μ is crucial for this purpose:

$$\widehat{I}_j^N \overline{ab} := 1 - kd(\overline{ab}, I_j^N) \text{ for } j \leq N, \quad \widehat{I}_j^N \overline{ab} = 0 \text{ for } j > N,$$

where k is an integer such that $\frac{1}{k} \leq m_\varphi(\varepsilon/6)$ for all $\varphi \in \mu^*$. By construction, all the predicates $\widehat{I}_j^N$ are k-Lipschitz:

$$|\widehat{I}_j^N \overline{ab} - \widehat{I}_j^N \overline{a'b}'| \leq |kd(\overline{ab}, I_j^N) - kd(\overline{a'b}', I_j^N)| \leq kd(\overline{ab}, \overline{a'b}').$$

and this constant is independent of N. We must verify again axioms A1-A5 for $j \leq N$ in the now continuous structure

$$\widehat{M}_N = ([\mathfrak{A}, \mathfrak{B}], \widehat{I}_0^N, ..., \widehat{I}_j^N, \widehat{I}_{j+1}^N, ...).$$

A0 and A1 are obvious.
- A2. If $P(a_i) < 1$, then $P(a_i) = 0$ and thus $a_i \in B$. This implies $d(\overline{ab}, I_j^N) = 1$ and thus $\widehat{I}_j^N \overline{ab} = 0 \leq P(a_i)$, etc.
- A3. Given $a \in A$, there is b such that $(a, b) \in I_1^N$ hence $\widehat{I}_1^N(a, b) = 1 = (\exists xy I_1 xy)^{\widehat{M}_N}$.
- A4. $\widehat{I}_j^N \overline{ab} \geq r > 0$ implies for small enough $\delta > 0$ the existence of $(\overline{a'b}') \in I_j^N$ such that $d(\overline{ab}, \overline{a'b}') \leq \frac{1-r+\delta}{k} < 1$. Moreover, given $a \in A$ there is c such that $(\overline{a'ab}'c) \in I_{j+1}^N$. Then $d(\overline{aabc}, I_{j+1}^N) \leq d(\overline{aabc}, \overline{a'ab}'c) = d(\overline{ab}, \overline{a'b}') \leq \frac{1-r+\delta}{k}$, and thus $(\exists w I_{j+1} \overline{aabw}')^{\widehat{M}_N} \geq 1 - kd(\overline{aabc}, I_{j+1}^N) \geq r - \delta$. As δ is arbitrarily small, $(\exists w I_{j+1} \overline{aabw}')^{\widehat{M}_N} \geq r$, showing that $\widehat{I}_j^N \overline{ab} \leq (\exists w I_{j+1} \overline{aabw}')^{\widehat{M}_N}$. The other case is similar.
- A5. As in A4, $\widehat{I}_j^N \overline{ab} \geq r > 0$ means that $d(\overline{ab}, \overline{a'b}') \leq \frac{1-r+\delta}{k} < \frac{1}{k}$ for some

$(\overline{a'}\overline{b'}) \in I_j^N$ and small enough δ; hence, $d(a_i, a_i'), d(b_i, b_i') < \frac{1}{k} < m_\varphi(\varepsilon/6)$ and thus

$$|\varphi^P[\overline{a}] - \varphi'^{P'}[\overline{b}]| \leq |\varphi^P[\overline{a}] - \varphi^{P'}[\overline{a'}]| + |\varphi^P[\overline{a'}] - \varphi^{P'}[\overline{b'}]| + |\varphi^{P'}[\overline{b'}] - \varphi^{P'}[\overline{b}]|$$
$$\leq \frac{\varepsilon}{6} + \frac{\varepsilon}{6} + \frac{\varepsilon}{6} = \varepsilon/2,$$

the first and last bounds $\frac{\varepsilon}{6}$ due to uniform continuity of φ and the definition of k, and the middle one because $(\overline{a'}\overline{b'}) \in I_j^N$. Thus $\widehat{I_j^N}\overline{a}\overline{b} \leq (|\varphi^P[\overline{x}] - \varphi'^{P'}[\overline{x}]|_{\leq \varepsilon/2}) = 1$.

By construction, the continuity moduli of each predicate in $\widehat{M_N}$ is independent of N since $[\mathfrak{A}, \mathfrak{B}]$ remains fixed. Then we may apply compactness to obtain a continuous S-model of the full theory A1-A5:

$$M_{\mu\varepsilon} = ([\mathfrak{A}^\circ, \mathfrak{B}^\circ], \widehat{I}_1, ..., \widehat{I}_j, \widehat{I}_{j+1}, ...),$$

which by the (approximate) downward Löwenheim-Skolem theorem for L we may assume is ε-separable. Moreover, we have

1. By A3(b), there is (a_1^*, b_1^*) such that $\widehat{I}_1(a_1^*, b_1^*) \geq 1 - \frac{\varepsilon}{2^2}$.

2. By A4, if $\widehat{I}_j(\overline{a}, \overline{b}) \geq 1 - \rho$ and $a \in A$ ($b \in B$) there is $b \in B$ ($a \in A$) such that $\widehat{I}_{j+1}(\overline{a}a, \overline{b}b) \geq 1 - \rho - \frac{\varepsilon}{2^{n+2}}$.

3. By A5, if $\widehat{I}_j(\overline{a}, \overline{b}) \geq 1 - \rho$ then $|\varphi^{\mathfrak{A}^\circ}[\overline{a}] - \varphi^{\mathfrak{B}^\circ}[\overline{b}]| \leq \frac{\varepsilon}{2} + \rho$ for all atomic $\varphi \in \mu^*$.

Choose listings $\{a_1, a_2...\}$, $\{b_1, b_2...\}$ of an ε-dense subsets of A° and B°, respectively, and construct a relation

$$R = \{(a_i^*, b_i^*) : i \in \omega\} \subseteq A^\circ \times B^\circ$$

starting with (1) and utilizing (2) in a back-and-forth manner to choose inductively (a_n^*, b_n^*) so that the domain and range of R contain $\{a_i\}_i$ and $\{b_i\}$, respectively, and

$$\widehat{I}_n(a_1^*...a_n^*, b_1^*...b_n^*) \geq 1 - (\frac{\varepsilon}{2^2} + ... + \frac{\varepsilon}{2^{n+1}})$$

for all $n \geq 1$. By (3), $|\varphi^{\mathfrak{A}^\circ} a_1^*...a_n^* - \varphi^{\mathfrak{B}^\circ} b_1^*...b_n^*| \leq \frac{\varepsilon}{2} + (\frac{\varepsilon}{4} + ... + \frac{\varepsilon}{2^n}) \leq \varepsilon$ for all atomic formulas in μ^*. Hence, R is a $\mu\varepsilon$-approximation from $\mathfrak{A}^\circ$ to $\mathfrak{B}^\circ$, which finishes the proof. $\square$

Theorem 6.3. (Maximality of Łukasiewicz logic) *Let $L \geq L\forall$ be a L-regular compact logic with the approximate Löwenheim-Skolem property. Then any $\varphi \in L_\tau$ is equivalent with respect to satisfaction to a countable theory of $L\forall$, in any equicontinuous class of complete structures.*

Proof. Consider $\varphi \in L_\tau$ and $\mathfrak{A} \notin Mod_\tau(\varphi)$ then $\mathfrak{A} \in C = Mod_\tau(\varphi_{\leq \frac{n}{n+1}})$ for some n. By Proposition 6.2, given S, there is a sentence $\varphi_n \in L\forall_\tau$ such that $Mod_{\tau,S}(\varphi) \subseteq Mod_{\tau,S}(\varphi_n)$ and $C \cap Mod_{\tau,S}(\varphi_n) = \emptyset$. Hence, $\mathfrak{A} \notin Mod_{\tau,S}(\varphi_n)$ and thus $Mod_{\tau,S}(\varphi) = \cap_n Mod_{\tau,S}(\varphi_n) = Mod_{\tau,S}(\{\varphi_n : n \in \omega\})$. $\square$

Corollary 6.4. *CL and Ł∀ have the same axiomatizability strength in any equicontinuous class of (not necessarily complete) structures.*

Proof. Any continuous structure is CL-equivalent to its completion, thus two CL–theories have the same continuous models if and only if they have the same complete models. ∎

Theorem 6.5. (**Maximality of continuous logic**) *Let $L \geq Ł∀$ be a Ł-regular compact logic with the approximate Löwenheim-Skolem property. Then any $\varphi \in L_\tau$ is equivalent to a sentence of CL, in any equicontinuous class of complete structures.*

Proof. We utilize the following version of the Stone-Weirstrass theorem which follows from Lemma 16.3 in [16]:

If X is a compact Hausdorff space and L is a sublattice of $C(X, [0,1])$ such that for any two distinct elements $x, y \in X$, and $a, b \in [0,1]$ there exists $f \in L$ with $f(x) = a$ and $f(y) = b$. Then L is dense in $C(X, [0,1])$.

For $\varphi \in L_\tau$ the map $F_\varphi : St_{\tau,S} \to [0,1]$, $A \longmapsto \varphi^{\mathfrak{A}}$, is continuous in the Ł∀-topology because the inverse image of each closed subbasic: $F_\varphi^{-1}([r,s]) = Mod(\varphi_{\geq r} \wedge \varphi_{\leq s})$ is L-closed and thus Ł∀-closed by Theorem 6.3. By continuity, F_φ factors through the quotient $St_{\tau/\equiv_{Ł∀}}$, which is compact and Hausdorff by regularity of the space $St_\tau(Ł∀)$:

$$
\begin{array}{ccc}
St_\tau & \xrightarrow{F_\varphi} & [0,1] \\
{\scriptstyle \eta} \searrow & & \nearrow {\scriptstyle \widehat{F}_\varphi} \\
& St_{\tau/\equiv_{Ł∀}} &
\end{array}
$$

Now, this is true also for the maps F_θ with $\theta \in$ CL. Let $\mathcal{F} = \{\widehat{F}_\theta : \theta \in \text{CL}\} \subseteq \mathcal{C}(St_{\tau/\equiv_{Ł∀}}, [0,1])$. Obviously, $\mathcal{F}$ is closed under composition with all continuous connectives and is a lattice thanks to the presence of $\wedge, \vee$. Given distinct points $M_{/\equiv_{Ł∀}}$, $N_{/\equiv_{Ł∀}}$ in $St_{\tau/\equiv_{Ł∀}}$ there must exist a Ł∀-sentence θ such that $\theta^M < \theta^N$ and given values $a, b \in [0,1]$ we may find a continuous connective c (actually in Ł∀($\mathbb{Q}$)) such that $c(\theta^M) = a$, $c(\theta^N) = b$, then the map $c \circ \widehat{F}_\theta = \widehat{F}_{c\theta}$ satisfies the hypothesis of the Stone-Weirstrass theorem granting that the uniform closure of $\mathcal{F}$ is $\mathcal{C}$. Therefore, $\widehat{F}_\varphi$ is the uniform limit of a sequence $\{\widehat{F}_{\theta_n}\}_n$ with $\theta_n \in \mathcal{F}$. Thus $F_\varphi = \widehat{F}_\varphi \circ \eta$ is the uniform limit of $F_{\theta_n} = \widehat{F}_{\theta_n} \circ \eta$ and by Lemma 2.3 $F_\varphi = F_\theta$ with $\theta \in$ CL.

Example $L = \Sigma_1^1(\text{CL})$ inherits from CL compactness, and the downward Löwenheim-Skolem property and thus it satisfies Proposition 6.2; however, it does not satisfy the conclusion of Theorems 6.3 and 6.5, because the crisp structures $(\mathbb{Q}, <)$ and $(\mathbb{R}, <)$ are complete for the discrete metric and equivalent in Ł∀ (which collapses to classical logic in these structures) and thus in CL, but the order incompleteness of the first structure is expressible in $\Sigma_1^1(Ł∀)$. We conclude that $\Sigma_1^1(\text{CL})$ cannot be closed under both $\to$ and $\neg$, and this closure hypothesis cannot be eliminated from Theorem 6.3. Actually, $\Sigma_1^1(\text{CL})$ is closed under $\wedge, \vee, \oplus, \odot$ and $(\ )_{\geq r}$ but not under, or $(\ )_{\leq r}$.

7   Comments

The countable theory given in Theorem 6.3 and its corollary and the sentence obtained in Theorem 6.5 depend on the uniform continuity moduli system S. Is this necessarily so?

Do theorems 6.3 and 6.5 hold for arbitrary continuous structures? Otherwise, under which conditions on the extension L will they hold if we allow incomplete continuous structures? An obvious condition is asking L to have the property that any satisfiable theory has a complete model (as in [9]) because in such case two theories or sentences are equivalent in continuous structures if and only if they are equivalent in complete ones, thus we may restrict the logic to the latter and apply the maximality results.

Lemmas 2.2 and 2.3 say that the uniform closure of Ł$\forall(\mathbb{Q})$ in [0,1]-valued structures is CL. Which is the uniform closure of Ł$\forall$?

8   Acknowledgments

This research was partially supported by a 2015-16 Seed Project of the Faculty of Sciences, Universidad de los Andes, and it was partially written during a visit to the University of Texas at San Antonio, supported by the NSF grant DMS-0819590. I thank José Iovino for his insistence that I write these results.

Bibliography

[1] L. P. Belluce, *Further results on infinite valued predicate logic*, J. Symb Logic **29** (1964) 69–78.

[2] L. P. Belluce and C. C. Chang, *A weak completeness theorem on infinite valued predicate logic*, J. Symb Logic **28** (1963) 43–50.

[3] I. Ben Yaacov, A. Berenstein, C. Ward Henson, and A. Usvyatsov, *Model theory for metric structures*, in: Model theory with applications to algebra and analysis, vol. 2, London Math. Soc. Lecture Note Ser., vol. 350, Cambridge Univ. Press, 2008, pp. 315–427.

[4] I. Ben Yaacov, J. Iovino, *Model theoretic forcing in analysis*, Annals of Pure and Applied Logic **158** (2009) 163–174.

[5] I. Ben Yaacov and A. Usvyatsov, *On d-finiteness in continuous structures*, Fundamenta Mathematicae **194** (2007) 67–88.

[6] I. Ben Yaacov and A. Usvyatsov, *Continuous first order logic and local stability*, Trans. Amer. Math. Soc. 362, **10** (2010) 5213–5259.

[7] I. Ben Yaacov, A. Nies, and T. Tsankov, *A López-Escobar Theorem for continuous logic*, preprint: arXiv:1407.7102v1.

[8] X. Caicedo, *Lindström's theorem for positive logics, a topological view*, in: Logic Without Borders, Essays on Set Theory, Model Theory, Philosophical Logic and Philosophy of Mathematics, Roman Kossak, Juha Kontinen, Åsa Hirvonen, Andrés Villaveces, eds., Walter de Gruyter 2015, pp. 73–90.

[9] X. Caicedo and J. Iovino, *Omitting uncountable types, and the strength of [0,1]-valued logics*, Annals of Pure and Applied Logic **165** (2014) 1169–1200.

[10] C. C. Chang, *Algebraic analysis of many valued logics*, Trans. Amer. Math. Soc, **88** (1958) 467–490.

[11] C.C. Chang, *Theory of models of infinite valued logic*, I, II, III, Notices Amer. Math. Soc. 8 (1961), 68–69.

[12] C.C. Chang and H. J. Keisler, *Model theories with truth values in a uniform space*, Bull. Amer. Math. Soc. **68** (1962), 107–109.

[13] C. C. Chang and H. J. Keisler, *Continuous Model Theory*, Annals of Mathematics Studies, No. 58, Princeton Univ. Press, Princeton, N.J., 1966.

[14] Ch. J Eagle. *Omitting types for infinitary* $[0, 1]$-*valued logic*. Annals of Pure and Applied Logic, **165**(3) (2014) 913-932.

[15] P. Hájek, *Metamathematics of fuzzy logic*, Trends in Logic—Studia Logica Library, vol. 4, Kluwer Academic Publishers, Dordrecht, 1998.

[16] L. Gillman and M. Jerison, *Rings of continuous functions*, Springer-Verlag, New York, 1976 (Reprint of the 1960 edition), Graduate Texts in Mathematics, No. 43.

[17] C. W. Henson. *Nonstandard hulls of Banach spaces*. Israel J. Math., **25**(1-2) (1976) 108–144.

[18] C. W. Henson and J. Iovino, *Ultraproducts in analysis*, Analysis and logic (Mons, 1997), London Math. Soc. Lecture Note Ser., vol. 262, Cambridge Univ. Press, Cambridge, 2002, pp. 1–110.

[19] J. Iovino, *On the maximality of logics with approximations*, J. Symbolic Logic **66**(4) (2001) 1909–1918.

[20] J. Iovino, *Applications of model theory to functional analysis*. Dover Publications 2014. Revised translation of *Ultraproductos en Análisis,* XV Escuela Venezolana de Matemáticas, Sept. 2002, Mérida, Venezuela.

[21] M. Katz, *Real valued models with metric equality and uniformly continuous predicates*, The Journal of Symbolic Logic **47**(4) (1982), 772–792.

[22] J. L. Krivine, *Sous-espaces de dimension finie des espaces de Banach réticulés*, Annals of Mathematics **104** (1976) 1–29.

[23] J. L. Krivine and B. Maurey. *Espaces de Banach stables*. Israel J. Math., 39, 4 (1981) 273–295.

[24] P. Lindström, *On extensions of elementary logic*, Theoria **35** (1969) 1–11.

[25] R. McNaughton. *A theorem about infinite-valued sentential logic*, J. Symbolic Logic, **16** (1951) 1–13.

[26] D. Mundici, *A compact* $[0,1]$-*valued first-order Łukasiewicz logic with identity on Hilbert space*, J. Logic and Computation, 21, 3 (2011) 509–525.

[27] D. Mundici, *Avanced Łukasiewicz calculus and MV algebras*, Springer, Trends in Logic. Vol. 35, 2001.

[28] S. Willard, *General Topology*. Addison-Wesley, 1970.

Chapter 5

Model theory and metric convergence I: Metastability and dominated convergence

Eduardo Dueñez
The University of Texas at San Antonio
San Antonio, Texas, USA

José Iovino
The University of Texas at San Antonio
San Antonio, Texas, USA

Dedicated to Ward Henson with gratitude on the occasion of his retirement.

This research was partially funded by NSF grant DMS-1500615.

The concept of convergence in metric spaces is fundamental in analysis. The present chapter is the first of a series of articles focusing on results, both classical and new, in which the convergence of some sequence(s)—or, more generally, some nets—follows from suitable hypotheses. We shall use the loose nomenclature "convergence theorem" for any result of this kind; the best known such results are the classical Monotone and Dominated Convergence theorems, as well as the ergodic convergence theorems of von Neumann and Birkhoff. For simplicity, we assume that all metric spaces under consideration are complete (an alternative perspective would be the study of theorems about Cauchy sequences and nets in metric spaces not necessarily complete).

Given a convergence theorem, it is natural to ask whether it admits refinements whereby the conclusion states a stricter mode of convergence. When the statement of a convergence theorem involves a collection of sequences, the classical refinement of the property of simultaneous convergence is that of uniform convergence. However, few convergence theorems in analysis admit a natural refinement implying uniform convergence. Furthermore, even in the rare cases when uniform convergence of a family of sequences is implied, the parameters of uniform convergence are rarely universal: Typically, uniform convergence will hold in every structure (consisting of the ambient metric space, sequences therein, plus any other necessary ingredients) satisfying suitable hypotheses, but not uniformly across all such structures. In fact, even if a convergence theorem should refer to the convergence of a single sequence, our focus shall be on the entirety of structures to which the theorem applies, and hence on all instances of relevant sequences. Thus, even when a "single" sequence is under immediate discussion, we ask whether its mode of convergence admits parameters that are uniform over all instances.

Tao [Tao08] introduced the notion of *metastable convergence* (Definition 1.2), which is an equivalent formulation of the usual Cauchy property (Proposition 1.5). The metastable-convergence viewpoint leads to the notion of *uniformly metastable convergence* (Definition 1.7), which is not only a metastable analogue of the classical property of uniform convergence of a family of sequences, but also a generalization thereof (Remark 1.8). Tao obtains a metastable version of the classical Dominated Convergence Theorem that holds with metastable rates that are universal; this result plays a crucial role in the proof his remarkable result on the convergence of ergodic averages for polynomial abelian group actions [Tao08]. Walsh's subsequent generalization [Wal12] of Tao's theorem to polynomial nilpotent group actions relies on a similar convergence theorem.

In this chapter we prove that metastability with a given rate is the only formulation of metric convergence that can be captured by a theory in continuous first-order logic (Proposition 2.3). This is a precise statement of Tao's observation that metastable convergence with a prescribed rate is a "finitary" property [Tao]. The conceptual backbone of the manuscript is the Uniform Metastability Principle (Proposition 2.4), which may be formulated as the following meta-theorem: *"If a classical statement about convergence in metric structures is refined to a statement about metastable convergence with some rate, then the validity of the original statement implies the validity of its metastable version."* As an instance of this phenomenon, we obtain a soft proof of a version of the Metastable Dominated Convergence Theorem. Our proof depends on neither infinitary arguments *à la* Tao [Tao08], nor recursive arguments and constructive analysis *à la* Avigad *et al.* [ADR12].

We show that the Uniform Metastability Principle follows directly from the fundamental theorem of model theory of metric structures, namely, the Compactness Theorem. We believe that, in spite of its simplicity, this principle captures a certain philosophical view revealing the scope of applicability of model-theoretic methods to the study of convergence in metric spaces. Although anticipated by Avigad and Iovino [AI13], we are not aware of a purely model-theoretical formulation of this principle hitherto.

In order to make the results of this chapter accessible to readers with no prior background in logic, particularly to researchers in analysis, the last section (Section 6) is a self-contained tutorial on the basics of model theory of metric structures. The literature contains several equivalent formulations of this theory; the one in current widespread use is the framework of first-order continuous logic developed by Ben Yaacov and Usvyatsov [BYU10, BYBHU08] building upon ideas of Chang-Keisler [CK62, CK66] and Henson [HI02]. We use the language of continuous approximations originally developed by Henson, as we feel that it is simpler and more natural for the applications at hand. The tutorial of Section 6 parallels portions of earlier introductions given by Henson and the second author that emphasize Banach spaces [HI02]; however, the present exposition places greater emphasis on metric structures and on topics of direct interest for the study of convergence theorems, particularly structures that are hybrid in the sense that they include metric sorts alongside discrete sorts. Readers already comfortable with model theory of metric structures from this perspective need to consult Section 6 only for reference.

Section 1, which does not depend on model theory at all, is an introduction to the metastable viewpoint of convergence. As an attempt to shed light into the finitary nature of uniformly metastable convergence, Subsection 1-I discusses the relation between the usual Cauchy criterion and Tao's notion of metastable convergence. In Subsection 1-II we define various notions of metastability, oscillation and uniform metastability for arbitrary sequences or nets in a metric space, and show how these notions relate to classical ones. Proposition 1.5, in particular, states that metastable convergence is equivalent to usual convergence (always in complete spaces).

In Section 2, we begin the discussion of sequences and nets from the viewpoint of model theory. We exhibit a (continuous) first-order axiomatization of the property of metastable convergence with a given uniform rate. Subsequently, we state and prove the Uniform Metastability Principle and a useful corollary thereof (Proposition 2.4 and Corollary 2.5).

In Section 3, we study finite measure structures from a model-theoretic viewpoint. We introduce a certain class of metric structures, which we call

Loeb structures due to their close relation to Loeb probability spaces from nonstandard analysis [Loe75].[1] We start by defining the notion of pre-Loeb structure. Roughly speaking, a pre-Loeb structure is a metric structure that satisfies all the first-order properties of probability spaces $(\Omega, \mathcal{A}, \mathbf{P})$, where $\mathbf{P}$ is a probability measure on a Boolean algebra $\mathcal{A}$ of subsets of Ω. Loeb structures are then defined as pre-Loeb structures that satisfy a saturation hypothesis. By basic model theory (Section 6-X), every metric structure can be extended naturally to a saturated structure. As a by-product of saturation, Loeb structures possess the infinitary properties (i.e., countable additivity) of *bona fide* classical probability spaces (Proposition 3.4).[2]

In Section 4, we study integration from a model-theoretic viewpoint. We begin by introducing the class of pre-integration structures. True to their name, these are structures satisfying the first-order axioms of the space $\mathcal{L}_{\Omega}^{\infty}$ of essentially bounded real functions on, say, a probability space $(\Omega, \mathcal{A}, \mathbf{P})$, endowed with the integration linear functional $I : \mathcal{L}_{\Omega}^{\infty} \to \mathbb{R}$ mapping f to $\int_{\Omega} f(\omega) \mathrm{d}\mathbf{P}(\omega)$. (In particular, a pre-integration structure is a pre-Loeb structure.) Of course, most models of those axioms will not correspond to *bona fide*, countably additive probability spaces, nor will I correspond to the operation of integration with respect to a measure in the usual sense. Nevertheless, a Riesz Representation Theorem (Theorem 4.7) for integration structures (that is, for saturated pre-integration structures), asserts that the space $\mathcal{L}_{\Omega}^{\infty}$ (whose elements correspond to bounded functions on the induced Loeb probability space) is endowed with a positive functional extending I and given by the usual operation of integration with respect to Loeb measure.

In Section 5, we introduce directed pre-integration structures. In essence, these are pre-integration structures $(\Omega, \mathcal{A}, \mathbf{P}, \mathcal{L}_{\Omega}^{\infty}, I)$ further endowed, say, for simplicity, with a bounded sequence $\varphi = (\varphi_n : n \in \mathbb{N})$ of elements of $\mathcal{L}_{\Omega}^{\infty}$ (identified with bounded functions on Ω). Directed integration structures are defined as saturated directed pre-integration structures, as expected. By the Riesz Representation Theorem (Theorem 4.7), under the saturation hypothesis, I corresponds to a classical integration operator on $\mathcal{L}_{\Omega}^{\infty}$; therefore, the usual proof of the Dominated Convergence Theorem applies verbatim (Proposition 5.3). Obviously, the conclusion of the Dominated Convergence Theorem continues to hold if an additional hypothesis of uniform metastable pointwise convergence is imposed on the family φ; moreover, this hypothesis is axiomatizable. It follows from the Uniform Metastability Principle that, under the additional hypothesis of uniform metastable pointwise convergence, the conclusion of the Dominated Convergence Theorem must admit a strengthening to convergence with a certain metastable rate. This yields an immediate proof of Tao's Metastable Dominated Convergence Theorem (Corollary 5.4).

In the Appendix to this manuscript we state the viewpoint that a property of metric spaces should be considered finitary when it is equivalent to

---

[1]This manuscript makes no use of results or notions from nonstandard analysis.

[2]More precisely, every Loeb structure induces a classical probability space $(\Omega, \mathcal{A}_{\mathrm{L}}, \mathbf{P}_{\mathrm{L}})$ on the same underlying sample space Ω (Proposition 3.4).

the satisfaction of a collection of axioms in a first-order language for metric structures.

We conclude this introduction by mentioning that Tao has formulated a nonstandard-analysis version of his Metastable Dominated Convergence Theorem in a blog post on *Walsh's ergodic theorem, metastability, and external Cauchy convergence* [Tao]. Tao's insightful post served as philosophical motivation for the model-theoretic perspective adopted in the current manuscript.

We gratefully acknowledge support for this research from the National Science Foundation through grant DMS-1500615. We thank the administration and sponsors of the Banff International Research Station for providing an excellent research environment in June 2016 during the meeting of the focused research group on Topological Methods in Model Theory (16frg676). We also thank Xavier Caicedo, Chris Eagle, and Frank Tall for valuable discussions and comments that helped refine some of the ideas and content of this manuscript.

1 A finitary formulation of convergence

I Motivation

Given $\epsilon \geq 0$, we say that a sequence $a = (a_n : n \in \mathbb{N})$ in some metric space X is $[\epsilon]$-*Cauchy* if at least one of its tails $a_{\geq N} = (a_n : n \geq N)$ satisfies the inequality $\mathrm{osc}_{\geq N}(a) := \sup_{m,n \geq N} \mathrm{d}(a_m, a_n) \leq \epsilon$. The $[\epsilon]$-Cauchy condition is infinitary in the sense that $\mathrm{osc}_{\geq N}(a)$ depends on the values a_n as n varies over the infinite set $\{N, N+1, \ldots\}$. Tao's criterion for "metastable convergence" [Tao08] imposes small oscillation conditions only on *finite* segments $a_{[N,N']} = (a_n : N \leq n \leq N')$ of a. For every fixed choice of a strictly increasing function $F : \mathbb{N} \to \mathbb{N}$ we regard the collection $\eta = ([N, F(N)] : N \in \mathbb{N})$ as a "sampling" of $\mathbb{N}$, one finite segment at a time; abusing nomenclature, we use the name *sampling* to refer to either the collection η or the function F defining it. In Tao's nomenclature, for $\epsilon > 0$ and sampling F, the sequence a is $[\epsilon, F]$-*metastable* if the inequality $\mathrm{osc}_{[N,F(N)]}(a) := \sup_{N \leq m,n \leq F(N)} \mathrm{d}(a_m, a_n) \leq \epsilon$ holds for some N. For fixed $\epsilon > 0$ and sampling F, "$[\epsilon, F]$-*metastable*" is a weaker property than "$[\epsilon]$-*Cauchy*" inasmuch as the former involves only the values of a on the subsets $a_{[N,F(N)]}$ of the tails $a_{\geq N}$. However, when F varies over all samplings of $\mathbb{N}$, the conjunction of the corresponding properties of $[\epsilon, F]$-metastability of a implies that a is $[\epsilon]$-Cauchy. This leads to Tao's characterization of convergence (i.e., of the Cauchy property in complete spaces) as $[\epsilon, F]$-metastability for all $\epsilon > 0$ and all samplings F (Proposition 1.5).

The metastable characterization of convergence is still not quite finitary because the existential statement on $N \in \mathbb{N}$ is infinitary. (We use the term "finitary" as a synonym for "axiomatizable in first-order continuous logic" per the Appendix to this manuscript.) Tao's concept of *metastability with a given rate* arises by restricting this existential statement to a bounded (finitary) one. After Tao, for fixed $\epsilon > 0$ and sampling F, we call $M \in \mathbb{N}$ an *(upper bound on the) rate of $[\epsilon, F]$-metastable convergence of a* if $\mathrm{osc}_{[N,F(N)]}(a) \leq \epsilon$ holds for some $N \leq M$. Any collection $M_\bullet = (M_{\epsilon,F}) \subset \mathbb{N}$ of natural numbers, one for each $\epsilon > 0$ and sampling F, is an *(upper bound on the) rate of metastability of a* if $M_{\epsilon,F}$ bounds the $[\epsilon, F]$-rate of metastable convergence of a for all ϵ, F. Evidently, given ϵ, F and $M \in \mathbb{N}$, the property "M is a rate of $[\epsilon, F]$-metastability for a" is finitary.

Given arbitrary ϵ, F and any Cauchy sequence a, it is clear that a admits *some* bound $M_\bullet$ on its rate of metastability—one may take $M_{\epsilon,F}$ to be any N satisfying $\mathrm{osc}_{\geq N}(a) \leq \epsilon$. However, no choice of $M_\bullet$ applies uniformly to *all* Cauchy sequences.

In Subsection II below, we define various notions of metastable convergence for nets in metric spaces. In particular, the notion of metastable convergence with a given rate (Definition 1.7) is the natural finitary notion of convergence of nets extending Tao's (for sequences).

II   Metastable convergence of nets in metric spaces

Throughout this section we fix a nonempty directed set $(\mathcal{D}, \leq)$ (that is, $\leq$ is a reflexive, antisymmetric and transitive binary relation on $\mathcal{D}$ such that every pair of elements has an upper bound). We denote by $\mathcal{D}_{\geq i}$ the final segment $\{j \in \mathcal{D} : j \geq i\} \subset \mathcal{D}$.

We recall that a $\mathcal{D}$-net $a_\bullet = (a_i : i \in \mathcal{D})$ in a topological space X *converges to $b \in X$* if, for every neighborhood B of b in X, there exists $i \in \mathcal{D}$ such that $a_i \in B$ for all $j \geq i$. If (X, d) is a metric space and $\epsilon \geq 0$, we will say that the $\mathcal{D}$-net $a_\bullet$ is *$[\epsilon]$-Cauchy* if there exists $i \in \mathcal{D}$ such that $\mathrm{d}(a_j, a_{j'}) \leq \epsilon$ for all $j, j' \geq i$, and that $a_\bullet$ is *(ϵ)-Cauchy* (or *ϵ-Cauchy*) if $a_\bullet$ is $[\epsilon']$-Cauchy for all $\epsilon' > \epsilon$. The net $a_\bullet$ is Cauchy in the usual sense when it is 0-Cauchy; in this case, $a_\bullet$ converges to b if for all $\epsilon > 0$ there is $i \in \mathcal{D}$ such that $\mathrm{d}(a_i, b) \leq \epsilon$ for all $j \geq i$. Every Cauchy net in a complete metric space X converges to some (necessarily unique) element $b \in X$. All metric spaces under consideration shall be complete, so we use the term "convergent" as a synonym of "Cauchy".

Definition 1.1. A *sampling* of the directed set $(\mathcal{D}, \leq)$ is any collection $\eta = (\eta_i : i \in \mathcal{D})$ of finite subsets of $\mathcal{D}$ (indexed by $\mathcal{D}$ itself), such that η_i is a nonempty finite subset of $\mathcal{D}_{\geq i}$ for each $i \in \mathcal{D}$.[3] The collection of all samplings of $\mathcal{D}$ will be denoted $\mathrm{Smpl}(\mathcal{D})$.

Definition 1.2. Fix a metric space (X, d) and directed set $(\mathcal{D}, \leq)$. For $\eta \in \mathrm{Smpl}(\mathcal{D})$ and $\epsilon > 0$, a $\mathcal{D}$-net $a_\bullet = (a_i : i \in \mathcal{D})$ in X is:

· *strict ϵ, η-metastable*, or *$[\epsilon, \eta]$-metastable* (note the square brackets), if

---

[3]The condition "η_i is nonempty" is not logically necessary. We impose it for heuristic reasons only.

there exists $i \in \mathcal{D}$ such that $\mathrm{d}(a_j, a_{j'}) \leq \epsilon$ for all $j, j' \in \eta_i$. Any such i is a *witness* of the (strict) $[\epsilon, \eta]$-metastability of $a_\bullet$.

· *lax ϵ, η-metastable*, or *(ϵ, η)-metastable* (note the round parentheses), if $a_\bullet$ is $[\epsilon', \eta]$-metastable for all $\epsilon' > \epsilon$,

· *η-metastable*, if $a_\bullet$ is $(0, \eta)$-metastable,

· *ϵ-metastable*, if $a_\bullet$ is (ϵ, η')-metastable for all $\eta' \in \mathrm{Smpl}(\mathcal{D})$,

· *metastable* if $a_\bullet$ is 0-metastable.

For fixed ϵ, η, (ϵ, η)-metastability (resp., $[\epsilon, \eta]$-metastability, ϵ-metastability) implies (ϵ', η)-metastability (resp., $[\epsilon', \eta]$-metastability, ϵ'-metastability) for all $\epsilon' \geq \epsilon$. It is also clear that strict $[\epsilon, \eta]$-metastability implies lax (ϵ, η)-metastability, that lax (ϵ', η)-metastability for all $\epsilon' > \epsilon \geq 0$ implies lax (ϵ, η)-metastability, and consequently that ϵ'-metastability for all $\epsilon' > \epsilon \geq 0$ implies ϵ-metastability. We also remark that any net $a_\bullet$ is necessarily C-metastable if C is an upper bound on the distances $\mathrm{d}(a_i, a_j)$ $(i, j \in \mathcal{D})$; this is the case, in particular, if $\mathrm{d}(a_i, x_0) \leq C/2$ for some fixed $x_0 \in X$ and all $i \in \mathcal{D}$.

Definition 1.3. Let $a_\bullet = (a_i : i \in \mathcal{D})$ be an arbitrary bounded $\mathcal{D}$-net on a metric space (X, d).

For every $\eta \in \mathrm{Smpl}(\mathcal{D})$, the *$\eta$-oscillation* $\mathrm{osc}_\eta(a_\bullet)$ *of $a_\bullet$* is

$$\mathrm{osc}_\eta(a_\bullet) = \inf\{\epsilon \geq 0 : a_\bullet \text{ is } [\epsilon, \eta]\text{-metastable}\}.$$

The *oscillation* $\mathrm{osc}(a_\bullet)$ *of $a_\bullet$* is

$$\mathrm{osc}(a_\bullet) = \sup\{\mathrm{osc}_\eta(a_\bullet) : \eta \in \mathrm{Smpl}(\mathcal{D})\}.$$

Note that the notations "osc", "osc_η" fail to exhibit the dependence of $\mathrm{osc}(a_\bullet)$, $\mathrm{osc}_\eta(a_\bullet)$ on $\mathcal{D}$. However, when these notations are used, the directed set $\mathcal{D}$ will be fixed, precluding ambiguous interpretations.

Proposition 1.4. *For any bounded $\mathcal{D}$-net $a_\bullet$, $\eta \in \mathrm{Smpl}(\mathcal{D})$ and $\epsilon \geq 0$:*

(1) $\mathrm{osc}_\eta(a_\bullet) \leq \epsilon$ *if and only if $a_\bullet$ is (ϵ, η)-metastable,*

(2) $\mathrm{osc}_\eta(a_\bullet) = \inf\{\epsilon \geq 0 : \mathrm{osc}_{\eta_i}(a_\bullet) \leq \epsilon \text{ for some } i \in \mathcal{D}\}$,

(3) $\mathrm{osc}_\eta(a_\bullet) = \min\{\epsilon \geq 0 : a_\bullet \text{ is } (\epsilon, \eta)\text{-metastable}\}$,

(4) $\mathrm{osc}_\eta(a_\bullet) = \inf\{\mathrm{osc}_{\eta_i}(a_\bullet) : i \in \mathcal{D}\}$,

(5) $\mathrm{osc}(a_\bullet) = \min\{\epsilon \geq 0 : a_\bullet \text{ is } \epsilon\text{-metastable}\}$,

(6) $\mathrm{osc}(a_\bullet) = \sup\{\epsilon \geq 0 : \text{for all } i \in \mathcal{D} \text{ there exist } j, j' \geq i \text{ with } \mathrm{d}(a_j, a_{j'}) \geq \epsilon\}$,

(7) $a_\bullet$ *is ϵ-Cauchy if and only if $\mathrm{osc}(a_\bullet) \leq \epsilon$,*

(8) $a_\bullet$ *is ϵ-Cauchy if and only if $a_\bullet$ is ϵ-metastable.*

It is customary to use (6) above as the definition of $\mathrm{osc}(a_\bullet)$.

Proof. 1. Let $r = \mathrm{osc}_\eta(a_\bullet)$. For any $t > r$ the definition of osc_η implies that $a_\bullet$ is $[s, \eta]$-metastable for some $s \in (r, t)$, hence also $[t, \eta]$-metastable; therefore, $a_\bullet$ is (r, η)-metastable, hence *a fortiori* (ϵ, η)-metastable for all $\epsilon \geq r$. Conversely, if $a_\bullet$ is (ϵ, η)-metastable, then it is $[\epsilon', \eta]$-metastable for all $\epsilon' \geq \epsilon$, so the definition of osc_η implies $r \leq \epsilon$.

2. Clearly, $a_\bullet$ is $[\epsilon, \eta]$-metastable iff $\mathrm{osc}_{\eta_i}(a_\bullet) \leq \epsilon$ for some $i \in \mathcal{D}$.

3. This follows immediately from (1).

4. Let $r = \mathrm{osc}_\eta(a_\bullet)$ and $s = \inf\{\mathrm{osc}_{\eta_i}(a_\bullet) : i \in \mathcal{D}\}$. By part (2), we have $r \leq s$. Conversely, if $\epsilon \geq \mathrm{osc}_{\eta_i}(a_\bullet)$ for some $i \in \mathcal{D}$, then $\epsilon \geq s$, hence $s \leq r$.

5. Clearly, the set $\{\epsilon \geq 0 : a_\bullet \text{ is } \epsilon\text{-metastable}\}$ has a least element, say s, since ϵ'-metastability for all $\epsilon' > \epsilon$ is equivalent to ϵ-metastability. Let $r = \mathrm{osc}(a_\bullet)$. If $a_\bullet$ is ϵ-metastable, then it is (ϵ, η)-metastable for all $\eta \in \mathrm{Smpl}(\mathcal{D})$, hence $\mathrm{osc}_\eta(a_\bullet) \leq \epsilon$ by (1), so $r \leq s$ by (3). Conversely, for all $\eta \in \mathrm{Smpl}(\mathcal{D})$, $a_\bullet$ is (r, η)-metastable by (1) and the definition of osc, and thus $a_\bullet$ is r-metastable, so $r \geq s$.

6. Let $r = \mathrm{osc}(a_\bullet)$ and

$$s = \sup\{\epsilon \geq 0 : \text{for all } i \in \mathcal{D} \text{ there exist } j, j' \geq i \text{ with } \mathrm{d}(a_j, a_{j'}) \geq \epsilon\}.$$

If $0 \leq t < s$, then for all $i \in \mathcal{D}$ there exists $\eta_i = \{j, j'\} \subset \mathcal{D}_{\geq i}$ such that $\mathrm{d}(a_j, a_{j'}) > t$, hence $a_\bullet$ is not $[t, \eta]$-metastable for $\eta = (\eta_i : i \in \mathcal{D})$. It follows that $t \leq \mathrm{osc}_\eta(a_\bullet) \leq r$. As this holds for all positive $t < s$, we have $s \leq r$. Conversely, if $0 \leq t < r$, then $\mathrm{osc}_\eta(a_\bullet) > t$ for some $\eta \in \mathrm{Smpl}(\mathcal{D})$. By (1), $a_\bullet$ is not (t, η)-metastable. Hence, there is $t' > t$ such that for all $i \in \mathcal{D}$ there exist $j, j' \in \eta_i \subset \mathcal{D}_{\geq i}$ with $\mathrm{d}(a_j, a_{j'}) > t' > t$; hence, $t \leq s$. As this holds for all $t < r$, we have $r \leq s$.

7. Let $r = \mathrm{osc}(a_\bullet)$. It follows from (6) that $a_\bullet$ is r-Cauchy, thus also ϵ-Cauchy for all $\epsilon \geq r$, Conversely, if $0 \leq \epsilon < r$, then by (6) there exists $\epsilon' \in (\epsilon, r)$ with the property that for all $i \in \mathcal{D}$ there exist $j, j' \geq i$ with $\mathrm{d}(a_j, a_{j'}) \geq \epsilon'$. Then $a_\bullet$ is not ϵ-Cauchy. □

8. This follows immediately from (5) and (7), plus the remarks following Definition 1.2.

Proposition 1.5 (Metastable characterization of the Cauchy property). *A net in a metric space X is Cauchy if and only if it is metastable.*

Proof. This is the particular case $\epsilon = 0$ of (8) in Proposition 1.4. □

Remark 1.6. Propositions 1.4 and 1.5 remain true if we consider only samplings η such that η_i consists of no more than two elements of $\mathcal{D}_{\geq i}$ for all $i \in \mathcal{D}$. This more restrictive definition of sampling could be used in all further developments without any essential changes.

Definition 1.7. Fix a directed set $\mathcal{D}$. The collection of all finite subsets of $\mathcal{D}$ will be denoted $\mathcal{P}_{\text{fin}}(\mathcal{D})$. Let $a_\bullet = (a_i : i \in \mathcal{D})$ be a $\mathcal{D}$-net in a metric space X.

Given $\epsilon > 0$ and $\eta \in \text{Smpl}(\mathcal{D})$, a set $E \in \mathcal{P}_{\text{fin}}(\mathcal{D})$ is called *a (bound on the) rate of $[\epsilon, \eta]$-metastability of $a_\bullet$* if there exists a witness $i \in E$ of the $[\epsilon, \eta]$-metastability of $a_\bullet$. (No sequence $a_\bullet$ has this property if E is empty.)

For $\epsilon > 0$, a collection $E_\bullet = (E_\eta : \eta \in \text{Smpl}(\mathcal{D})) \subset \mathcal{P}_{\text{fin}}(\mathcal{D})$ is called a *(bound on the) rate of $[\epsilon]$-metastability of $a_\bullet$* if E_η is a rate of $[\epsilon, \eta]$-metastability of $a_\bullet$ for all $\eta \in \text{Smpl}(\mathcal{D})$.

For $r \geq 0$ and $\eta \in \text{Smpl}(\mathcal{D})$, a collection $E_\bullet = (E_\epsilon : \epsilon > r)$ in $\mathcal{P}_{\text{fin}}(\mathcal{D})$ is called a *(bound on the) rate of (r, η)-metastability of $a_\bullet$* if E_ϵ is a rate of $[\epsilon, \eta]$-metastability of $a_\bullet$ for all $\epsilon > r$.

For $r \geq 0$, a collection $E_\bullet = (E_{\epsilon,\eta} : \eta \in \text{Smpl}(\mathcal{D}), \epsilon > r)$ in $\mathcal{P}_{\text{fin}}(\mathcal{D})$ is called a *(bound on the) rate of r-metastability of $a_\bullet$*, if $E_{\cdot,\eta}$ is a rate of (r, η)-metastability of $a_\bullet$ for all $\eta \in \text{Smpl}(\mathcal{D})$, where $E_{\cdot,\eta} = (E_{\epsilon,\eta} : \epsilon > r)$. When $r = 0$, we say simply that $E_\bullet$ is a (bound on the) rate of metastability of $a_\bullet$.

If $\mathcal{C}$ is any collection of $\mathcal{D}$-nets in X, we say that:

· E is a *uniform (bound on the) rate of $[\epsilon, \eta]$-metastability for $\mathcal{C}$*, or $\mathcal{C}$ is *E-uniformly $[\epsilon, \eta]$-metastable*, if $a_\bullet$ is E-uniformly $[\epsilon, \eta]$-metastable for all $a_\bullet \in \mathcal{C}$;

· $E_\bullet$ is a *uniform (bound on the) rate of $[\epsilon]$-metastability for $\mathcal{C}$*, or $\mathcal{C}$ is *$E_\bullet$-uniformly $[\epsilon]$-metastable*, if $a_\bullet$ is $E_\bullet$-uniformly $[\epsilon]$-metastable for all $a_\bullet \in \mathcal{C}$;

· $E_\bullet$ is a *uniform (bound on the) rate of r-metastability for $\mathcal{C}$*, or $\mathcal{C}$ is *$E_\bullet$-uniformly r-metastable*, if $a_\bullet$ is $E_\bullet$-uniformly r-metastable for all $a_\bullet \in \mathcal{C}$. (When $r = 0$, we usually omit it.)

Remark 1.8. We show that the concept of uniform metastability generalizes that of uniform convergence. Let $\mathcal{C}$ be any collection of $\mathcal{D}$-nets. Given $M \in \mathcal{D}$ and $\epsilon > 0$, the collection $\mathcal{C}$ is *M-uniformly $[\epsilon]$-Cauchy* if $\text{osc}_{\geq M}(a_\bullet) := \sup_{j,j' \geq M} d(a_j, a_{j'}) \leq \epsilon$ for each $\epsilon > 0$ and $a_\bullet \in \mathcal{C}$. Given $M_\bullet = (M_\epsilon : \epsilon > 0) \subset \mathcal{D}$, the collection $\mathcal{C}$ is *$M_\bullet$-uniformly Cauchy* if every $a_\bullet \in \mathcal{C}$ is M_ϵ-uniformly $[\epsilon]$-Cauchy for each $\epsilon > 0$. Corresponding to $M_\bullet$ there is a rate of metastability $E_\bullet = (E_{\epsilon,\eta})$ defined by $E_{\epsilon,\eta} = \{M_\epsilon\} \in \mathcal{P}_{\text{fin}}(D)$. Under this identification, the collection $\mathcal{C}$ is $M_\bullet$-uniformly Cauchy if and only if it is $E_\bullet$-metastable.

Clearly, metastability rates $E_\bullet$ obtained from a collection $M_\bullet$ as above are very special. The following example, due to Avigad *et al.* [ADR12], exhibits a family of uniformly metastable sequences that are not uniformly convergent. Every monotonically increasing sequence in $[0, 1]$ is convergent. Let $\mathcal{C}$ be the collection of all such sequences. Clearly, there is no rate $M_\bullet$ such that all sequences $a_\bullet \in \mathcal{C}$ are $M_\bullet$-uniformly Cauchy; in fact, for $\epsilon \in (0,1)$, the sequence $a_\bullet$ with $a_m = 0$ for $m \leq M_\epsilon$ and $a_m = 1$ for $m > M_\epsilon$ satisfies $\mathrm{osc}_{\geq E_\epsilon}(a_\bullet) = 1 > \epsilon$. On the other hand, for any $\epsilon > 0$ and a sampling of $\mathbb{N}$ given as a strictly increasing function $F : \mathbb{N} \to \mathbb{N}$ (as per Section 1.I), let $E_{\epsilon,F} = \{m \in \mathbb{N} : m \leq F^{(k)}(0)\}$ where $k = \lceil \epsilon^{-1} \rceil$ is the smallest integer no smaller than ϵ^{-1} and $F^{(k)}(0)$ is the k-fold iterate of F applied to 0. Since $k\epsilon \geq 1$, at least one of the k differences $a_{F^{(j+1)}(0)} - a_{F^{(j)}(0)}$ $(j = 0, 1, \ldots, k-1)$ must not exceed ϵ whenever $a_\bullet \in \mathcal{C}$, hence $\mathcal{C}$ is $E_{\epsilon,F}$-uniformly $[\epsilon]$-metastable. The collection $E_\bullet = (E_{\epsilon,F})$ is a uniform metastability rate for $\mathcal{C}$.

The concept of uniform metastability is crucial to our applications. As discussed in Remark 1.8 above, uniform metastability is a proper generalization of the classical notion of uniform convergence. Moreover, uniform metastability with a given rate is axiomatizable in the logic of metric structures (Proposition 2.3 below). This allows for powerful applications of model theory, particularly of compactness (e.g., the Uniform Metastability Principle, Proposition 2.4). We believe that many convergence results in analysis follow from hypotheses captured by the semantics of first-order logic for metric structures; consequently, such results ought to admit refinements to convergence with a metastability rate—and moreover the rate ought to be universal, i.e., independent of the structure to which the theorem is applied. Tao's Metastable Dominated Convergence Theorem, as stated and proved in Section 5 below, is but one example of this philosophy.

2　Convergence of nets in metric structures and uniform metastability

Throughout Sections 2 to 5, we will assume that the reader is familiar with the material presented in Section 6. In particular, we assume familiarity with the notions of Henson metric structure, signature, positive bounded formula, approximate satisfaction, the Compactness Theorem 6.31, and saturated structures. We will deal with multi-sorted structures that contain discrete sorts alongside nondiscrete ones. Recall that $\mathbb{R}$, equipped with its ordered field structure and a constant for each rational number, occurs tacitly as a sort of every metric structure, and that discrete predicates in a structure are seen as $\{0,1\}$-valued functions. If (M, d, a) is a pointed metric space and $C \geq 0$, the set $\{x \in M \mid d(x, a) \leq C\}$ will be denoted $M^{[C]}$.

Throughout the chapter, L will be a many-sorted signature with sorts $(\mathbb{S}_i : i < \alpha)$ (for some ordinal $\alpha > 1$), and $\mathbb{S}_0$ will be the special sort designated for $\mathbb{R}$. For notational convenience, we will identify $\mathbb{S}_0$ with $\mathbb{R}$.

Hereafter, $(\mathcal{D}, \leq)$ will denote a directed set with least element j_0. We will regard $(\mathcal{D}, \leq)$ as a discrete metric structure with sorts $\mathcal{D}$ and $\mathbb{R}$, and the point

j_0 will be regarded as the anchor for the sort $\mathcal{D}$. We will refer to $(\mathcal{D}, \leq, j_0)$ as a *pointed directed set*.

Definition 2.1 (Directed structure). Fix a pointed directed set $(\mathcal{D}, \leq, j_0)$. Let L be a many-sorted signature with sorts $(\mathbb{S}_i : i < \alpha)$ (for some ordinal $\alpha > 1$), where $\mathbb{S}_0 = \mathbb{R}$. The sort $\mathbb{S}_1$ will be called the *directed sort*; it will be denoted by $\mathbb{D}$ henceforth. Let L include a symbol $[\![\cdot \leq \cdot]\!]$ for a function $\mathbb{D} \times \mathbb{D} \to \{0, 1\}$. Let L also include distinct constant symbols c_j, one for each $j \in \mathcal{D}$. In addition to the function symbols for the sort metrics and the operations on $\mathbb{R}$ plus constants for rational numbers, L may include other function and constant symbols, as well as any other sorts than those already mentioned. A $\mathcal{D}$-*directed structure* is any metric L-structure $\mathcal{M}$ such that

- the sort $\mathbb{D}^{\mathcal{M}}$ is discrete,

- the interpretation of $[\![\cdot \leq \cdot]\!]$ induces an order on $\mathbb{D}^{\mathcal{M}}$, denoted $\leq$ (by a slight abuse of notation), such that $(\mathbb{D}^{\mathcal{M}}, \leq, c_{j_0}^{\mathcal{M}})$ is an anchored directed set and, for all $i, j \in \mathbb{D}^{\mathcal{M}}$, $i \leq j$ holds precisely when $[\![i \leq j]\!] = 1$,

- the map $\mathcal{D} \to \mathbb{D}^{\mathcal{M}}$ defined by $j \mapsto c_j^{\mathcal{M}}$ is an order-preserving injection;

It should be clear that the class of $\mathcal{D}$-directed structures is axiomatizable in the semantics of approximate satisfaction of positive bounded L-formulas. We note that the embedding $\mathcal{D} \hookrightarrow \mathbb{D}^{\mathcal{M}}$ is usually not surjective.

In order to discuss nets in $\mathcal{D}$-directed structures, we need to extend the language L with a function symbol $\mathbf{s} : \mathbb{D} \to \mathbb{S}_\iota$ interpreted as a function from the directed sort $\mathbb{D}$ into some other sort $\mathbb{S}_\iota$ of L. In this context, T shall denote any fixed uniform $L[\mathbf{s}]$-theory extending the theory of $\mathcal{D}$-directed L-structures. (Extensions of a Henson language are discussed in Section 6.XI.)

Definition 2.2. Fix a $\mathcal{D}$-directed L-structure $\mathcal{M}$, a new function symbol $\mathbf{s} : \mathbb{D} \to \mathbb{S}_\iota$, and a uniform $L[\mathbf{s}]$-theory T. An *(external) net* in $\mathcal{M}$ is any $\mathcal{D}$-net $s_\bullet = (s_i : i \in \mathcal{D})$ taking values in some sort of L. An *internal net (modulo T)* is a function $s : \mathbb{D}^{\mathcal{M}} \to \mathbb{S}_\iota^{\mathcal{M}}$ such that $(\mathcal{M}, s)$ is a model of T.

Note that an internal net s yields an external $\mathcal{D}$-net $s_\bullet = (s_i : i \in \mathcal{D})$, letting $s_i = s(i)$ for $i \in \mathcal{D}$.

For any rational $\epsilon > 0$, $\eta \in \mathrm{Smpl}(\mathcal{D})$ and internal net $s = \mathbf{s}^{\mathcal{M}}$, the $\mathcal{D}$-net $s_\bullet$ is strict $[\epsilon, \eta]$-*metastable* if and only if there exists $i \in \mathcal{D}$ such that

$$\mathcal{M} \not\approx \xi_i^\eta(\epsilon) \qquad (\text{equivalently}, \quad \mathcal{M} \models \xi_i^\eta(\epsilon)),$$

where ξ_i^η is the positive bounded L-formula

$$\xi_i^\eta(\mathbf{t}) : \bigwedge_{j, j' \in \eta_i} \left(\mathrm{d}_\iota(\mathbf{s}(j), \mathbf{s}(j')) \leq \mathbf{t} \right), \tag{5.1}$$

as follows from the semantics of approximate satisfaction and part (3) of

Proposition 1.4.[4] We emphasize that ξ_i^η is a *bona fide* L-formula since it is a finite conjunction of atomic formulas. Nevertheless, the property "$s_\bullet$ *is* $[\epsilon, \eta]$-*metastable*" is not L-axiomatizable since the asserted existence of the witness $i \in \mathcal{D}$ amounts to an infinite disjunction of formulas when $\mathcal{D}$ is itself infinite.

For fixed $\eta \in \mathrm{Smpl}(\mathcal{D})$ and $\epsilon > 0$, call $s_\bullet$ an $[\epsilon, \eta]$-*unstable* $\mathcal{D}$-net if, for all $i \in \mathcal{D}$, there exist $j, j' \in \eta_i$ with $\mathrm{d}_\iota(s_j, s_{j'}) \geq \epsilon$. For rational $\epsilon \geq 0$, the assertion "$s_\bullet$ *is* $[\epsilon, \eta]$-*unstable*" is equivalent to

$$\mathcal{M} \not\approx {}^{\text{\tiny W}}\xi_i^\eta(\epsilon) \qquad \text{for all } i \in \mathcal{D}, \quad (\text{alternatively,} \quad \mathcal{M} \models {}^{\text{\tiny W}}\xi_i^\eta(\epsilon))$$

where

$$ {}^{\text{\tiny W}}\xi_i^\eta(\mathbf{t}) : \bigvee_{j, j' \in \eta_i} \left(\mathrm{d}_\iota(\mathbf{s}(j), \mathbf{s}(j')) \geq \mathbf{t} \right). \tag{5.2}$$

Note that $[\epsilon, \eta]$-metastability is consistent with $[\epsilon', \eta]$-instability precisely when $\epsilon' \leq \epsilon$. In contrast to the property of strict metastability, the property of $[\epsilon, \eta]$-instability for given $\eta \in \mathcal{D}$ and $\epsilon \geq 0$ is axiomatized by the collection

$$\{ {}^{\text{\tiny W}}\xi_i^\eta(\epsilon') : \text{rational } \epsilon' < \epsilon \text{ and } i \in \mathcal{D} \}$$

of L-sentences, in the semantics of approximate (or discrete) satisfaction.

Recall that if $\mathcal{M}$ is a structure, the complete theory $\mathrm{Th}(\mathcal{M})$ of $\mathcal{M}$ is the set of all sentences satisfied by $\mathcal{M}$. It should be clear from the preceding observations that if $s_\bullet$ is a $\mathcal{D}$-net underlying an internal net s, then classical properties of $s_\bullet$, including: "$s_\bullet$ *is* $[\epsilon, \eta]$-*metastable*", "$s_\bullet$ *is* (ϵ, η)-*metastable*", "$s_\bullet$ *is* η-*metastable*", "$r \leq \mathrm{osc}(s_\bullet) \leq s$", etc., depend only on the complete theory T of $\mathcal{M}$, though some of these properties are not axiomatizable.

Proposition 2.3 (Axiomatizability of uniform metastability). *Fix a pointed directed set* $(\mathcal{D}, \leq, j_0)$, *a language* L *for* $\mathcal{D}$-*directed structures, a new function symbol* $\mathbf{s} : \mathbb{D} \to \mathbb{S}_\iota$ *and a uniform* $L[\mathbf{s}]$-*theory* T. *For fixed reals* $\epsilon > r \geq 0$ *and sampling* $\eta \in \mathrm{Smpl}(\mathcal{D})$, *the following properties of the* $\mathcal{D}$-*net* $s_\bullet$ *induced by an internal net* $s = \mathbf{s}^{\mathcal{M}}$ *are* $L[\mathbf{s}]$-*axiomatizable (modulo* T):

- *"E is a rate of $[\epsilon, \eta]$-metastability for $s_\bullet$", for any fixed $E \in \mathcal{P}_{\mathrm{fin}}(\mathcal{D})$;*

- *"$E_\bullet$ is a rate of (r, η)-metastability for $s_\bullet$", for any family $E_\bullet = (E_\epsilon : \epsilon > r)$ in $\mathcal{P}_{\mathrm{fin}}(\mathcal{D})$;*

- *"$E_\bullet$ is a rate of r-metastability for $s_\bullet$", for any family $E_\bullet = (E_{\epsilon, \eta} : \eta \in \mathrm{Smpl}(\mathcal{D}), \epsilon > r)$ in $\mathcal{P}_{\mathrm{fin}}(\mathcal{D})$.*

Hence, each of the preceding properties characterizes an axiomatizable subclass of models of T. *In fact, in each case the collection of axioms is independent of* T *and* L: *It depends only on the underlying directed set* $\mathcal{D}$.

---

[4]For notational convenience, in (5.1) we write $\mathbf{s}_k$ for $\mathbf{s}(\mathbf{c}_k)$, where $\mathbf{c}_k$ is the constant denoting the element $k \in \mathcal{D}$. Similar simplifications will be usually made without comment whenever the intended strict syntax is otherwise clear.

Proof. For $E \in \mathcal{P}_{\mathrm{fin}}(\mathcal{D})$ and $\eta \in \mathcal{D}$, let

$$\xi_E^{\eta} = \bigvee_{i \in E} \xi_i^{\eta} \quad \text{with } \xi_i^{\eta} \text{ given by (5.1).}$$

The first stated property is axiomatized by the formulas $\xi_E^{\eta}(\epsilon')$ for rational $\epsilon' > \epsilon$, the second by all formulas $\xi_{E_\epsilon}^{\eta}(\epsilon')$ for rational $\epsilon' > \epsilon$, and the third by all formulas $\xi_{E_{\epsilon,\eta}}^{\eta}(\epsilon')$ for $\epsilon > r$, rational $\epsilon' > \epsilon$, and all $\eta \in \mathrm{Smpl}(\mathcal{D})$. □

Proposition 2.4 (Uniform Metastability Principle). *Fix a pointed directed set $(\mathcal{D}, \leq, j_0)$, a language L for $\mathcal{D}$-directed structures, a new function symbol $\mathbf{s} : \mathbb{D} \to \mathbb{S}_\iota$ and a uniform $L[\mathbf{s}]$-theory T. Let $\eta \in \mathrm{Smpl}(\mathcal{D})$ and $r \geq 0$. If $s_{\bullet}$ is (r, η)-metastable whenever $(\mathcal{M}, s) \succapprox T$, then there exists a collection $E_{\bullet} = (E_\epsilon : \epsilon > r)$ in $\mathcal{P}_{\mathrm{fin}}(\mathcal{D})$ such that $E_{\bullet}$ is a rate of (r, η)-metastability for $s_{\bullet}$, uniformly over all models $(\mathcal{M}, s)$ of T.*

Loosely speaking, if all internal sequences are metastable, then they are uniformly metastable. (Of course the notion of internal sequence is in reference to a fixed uniform theory T.) On the other hand, the uniform rate $E_{\bullet}$ is certainly dependent on T. The uniform theory T implies an upper bound $C \geq 0$ such that $s(j) \in (\mathbb{S}_\iota)^{[C]}$ for all $j \in \mathbb{D}^{\mathcal{M}}$ whenever $(\mathcal{M}, s) \succapprox T$.

Proof. It is enough to show that a bound E_ϵ on the rate of $[\epsilon, \eta]$-metastability exists for each rational $\epsilon > r$, uniformly for all $s_{\bullet}$ arising from $\mathbf{s}^{\mathcal{M}}$ for arbitrary $\mathcal{M} \succapprox T$. Assume no such E_ϵ exists for some $\epsilon > r$. With the notation of Proposition 2.3, given $E \in \mathcal{P}_{\mathrm{fin}}(\mathcal{D})$ there exists $(\mathcal{M}, s) \succapprox T$ and some rational $\epsilon' > \epsilon$ such that $(\mathcal{M}, s) \not\succapprox \xi_E^{\eta}(\epsilon')$. Consequently, $(\mathcal{M}, s) \succapprox {}^{\mathrm{w}}\xi_E^{\eta}(\epsilon)$: $\bigwedge_{i \in E} {}^{\mathrm{w}}\xi_i^{\eta}(\epsilon)$, with ${}^{\mathrm{w}}\xi_i^{\eta}$ as in (5.2). Thus, the collection

$$X_\epsilon^{\eta} = \{ {}^{\mathrm{w}}\xi_i^{\eta}(\epsilon) : i \in \mathcal{D} \}$$

is finitely jointly satisfiable with T. By the Compactness Theorem 6.31, there exists a model $(\mathcal{M}, s)$ of $T \cup X_\epsilon^{\eta}$. On the one hand, $s_{\bullet}$ is $[\epsilon, \eta]$-unstable because $\mathcal{M} \succapprox X_\epsilon^{\eta}$; on the other hand, $s_{\bullet}$ is (r, η)-metastable by hypothesis, since $(\mathcal{M}, s) \succapprox T$. This is a contradiction since $r < \epsilon$. □

The Uniform Metastability Principle may be formulated as the following dichotomy: For a fixed uniform $L[\mathbf{s}]$-theory T, real $r \geq 0$ and $\eta \in \mathrm{Smpl}(\mathcal{D})$,

· Either: There exists a rate $E_{\bullet} = (E_\epsilon : \epsilon > r)$ of (r, η)-metastability for $s_{\bullet}$ valid uniformly for all models $(\mathcal{M}, s)$ of T;

· Or else: There exists a model $(\mathcal{M}, s)$ of T such that $s_{\bullet}$ fails to be (r, η)-metastable.

Below, we state a form of the Uniform Metastability Principle that applies when many internal nets are realized in the same structure. Fix a pointed directed set $(\mathcal{D}, \leq, j_0)$ and a signature L for $\mathcal{D}$-directed structures. Extend L

with function symbols $\mathbf{s} : \mathbb{D} \to \mathbb{S}_\iota$, $\boldsymbol{\sigma} : \mathbb{A} \times \mathbb{D} \to \mathbb{S}_\iota$ where $\mathbb{A}$ is of the form $\mathbb{S}_{i_1} \times \cdots \times \mathbb{S}_{i_n}$ (a Cartesian product of sorts of L) and $\mathbb{S}_\iota$ is any sort of L. We will treat $\mathbb{A}$ itself as a sort, writing $\mathbb{A}^{\mathcal{M}}$ for $\mathbb{S}_{i_1}^{\mathcal{M}} \times \cdots \times \mathbb{S}_{i_n}^{\mathcal{M}}$ in any L-structure $\mathcal{M}$. If $(\mathcal{M}, \sigma, s)$ is an $L[\boldsymbol{\sigma}, \mathbf{s}]$-structure, we will say that s *admits σ-parameters of size $\leq C$* if there exists $\bar{a} \in (\mathbb{A}^{\mathcal{M}})^{[C]}$ such that $s(j) = \sigma(\bar{a}, j)$ for all $j \in \mathbb{D}^{\mathcal{M}}$. The slightly more general notion that s *approximately admits σ-parameters of size $\lesssim C$* means that, for all $\epsilon > 0$, there exists $\bar{a} \in (\mathbb{A}^{\mathcal{M}})^{[C+\epsilon]}$ such that $\mathrm{d}(s(j), \sigma(\bar{a}, j)) \leq \epsilon$ for all $j \in \mathbb{D}^{\mathcal{M}}$. If C is rational, the latter property of $L[\boldsymbol{\sigma}, \mathbf{s}]$-structures $(\mathcal{M}, \sigma, s)$ is captured by a single axiom

$$\upsilon_C : (\exists_C \bar{a})(\forall j)(\mathrm{d}(\mathbf{s}(j), \boldsymbol{\sigma}(\bar{a}, j)) \leq 0). \tag{5.3}$$

In general, the property is axiomatized by the scheme $\{\upsilon_D : \text{rational } D > C\}$.

Corollary 2.5 (Uniform metastability of parametrized sequences). *Fix a pointed directed set $(\mathcal{D}, \leq, j_0)$ and a signature L for $\mathcal{D}$-directed structures. Extend L with function symbols $\mathbf{s} : \mathbb{D} \to \mathbb{S}_\iota$ and $\boldsymbol{\sigma} : \mathbb{A} \times \mathbb{D} \to \mathbb{S}_\iota$ where $\mathbb{A}$ is of the form $\mathbb{S}_{i_1} \times \cdots \times \mathbb{S}_{i_n}$ (a Cartesian product of sorts of L) and $\mathbb{S}_\iota$ is any sort of L. Fix a rational $C \geq 0$ and a uniform $L[\boldsymbol{\sigma}, \mathbf{s}]$-theory T such that s approximately admits σ-parameters of size $\lesssim C$ for every model $(\mathcal{M}, \sigma, s) \approx T$ (i.e., $T \approx \upsilon_D$ for all rational $D > C$ with υ_D defined in (5.3) above). For some $r \geq 0$ and $\eta \in \mathrm{Smpl}(\mathcal{D})$, assume that the external $\mathcal{D}$-net $s_\bullet$ is (r, η)-metastable whenever $(\mathcal{M}, \sigma, s) \approx T$. Then there exists a collection $E_\bullet^r = (E_\epsilon : \epsilon > r)$ in $\mathcal{P}_{\mathrm{fin}}(\mathcal{D})$ such that $s_\bullet$ is $E_\bullet$-uniformly (r, η)-metastable whenever $(\mathcal{M}, \sigma, s) \approx T$. Moreover, $E_\bullet$ depends only on T.*

As a simple application of saturation, the notions "*s admits σ-parameters of size $\leq C$*" and "*s approximately admits σ-parameters of size $\lesssim C$*" are seen to be equivalent in any ω^+-saturated $L[\boldsymbol{\sigma}, \mathbf{s}]$-structure $(\mathcal{M}, \sigma, s)$. Since every model of T admits an ω^+-saturated elementary extension, the hypotheses of Corollary 2.5 may be relaxed to state that the external $\mathcal{D}$-nets $s_\bullet$ are (r, η)-metastable whenever s admits (exact) parameters of size $\leq C$ via σ (without changing the conclusion).

Proof. By the Uniform Metastability Principle 2.4, if the collection $E_\bullet$ did not exist, there would be a model $(\mathcal{M}, \sigma, s)$ of T such that $s_\bullet$ is not (r, η)-metastable, contradicting the hypotheses. Hence, the metastability rate $E_\bullet$ must exist, and depends only on T. (Note that $E_\bullet$ implicitly depends on C.) $\square$

---

3 Loeb structures

In this section, L will be a many-sorted signature with sorts $(\mathbb{S}_i : i < \alpha)$ (for some ordinal α), where $\mathbb{S}_0 = \mathbb{R}$ (as convened), $\mathbb{S}_1 = \Omega$, and $\mathbb{S}_2 = \mathcal{A}$.

The sort Ω will be interpreted as a set with anchor point ω_0, and $\mathcal{A}$ will be interpreted as a Boolean algebra with its least element as anchor point.

Recall that, by default, L comes equipped with the following symbols:

- a binary function symbol $d_i : \mathbb{S}_i \times \mathbb{S}_i \to \mathbb{R}$ $(i < \alpha)$ for the metric of $\mathbb{S}_i$ (we may dispense with d_0 since $d_0(x, y) = |y - x|$ for $x, y \in \mathbb{R}$, and we shall denote d_1 by d_Ω, d_2 by $d_\mathcal{A}$);

- binary symbols $+_\mathbb{R}$, $-_\mathbb{R}$, $\cdot_\mathbb{R}$ for the operations of addition, subtraction, multiplication of $\mathbb{R}$ and a monadic function symbol $|\cdot|_\mathbb{R}$ for the absolute value of $\mathbb{R}$;

- binary symbols $\wedge_\mathbb{R}$ ("minimum") and $\vee_\mathbb{R}$ ("maximum") for the lattice operations of $\mathbb{R}$;

- an $\mathbb{R}$-valued constant c_r for each rational number $r \in \mathbb{R}$.

We will require that, in addition to these symbols, the signature L include the following:

- a symbol for the *measure*:

$$\mu : \mathcal{A} \to \mathbb{R};$$

- function symbols

$$\cup : \mathcal{A} \times \mathcal{A} \to \mathcal{A}, \qquad \cap : \mathcal{A} \times \mathcal{A} \to \mathcal{A}, \qquad {}^c : \mathcal{A} \to \mathcal{A}$$

for the union (join), intersection (meet) and complementation operations of the interpretation of $\mathcal{A}$;

- a symbol $[\![\cdot \in \cdot]\!] : \Omega \times \mathcal{A} \to \{0, 1\}$;

- $\mathcal{A}$-valued constants $\varnothing$, Ω for the zero (null element) and unity (universal element) of the interpretation of $\mathcal{A}$;

- an Ω-valued constant c_{ω_0} for the distinguished element ω_0 of the interpretation of Ω.

The signature L may include other function and constant symbols, as well as many other sorts than those already mentioned.

In order to simplify the notation, we normally omit the subscripts in the operations $+_\mathbb{R}$, $-_\mathbb{R}$, $\cdot_\mathbb{R}$, $\wedge_\mathbb{R}$, $\vee_\mathbb{R}$, and $|\cdot|_\mathbb{R}$ of $\mathbb{R}$. Also, for a rational number r, we denote the constant c_r simply as r.

If $\mathcal{M}$ is a fixed L-structure, we will denote the sorts of $\mathcal{M}$ corresponding to $\mathcal{A}$ and Ω also as $\mathcal{A}$ and Ω. Moreover, if the context allows it, we will identify the symbols of L with their interpretation in $\mathcal{M}$; thus, for instance, we denote $\cup^\mathcal{M}$ and $\Omega^\mathcal{M}$ simply as $\cup$ and Ω (respectively).

Definition 3.1 (Loeb structure). A signature L as described above will henceforth be called a *Loeb signature*. A *pre-Loeb finite measure structure* is an L-structure $\mathcal{M}$ such that:

- the metrics d_Ω on Ω and $d_\mathcal{A}$ on $\mathcal{A}$ are discrete;

- $[\![\cdot \in \cdot]\!]$ is a $\{0,1\}$-valued function identifying $(\mathcal{A}, \varnothing, \Omega, \cup, \cap, \cdot^c)$ with an algebra of subsets of Ω, i.e., for all $A, B \in \mathcal{A}$ and $\omega \in \Omega$:

  - $d_\mathcal{A}(A, B) = \sup_{x \in \Omega} \big| [\![x \in A]\!] - [\![x \in B]\!] \big|$;
  - $[\![\omega \in \varnothing]\!] = 0$;
  - $[\![\omega \in \Omega]\!] = 1$;
  - $[\![\omega \in A \cup B]\!] = [\![\omega \in A]\!] \vee [\![\omega \in B]\!]$;
  - $[\![\omega \in A \cap B]\!] = [\![\omega \in A]\!] \wedge [\![\omega \in B]\!]$;
  - $[\![\omega \in A]\!] + [\![\omega \in A^c]\!] = 1$;

- the interpretation μ of $\boldsymbol{\mu}$ is a finitely additive measure on $\mathcal{A}$. For some $C \geq 0$:

  - $\mu(\varnothing) = 0$ and $0 \leq \mu(A) \leq \mu(\Omega) =: \|\mu\| \leq C$ for all $A \in \mathcal{A}$;
  - $\mu(A \cup B) + \mu(A \cap B) = \mu(A) + \mu(B)$ for all $A, B \in \mathcal{A}$.

In a *pre-Loeb signed measure structure* $\mathcal{M}$, the last axioms are replaced by:

- For some $C \geq 0$, $\mu = \boldsymbol{\mu}^{\mathcal{M}}$ is a signed measure on $\mathcal{A}$ with total variation at most C:

  - $\mu(\varnothing) = 0$ and $\mu(A \cup B) + \mu(A \cap B) = \mu(A) + \mu(B)$ for all $A, B \in \mathcal{A}$;
  - $\|\mu\| := \sup\{|\mu(A)| + |\mu(B)| - |\mu(A \cap B)| : A, B \in \mathcal{A}\} \leq C$.

A *pre-Loeb probability structure* is a pre-Loeb finite measure structure $\mathcal{M}$ with $\|\mu\| = 1$. A *Loeb probability (finite measure, signed measure) structure* is any λ^+-saturated pre-Loeb probability (finite measure, signed measure) structure, where $\lambda = \text{card}(L)$.

Clearly, a positive pre-Loeb structure is a signed pre-Loeb structure. It can be easily verified that the equality $\|\mu\| = \sup\{|\mu(A)| + |\mu(B)| - |\mu(A \cap B)| : A, B \in \mathcal{A}\}$ holds in any pre-Loeb measure structure (probability, finite or positive). On the other hand, in signed measure structures only the inequality $|\mu(\Omega)| \leq \|\mu\|$ holds in general.

Recall (Definition 6.22) that if L' is a signature, a class $\mathcal{C}$ of metric structures is said to be L'-axiomatizable if it consists of the models of a (positive bounded) L'-theory.

Proposition 3.2. *Let L be a signature for Loeb structures. For every fixed $C \geq 0$, the class of all pre-Loeb probability (finite measure, signed measure) L-structures $\mathcal{M}$ such that $\|\mu\| \leq C$ is axiomatizable.*

(The assumption $C = 1$ in the case of probability structures is tacit.)

Proof. Given $C \geq 0$, we have to show that the clauses of Definition 3.1 are equivalent to the (approximate) satisfaction of positive bounded L-sentences. Clearly, it suffices to do so under the assumption that C is rational (if C is irrational, simply take the union of all axiom schemes for rational $D > C$). This is a routine exercise, so we give only one example. The condition $\|\mu\| \leq C$ amounts to the approximate satisfaction of the sentence[5]

$$(\forall A, B)(|\boldsymbol{\mu}(A)| + |\boldsymbol{\mu}(B)| - |\boldsymbol{\mu}(A \cap B)| \leq C).$$

The reader is invited to write down formulas axiomatizing the remaining clauses. □

Definition 3.3. For any pre-Loeb structure $\mathcal{M}$, define

$$\mathcal{A}^{\mathcal{M}} \to \mathcal{P}(\Omega^{\mathcal{M}})$$
$$A \mapsto [A]^{\mathcal{M}} = \{\omega \in \Omega^{\mathcal{M}} : [\![\omega \in A]\!] = 1\}.$$

The collection

$$[\mathcal{A}]^{\mathcal{M}} = \{[A]^{\mathcal{M}} : A \in \mathcal{A}\}$$

of subsets of $\Omega^{\mathcal{M}}$ is the *induced (external) algebra in* $\Omega^{\mathcal{M}}$. It follows from the definition of pre-Loeb structure that $[\mathcal{A}]^{\mathcal{M}}$ is an algebra of subsets of $\Omega^{\mathcal{M}}$ (the map $A \mapsto [A]^{\mathcal{M}}$ is an isomorphism between the Boolean algebras $\mathcal{A}$ and $[\mathcal{A}]^{\mathcal{M}}$).

The *(external) measure induced by* μ *on* $\mathcal{M}$ is the real-valued function

$$[\mu]^{\mathcal{M}} : [\mathcal{A}]^{\mathcal{M}} \to \mathbb{R}$$
$$[A]^{\mathcal{M}} \mapsto \mu(A).$$

$[\mu]^{\mathcal{M}}$ is well defined because $A \mapsto [A]^{\mathcal{M}}$ is injective.

When $\mathcal{M}$ is fixed, we usually write $[\mu], [\mathcal{A}], [A]$ for $[\mu]^{\mathcal{M}}, [\mathcal{A}]^{\mathcal{M}}, [A]^{\mathcal{M}}$. We may also write $[\Omega]$ instead of $\Omega^{\mathcal{M}}$, for emphasis.

Proposition 3.4. *Fix a Loeb structure $\mathcal{M}$ and let $\mathcal{A}_{\mathrm{L}}$ be the σ-algebra of subsets of Ω generated by $[\mathcal{A}]$.*

(1) *The function $[\mu]$ extends to a unique countably-additive positive [signed] measure μ_{L} on $\mathcal{A}_{\mathrm{L}}$ with total variation $\mathrm{var}(\mu_{\mathrm{L}}) = \|\mu\|$.*

(2) *For all $S \in \mathcal{A}_{\mathrm{L}}$ and $\epsilon > 0$ there exist $A, B \in \mathcal{A}$ such that $[A] \subset S \subset [B]$ and $\|\mu{\restriction}(B \setminus A)\| \leq \epsilon$, where*

$$\mu{\restriction}X : \mathcal{A} \to \mathbb{R}$$
$$Y \mapsto \mu(Y \cap X)$$

for any $X \in \mathcal{A}$.

---

[5]We remark that the value $D = \|\mu\|$ is itself uniquely characterized by the formula $(\forall A, B)(|\boldsymbol{\mu}(A)| + |\boldsymbol{\mu}(B)| - |\boldsymbol{\mu}(A \cap B)| \leq D) \wedge (\exists A, B)(|\boldsymbol{\mu}(A)| + |\boldsymbol{\mu}(B)| - |\boldsymbol{\mu}(A \cap B)| \geq D).$

(3) *For every $S \in \mathcal{A}_L$ there exists $A \in \mathcal{A}$ such that $|\mu_L|(S\triangle[A]) = 0$, where $X\triangle Y = (X \setminus Y) \cup (Y \setminus X)$ is the symmetric set difference and $|\mu_L|$ is the (nonnegative) absolute measure of μ_L.*

Proof. We prove the statements for μ a positive finite measure, leaving the case of a signed measure to the reader.

1. Clearly, $[\mu]$ is a finitely additive nonnegative measure on $([\Omega], [\mathcal{A}])$ with finite total variation $\mathrm{var}([\mu]) = \mu(\Omega) = \|\mu\|$. Suppose that $([A_i] : i \in \mathbb{N})$ is a descending chain in $[\mathcal{A}]$ (for some descending chain (A_i) in $\mathcal{A}$). Assume that, for every $i \in \mathbb{N}$, $[A_i]$ is nonempty. Choose $\omega_i \in [A_i]$. By saturation, since (A_i) is descending, we have $[\![\omega \in A_i]\!] = 1$ for some ω and all $i \in \mathbb{N}$. Certainly, $\omega \in \bigcap_{i \in \mathbb{N}}[A_i]$. Therefore, if $\bigcap_{i \in \mathbb{N}}[A_i] = \emptyset$, we must have $A_j = \emptyset$ for some $j \in \mathbb{N}$, hence $\inf_{i \in \mathbb{N}}[\mu]([A_i]) = [\mu]([A_j]) = \mu(A_j) = \mu(\emptyset) = 0$. By taking relative complements, the preceding argument implies that $[\mu]$ is countably additive on $[\mathcal{A}]$, hence a premeasure thereon (in fact, if a set $B \in [\mathcal{A}]$ is a countable union of sets in $[\mathcal{A}]$, then B is necessarily a finite union of such sets). By the Carathéodory Extension Theorem, $[\mu]$ admits an extension to a countably additive measure on $\sigma[\mathcal{A}] = \mathcal{A}_L$ with total variation $\mathrm{var}(\mu_L) = \mathrm{var}([\mu]) = \|\mu\|$. The extension is unique because $[\mu]$ has finite total variation.

2. This assertion also follows from Carathéodory's theorem.

3. Let $S \in \mathcal{A}_L$. By part (2), for each $n \in \mathbb{N}$ we may choose $A_n, B_n \in \mathcal{A}$ with $[A_n] \subset S \subset [B_n]$ and $\mu(B_n \setminus A_n) \le 1/(n+1)$. Clearly, $\inf_n \mu(B_n \setminus A_n) = 0$. Without loss of generality, (A_n) is increasing and (B_n) decreasing. By saturation, there exists $A \in \mathcal{A}$ such that $A_n \subset A \subset B_n$ for all $n \in \mathbb{N}$. Since (A_n) is increasing and (B_n) decreasing, $U = \bigcup_{n \in \mathbb{N}}[A_n] \subset [A] \subset \bigcap_{n \in \mathbb{N}}[B_n] = V$, and also $U \subset S \subset V$. We have $\mu_L(S\triangle[A]) \le \mu_L(V \setminus U) = \inf_n \mu(B_n \setminus A_n) = 0$, so A is as required. $\qquad\square$

Definition 3.5. We call $\mathcal{A}_L$ the *Loeb algebra* of the Loeb structure $\mathfrak{M}$, and μ_L the *Loeb measure* on $[\Omega]$ (i.e., on $\Omega^{\mathfrak{M}}$).

Remarks 3.6.

1. The proof of Proposition 3.4 given above is an adaptation of the classical construction of Loeb measures [Loe75] (see also the articles by Cutland [Cut83, Cut00], and Ross [Ros97], on which we base our approach). Our context differs from the classical one in the sense that we do not need to use nonstandard universes or hyperreals, and our measures need not be probability measures.

2. Our definition of Loeb algebra differs from the classical definition, according to which the Loeb algebra is the completion $\overline{\mathcal{A}_L}$ of $\mathcal{A}_L$ relative to μ_L and every subset U of an μ_L-null set is declared to be $\overline{\mathcal{A}_L}$-measurable and null. While the classical definition has the advantage

that the converse statements to (2) and (3) of Proposition 3.4 hold, we prefer to avoid completing $\mathcal{A}_L$ so its definition is independent of μ.

---

4 Integration structures

In this section, L will denote a signature for Loeb structures (with sorts $\mathbb{R}$, Ω, $\mathcal{A}$) such that, in addition to all the constant and function symbols required for Loeb structures, L includes a sort $\mathcal{L}_\Omega^\infty$ (whose interpretation will be a Banach lattice-algebra) and the following function symbols:

· a monadic symbol $\|\cdot\|_{\mathcal{L}_\Omega^\infty} : \mathcal{L}_\Omega^\infty \to \mathbb{R}$ (to be interpreted as the Banach norm of sort $\mathcal{L}_\Omega^\infty$);

· function symbols $\sup : \mathcal{L}_\Omega^\infty \to \mathbb{R}$ and $\inf : \mathcal{L}_\Omega^\infty \to \mathbb{R}$;

· binary symbols

$$+_{\mathcal{L}_\Omega^\infty} : \mathcal{L}_\Omega^\infty \times \mathcal{L}_\Omega^\infty \to \mathcal{L}_\Omega^\infty,$$
$$-_{\mathcal{L}_\Omega^\infty} : \mathcal{L}_\Omega^\infty \times \mathcal{L}_\Omega^\infty \to \mathcal{L}_\Omega^\infty,$$
$$\cdot_{\mathcal{L}_\Omega^\infty} : \mathcal{L}_\Omega^\infty \times \mathcal{L}_\Omega^\infty \to \mathcal{L}_\Omega^\infty,$$

and $\bullet_{\mathcal{L}_\Omega^\infty} : \mathbb{R} \times \mathcal{L}_\Omega^\infty \to \mathcal{L}_\Omega^\infty$ (to be interpreted as the algebra operations of $\mathcal{L}_\Omega^\infty$ and the scalar multiplication of the interpretation of $\mathcal{L}_\Omega^\infty$, respectively);

· binary symbols

$$\wedge_{\mathcal{L}_\Omega^\infty} : \mathcal{L}_\Omega^\infty \times \mathcal{L}_\Omega^\infty \to \mathcal{L}_\Omega^\infty, \qquad \vee_{\mathcal{L}_\Omega^\infty} : \mathcal{L}_\Omega^\infty \times \mathcal{L}_\Omega^\infty \to \mathcal{L}_\Omega^\infty$$

(to be interpreted as the lattice operations of the interpretation of $\mathcal{L}_\Omega^\infty$);

· monadic symbols $\sup_\Omega : \mathcal{L}_\Omega^\infty \to \mathbb{R}$ and $\inf_\Omega : \mathcal{L}_\Omega^\infty \to \mathbb{R}$;

· a binary symbol $\mathrm{ev}_\Omega : \mathcal{L}_\Omega^\infty \times \Omega \to \mathbb{R}$;

· a monadic symbol $\chi : \mathcal{A} \to \mathcal{L}_\Omega^\infty$;

· $I : \mathcal{L}_\Omega^\infty \to \mathbb{R}$;

· $\mathcal{L}_\Omega^\infty$-valued constant symbols $0_{\mathcal{L}_\Omega^\infty}$ and $1_{\mathcal{L}_\Omega^\infty}$.

A signature including the sorts and symbols above will be called a *signature for integration structures*. Such a signature may include many other sorts, function symbols, and constant symbols.

Notation 4.1. If L is a signature for integration structures, $\mathcal{M}$ is an L-structure, $f \in (\mathcal{L}_\Omega^\infty)^\mathcal{M}$, $\omega \in \Omega^\mathcal{M}$ and $A \in \mathcal{A}^\mathcal{M}$, we write $\mathrm{ev}_\Omega^\mathcal{M}(f, \omega)$ for $f(\omega)$, χ_A for $\chi(A)$, and If for $I(f)$.

If $\mathcal{M}$ is an L-structure and $\mathcal{M}$ is fixed by the context, in order to simplify the notation, we will denote the sort corresponding to $\mathcal{L}_\Omega^\infty$ in $\mathcal{M}$ as $\mathcal{L}_\Omega^\infty$, rather than $(\mathcal{L}_\Omega^\infty)^\mathcal{M}$. Carrying this simplification one step further, we will remove the $\mathcal{M}$-superscript from the interpretations of the function symbols in $\mathcal{M}$; thus, for instance, we write $\|\cdot\|_{\mathcal{L}_\Omega^\infty}$ and $+_{\mathcal{L}_\Omega^\infty}$ instead of $(\|\cdot\|_{\mathcal{L}_\Omega^\infty})^\mathcal{M}$ and $(+_{\mathcal{L}_\Omega^\infty})^\mathcal{M}$, respectively.

Definition 4.2 (Integration structure). Let L be a signature for integration structures. A *positive pre-integration structure* is an L-structure $\mathcal{M}$ such that:

· The interpretation of $(\mathcal{L}_\Omega^\infty, +_{\mathcal{L}_\Omega^\infty}, -_{\mathcal{L}_\Omega^\infty}, \cdot_{\mathcal{L}_\Omega^\infty}, \cdot_{\mathcal{L}_\Omega^\infty}, \wedge_{\mathcal{L}_\Omega^\infty}, \vee_{\mathcal{L}_\Omega^\infty}, \|\cdot\|_{\mathcal{L}_\Omega^\infty}, 0_{\mathcal{L}_\Omega^\infty})$ in $\mathcal{M}$ is a Banach algebra and Banach lattice with anchor $0_{\mathcal{L}_\Omega^\infty}$ (the *zero function*);

· for all $f \in \mathcal{L}_\Omega^\infty$ and $\omega \in \Omega$, we have:

  – the metric on $\mathcal{L}_\Omega^\infty$ is that induced by the norm $\|\cdot\|_{\mathcal{L}_\Omega^\infty}$;

  – $\|f\|_{\mathcal{L}_\Omega^\infty} = \sup_{\omega' \in \Omega} |f(\omega')|$;

  – $\mathrm{ev}_\Omega(\cdot, \omega)$ is an algebra and lattice homomorphism;

  – $\sup f = \sup_{\omega' \in \Omega} f(\omega')$ and $\inf f = \inf_{\omega' \in \Omega} f(\omega')$;

  – $0_{\mathcal{L}_\Omega^\infty}(\omega) = 0$ and $1_{\mathcal{L}_\Omega^\infty}(\omega) = 1$;

· each $f \in \mathcal{L}_\Omega^\infty$ is *approximately $\mathcal{A}$-measurable* ($\mathcal{A}\approx$*measurable*) in the following sense: For any reals $u < v$ there exists $A \in \mathcal{A}$ such that $f(\omega) \le v$ if $[\![\omega \in A]\!] = 1$, and $f(\omega) \ge u$ if $[\![\omega \in A]\!] = 0$;

· $\chi_A(\omega) = [\![\omega \in A]\!]$ for all $A \in \mathcal{A}$ and $\omega \in \Omega$;

· for some $C \ge 0$:

  – $\mathcal{M}$ is a pre-Loeb finite measure structure satisfying $\mu(A) = I\chi_A$ for all $A \in \mathcal{A}$, and $\|\mu\| := \mu(\Omega) \le C$ (we also define $\|I\| = \|\mu\|$);

  – the *integration operation* I is a $\|\mu\|$-Lipschitz linear functional $\mathcal{L}_\Omega^\infty \to \mathbb{R}$;

  – $\|\mu\| \inf f \le If \le \|\mu\| \sup f$ for all $f \in \mathcal{L}_\Omega^\infty$ (in particular, I is a positive functional: $If \ge 0$ if $f \ge 0$).

In a *signed pre-integration structure*, the last axiom becomes:

· For some $C \ge 0$:

  – $\mathcal{M}$ is a pre-Loeb signed measure structure satisfying $\|\mu\| \le C$ and $\mu(A) = I\chi_A$ for all $A \in \mathcal{A}$ (we define $\|I\| := \|\mu\|$);

 − I is a $\|\mu\|$-Lipschitz linear functional $\mathcal{L}_\Omega^\infty \to \mathbb{R}$;
 − $|If| \leq \|\mu\|\|f\|_{\mathcal{L}_\Omega^\infty}$ for all $f \in \mathcal{L}_\Omega^\infty$.

A *probability pre-integration structure* is a positive integration structure with $\|\mu\| = 1$. A *probability* (resp., *positive, signed*) *integration structure* is any λ^+-saturated probability (resp., positive, signed) pre-integration structure, where $\lambda = \mathrm{card}(L)$.

For $f \in \mathcal{L}_\Omega^\infty$, we let $f_+ = (f \vee 0)$, $f_- = (-f)_+$ (so $f = f_+ + f_-$ with $f_+, f_- \geq 0$), and $|f| = f_+ + f_-$ (so $|f|(\omega) = |f(\omega)|$ for all $\omega \in \Omega$). For syntactic convenience, we treat the order of the lattice $\mathcal{L}_\Omega^\infty$ as part of the language, so we write $f \leq g$ to mean $\|(f \wedge g) - f\|_{\mathcal{L}_\Omega^\infty} \leq 0$.

Remark 4.3. The usual definition of measurability suggests postulating that for every $f \in \mathcal{L}_\Omega^\infty$ and every interval $J \subset \mathbb{R}$ there shall exist $A \in \mathcal{A}$ such that $\omega \in \mathcal{A}$ if and only if $f(\omega) \in J$. However, exact measurability in this sense is not axiomatizable by positive bounded L-sentences. On the other hand, approximate measurability is axiomatizable as shown below in Proposition 4.6. Perhaps surprisingly, this postulate fails even in (saturated) probability structures. By working in saturated probability spaces and externally enlarging $[\mathcal{A}]$ to a σ-algebra $\sigma[\mathcal{A}]$ of subsets of Ω, the $\mathcal{A}{\approx}$measurability axiom implies the exact $\sigma[\mathcal{A}]$-measurability of all functions $[f] : \Omega \to \mathbb{R}$ for $f \in \mathcal{L}_\Omega^\infty$. See Proposition 4.6 below.

Proposition 4.4. *Let L be a language for pre-measure structures. For every $C \geq 0$, the class of all (probability, positive, or signed) pre-measure L-structures $\mathcal{M}$ with $\|\mu^{\mathcal{M}}\| \leq C$ is axiomatizable in the logic of approximate satisfaction of positive bounded formulas.*

Proof. It is a routine exercise to verify that the axioms in the definition of integration structure can be written as sets of positive bounded L-sentences. As an example, the $\mathcal{A}{\approx}$measurability condition amounts to the axiom schema

$$(\forall_D f)(\exists A)(\forall \omega)(r\chi_{A^c} - D\chi_A \leq f(\omega) \leq r\chi_A + D\chi_{A^c})$$

for all rational $D > 0$ and r, as easily seen from the semantics of approximate satisfaction. The remaining axioms are handled similarly. □

Notation 4.5. Given any function $G : \mathbb{R} \to \mathbb{R}$, an element $f \in \mathcal{L}_\Omega^\infty$, and a real number $t \in \mathbb{R}$, we let

$$\{G(f) \leq t\} := \{\omega \in \Omega : G(f(\omega)) \leq t\},$$

with similar definitions for $\{G(f) < t\}$, $\{G(f) \geq t\}$, and $\{G(f) > t\}$.

Proposition 4.6. *Let $\mathcal{M}$ be an integration L-structure. For every $f \in \mathcal{L}_\Omega^\infty$, the function $[f] = [f]^{\mathcal{M}} : \Omega \to \mathbb{R}$ defined by $\omega \mapsto f(\omega)$ is $\mathcal{A}_L$-measurable.*

Proof. Since $\mathcal{A}_L$ is a σ-algebra, it suffices to show that $\{f \leq t\} \in \mathcal{A}_L$ for any fixed real number t.

Let $(u_n : n \in \mathbb{N})$ be a strictly decreasing sequence of rational numbers such that $\inf_n u_n = t$. By $\mathcal{A}\approx$measurability and saturation, for each $n \in \mathbb{N}$ there exists $A_n \in \mathcal{A}$ such that $f(\omega) \leq u_n$ if $\omega \in [A_n]$, and $f(\omega) \geq u_n$ if $\omega \notin [A_n]$. Clearly, $\{f \leq t\} = \bigcap_{n \in \mathbb{N}}[A_n] \in \mathcal{A}_L$. $\square$

Theorem 4.7 (Riesz Representation Theorem for integration structures). *Let $\mathcal{M}$ be an integration L-structure (positive or signed). For every $f \in \mathcal{L}_\Omega^\infty$, $[f]$ is Loeb-integrable and*

$$If = \int [f] \, d\mu_L.$$

Proof. We assume that I is positive, leaving the signed case to the reader. Let $C = \|I\| = \|\mu\|$ and $D = \|f\|_{\mathcal{L}_\Omega^\infty}$. $[f]$ is Loeb integrable because it is $\mathcal{A}_L$-measurable (Proposition 4.6) and bounded (by D). The Loeb measure μ_L is also positive with total variation $\text{var}(\mu_L) = \mu_L(\Omega) = \mu(\Omega) = \|\mu\| = C$. Let us write $\int F$ for $\int F d\mu_L$. If $A \in \mathcal{A}$, we have $I\chi_A = \mu(A) = \mu_L([A]) = \int[\chi_A]$. Fix $\epsilon > 0$ and let $(J_i : i < k)$ be any finite collection of disjoint intervals, each having length at most ϵ, such that $\bigcup_{i<k} J_i \supset [-D, D]$. For $i < k$, let $S_i = \{f \in J_i\} \subset \Omega$. By measurability of $[f]$ (Proposition 4.6), the collection $(S_i : i < k)$ is an $\mathcal{A}_L$-measurable disjoint cover of Ω, hence $\sum_{i<k} \mu_L(S_i) = \mu_L(\Omega) = C$. Choose rational numbers r_i $(i < k)$ such that $J_i \subset [r_i, r_i + 2\epsilon]$. The $\mathcal{A}_L$-simple function $F_\epsilon = \sum_{i<k} r_i \chi_{S_i} : \Omega \to \mathbb{R}$ satisfies $F_\epsilon \leq [f] \leq F_\epsilon + 2\epsilon$, hence $\int F_\epsilon \leq \int[f] \leq \int F_\epsilon + 2C\epsilon$. By internal approximability (Proposition 3.4(3)), there exist $A_i \in \mathcal{A}$ $(i < k)$ such that $\mu_L([A_i]\triangle S_i) = 0$; thus $\sum_{i<k} \mu(A_i) = \sum_{i<k} \mu_L(S_i) = C$. Without loss of generality we may assume that the sets A_i are pairwise disjoint. Let $B = \left(\bigcup_{i<k} A_i\right)^c$, so $\mu(B) = 0$. Let $f_\epsilon = f\chi_B + \sum_{i<k} r_i \chi_{A_i}$. Certainly, $f_\epsilon \leq f \leq f_\epsilon + 2\epsilon$, so $If_\epsilon \leq If \leq If_\epsilon + 2C\epsilon$. Since $\mu(B) = 0$, we have $If_\epsilon = \sum_{i<k} r_i \mu(A_i) = \sum_{i<k} r_i \mu_L(S_i) = \int F_\epsilon$. Thus, If and $\int[f]$ both lie in $[If_\epsilon, If_\epsilon + 2C\epsilon]$, so $|If - \int[f]| \leq 2C\epsilon$. Since C is fixed and this holds for all $\epsilon > 0$, we conclude that $If = \int[f]$. $\square$

5   Directed integration structures

Definition 5.1 (Directed integration structure). Fix a pointed directed set $(\mathcal{D}, \leq, j_0)$. Let L be both a signature for integration structures (with sorts $\mathbb{R}$, Ω, $\mathcal{A}$, $\mathcal{L}_\Omega^\infty$) and also for $\mathcal{D}$-directed structures (with directed sort $\mathbb{D}$). In addition, assume that L has a function symbol

$$\varphi : \mathbb{D} \to \mathcal{L}_\Omega^\infty.$$

Such L will be called a *signature for directed integration structures* (and it may include any other sorts and symbols than those named).

A $\mathcal{D}$-*directed pre-integration structure* is an L-structure $\mathcal{M}$ that is both a $\mathcal{D}$-directed L-structure and a pre-integration L-structure (positive or signed). A $\mathcal{D}$-*directed integration structure* is an λ^+-saturated $\mathcal{D}$-directed pre-integration structure, where $\lambda = \mathrm{card}(L)$.

Given any $\mathcal{D}$-directed pre-integration structure $\mathcal{M}$, let $\varphi = \boldsymbol{\varphi}^{\mathcal{M}}$. Note that the definition of L-structure implies that φ is uniformly bounded on the discrete sort $\mathbb{D}^{\mathcal{M}}$. We define

$$\|\varphi\| := \sup_{j \in \mathbb{D}} \|\varphi(j)\|_{\mathcal{L}_\Omega^\infty}.$$

For $j \in \mathbb{D}$, we will denote $\varphi(j)$ by φ_j. Each $\omega \in \Omega$ defines an external $\mathcal{D}$-net $\varphi_\bullet(\omega) := (\varphi_j(\omega) : j \in \mathcal{D})$ in $\mathcal{L}_\Omega^\infty$. We also have a real-valued $\mathcal{D}$-net $I\varphi_\bullet := (I(\varphi_j) : j \in \mathcal{D})$.

Proposition 5.2. *Let L be a language for directed integration structures and $C \geq 0$. The class of all (positive or signed) $\mathcal{D}$-directed pre-integration L-structures $\mathcal{M}$ such that $\|\varphi\| \leq C$ and $\|\mu\| \leq C$ is axiomatizable in the logic of approximate satisfaction.*

Proof. This follows from Propositions 3.2 and 4.4, upon remarking that the condition $\|\varphi\| \leq C$ is equivalent to the approximate satisfaction of the axioms[6]

$$(\forall j)\left(\|\boldsymbol{\varphi}_j\|_{\mathcal{L}_\Omega^\infty} \leq r\right) \qquad \text{for all rational } r > C. \quad \square$$

Proposition 5.3 (Dominated Convergence Theorem). *Assume that the directed set $\mathcal{D}$ is countable. Let $\mathcal{M}$ be a (saturated) $\mathcal{D}$-directed integration L-structure, for a suitable language L. Then*

$$\mathrm{osc}(I\varphi_\bullet) \leq \|I\| \sup_{\omega \in \Omega} \mathrm{osc}(\varphi_\bullet(\omega)) \qquad \text{for all } \omega \in \Omega.$$

In particular, $I\varphi_\bullet$ is convergent if $\varphi_\bullet(\omega)$ is convergent for each $\omega \in \Omega$.

Proof. We prove the inequality $\mathrm{osc}(I\varphi_\bullet) \leq \sup_{\omega \in \Omega} \mathrm{osc}(\varphi_\bullet(\omega))$ in the case when $\mathcal{M}$ is a probability integration structure (with $\|I\| = 1$), leaving the general case to the reader. Let $r = \sup_{\omega \in \Omega} \mathrm{osc}(\varphi_\bullet(\omega))$.

For $\epsilon > 0$ and $i, j, j' \in \mathcal{D}$, let

$$\Omega_\epsilon^{j,j'} = \{\omega \in \Omega : |\varphi_{j'}(\omega) - \varphi_j(\omega)| \leq r + \epsilon\}$$

and

$$\Omega_\epsilon^i = \bigcap_{j,j' \in \mathcal{D}_{\geq i}} \Omega_\epsilon^{j,j'} = \{\omega \in \Omega : |\varphi_{j'}(\omega) - \varphi_j(\omega)| \leq r + \epsilon \quad \text{for all } j, j' \in \mathcal{D}_{\geq i}\}.$$

---

[6]The value $D = \|\varphi\|$ is characterized by the approximate satisfaction of the formula

$$(\forall j)\left(\|\boldsymbol{\varphi}_j\|_{\mathcal{L}_\Omega^\infty} \leq D\right) \wedge (\exists j)\left(\|\boldsymbol{\varphi}_j\|_{\mathcal{L}_\Omega^\infty} \geq D\right).$$

Beyond First Order Model Theory

The functions $\varphi_j(\cdot)$ and $\varphi_{j'}(\cdot)$ are $\mathcal{A}_{\mathrm{L}}$-measurable, by Proposition 4.6. It is clear that each set $\Omega_\epsilon^{j,j'}$ is also $\mathcal{A}_{\mathrm{L}}$-measurable, and so is the (countable) intersection Ω_ϵ^i. Clearly, $\Omega_\epsilon^i \subset \Omega_\epsilon^{i'}$ for $i \leq i'$. Since $\omega \in \Omega_\epsilon^i$ implies $\mathrm{osc}_{\eta_i}(\varphi_\bullet(\omega)) \leq r+\epsilon$, while $\mathrm{osc}(\varphi_\bullet(\omega)) \leq r$ for all ω by hypothesis, part (2) of Proposition 1.4 gives:

$$\bigcup_{i \in \mathcal{D}} \Omega_\epsilon^i = \Omega.$$

Since $\mathcal{D}$ is countable and μ_{L} is a probability measure, we have $\sup_{i \in \mathcal{D}} \mu_{\mathrm{L}}(\Omega_\epsilon^i) = \mu_{\mathrm{L}}(\Omega) = 1$, hence $\mu(\Omega_\epsilon^{i_0}) \geq 1 - \epsilon/(\|\varphi\|+1)$ for some $i_0 \in \mathcal{D}$. For $j, j' \geq i_0$, we have:

$$|I(\varphi_{j'}) - I(\varphi_j)| \leq I(|\varphi_{j'} - \varphi_j|) = \int |\varphi_{j'} - \varphi_j| \chi_{\Omega_\epsilon^{i_0}} + \int |\varphi_{j'} - \varphi_j| \chi_{(\Omega_\epsilon^{i_0})^c}$$
$$\leq (r+\epsilon)\mu_{\mathrm{L}}(\Omega_\epsilon^{i_0}) + 2\|\varphi\|\mu_{\mathrm{L}}((\Omega_\epsilon^{i_0})^c) \leq r + 3\epsilon.$$

Therefore, $\mathrm{osc}(I\varphi_\bullet) \leq \mathrm{osc}_{\eta_i}(I\varphi_\bullet) \leq r + 3\epsilon$ (again, by part (2) of Proposition 1.4). Since ϵ is an arbitrary positive number, $\mathrm{osc}(I\varphi_\bullet) \leq r = \sup_{\omega \in \Omega} \mathrm{osc}(\varphi_\bullet(\omega))$. $\qquad\square$

Note that the proof above uses only the standard theory of integration.

Corollary 5.4 (Metastable Dominated Convergence Theorem). *Let $(\mathcal{D}, \leq, j_0)$ be a directed set with $\mathcal{D}$ countable. Let $\mathcal{M}$ be a (not necessarily saturated) $\mathcal{D}$-directed pre-integration L-structure, for a suitable signature L. Fix a real number $s \geq 0$. Let T be any uniform theory including the axiom $\|I\| \leq s$ (that is, the axioms $\|I\| \leq u$ for all rationals $u > s$) and extending the theory of $\mathcal{D}$-directed pre-integration L-structures. Given $r \geq 0$ and any collection $E_\bullet^r = (E_{\epsilon,\eta} : \eta \in \mathrm{Smpl}(\mathcal{D}), \epsilon > r)$ in $\mathcal{P}_{\mathrm{fin}}(\mathcal{D})$, there exists another collection $\widetilde{E}_\bullet^{rs} = (\widetilde{E}_{\epsilon,\eta} : \eta \in \mathrm{Smpl}(\mathcal{D}), \epsilon > rs)$ such that every model $\mathcal{M}$ of T satisfies the following property:*

If every external $\mathcal{D}$-net in the collection $\mathcal{C} = (\varphi_\bullet(\omega) : \varphi \in (\mathcal{L}_\Omega^\infty)^{[1]}, \omega \in \Omega)$ is $E_\bullet^r$-uniformly r-metastable, then $\widetilde{E}_\bullet^{rs}$ is a rate of rs-metastability for the collection $(I\varphi_\bullet : \varphi \in (\mathcal{L}_\Omega^\infty)^{[1]})$.

In fact, one such $\widetilde{E}_\bullet^{rs}$ may be found depending only on r, s, and $E_\bullet^r$.

Proof. Let $r, s \geq 0$ and $E_\bullet^r$ be given. Restrict the signature L so it only names the sorts and symbols strictly required for a signature of $\mathcal{D}$-directed pre-integration structures. Let $\widetilde{T} = T[\sigma, \mathsf{a}]$ where $\mathsf{a} : \mathbb{D} \to \mathbb{R}$ and $\sigma : \mathcal{L}_\Omega^\infty \times \Omega \times \mathbb{D} \to \mathbb{R}$ are new function symbols, and let T be the theory of pre-integration L-structures augmented with the following $\widetilde{L}$-axioms:

· $\|\mathtt{I}\| \leq C$, for all rational $C > s$;

· $(\forall_C \varphi)(\forall j)(\forall \omega)(|\sigma(\varphi, \omega, j) - \varphi_j(\omega)| \leq 0)$, for each rational $C \geq 0$;

· $(\exists_1 \varphi)(\exists \omega)(\forall j)(|\mathsf{a}(j) - \sigma(\varphi, \omega, j)| \leq 0)$.

Clearly, T is a uniform $\widetilde{L}$-theory, and σ is $(T{\restriction}L[\boldsymbol{\sigma}])$-definable in L; in fact, every model $\mathcal{M}$ of $T{\restriction}L$ admits a unique expansion to a model $(\mathcal{M}, \sigma)$ of $T{\restriction}L[\boldsymbol{\sigma}]$, and any such model admits some expansion $(\mathcal{M}, \sigma, a)$ to a model of T via any $a : \mathbb{D}^{\mathcal{M}} \to \mathbb{R}$ that approximately admits σ-parameters of size $\lesssim 1$. Using Proposition 2.3, extend T to a (necessarily uniform) $\widetilde{L}$-theory $\widetilde{T}$ such that $a_{\bullet}$ is $E_{\bullet}^r$-uniformly r-metastable for all models $(\mathcal{M}, \sigma, a)$ of $\widetilde{T}$. Henceforth all L'-structures for a language $L' \subset \widetilde{L}$ are assumed to be models of $\widetilde{T}{\restriction}L'$. If $(\mathcal{M}, \sigma) \approx \widetilde{T}{\restriction}L[\boldsymbol{\sigma}]$, every external sequence $\varphi_{\bullet}(\omega)$ with $\|\varphi\| \leq 1$ admits σ-parameters (φ, ω) of size ≤ 1, and every such $\varphi_{\bullet}(\omega)$ is of the form $a_{\bullet}$ for some expansion $(\mathcal{M}, \sigma, a) \approx \widetilde{T}$ of $(\mathcal{M}, \sigma)$. By Proposition 1.4, $\mathrm{osc}(a_{\bullet}) \leq r$, hence $\mathrm{osc}(\varphi_{\bullet}(\omega)) \leq r$ whenever $\|\varphi\| \leq 1$. Since $\|I\| \leq s$, we have $\mathrm{osc}(I\varphi_{\bullet}) \leq rs$ by Proposition 5.3. Let $\mathsf{b} : \mathbb{D} \to \mathbb{R}$ and $\boldsymbol{\tau} : \mathcal{L}_{\Omega}^{\infty} \times \mathbb{D} \to \mathbb{R}$ be new function symbols. Expand $\widetilde{T}$ to an $\widetilde{L}[\boldsymbol{\tau}, \mathsf{b}]$-theory $\widetilde{T}'$ by adding the axioms

- $(\forall_C \varphi)(\forall j)(|\boldsymbol{\tau}(\varphi, j) - I\varphi_j| \leq 0)$, for each rational $C \geq 0$;

- $(\exists_1 \varphi)(\forall j)(|\mathsf{b}(j) - \boldsymbol{\tau}(\varphi, j)| \leq 0)$.

Clearly, $\widetilde{T}'$ is uniform, τ is $(\widetilde{T}'{\restriction}\widetilde{L}[\boldsymbol{\tau}])$-definable in $\widetilde{L}$, and an expansion $(\mathcal{M}, \tau, b)$ of a model $(\mathcal{M}, \tau)$ of $\widetilde{T}'{\restriction}\widetilde{L}[\boldsymbol{\tau}]$ to a model of $\widetilde{T}'$ is via $b : \mathbb{D}^{\mathcal{M}} \to \mathbb{R}$ that approximately admits τ-parameters of size $\lesssim 1$. Henceforth, all L'-structures for any signature $L' \subset \widetilde{L}[\boldsymbol{\tau}, \mathsf{b}]$ are assumed to be models of $\widetilde{T}'{\restriction}L'$. Since $\widetilde{T}$ implies that $\mathrm{osc}(I\varphi_{\bullet}) \leq rs$ whenever $\|\varphi\| \leq 1$, $\widetilde{T}'$ implies[7] that $\mathrm{osc}(b_{\bullet}) \leq rs$. By uniform metastability of parametrized sequences (Corollary 2.5), there exists a collection $\widetilde{E_{\bullet}^{rs}}$ depending only on $\widetilde{T}'$ such that all sequences $b_{\bullet}$, and hence all sequences $I\varphi_{\bullet}$ for $\|\varphi\| \leq 1$, are $\widetilde{E_{\bullet}^{rs}}$-uniformly rs-metastable. Moreover, $\widetilde{E_{\bullet}^{rs}}$ depends only on $\widetilde{T}'$, hence only on r, s and the given rate $E_{\bullet}^r$. $\square$

6   Background on metric model theory

This section describes the general framework for the classes of structures that are the focus of the chapter. We refer to these structures as *metric structures*.

---

[7] *A priori*, $\widetilde{T}'$ only implies that $\mathrm{osc}(b_{\bullet}) \leq rs$ when $b_{\bullet}$ admits τ-parameters of size ≤ 1, so $b_{\bullet} : j \mapsto \tau(\varphi, j) = I\varphi_j$ for some $\|\varphi\| \leq 1$. If this is the case, however, then any $b_{\bullet}$ that approximately admits τ-parameters of size $\lesssim 1$ must still satisfy $\mathrm{osc}(b_{\bullet}) \leq rs$. Alternatively, the rest of the proof may proceed applying Corollary 2.5 in the stronger version that only assumes metastability when $b_{\bullet}$ admits exact parameters of size ≤ 1.

I   Henson metric structures

Recall that a *pointed metric space* is a triple (M, d, a), where (M, d) is a metric space and a is a distinguished element of M called the *anchor* of M. If (M, d, a) is a pointed metric space, the closed ball of radius r around the anchor point a will be denoted $B_M[r]$, or simply $B[r]$ if the ambient space M is clear from the context; the corresponding open ball will be denoted $M^{(r)}$ or $B(r)$. If $(M_1, d_1, a_1), \ldots, (M_n, d_n, a_n)$ are pointed metric spaces, we regard the product $\prod_{i=1}^{n}(M_i, d_i, a_i)$ tacitly as a pointed metric space by taking $(a_1, \ldots, a_n)$ as its anchor and using the supremum metric.

Definition 6.1. A *metric* (or *Henson*) *structure* $\mathcal{M}$ (often just called a structure in this manuscript) consists of the following items:

· A family $(M^{(s)} \mid s \in \mathbf{S})$ of pointed metric spaces,

· A collection of functions of the form

$$F : M^{(s_1)} \times \cdots \times M^{(s_n)} \to M^{(s_0)},$$

each of which is locally uniformly continuous, i.e., uniformly continuous on each bounded subset of its domain.

The spaces $M^{(s)}$ are called the *sorts* of $\mathcal{M}$. We say that $\mathcal{M}$ *is based on* the collection $(M^{(s)} \mid s \in \mathbf{S})$ of its sorts.

We do require that every metric structure contain the set $\mathbb{R}$ of real numbers, equipped with the usual distance and 0 as an anchor point, as a distinguished sort. We also require that the given metric on each sort of $\mathcal{M}$ be included in the list of functions of $\mathcal{M}$, and that the anchor of each sort be included as a (constant) function.

If $\mathcal{M}$ is based on $(M^{(s)} \mid s \in \mathbf{S})$ an element of $M^{(s)}$ will be called an *element* of $\mathcal{M}$ of sort s. The *cardinality* of $\mathcal{M}$, denoted card$(\mathcal{M})$, is defined as $\sum_{s \in S} \mathrm{card}(M^{(s)})$.

Some of the sorts M of a structure may be discrete metric spaces, with the respective metric d $: M \times M \to \{0, 1\}$ taking the value 1 at every pair of distinct points. If all the sorts of $\mathcal{M}$ are discrete, we will say that $\mathcal{M}$ is a *discrete* structure. Similarly, if the sorts of $\mathcal{M}$ are bounded, we will say that $\mathcal{M}$ is a *bounded* metric structure.

Some of the functions of a structure $\mathcal{M}$ may have arity 0. Such functions correspond to distinguished elements of $\mathcal{M}$. We will call these elements the *constants* of the structure. If F is a $\{0, 1\}$-valued function of $\mathcal{M}$, we will identify F with a subset of its domain, namely, $F^{-1}(1)$. Such a function will be called a *relation*, or a *predicate*, of $\mathcal{M}$.

We will require that the special sort $\mathbb{R}$ should come equipped with the field operations of $\mathbb{R}$, the order relation and the lattice operations ($\max(x, y)$ and $\min(x, y)$), plus a constant for each rational number.

If a structure $\mathcal{M}$ is based on $(M^{(s)} \mid s \in \mathbf{S})$ and $(F^i \mid i \in I)$ is a list of the functions of $\mathcal{M}$, we write

$$\mathcal{M} = (M^{(s)}, F_i \mid s \in \mathbf{S}, i \in I).$$

For notational simplicity, the real sort $\mathbb{R}$, the metrics on the sorts of $\mathcal{M}$, and their respective anchors need not be listed explicitly in this notation. We will only list them when needed for emphasis.

The structures that we will be dealing with are "hybrid" in the sense that some of their sorts are discrete, while others are genuine metric spaces. Typically, the nondiscrete structures will be Banach algebras or Banach lattices; in these the natural anchor point is 0. The discrete sorts that we will encounter include partial orders and purely algebraic structures; in structures of this type, the particular choice of anchor point is often inconsequential.

II   Henson signatures and structure isomorphisms

We will need a formal way to index the sorts and functions of any given structure $\mathcal{M}$. This is accomplished through the concept of *signature* of a metric structure.

Definition 6.2. Let $\mathcal{M}$ be a structure based on $(M^{(s)} \mid s \in \mathbf{S})$. A *Henson signature* L for $\mathcal{M}$ consists of the following items:

· A sort index set $\mathbf{S}$,

· A special element $s_{\mathbb{R}} \in \mathbf{S}$ such that $M^{(s_{\mathbb{R}})} = \mathbb{R}$,

· For each function $F : M^{(s_1)} \times \cdots \times M^{(s_n)} \to M^{(s_0)}$, a triple of the form

$$((s_1, \ldots, s_n), f, s_0),$$

where f is a purely syntactic symbol called a *function symbol* for F. We write $F = f^{\mathcal{M}}$ and call F the *interpretation* of f in $\mathcal{M}$. We call $s_1 \times \cdots \times s_n$ and s_0 the *domain* and *range* of f, respectively. We express this by writing (purely formally)

$$f : s_1 \times \cdots \times s_n \to s_0.$$

The number m is called the *arity* of the function symbol f. If $m = 0$ and the constant value of $f^{\mathcal{M}}$ in $M^{(s_0)}$ is c, we call f a *constant symbol* for c.

We express the fact that L is a signature for $\mathcal{M}$ by saying that $\mathcal{M}$ is an *L-structure*. A structure $\mathcal{M}$ for some Henson signature L will also be called a *Henson structure*. The *cardinality* of a signature L, denoted $\mathrm{card}(L)$, is defined as

$$\mathrm{card}(\mathbf{S}) + \mathrm{card}(\{ f \mid f \text{ is a function symbol of } L \}) + \aleph_0,$$

where $\mathbf{S}$ is the sort index set of L.

Definition 6.3. If L and L' are signatures, we say that L is a *subsignature* of L' (or that L' is an *extension* of L), and write $L \subseteq L'$, if the following conditions hold:

· The sort index set of L is a subset of the sort index set of L',

· Every triple of the form $((s_1, \ldots, s_n), f, s_0)$ that is in L is also in L'.

If L, L' are signatures, we say that L' is an *extension by constants* of L if L and L' have the same sort index set and every function symbol of L' that is not in L is a constant symbol. If the set of such constant symbols is C, we denote L' as $L[C]$.

Definition 6.4. Let L be a signature and let $\mathcal{M}$ and $\mathcal{N}$ be L-structures based on $(M^{(s)} \mid s \in \mathbf{S})$ and $(N^{(s)} \mid s \in \mathbf{S})$, respectively.

1. $\mathcal{M}$ is a *substructure* of $\mathcal{N}$ if $M^{(s)} \subseteq N^{(s)}$ and, for each function symbol f, the interpretation $f^{\mathcal{N}}$ of f in $\mathcal{N}$ is an extension of $f^{\mathcal{M}}$.

2. $\mathcal{M}$ and $\mathcal{N}$ are *isomorphic* if there exists a family $\mathcal{I} = (\mathcal{I}^{(s)} \mid s \in \mathbf{S})$ of maps (called an *isomorphism* from $\mathcal{M}$ into $\mathcal{N}$) such that for each $s \in \mathbf{S}$, $\mathcal{I}^{(s)} : M^{(s)} \to N^{(s)}$ is a bijection that commutes with the interpretations of the function symbols of L, in the sense that if $f : s_1 \times \cdots \times s_n \to s_0$, then $\mathcal{I}^{(s_0)}(f^{\mathcal{M}}(a_1), \ldots, f^{\mathcal{M}}(a_n)) = f^{\mathcal{N}}(\mathcal{I}^{(s_1)}(a_1), \ldots, \mathcal{I}^{(s_n)}(a_n))$. If a is an element of $M^{(s)}$ and the sort index s need not be made specific, we may write $\mathcal{I}(a)$ instead of $\mathcal{I}^{(s)}(a)$.

3. An *automorphism* of $\mathcal{M}$ is an isomorphism between $\mathcal{M}$ and $\mathcal{M}$.

III   Uniform classes and ultraproducts of metric structures

Recall that a *filter* on a nonempty set Λ is a collection $\mathcal{F}$ of subsets of Λ such that (i) $\Lambda \in \mathcal{F}$ and $\emptyset \notin \mathcal{F}$, (ii) $A \cap B \in \mathcal{F}$ if $A, B \in \mathcal{F}$, and (iii) $A \in \mathcal{F}$ if $B \in \mathcal{F}$ and $A \supset B$. An *ultrafilter* on Λ is a maximal filter $\mathcal{U}$ on Λ; equivalently, $\mathcal{U}$ is a filter such that (iv) $A \in \mathcal{U}$ or $\Lambda \setminus A \in \mathcal{U}$ for all $A \subset \Lambda$. If Λ is an index set and $\mathcal{F}$ is a filter on Λ, we will say that a subset of Λ is $\mathcal{F}$-*large* if it is in $\mathcal{F}$. An ultrafilter $\mathcal{U}$ on Λ is *principal* if there exists $\lambda_0 \in \Lambda$ such that $A \in \mathcal{U}$ iff $A \ni \lambda_0$ for all $A \subset \Lambda$; otherwise, $\mathcal{U}$ is *nonprincipal*. If X is a topological space, $(x_\lambda)_{\lambda \in \Lambda}$ is a family of elements of X, and $\mathcal{F}$ is a filter on Λ, we will say that $(x_\lambda)_{\lambda \in \Lambda}$ *converges to* an element $y \in X$ with respect to $\mathcal{F}$ if for every neighborhood U of y, the set $\{\lambda \in \Lambda \mid x_\lambda \in U\}$ is $\mathcal{F}$-large. If X is compact Hausdorff, then for every family $x_\bullet = (x_\lambda)_{\lambda \in \Lambda}$ and every ultrafilter $\mathcal{U}$ on Λ there exists a unique $y \in X$ such that $(x_\lambda)_{\lambda \in \Lambda}$ converges to y with respect to $\mathcal{U}$; this element y is called the $\mathcal{U}$-*limit* of $(x_\lambda)_{\lambda \in \Lambda}$ and is denoted $\mathcal{U}\lim x_\bullet$ or $\mathcal{U}\lim_\lambda x_\lambda$.

Let $(X_\lambda, d_\lambda)_{\lambda \in \Lambda}$ be a family of metric spaces and let $\mathcal{U}$ be an ultrafilter on Λ. The $\mathcal{U}$-*ultraproduct* of $(X_\lambda, d_\lambda)_{\lambda \in \Lambda}$ is the metric space defined in the following manner. Let $\ell^\infty(X_\lambda, d_\lambda)_\Lambda := \ell^\infty(X_\lambda, d_\lambda \mid \lambda \in \Lambda)$ be the set of all elements of $\prod_{\lambda \in \Lambda} X_\lambda$ that are bounded (when regarded as families indexed by Λ in the natural way). For $x = (x_\lambda)_{\lambda \in \Lambda}$, $y = (y_\lambda)_{\lambda \in \Lambda}$ in $\ell^\infty(X_\lambda, d_\lambda)_\Lambda$, and an ultrafilter $\mathcal{U}$ on Λ, define

$$d(x, y) = \mathcal{U}\lim_\lambda d_\lambda(x_\lambda, y_\lambda).$$

Since elements of $\ell^\infty(X_\lambda, d_\lambda)_\Lambda$ are bounded families, it is clear that d is well defined. It is also easy to verify that d is a pseudometric on $\ell^\infty(X_\lambda, d_\lambda)_\Lambda$. Now we can turn d into a metric in the usual way, namely by identifying any two elements $x, y \in \ell^\infty(X_\lambda, d_\lambda)_\Lambda$ such that $d(x, y) = 0$. For $x \in \ell^\infty(X_\lambda, d_\lambda)_\Lambda$, we let $(x)_\mathcal{U}$ denote the equivalence class of x under this identification, and for any two equivalence classes $(x)_\mathcal{U}$, $(y)_\mathcal{U}$, we define $d((x)_\mathcal{U}, (y)_\mathcal{U})$ as $d(x, y)$. The resulting metric space is called the $\mathcal{U}$-*ultraproduct* of the family $(X_\lambda, d_\lambda)_{\lambda \in \Lambda}$. It will be denoted $(\prod_{\lambda \in \Lambda} X_\lambda)_\mathcal{U}$.

If the spaces (X_λ, d_λ) are identical to the same space (X, d) for $\lambda \in \Lambda$, the $\mathcal{U}$-ultraproduct $(\coprod_{\lambda \in \Lambda} X_\lambda)_\mathcal{U}$ is called the $\mathcal{U}$-*ultrapower* of (X, d), denoted $(X)_\mathcal{U}$. Note that the map from X into $(X)_\mathcal{U}$ that assigns to each $x \in X$ the equivalence class of the constant family $(x \mid \lambda \in \Lambda)$ is an isometric embedding. When the ultrafilter $\mathcal{U}$ is principal or the space (X, d) is compact this map is surjective, though it is not so in general. The verification of these statements is left to the reader.

In the definition of ultraproduct, we lifted the metrics from the family $(X_\lambda, d_\lambda)_{\lambda \in \Lambda}$ to $(\prod_{\lambda \in \Lambda} X_\lambda)_\mathcal{U}$ by taking $\mathcal{U}$-limits. Doing the same for more general functions requires additional hypotheses. Let us introduce the concept of *uniform family of functions*.

Definition 6.5. Suppose that (X, d, a) and (Y, ρ, b) are pointed pseudometric spaces, B is a subset of X, and $F : X \to Y$ is uniformly continuous and bounded on B.

1. A *bound for F on B* is a number $\Omega \geq 0$ such that

$$x \in B \quad \Rightarrow \quad F(x) \in B_Y(\Omega).$$

2. A *modulus of uniform continuity for F on B* is a function $\Delta : (0, \infty) \to [0, \infty)$ such that, for all $x, y \in B$ and $\epsilon > 0$,

$$d(x, y) < \Delta(\epsilon) \quad \Rightarrow \quad \rho(F(x), F(y)) \leq \epsilon.$$

Definition 6.6. Let L be a signature and let $\mathscr{C}$ be a class of L-structures. We will say that $\mathscr{C}$ is a *uniform class* if the following two conditions hold for every function symbol $f : s_1 \times \cdots \times s_n \to s_0$ of L and every $r > 0$:

1. (Local equiboundedness condition for $\mathscr{C}$.) There exists $\Omega = \Omega_{f,r} \in [0, \infty)$ such that, for every structure $\mathfrak{M}$ of $\mathscr{C}$, the number Ω is a bound for $f^{\mathfrak{M}}$ on $B_{M_\lambda^{(s_1)}}(r) \times \cdots \times B_{M_\lambda^{(s_n)}}(r)$.

2. (Local equicontinuity condition for $\mathscr{C}$.) There exists $\Delta = \Delta_{f,r}$: $(0, \infty) \to [0, \infty)$ such that for every structure M of $\mathscr{C}$, the function Δ is a modulus of uniform continuity for $f^{\mathcal{M}}$ on $B_{M_\lambda^{(s_1)}}(r) \times \cdots \times B_{M_\lambda^{(s_n)}}(r)$.

Any collection $(\Omega_{r,f}, \Delta_{r,f} \mid r > 0)$ will be called a family of *moduli of local uniform continuity* for f. A collection $\mathbb{U} = (\Omega_{r,f}, \Delta_{r,f} \mid r > 0, f \in \mathbf{F})$, with f ranging over the collection $\mathbf{F}$ of function symbols of L, will be called a *modulus of uniformity* for L-structures.

Remark 6.7. Clearly, any single L-structure M admits some modulus of uniformity $\mathbb{U}$; however, no single such $\mathbb{U}$ is a modulus of uniformity for every L-structure. This is quite analogous to the fact that every Cauchy sequence is metastable with some uniform rate $E_\bullet$, but no single such rate of uniform metastability applies to all Cauchy sequences (refer to the discussion in Section 1-I).

Let $\mathscr{C}$ be a uniform class of L-structures. Let $(\mathcal{M}_\lambda)_{\lambda \in \Lambda}$ be family of structures in $\mathscr{C}$ such that $\mathcal{M}_\lambda$ is based on $(M_\lambda^{(s)} \mid s \in \mathbf{S})$ for each $\lambda \in \Lambda$. If $f : s_1 \times \cdots \times s_n \to s_0$ is a function symbol of L, then for any ultrafilter $\mathcal{U}$ on Λ we define a function

$$\Big(\prod_{\lambda \in \Lambda} f^{\mathcal{M}_\lambda} \Big)_{\mathcal{U}} : \Big(\prod_{\lambda \in \Lambda} M_\lambda^{(s_1)} \Big)_{\mathcal{U}} \times \cdots \times \Big(\prod_{\lambda \in \Lambda} M_\lambda^{(s_n)} \Big)_{\mathcal{U}} \to \Big(\prod_{\lambda \in \Lambda} M_\lambda^{(s_0)} \Big)_{\mathcal{U}}$$

naturally as follows: If $(x_\lambda^i)_{\lambda \in \Lambda} \in \ell^\infty(M_\lambda, d_\lambda^{(s_i)})_\Lambda$ for $i = 1, \ldots, n$, we let

$$\Big(\prod_{\lambda \in \Lambda} f^{\mathcal{M}_\lambda} \Big)_{\mathcal{U}} \big(((x_\lambda^1)_{\lambda \in \Lambda})_{\mathcal{U}}, \ldots, ((x_\lambda^m)_{\lambda \in \Lambda})_{\mathcal{U}} \big) = \big((f(x_\lambda^1, \ldots, x_\lambda^m))_{\lambda \in \Lambda} \big)_{\mathcal{U}}.$$

$$(5.1)$$

The uniformity of $\mathscr{C}$ implies that if $(x_\lambda^i)_{\lambda \in \Lambda} \in B_r\big(\ell^\infty(M_\lambda, d_\lambda^{(s_i)})_\Lambda\big)$, then $(f(x_\lambda^1, \ldots, x_\lambda^m))_{\lambda \in \Lambda} \in B_\Omega\big(\ell^\infty(M_\lambda, d_\lambda^{(s_i)})_\Lambda\big)$ for some $\Omega > 0$, hence the right-hand side of (5.1) is an element of $(\prod_{\lambda \in \Lambda} X_\lambda)_{\mathcal{U}}$. Thus, if Ω is a uniform bound for $f^{\mathcal{M}_\lambda}$ on $B_{M_\lambda^{(s_1)}}(r_1) \times \cdots \times B_{M_\lambda^{(s_n)}}(r_n)$ for all $\lambda \in \Lambda$, then Ω is also a bound for $(\prod_{\lambda \in \Lambda} f^{\mathcal{M}_\lambda})_{\mathcal{U}}$ on $B_{(\prod_{\lambda \in \Lambda} M_\lambda^{(s_1)})_{\mathcal{U}}}(r_1) \times \cdots \times B_{(\prod_{\lambda \in \Lambda} M_\lambda^{(s_n)})_{\mathcal{U}}}(r_n)$. It is also trivial to verify that, if Δ is a uniform continuity modulus for $f^{\mathcal{M}_\lambda}$ on $B_{M_\lambda^{(s_1)}}(r_1) \times \cdots \times B_{M_\lambda^{(s_n)}}(r_n)$ for all $\lambda \in \Lambda$, then Δ is also a modulus of uniform continuity for $(\prod_{\lambda \in \Lambda} f^{\mathcal{M}_\lambda})_{\mathcal{U}}$ on $B_{(\prod_{\lambda \in \Lambda} M_\lambda^{(s_1)})_{\mathcal{U}}}(r_1) \times \cdots \times B_{(\prod_{\lambda \in \Lambda} M_\lambda^{(s_n)})_{\mathcal{U}}}(r_n)$. Thus, equation (5.1) defines the interpretation of functions in $(\prod_{\lambda \in \Lambda} X_\lambda)_{\mathcal{U}}$ well. The following proposition summarizes the preceding discussion.

Proposition 6.8. *Let $\mathscr{C}$ be a uniform class of L-structures and let $(\mathcal{M}_\lambda)_{\lambda \in \Lambda}$ be a family of structures in $\mathscr{C}$ such that for each $\lambda \in \Lambda$ the structure $\mathcal{M}_\lambda$ is based on $(M_\lambda^{(s)} \mid s \in \mathbf{S})$. If $\mathcal{U}$ is an ultrafilter on Λ, we obtain an L-structure $(\prod_{\lambda \in \Lambda} \mathcal{M}_\lambda)_{\mathcal{U}}$ based on $((\prod_{\lambda \in \Lambda} M_\lambda^{(s)})_{\mathcal{U}} \mid s \in \mathbf{S})$ by interpreting any function symbol f of L in $(\prod_{\lambda \in \Lambda} \mathcal{M}_\lambda)_{\mathcal{U}}$ as $(\prod_{\lambda \in \Lambda} f^{\mathcal{M}_\lambda})_{\mathcal{U}}$.*

Furthermore, any modulus of uniformity for $\mathscr{C}$ is also a modulus of uniformity for $(\prod_{\lambda \in \Lambda} \mathcal{M}_\lambda)_{\mathcal{U}}$.

Definition 6.9. The structure $(\prod_{\lambda \in \Lambda} \mathcal{M}_\lambda)_{\mathcal{U}}$ in Proposition 6.8 is called the $\mathcal{U}$-*ultraproduct* of the family $(\mathcal{M}_\lambda)_{\lambda \in \Lambda}$.

Note that the hypothesis that a class $\mathscr{C}$ is uniform asserts that the collection $(f^{\mathcal{M}} \mid \mathcal{M} \in \mathscr{C})$ of interpretations of a given function symbol f is an equicontinuous and equibounded family, precisely as in the statement of the Arzelà-Ascoli theorem. This classical result is thus subsumed under the fact that any ultraproduct of a uniform family of L-structures is itself an L-structure. Just as the hypotheses of equicontinuity and equiboundedness are both necessary for the conclusion of the Arzelà-Ascoli theorem to hold, it should be clear that ultraproduct structures are defined, in general, only for subfamilies of some *uniform* class of structures—otherwise, the right-hand side of equation (5.1) may fail to define an element lying at finite distance from the anchor $((a_\lambda)_{\lambda \in \Lambda})_{\mathcal{U}}$, or else the function so defined may fail to be continuous.

IV Henson languages and semantics: Formulas and satisfaction

We now focus our attention on the precise connection between metric structures and their ultrapowers and, more generally, between families of metric structures and their ultraproducts. These connections are intimately connected to notions from model theory, a branch of mathematical logic.

In the current literature, there are two formally different but equivalent logical frameworks to study metric structures from a model-theoretic perspective. One of these frameworks is that of *continuous model theory* [BYU10, BYBHU08], which uses real-valued logic, and the other is the logic of approximate truth, introduced in the 1970s by C. W. Henson [Hen75, Hen76] and developed further by Henson and the second author [HI02, Iov14]. We have adopted the latter because, despite its less widespread use in the current literature, it has strong syntactic advantages, as it allows dealing with unbounded metric structures such as Banach spaces in a natural fashion, without having to replace the metric by an equivalent bounded metric.[8]

Let L be a fixed signature. In analogy with languages of traditional (discrete) first-order logic, we construct a language, called a *Henson language*, which is suitable for discussing properties of metric structures. The language consists of syntactic expressions called *positive bounded formulas* of L, or L-*formulas*. These are strings or symbols built from a basic alphabet that includes the following symbols:

· The function symbols of the signature L,

---

[8]For a proof of the equivalence among various formulations, see [Iov01, Iov09, CI14, Cai].

- For each sort index $s \in \mathbf{S}$ of L, a countable collection of symbols called *variables of sort s*, or variables *bound to the sort s*.

- Logical connectives $\vee$ and $\wedge$, and for each positive rational number r, quantifiers $\forall_r$ and $\exists_r$.

- Parentheses and the comma symbol.

First we define the concept of L-term. Intuitively, a term is a string of symbols that may be interpreted by elements of L-structures. Since elements of structures occur inside sorts, each term must have a sort associated with it. Thus we define the concept of *s-valued term*:

Definition 6.10. An *s-valued L-term* is any finite string of symbols that can be obtained by finitely many applications of the following rules of formation:

1. Every variable of sort s is an s-valued term,

2. If f is a function symbol with $f : s_1 \times \cdots \times s_n \to s$ and $t_1, \ldots, t_n$ are such that t_i is an s_i-valued for $i = 1, \ldots, n$, then $f(t_1, \ldots, t_n)$ is an s-valued term.

If t is a term and $x_1, \ldots, x_n$ is a list of variables that contains all the variables occurring in t, we write t as $t(x_1, \ldots, x_n)$.

A *real-valued term* is an $s_\mathbb{R}$-valued term. A *term* is string that is an s-valued term for some $s \in \mathbf{S}$.

Definition 6.11. Let $\mathcal{M}$ be an L-structure based on $(M^{(s)} \mid s \in \mathbf{S})$ and let $t(x_1, \ldots, x_n)$ be an L-term, where s_i is a variable of sort s_i, for $i = 1, \ldots, n$. If $a_1, \ldots, a_n$ are elements of $\mathcal{M}$ such that a_i is of sort s_i, for $i = 1, \ldots, n$, the *evaluation* of t in $\mathcal{M}$ at $a_1, \ldots, a_n$, denoted $t^{\mathcal{M}}[a_1, \ldots, a_n]$, is defined by induction on the length of t as follows:

1. If t is x_i, then $t^{\mathcal{M}}[a_1, \ldots, a_n]$ is a_i,

2. If t is $f(t_1, \ldots, t_n)$, where f is a function symbol and $t, \ldots, t_n$ are terms of lower length, then $t^{\mathcal{M}}[a_1, \ldots, a_n]$ is

$$f^{\mathcal{M}}(t_1^{\mathcal{M}}[a_1, \ldots, a_n], \ldots, t_n^{\mathcal{M}}[a_1, \ldots, a_n]).$$

As an addendum to (2), by a slight abuse of notation, if f is a nullary $M^{(s)}$-valued function symbol (i.e., a constant symbol) we usually interpret f as be the (unique) element $a \in M^{(s)}$ in the range of the function $f^{\mathcal{M}}$.

Notation 6.12. Recall from the definition of signature that every signature L must include a special sort index $s_\mathbb{R}$ and constant symbol for each rational number. Informally we will identify each rational number with its constant symbol in L. More generally, since L includes function symbols for the addition and multiplication in $\mathbb{R}$, for every polynomial $p(x_1, \ldots, x_n) \in \mathbb{Q}[x_1, \ldots, x_n]$

there exists a real-valued L-term $t(x_1, \ldots, x_n)$ such that $t^{\mathcal{M}}[a_1, \ldots, a_n] = p(a_1, \ldots, a_n)$ for any L-structure $\mathcal{M}$ and $a_1, \ldots, a_n \in \mathbb{R}$. We will identify t and p. Thus, if $t_1, \ldots, t_n$ are L terms and $p(x_1, \ldots, x_n) \in \mathbb{Q}[x_1, \ldots, x_n]$, we may refer to the L-term $p(t_1, \ldots, t_n)$.

Definition 6.13. A *positive bounded L-formula* (or simply an *L-formula*) is any finite string of symbols that can be obtained by finitely many applications of the following rules of formation:

1. If t is a real-valued L-term and r is a rational number, then the expressions

$$t \leq r \qquad \text{and} \qquad t \geq r$$

   are L-formulas. These are the *atomic L-formulas*.

2. If φ and ψ are positive bounded L-formulas, then the expressions

$$(\varphi \wedge \psi) \qquad \text{and} \qquad (\varphi \vee \psi)$$

   are positive bounded L-formulas. These are the *conjunction* and *disjunction*, respectively, of φ and ψ.

3. If φ is positive bounded L-formula, r is a positive rational, and x is a variable, then the expressions

$$\exists_r x \, \varphi \qquad \text{and} \qquad \forall_r x \, \varphi$$

   are positive bounded L-formulas.

Notation 6.14. Whenever possible, we shall omit parentheses according to the usual syntactic simplification rules. If t is a real-valued term and r_1, r_2 are rational numbers, we will write $r_1 \leq t \leq r_2$ as an abbreviation of the conjunction $(r_1 \leq t \wedge t \leq r_2)$. Similarly, we regard $t = r$ as an abbreviation of the conjunction $(t \leq r \wedge t \geq r)$. If t_1 and t_2 are real-valued terms, we regard $t_1 \leq t_2$ as an abbreviation of $0 \leq t_2 - t_1$ and, if t_1, t_2 are s-valued terms, we regard the expression $t_1 = t_2$ as an abbreviation of $d(t_1, t_2) \leq 0$, where d is the function symbol designating the metric of the sort indexed by s. If $\varphi_1, \ldots, \varphi_n$ are formulas, we may write $\bigwedge_{i=1}^n \varphi_i$ and $\bigvee_{i=1}^n \varphi_i$ as abbreviations of $\varphi_1 \wedge \cdots \wedge \varphi_n$ and $\varphi_1 \vee \cdots \vee \varphi_n$, respectively. If t is an s-valued term and d, a are the designated function symbol and constant symbol, respectively, for the metric and the anchor of this sort, we shall regard the expression $t \in B_r$ as an abbreviation of the formula $d(t, a) \leq r$.

Definition 6.15. A *subformula* of a formula φ is a substring of φ that is itself a formula. If φ is a formula and x is a variable, we say that x occurs *free* in φ if there is at least one occurrence of x in φ that is not within any subformula of the form $\forall_r x \, \varphi$ or $\exists_r x \, \varphi$. If $x_1, \ldots, x_n$ are variables, we write φ as $\varphi(x_1, \ldots, x_n)$ if all the free variables of φ are among $x_1, \ldots, x_n$. A positive bounded *L-sentence* is a positive bounded formula without any free variables.

The definition below introduces the most basic concept of model theory, namely, the satisfaction relation $\models$ between structures and formulas. Intuitively, if $\mathcal{M}$ is an L structure, $\varphi(x_1, \ldots, x_n)$ is an L- formula and $a_1, \ldots, a_n$ are elements of $\mathcal{M}$,

$$\mathcal{M} \models \varphi[a_1, \ldots, a_n]$$

means that φ is true $\mathcal{M}$ if $x_1, \ldots x_n$ are interpreted as $a_1, \ldots, a_n$, respectively. Evidently, for this to be meaningful, the variable x_i must be of the same sort as the element a_i, for $i = 1, \ldots, n$.

Definition 6.16. Let $\mathcal{M}$ be an L-structure based on $(M^{(s)} \mid s \in \mathbf{S})$ and let $\varphi(x_1, \ldots, x_n)$ be an L-formula, where s_i is a variable of sort s_i, for $i = 1, \ldots, n$. If $a_1, \ldots, a_n$ are elements of $\mathcal{M}$ such that a_i is of sort s_i, for $i = 1, \ldots, n$, the *(discrete) satisfaction relation* $\mathcal{M} \models \varphi[a_1, \ldots, a_n]$ is defined inductively as follows:

1. If $\varphi(x_1, \ldots, x_n)$ is $t \leq r$, where $t = t(x_1, \ldots, x_n)$ is a real-valued term and r is rational, then $\mathcal{M} \models \varphi[a_1, \ldots, a_n]$ if and only if $t^{\mathcal{M}}[a_1, \ldots, a_n] \leq r$.

2. If $\varphi(x_1, \ldots, x_n)$ is $t \geq r$, where $t = t(x_1, \ldots, x_n)$ is a real-valued term and r is rational, then $\mathcal{M} \models \varphi[a_1, \ldots, a_n]$ if and only if $t^{\mathcal{M}}[a_1, \ldots, a_n] \geq r$.

3. If $\varphi(x_1, \ldots, x_n)$ is $(\psi_1 \wedge \psi_2)$, where ψ_1 and ψ_2 are L-formulas, then $\mathcal{M} \models \varphi[a_1, \ldots, a_n]$ if and only if

$$\mathcal{M} \models \psi_1[a_1, \ldots, a_n] \quad \text{and} \quad \mathcal{M} \models \psi_2[a_1, \ldots, a_n].$$

4. If $\varphi(x_1, \ldots, x_n)$ is $(\psi_1 \vee \psi_2)$, where ψ_1 and ψ_2 are L-formulas, then $\mathcal{M} \models \varphi[a_1, \ldots, a_n]$ if and only if

$$\mathcal{M} \models \psi_1[a_1, \ldots, a_n] \quad \text{or} \quad \mathcal{M} \models \psi_2[a_1, \ldots, a_n].$$

5. If $\varphi(x_1, \ldots, x_n)$ is $\exists_r x\, \psi(x, x_1, \ldots, x_n)$, where r is a positive rational, x is a variable of sort s, and $\psi(x, x_1, \ldots, x_n)$ is an L-formula, then $\mathcal{M} \models \varphi[a_1, \ldots, a_n]$ if and only if

$$\mathcal{M} \models \psi[a, a_1, \ldots, a_n] \quad \text{for some } a \in B_{M^{(s)}}[r].$$

6. If $\varphi(x_1, \ldots, x_n)$ is $\forall_r x\, \psi(x, x_1, \ldots, x_n)$, where r is a positive rational, x is a variable of sort s, and $\psi(x, x_1, \ldots, x_n)$ is an L-formula, then $\mathcal{M} \models \varphi[a_1, \ldots, a_n]$ if and only if

$$\mathcal{M} \models \psi[a, a_1, \ldots, a_n] \quad \text{for every } a \in B_{M^{(s)}}(r).$$

If $\mathcal{M} \models \varphi[a_1, \ldots, a_n]$, we say that $\mathcal{M}$ *satisfies* φ at $a_1, \ldots, a_n$.

Note that universal (resp., existential) quantification is interpreted in open (resp., closed) balls.

Definition 6.17. If φ is a set of formulas, we denote it by $\varphi(x_1, \ldots, x_n)$ if all the free variables of all the formulas in φ are among $x_1, \ldots, x_n$. If $\mathcal{M}$ is a structure and $a_1, \ldots, a_n$ are elements of $\mathcal{M}$, we write $\mathcal{M} \models \varphi[a_1, \ldots, a_n]$ if $\mathcal{M} \models \varphi[a_1, \ldots, a_n]$ for every $\varphi \in \varphi$.

V Approximations and approximate satisfaction

We begin this subsection by defining a strict partial ordering of positive bounded L-formulas, namely the relation "ψ *is an approximation of* φ", denoted $\varphi \Rightarrow \psi$ (or $\psi \Leftarrow \varphi$). Roughly speaking, this means that ψ arises when every estimate occurring in φ is relaxed. The formal definition of the approximation relation is by induction on the complexity of φ, as given by the following table.

If φ is:	The approximations of φ are:
$t \le r$	$t \le r'$ where $r' > r$
$t \ge r$	$t \ge r'$ where $r' < r$
$(\xi \wedge \psi)$	$(\xi' \wedge \psi')$ where $\xi \Rightarrow \xi'$ and $\psi \Rightarrow \psi'$
$(\xi \vee \psi)$	$(\xi' \vee \psi')$ where $\xi \Rightarrow \xi'$ and $\psi \Rightarrow \psi'$
$\exists_r x\, \psi$	$\exists_{r'} x\, \psi'$ where $\psi \Rightarrow \psi'$ and $r' > r$
$\forall_r x\, \psi$	$\forall_{r'} x\, \psi'$ where $\psi \Rightarrow \psi'$ and $r' < r$

Definition 6.18. Let $\mathcal{M}$ be an L-structure based on $(M^{(s)} \mid s \in \mathbf{S})$ and let $\varphi(x_1, \ldots, x_n)$ be an L-formula, where x_i is a variable of sort s_i, for $i = 1, \ldots, n$. If $a_1, \ldots, a_n$ are elements of $\mathcal{M}$ such that a_i is of sort s_i, for $i = 1, \ldots, n$, we say that $\mathcal{M}$ *approximately satisfies* φ at $a_1, \ldots, a_n$, and write

$$\mathcal{M} \approx\!\!\!\mid \varphi[a_1, \ldots, a_n],$$

if

$$\mathcal{M} \models \varphi'[a_1, \ldots, a_n], \quad \text{for every } \varphi' \Leftarrow \varphi.$$

If $\varphi(x_1, \ldots, x_n)$ is a set of formulas, we say that $\mathcal{M}$ *approximately satisfies* φ at $a_1, \ldots, a_n$, and write $\mathcal{M} \approx\!\!\!\mid \varphi[a_1, \ldots, a_n]$, if $\mathcal{M} \approx\!\!\!\mid \varphi[a_1, \ldots, a_n]$ for every $\varphi \in \varphi$.

Clearly, approximate satisfaction is a weaker notion of truth than discrete satisfaction. For nondiscrete metric space structures, the approximate satisfaction relation $\approx\!\!\!\mid$ is the "correct" notion of truth, in the sense that for these structures it is not the notion of discrete satisfaction, but rather that of approximate satisfaction, that yields a well-behaved model theory. (For discrete structures, the two relations are clearly equivalent.)

The negation connective ("$\neg$") is not allowed in positive bounded formulas. However, for every positive bounded formula φ there is a positive bounded formula $\overset{\mathrm{w}}{\neg}\, \varphi$, called the *weak negation of* φ, that plays a role analogous to that played by the negation of φ.

Definition 6.19. The unary pseudo-connective $\overset{\text{w}}{\neg}$ of *weak negation* of L-formulas is defined recursively as follows.

If φ is:	$\overset{\text{w}}{\neg}\varphi$ is:
$t \leq r$	$t \geq r$
$t \geq r$	$t \leq r$
$(\xi \wedge \psi)$	$(\overset{\text{w}}{\neg}\xi \vee \overset{\text{w}}{\neg}\psi)$
$(\xi \vee \psi)$	$(\overset{\text{w}}{\neg}\xi \wedge \overset{\text{w}}{\neg}\psi)$
$\forall_r x\, \psi$	$\exists_r x\, \overset{\text{w}}{\neg}\psi$
$\exists_r x\, \psi$	$\forall_r x\, \overset{\text{w}}{\neg}\psi.$

Remarks 6.20.

1. If φ, φ' are positive bounded formulas, then $\varphi \Rrightarrow \varphi'$ if and only if $\overset{\text{w}}{\neg}\varphi' \Rrightarrow \overset{\text{w}}{\neg}\varphi$.

2. Although languages for metric structures do not include a connective interpreted as the implication "*if φ then ψ*", a formula of the form "$\overset{\text{w}}{\neg}\varphi \vee \psi$" may be regarded as a weak conditional.

3. If $\mathcal{M}$ is an L-structure and $\varphi(x_1, \ldots, x_n)$ is a positive bounded L-formula such that $\mathcal{M} \not\models \varphi[a_1, \ldots, a_n]$, then $\mathcal{M} \models \overset{\text{w}}{\neg}\varphi[a_1, \ldots, a_n]$. If φ' is an approximation of φ such that $\mathcal{M} \models \overset{\text{w}}{\neg}\varphi'[a_1, \ldots, a_n]$, then $\mathcal{M} \not\models \varphi[a_1, \ldots, a_n]$.

Proposition 6.21. *Let $\mathcal{M}$ be an L-structure based on $(M^{(s)} \mid s \in \mathbf{S})$, let $\varphi(x_1, \ldots, x_n)$ be an L-formula, where x_i is a variable of sort s_i, for $i = 1, \ldots, n$, and let $a_1, \ldots, a_n$ be elements of $\mathcal{M}$ such that a_i is of sort s_i, for $i = 1, \ldots, n$. Then, $\mathcal{M} \not\approx \varphi[a_1, \ldots, a_n]$ if and only there exists a formula $\varphi' \Lleftarrow \varphi$ such that $\mathcal{M} \not\approx \overset{\text{w}}{\neg}\varphi'[a_1, \ldots, a_n]$.*

Proof. In order to simplify the nomenclature, let us suppress the lists $x_1, \ldots, x_n$ and $a_1, \ldots, a_n$ from the notation.

If $\mathcal{M} \not\approx \varphi$, there exists $\varphi' \Lleftarrow \varphi$ such that $\mathcal{M} \not\models \varphi'$. We have $\mathcal{M} \models \overset{\text{w}}{\neg}\varphi'$ and hence $\mathcal{M} \approx \overset{\text{w}}{\neg}\varphi'$. Conversely, assume that there exists $\varphi' \Lleftarrow \varphi$ such that $\mathcal{M} \approx \overset{\text{w}}{\neg}\varphi'$. Take sentences ψ, ψ' such that $\varphi \Rrightarrow \psi \Rrightarrow \psi' \Rrightarrow \varphi'$. Then $\mathcal{M} \models \overset{\text{w}}{\neg}\psi'$ (by Remark 6.20-(3)) and hence $\mathcal{M} \not\models \psi$, so $\mathcal{M} \not\approx \varphi$. $\square$

VI   Theories, elementary equivalence, and elementary substructures

In this section we include some basic definitions from model theory.

Definition 6.22. Let L be a signature.

1. An *L-theory* (or simply a *theory*) is a set of L-sentences.

2. If T is a theory and $\mathcal{M} \approx T$, we say that $\mathcal{M}$ is a *model* of T. A theory is *satisfiable* (or *consistent*) if it has a model. The class of all models of a theory T is denoted $\mathrm{Mod}(T)$.

3. An *axiomatizable class* $\mathscr{C}$ (or *elementary class*) is one that consists of all the models of a fixed theory T. We say that $\mathscr{C}$ *is L-axiomatizable*, or $\mathscr{C}$ *is axiomatizable by* T, when the language or the theory need to be specified.

4. An L-theory T is *uniform* if the class of all models of T is uniform. (See Proposition 6.24.)

5. The *complete L-theory of a structure* $\mathcal{M}$, denoted $\mathrm{Th}(\mathcal{M})$, is the set of all L-sentences φ such that $\mathcal{M} \approx \varphi$. A *complete L-theory* is the complete L-theory of any L-structure $\mathcal{M}$.

6. The *complete L-theory of class* $\mathscr{C}$ *of L-structures* is $\mathrm{Th}(\mathscr{C}) = \bigcap_{\mathcal{M} \in \mathscr{C}} \mathrm{Th}(\mathcal{M})$.

7. Two L-structures $\mathcal{M}, \mathcal{N}$ are *elementarily equivalent*, written $\mathcal{M} \equiv \mathcal{N}$, if they have the same complete theory, i.e., if

$$\mathcal{M} \approx \varphi \quad \Leftrightarrow \quad \mathcal{N} \approx \varphi, \qquad \text{for every } L\text{-sentence } \varphi.$$

8. If $\mathcal{M}$ and $\mathcal{N}$ are L-structures and $\mathcal{M}$ is a substructure of $\mathcal{N}$, we say that $\mathcal{M}$ is an *elementary substructure* of $\mathcal{N}$, and we write $\mathcal{M} \prec \mathcal{N}$, if whenever $a_1, \ldots, a_n$ are elements of $\mathcal{M}$ and $\varphi(x_1, \ldots, x_n)$ is an L-formula such that a_i is of the same sort as x_i, for $i = 1, \ldots, n$, we have

$$\mathcal{M} \approx \varphi[a_1, \ldots, a_n] \quad \Leftrightarrow \quad \mathcal{N} \approx \varphi[a_1, \ldots, a_n].$$

Remarks 6.23. 1. When $\mathcal{M}$ is an L-structure, the interpretation of each function symbol is locally bounded and locally uniformly continuous by definition, hence $\mathrm{Th}(\mathcal{M})$ is necessarily a uniform theory.

2. Any satisfiable theory T admits some extension to a uniform theory $T' = \mathrm{Th}(\mathcal{M})$, where $\mathcal{M}$ is any model of T. Neither the extension nor a modulus of uniformity thereof is uniquely determined by T in general.

3. If T' extends a $\mathbb{U}$-uniform theory T, then T' is also $\mathbb{U}$-uniform.

Proposition 6.24. *Let L be a signature. Given a modulus of uniformity $\mathbb{U}$, the class of L-structures $\mathcal{M}$ such that $\mathbb{U}$ is a modulus of uniformity for $\mathcal{M}$ is axiomatizable.*

Proof. Let $\mathbb{U} = (\Omega_{r,f}, \Delta_{r,f} \mid r > 0, f \in \mathbf{F})$ be a modulus of uniformity. For a function symbol $f : s_1 \times \cdots \times s_n \to s_0$ and rational numbers $u, v, w > 0$, consider the L-sentences

$$\chi_{f,u,v} \; : \; \forall_u x_1 \ldots \forall_u x_n \big(d(f(x_1, \ldots, x_n), a_0) \leq v \big),$$

$$\xi_{f,u,v,w} \; : \; \forall_u x_1 \forall_u y_1 \ldots \forall_u x_n \forall_u y_n \big(\overline{d}(\overline{x}, \overline{y}) \geq v \lor d(f(\overline{x}), f(\overline{y})) \leq w \big),$$

where $\overline{x}, \overline{y}$ denote the n-tuples $x_1, \ldots, x_n$ and $y_1, \ldots, y_n$, respectively, $\overline{d}$ denotes the supremum distance on $\mathcal{M}^{(s_1)} \times \cdots \times \mathcal{M}^{(s_n)}$, and a_0 the constant symbol for the anchor of the sort indexed by s_0. It should be clear that for any signature L, the class of L-structures $\mathcal{M}$ such that $\mathbb{U}$ is a modulus of uniformity for $\mathcal{M}$ is axiomatized by the union of the following sets of sentences:

$$\{ \chi_{f,u,v} \mid f \in \mathbf{F}, \, u, v \in \mathbb{Q}_+, \, u < r, \, v > \Omega_{r,f} \text{ for some } r \in (0, \infty) \},$$

$$\{ \xi_{f,u,v,w} \mid f \in \mathbf{F}, \, u, v \in \mathbb{Q}_+, \, u < r, \, v < t, \, w > \Delta_{r,f}(t) \text{ for some } r, t \in (0, \infty) \} \square$$

It is easy to construct examples showing that the meta-property "*the class $\mathscr{C}$ is uniform*" (without specifying a modulus of uniformity) is not axiomatizable in general.

The following proposition gives simple conditions to verify $\equiv$ and $\prec$.

Proposition 6.25. *Let $\mathcal{M}$ and $\mathcal{N}$ be L-structures.*

(1) $\mathcal{M} \equiv \mathcal{N}$ *if and only if for every L-sentence φ,*

$$\mathcal{M} \models \varphi \quad \Rightarrow \quad \mathcal{N} \models \varphi.$$

(2) (Tarski-Vaught test for $\prec$). *A substructure $\mathcal{M}$ of the structure $\mathcal{N}$ is an elementary substructure if and only if the following condition holds: If $\varphi(x_1, \ldots, x_n, y)$ is an L-formula, $a_1, \ldots, a_n$ are elements of $\mathcal{M}$ with a_i of the same sort as x_i for $i = 1, \ldots, n$ such that $\mathcal{N} \models \exists_r y \, \varphi[a_1, \ldots, a_n]$ for some $r > 0$, then $\mathcal{M} \models \exists_r y \, \varphi[a_1, \ldots, a_n]$.*

Proof. For part (1), the direct implication follows by definition of elementary equivalence, while the converse follows from Proposition 6.21. For (2), the direct implication is trivial, and the converse follows by induction on the complexity of φ. $\square$

The following is an immediate consequence of Proposition 6.25-(2):

Proposition 6.26 (Downward Löwenheim-Skolem Theorem). *Let $\mathcal{M}$ be a structure and let A be a set of elements of $\mathcal{M}$. Then there exists a substructure $\mathcal{M}_0$ of $\mathcal{M}$ such that*

- $\mathcal{M}_0 \prec \mathcal{M}$,

- *Every element of A is an element of $\mathcal{M}_0$,*

- $\mathrm{card}(\mathcal{M}_0) \leq \mathrm{card}(A) + \mathrm{card}(L)$.

Definition 6.27. An *elementary chain* is a family $(\mathcal{M}_i)_{i \in I}$ of structures, indexed by some linearly ordered set $(I, <)$, such that $\mathcal{M}_i \prec \mathcal{M}_j$ if $i < j$.

Another useful consequence of Proposition 6.25 is the elementary chain property:

Proposition 6.28 (The Elementary Chain Property). *If $(\mathcal{M}_i)_{i \in I}$ is an elementary chain, then $\bigcup_{i \in I} \mathcal{M}_i$ is an elementary extension of $\mathcal{M}_j$ for every $j \in I$.*

Proof. By Proposition 6.25. $\square$

VII   Łoś' Theorem

The following fundamental theorem, proved by J. Łoś in the 1950s [Łoś55], intuitively states that a formula φ is satisfied by an ultraproduct of a family $(\mathcal{M}_\lambda)_{\lambda \in \Lambda}$ of structures if and only if every approximation of φ is satisfied by almost all of the structures $\mathcal{M}_\lambda$. Łoś proved the theorem for discrete structures (i.e., traditional first-order logic), where approximations are not needed; however, essentially the same argument holds for arbitrary metric structures.

Theorem 6.29 (Łoś' Theorem for Metric Structures). *Let L be a signature and let $(\mathcal{M}_\lambda)_{\lambda \in \Lambda}$ be a family of L-structures in a uniform class such that for each $\lambda \in \Lambda$ the structure $\mathcal{M}_\lambda$ is based on $(M_\lambda^{(s)} \mid s \in \mathbf{S})$. Let $(a_{1,\lambda})_{\lambda \in \Lambda}, \ldots, (a_{n,\lambda})_{\lambda \in \Lambda}$ be such that of $(a_{i,\lambda})_{\lambda \in \Lambda} \in \ell^\infty(M_\lambda^{(s_i)})_\Lambda$ for $i = 1, \ldots, n$ and let $\varphi(x_1, \ldots, x_n)$ be an L-formula such that x_i is of sort s_i. Then, for any ultrafilter $\mathcal{U}$ on Λ,*

$$\Big(\prod_{\lambda \in \Lambda} \mathcal{M}_\lambda \Big)_\mathcal{U} \models \varphi[\,((a_{1,\lambda})_{\lambda \in \Lambda})_\mathcal{U}, \ldots, ((a_{n,\lambda})_{\lambda \in \Lambda})_\mathcal{U}\,]$$

if and only if for every approximation $\varphi' > \varphi$, the set

$$\{\, \lambda \in \Lambda \mid \mathcal{M}_\lambda \models \varphi'[a_{1,\lambda}, \ldots, a_{n,\lambda}] \,\}$$

is $\mathcal{U}$-large.

Proof. Using the definition of $\mathcal{U}$-ultraproduct of structures and the interpretation of function symbols therein (Proposition 6.8 and Definition 6.9), Łoś' Theorem follows by induction on the complexity of φ. $\square$

An important corollary of Łoś' theorem is the special case when all the structures $\mathcal{M}_\lambda$ equal the same structure $\mathcal{M}$. In this case, the $\mathcal{U}$-ultraproduct $(\prod_{\lambda \in \Lambda} \mathcal{M}_\lambda)_\mathcal{U}$ is the $\mathcal{U}$-ultrapower of $\mathcal{M}$. Hence we have the following:

Corollary 6.30. *Every metric structure is an elementary substructure of its ultrapowers.*

VIII Compactness

The Compactness Theorem is arguably the most distinctive theorem of first-order logic.

For a set φ of formulas, let $\varphi_\approx$ denote the set of all approximations of formulas in φ.

Theorem 6.31 (Compactness Theorem). *Let $\mathscr{C}$ be a uniform class of structures and let T be an L-theory. If every finite subset of $T_\approx$ has a model in $\mathscr{C}$ in the semantics of discrete satisfaction, then T has a model in the semantics of approximate satisfaction. This model may be taken to be an ultraproduct of structures in $\mathscr{C}$ that admits the same modulus of uniformity as the class $\mathscr{C}$.*

Proof. Let Λ be the set of finite subsets of $T_\approx$, and for each λ in Λ, let $\mathcal{M}_\lambda$ be a model in $\mathscr{C}$ of all the sentences in λ in the semantics of discrete satisfaction. For every finite subset φ of $T_\approx$, let $\Lambda_{\supseteq\varphi}$ be the set of all $\lambda \in \Lambda$ such that $\lambda \supseteq \varphi$. Then $\mathcal{M}_\lambda \models \varphi$ for every $\lambda \in \Lambda_{\supseteq\varphi}$. Note that the collection of subsets of Λ of the form $\Lambda_{\supseteq\varphi}$ is closed under finite intersections since $\Lambda_{\supseteq\varphi} \cap \Lambda_{\supseteq\Psi} = \Lambda_{\supseteq\varphi\cup\Psi}$. Let $\mathcal{U}$ be an ultrafilter on Λ extending this collection. Then, by Łoś' Theorem (Theorem 6.29), we have $(\prod_{\lambda\in\Lambda} \mathcal{M}_\lambda)_\mathcal{U} \approx T$. Furthermore, $(\prod_{\lambda\in\Lambda} \mathcal{M}_\lambda)_\mathcal{U}$ admits the same modulus of uniformity as the family $(\mathcal{M}_\lambda)_{\lambda\in\Lambda}$, by Proposition 6.8. $\square$

The following corollary amounts to a restatement of the Compactness Theorem that does not explicitly mention approximation of formulas.

Corollary 6.32. *If Υ is a uniform theory and T is any collection of sentences such that every finite subset of T is approximately satisfied by a model of Υ, then $\Upsilon \cup T$ has a model.*

Proof. Every finite subset φ of $(\Upsilon \cup T)_\approx$ must be a subset of $(\Upsilon \cup T')_\approx$ for some finite subset $T' \subseteq T$. By hypothesis, $\Upsilon \cup T'$ has a model, which is thus also a model of φ in the semantics of discrete satisfaction. By the Compactness Theorem (Theorem 6.31), $\Upsilon \cup T$ has a model. $\square$

Remark 6.33. Fix a signature L and a satisfiable uniform L-theory Υ. Let $\mathscr{C} = \mathrm{Mod}(\Upsilon)$ be the class of all models of Υ. For any set φ of L-formulas, let $\mathrm{Mod}_\mathscr{C}(\varphi)$ the subclass of $\mathscr{C}$ consisting of models of φ. Corollary 6.32 is equivalent to the following statement: If the collection $(\mathrm{Mod}_\mathscr{C}(\varphi) \mid \varphi \in T)$ has the finite intersection property, then

$$\bigcap_{\varphi\in T} \mathrm{Mod}_\mathscr{C}(\varphi) \neq \emptyset.$$

Thus, if we topologize $\mathscr{C}$ by letting the classes of the form $\mathrm{Mod}_\mathscr{C}(\varphi)$ with $\varphi \in T$ be the basic closed sets, then $\mathscr{C}$ is compact. Moreover, if $(\mathcal{M}_\lambda)_{\lambda\in\Lambda}$ is any family in $\mathscr{C}$ and $\mathcal{U}$ is an ultrafilter on Λ, then the ultraproduct $(\prod_{\lambda\in\Lambda} \mathcal{M}_\lambda)_\mathcal{U}$ is exactly the $\mathcal{U}$-limit of $(\mathcal{M}_\lambda)_{\lambda\in\Lambda}$ in this topology. This explains why the Compactness Theorem is so named, as well as the naturality of the ultraproduct construction. We emphasize, however, that the class $\mathscr{C}$ so topologized must be uniform in order to be compact. More generally, if $\mathscr{C}'$ is any uniform class endowed with the above topology, then $\mathscr{C}'$ is a relatively compact subset of the class $\mathscr{C} = \mathrm{Mod}(\Upsilon)$ axiomatized by the uniform theory $\Upsilon = \mathrm{Th}(\mathscr{C})$.

Now we present three useful applications of the Compactness Theorem. The first one (Corollary 6.34) gives a finitary condition for a theory to be of the form $\mathrm{Th}(\mathcal{M})$ for some structure $\mathcal{M}$. The second one (Corollary 6.37), states that any two models of a complete theory T can be jointly elementarily embedded in a single model of T.

Recall (Definition 6.22) that the complete theory of a structure $\mathcal{M}$ is denoted $\mathrm{Th}(\mathcal{M})$.

Corollary 6.34. *The following conditions are equivalent for a theory T:*

(1) *The theory T is complete, i.e., there exists a structure $\mathcal{M}$ such that $T = \mathrm{Th}(\mathcal{M})$.*

(2) (a) *There exists a uniform theory Υ such that every finite subset of $\Upsilon \cup T$ is satisfiable, and*

    (b) *For every L-sentence φ, if $\varphi \notin T$, then there exists $\varphi' \Leftarrow \varphi$ such that $\overset{w}{\neg} \varphi' \in T$.*

Proof. We only have to prove (2)$\Rightarrow$(1). By part (a) of (2) and Corollary 6.32, T has a model $\mathcal{M}$ whose theory $\mathrm{Th}(\mathcal{M})$ extends T. Part (b) of (2) gives $\mathrm{Th}(\mathcal{M}) \subseteq T$, proving (1). $\qquad\square$

Notation 6.35. If $\mathcal{M}$ is an L-structure and $A = (a_i)_{i \in I}$ is an indexed family of elements of $\mathcal{M}$, we denote by $(\mathcal{M}, a_i \mid i \in I)$ the expansion of $\mathcal{M}$ that results from adding a distinct constant for a_i, for each $i \in I$. In particular, if A is a set of elements of $\mathcal{M}$, we denote by

$$(\mathcal{M}, a \mid a \in A)$$

the expansion of $\mathcal{M}$ that results from adding a constant for each $a \in A$. If, in this case, C is a set of constant symbols not already in L that includes one constant symbol designating each constant $a \in A$, we informally refer to the signature $L[C]$ (see Definition 6.3) as $L[A]$. Thus, we informally call the preceding expansion of $\mathcal{M}$ an "$L[A]$-structure". If A consists of all the elements of $\mathcal{M}$, we write $(\mathcal{M}, a \mid a \in A)$ as $(\mathcal{M}, a \mid a \in \mathcal{M})$ and denote the expanded signature by $L[\mathcal{M}]$.

Definition 6.36. Let $\mathcal{M}$ and $\mathcal{N}$ be L-structures. An *elementary embedding* of $\mathcal{M}$ into $\mathcal{N}$ is a map e that assigns to each element a of $\mathcal{M}$ an element $e(a)$ of $\mathcal{N}$ such that, whenever $a_1, \ldots, a_n$ are elements of $\mathcal{M}$ and $\varphi(x_1, \ldots, x_n)$ is an L-formula such that a_i is of the same sort as x_i, for $i = 1, \ldots, n$, we have

$$\mathcal{M} \approx \varphi[a_1 \ldots, a_n] \quad \Leftrightarrow \quad \mathcal{N} \approx \varphi[e(a_1) \ldots, e(a_n)].$$

Note that e is an elementary embedding of $\mathcal{M}$ into $\mathcal{N}$ if and only if the $L[A]$-structures $(\mathcal{M}, a \mid a \in A)$ and $(\mathcal{N}, e(a) \mid a \in A)$ are elementarily equivalent.

Corollary 6.37. *The following conditions are equivalent for two L-structures $\mathcal{M}, \mathcal{M}'$.*

(1) $\mathcal{M} \equiv \mathcal{M}'$.

(2) *There exists a structure $\mathcal{N}$ such that $\mathcal{M} \prec \mathcal{N}$ and there is an elementary embedding of $\mathcal{M}'$ into $\mathcal{N}$. Moreover, $\mathcal{N}$ can be taken to be an ultrapower of $\mathcal{M}$.*

Proof. It suffices to prove the direct implication, since the inverse is clear. Assume $\mathcal{M} \equiv \mathcal{M}'$. Let T be the complete $L[\mathcal{M}]$-theory of $\mathcal{M}$ and T' the complete $L[\mathcal{M}']$-theory of $\mathcal{M}'$. Both T and T' are uniform theories, by Remark 6.23. First we show that every finite subset of $T'_{\approx} \cup T$ has a model. Since T' is closed under conjunctions, it suffices to show that $T \cup \{\psi\}$ has a model whenever $\psi \in T'_{\approx}$. Any given formula in $T'_{\approx}$ is of the form $\varphi'(c_{b_1}, \ldots, c_{b_n})$, where φ' is an approximation of a formula φ in T' and c_{b_i} is a constant for an element b_i of f $\mathcal{M}'$, for $i = 1, \ldots, n$. Since $\mathcal{M}' \approx \varphi[b_1, \ldots, b_n]$ by assumption, then there exists $r > 0$ such that $\mathcal{M}' \approx \exists_r x_1 \ldots \exists_r x_n \, \varphi(x_1, \ldots, x_n)$. Now, since $\mathcal{M} \equiv \mathcal{M}'$ and $\varphi' \Leftarrow \varphi$, the semantics of approximate satisfaction ensure the existence of elements $a_1, \ldots, a_n$ of $\mathcal{M}$ such that $\mathcal{M} \approx \varphi'[a_1, \ldots, a_n]$; hence, $(\mathcal{M}, a \mid a \in \mathcal{M})$ admits an expansion to a model $\widetilde{\mathcal{M}}$ of $T \cup \{\psi\}$ simply by letting $b_1^{\widetilde{\mathcal{M}}} := a_1$, $\ldots$, $b_n^{\widetilde{\mathcal{M}}} := a_n$. By the Compactness Theorem 6.31, $T \cup T'$ has a model. Let

$$(\mathcal{N}, \widetilde{a}, \widetilde{b} \mid a \in \mathcal{M}, b \in \mathcal{M}')$$

be a model of this theory. The maps $a \mapsto \widetilde{a}$ and $b \mapsto \widetilde{b}$ are elementary embeddings of $\mathcal{M}$ and $\mathcal{M}'$, respectively, into $\mathcal{N}$. Without loss of generality, we may assume both that $\mathcal{N}$ is an ultrapower of $\mathcal{M}$, and $\widetilde{a} = a$ for all elements a of $\mathcal{M}$. This proves (2). $\qquad\square$

Essentially the same argument used to prove Corollary 6.37 proves the following result:

Corollary 6.38. *Given any family $(\mathcal{M}_i)_{i \in I}$ of elementarily equivalent L-structures there exists an L-structure $\mathcal{N}$ such that, for each $i \in I$, there is an elementary embedding of $\mathcal{M}_i$ into $\mathcal{N}$.*

Proposition 6.39. *If $\mathcal{M}, \mathcal{N}$ are elementarily equivalent structures, then there exist structures $\hat{\mathcal{M}}, \hat{\mathcal{N}}$ such that $\mathcal{M} \prec \hat{\mathcal{M}}$, $\mathcal{N} \prec \hat{\mathcal{N}}$, and $\hat{\mathcal{M}}$ is isomorphic to $\hat{\mathcal{N}}$.*
Furthermore, if $(a_i)_{i \in I}$, $(b_i)_{i \in I}$ are elements of $\mathcal{M}$ and $\mathcal{N}$ such that

$$(\mathcal{M}, a_i \mid i \in I) \equiv (\mathcal{N}, b_i \mid i \in I),$$

then there exist elementary extensions $\hat{\mathcal{M}} \succ \mathcal{M}$ and $\hat{\mathcal{N}} \succ \mathcal{N}$ and an isomorphism $\mathfrak{I}$ from $\hat{\mathcal{M}}$ into $\hat{\mathcal{N}}$ such that $\mathfrak{I}(a_i) = b_i$ for all $i \in I$.

Proof. Inductively, for every ordinal $n < \omega$, use Corollary 6.37 to construct structures $\mathcal{M}_n, \mathcal{N}_n$, and maps e_n, f_n, such that

(i) $\mathcal{M}_0 = \mathcal{M}$, $\mathcal{N}_0 = \mathcal{N}$,

(ii) e_n is an elementary embedding of $\mathcal{M}_n$ into $\mathcal{N}_{n+1}$ and f_n is an elementary embedding of $\mathcal{N}_{n+1}$ into $\mathcal{M}_{n+1}$,

(iii) $f_{n+1}(e_n(a)) = a$ for every element a of the universe of $\mathcal{M}_n$.

Let $\hat{\mathcal{M}} = \bigcup_{n<\omega} \mathcal{M}_n$, $\hat{\mathcal{N}} = \bigcup_{n<\omega} \mathcal{M}_n$, and $e = \bigcup_{n<\omega} e_n$. The e is an isomorphism from $\hat{\mathcal{M}}$ into $\hat{\mathcal{N}}$.

The second part of the statement is given by the preceding construction, since e_0 can be chosen to map a_i to b_i for each $i \in I$. Alternatively, let $C = (c_i \mid i \in I)$ be a collection of new constants and apply the result just proved to the $L[C]$-structures $\widetilde{\mathcal{M}} = (\mathcal{M}, a_i \mid i \in I)$ and $\widetilde{\mathcal{N}} = (\mathcal{N}, b_i \mid i \in I)$. $\square$

IX Types

We begin this subsection defining the notion of *finite satisfiability* of a set of formulas.

Definition 6.40. If $\varphi(x_1, \ldots, x_n)$ is a set of L-formulas and $\mathcal{M}$ is an L-structure, we say that φ is *finitely satisfiable* in $\mathcal{M}$ if there exists r such that for every finite $\varphi_0 \subseteq \varphi$,

$$\mathcal{M} \approx \exists_r x_1 \ldots \exists_r x_n \bigwedge_{\varphi \in \varphi_0} \varphi(x_1, \ldots, x_n). \tag{*}$$

Note that if (*) holds, then for every finite $\varphi_0 \subseteq \varphi_\approx$ there exist $c_1, \ldots, c_n$ in $\mathcal{M}$ such that $\mathcal{M} \models \bigwedge_{\varphi \in \varphi_0} \varphi[c_1, \ldots, c_n]$ (discrete satisfaction). However, the tuple $(c_1, \ldots, c_n)$ depends on φ_0.

We now introduce one of the central concepts of model theory: that of *type*.

Definition 6.41. If $\mathcal{M}$ is an L-structure, A is a set of elements of $\mathcal{M}$, and $(c_1, \ldots, c_n)$ is a tuple of elements of $\mathcal{M}$, the *type of* $(c_1, \ldots, c_n)$ *over* A, denoted $\mathrm{tp}_A(c_1, \ldots, c_n)$, is the set of all $L[A]$-formulas $\varphi(x_1, \ldots, x_n)$ such that $(\mathcal{M}, a \mid a \in A) \approx \varphi[c_1, \ldots, c_n]$. We denote $\mathrm{tp}_\emptyset(c_1, \ldots, c_n)$ by $\mathrm{tp}_L(c_1, \ldots, c_n)$.

If T is a complete L-theory and $t = t(x_1, \ldots, x_n)$ is a set of $L[C]$-formulas, where C is a set of constant symbols not in L, we say that t is an *n-type of* T if there exists an $L[C]$-model $\mathcal{M}$ of T and elements $c_1, \ldots, c_n$ in $\mathcal{M}$ such that $t = \mathrm{tp}_A(c_1, \ldots, c_n)$ where A is the subset of $\mathcal{M}$ consisting of elements interpreting the constants in C. In this case, we say t *is realized* in $\mathcal{M}$, and that the n-tuple $(c_1, \ldots, c_n)$ *realizes* t in $\mathcal{M}$.

Remarks 6.42.

1. If T is a complete theory and t is a type for T, then $T \subseteq t$. In fact, if $\mathcal{M}$ is a model of T and t is a type over a set A of elements of $\mathcal{M}$, then the set of sentences in t is precisely the complete $L[A]$-theory $\mathrm{Th}(\mathcal{M}, a \mid a \in A)$, which extends $T = \mathrm{Th}(\mathcal{M})$.

2. The notation $\mathrm{tp}_A(c_1, \ldots, c_n)$ is imprecise in the sense that it does not make reference to the structure $\mathcal{M}$ where the elements $c_1, \ldots, c_n$ and the set A "live". However, since T is a complete L-theory by assumption, precise knowledge of $\mathcal{M}$ is to a large extent unnecessary. In fact, if we are given a family $(t_i)_{i \in I}$ of types of T such that, for $i \in I$, the type t_i is over a set of parameters A_i realized in an L-structure $\mathcal{M}_i$, then by Corollary 6.38 there exists a single structure $\mathcal{N}$ in which all the structures $\mathcal{M}_i$ are elementary embedded. In fact, as we shall see in Proposition 6.46 below, given any cardinal κ we can fix a "big" model $\mathcal{N}$ of T that is an elementary extension of every L-structure $\mathcal{M}$ with cardinality $\mathrm{card}(\mathcal{M}) < \kappa$, and such that $\mathcal{N}$ realizes every type over any of its subsets A with $\mathrm{card}(A) < \kappa$. Furthermore, the model $\mathcal{N}$ can be taken with the following additional homogeneity property: If $c_1, \ldots, c_n$ and $c_1', \ldots, c_n'$ are elements of $\mathcal{N}$ and A is a set of elements of $\mathcal{N}$ with $\mathrm{card}(A) < \kappa$, then $\mathrm{tp}_A(c_1, \ldots, c_n) = \mathrm{tp}_A(c_1', \ldots, c_n')$ if and only if there is an automorphism of $\mathcal{N}$ carrying $c_1, \ldots, c_n$ to $c_1', \ldots, c_n'$ and fixing A pointwise. This allows viewing types as orbits on the big model under the action of its group of automorphisms, and enables a Galois-theoretic viewpoint of complete theories. It also explains the use of the word "type". We will discuss this in more detail in part XII.

3. If T is a complete theory, $\mathcal{M}$ is an arbitrary model of T, A is a set of elements of $\mathcal{M}$, and t is a type of T over A, there is no guarantee that t is realized in $\mathcal{M}$. However:

   (a) The equivalence between (1) and (2) of Proposition 6.43 below shows that $\mathcal{M}$ has an elementary extension where t is realized. In particular, every elementary extension of $\mathcal{M}$ has a further elementary extension where t is realized.

   (b) t is finitely satisfiable in every model of $\mathrm{Th}(\mathcal{M}, a \mid a \in A)$.

Proposition 6.43. *Let T be a complete theory and let $t(x_1, \ldots, x_n)$ be a set of $L[C]$-formulas, where C is a set of constant symbols not in L. The following conditions are equivalent.*

(1) *There exists a model $\mathcal{M}$ of T, a set A of elements of $\mathcal{M}$, and elements $c_1, \ldots, c_n$ of $\mathcal{M}$ such that $t(x_1, \ldots, x_n) = \mathrm{tp}_A(c_1, \ldots, c_n)$.*

(2) *For every model $\mathcal{N}$ of T there exists an elementary extension $\mathcal{N}'$ of $\mathcal{N}$, a set A of elements of $\mathcal{N}'$, and elements $c_1, \ldots, c_n$ of $\mathcal{N}'$ such that $t(x_1, \ldots, x_n) = \mathrm{tp}_A(c_1, \ldots, c_n)$.*

(3) (a) *There exists a positive rational r such that, for every finite subset $\varphi \subset t$, the formula $\exists_r x_1 \ldots \exists_r x_n \bigwedge_{\varphi \in \varphi} \varphi(x_1, \ldots, x_n)$ is in t.*

   (b) *For every $L[C]$-formula $\varphi(x_1, \ldots, x_n)$, if $\varphi \notin t$, then there exists $\varphi' \Lleftarrow \varphi$ such that $\stackrel{w}{=} \varphi' \in t$.*

Proof. The equivalence between (1) and (3) is given by Corollary 6.34 (by replacing the variables $x_1, \ldots, x_n$ with constant symbols not already in $L[A]$). The equivalence between (3) and (2) is given by Corollary 6.37. □

Remark 6.44. Part (3) of Proposition 6.43 gives a purely syntactic criterion for a given set of formulas to be a type.

X   Saturated and homogeneous models

Strictly for reasons of notational simplicity, we will focus our attention on one-sorted structures throughout this subsection. When a structure $\mathcal{M}$ has only one sort, this sort is called the *universe* of $\mathcal{M}$. The *cardinality* of a one-sorted structure is defined as the cardinality of its universe.

The observations in Remark 6.43 lead naturally to the concept of *saturated structure*:

Definition 6.45. Let $\mathcal{M}$ is a structure with universe M and let κ be an infinite cardinal with $\kappa \leq \mathrm{card}(M)$. We say that $\mathcal{M}$ is κ-*saturated* if whenever t is a type of $\mathrm{Th}(\mathcal{M})$ over a subset of M of cardinality strictly less than κ, there is a realization of t in $\mathcal{M}$.

In this context, we may abuse notation and use ω as a synonym for $\aleph_0$, the cardinality of (infinite) countable sets. Thus, an ω^+-saturated structure is one that realizes types over any countable subset of the universe. The informal terminology "countable saturation" shall mean ω^+-saturation, and not ω-saturation.

Notice that an L-structure $\mathcal{M}$ is κ-saturated if and only if, whenever A is a subset of the universe of $\mathcal{M}$ of cardinality less than κ and $\varphi(x_1, \ldots, x_n)$ is a set of $L[A]$-formulas that is finitely satisfiable in $\mathcal{M}$, there exist $c_1, \ldots, c_n$ in $\mathcal{M}$ such that $\varphi(x_1, \ldots, x_n) \subseteq \mathrm{tp}_A(c_1, \ldots, c_n)$.

Proposition 6.46. *If $\mathcal{M}$ is κ-saturated and $\mathcal{N}$ is a structure of cardinality less than κ such that $\mathcal{N} \equiv \mathcal{M}$, then $\mathcal{N}$ can be elementarily embedded in $\mathcal{M}$.*

Proof. Assume that $\mathcal{M}$ is κ-saturated, $\mathcal{N} \equiv \mathcal{M}$, and $\mathrm{card}(\mathcal{N}) < \kappa$. If follows easily by induction that, if α is an ordinal satisfying $\alpha < \kappa$ and $(a_i)_{i<\alpha}$ is a list of elements of $\mathcal{N}$, then there exists $(a_i')_{i<\alpha}$ in $\mathcal{M}$ such that $(\mathcal{N}, a_i \mid i < \alpha) \equiv (\mathcal{M}, a_i' \mid i < \alpha)$. Thus, in the case when $(a_i)_{i<\alpha}$ lists all the elements of $\mathcal{N}$, the map $a_i \mapsto a_i'$ is an elementary embedding of $\mathcal{N}$ into $\mathcal{M}$. □

Proposition 6.47. *If* $\mathcal{M}$ *is an* $\aleph_1$*-saturated L-structure, then* $\models\!\approx$ *and* $\models$ *coincide on* $\mathcal{M}$, *i.e., for an L-formula* $\varphi(x_1, \ldots, x_n)$ *and elements* $a_1, \ldots, a_n$ *of suitable sorts, we have*

$$\mathcal{M} \models\!\approx \varphi[a_1, \ldots, a_n] \quad \Leftrightarrow \quad \mathcal{M} \models \varphi[a_1, \ldots, a_n].$$

Proof. By induction on the complexity of φ. □

Recall that, if κ is a cardinal, then κ^+ denotes the smallest cardinal larger than κ.

Proposition 6.48. *If* κ *is an infinite cardinal, then every structure has a* κ^+*-saturated elementary extension.*

Proof. Fix an infinite cardinal κ and an L-structure $\mathcal{M}$. Applying Remark 6.42-(2), we construct inductively, for every ordinal $i < \kappa^+$, a structure $\mathcal{M}_i$, such that

(i) $\mathcal{M}_0 = \mathcal{M}$,

(ii) $\mathcal{M}_i \prec \mathcal{M}_{i+1}$ and every type over a subset of the universe of $\mathcal{M}_i$ of cardinality less than κ is realized in $\mathcal{M}_{i+1}$,

(iii) If $j < \kappa^+$ is a limit ordinal, then $\mathcal{M}_j = \bigcup_{i<j} \mathcal{M}_i$.

It follows from the Elementary Chain Property (Proposition 6.28) that $(\mathcal{M}_i)_{i<\kappa^+}$ is an elementary chain, and that $\bigcup_{i<\kappa^+} \mathcal{M}_i$ is an elementary extension of $\mathcal{M}$, which is clearly κ^+-saturated. □

Suppose that $\mathcal{M}$ is κ-saturated and let α be an ordinal with $\alpha < \kappa$. It follows directly from the κ-saturation of $\mathcal{M}$ that if $(a_i)_{i<\alpha}$, $(a'_i)_{i<\alpha}$ are families in $\mathcal{M}$ such that $(\mathcal{M}, a_i \mid i < \alpha) \equiv (\mathcal{M}, a'_i \mid i < \alpha)$, then for every element b of $\mathcal{M}$ there exists an element b' such that

$$(\mathcal{M}, b, a_i \mid i < \alpha) \equiv (\mathcal{M}, b', a'_i \mid i < \alpha).$$

A structure $\mathcal{M}$ that has this extension property for all pairs of families $(a_i)_{i<\alpha}$, $(a'_i)_{i<\alpha}$ with $\alpha < \kappa$ is said to be κ-*homogeneous*. Note that if $\mathcal{M}$ is κ-homogeneous with $\kappa = \mathrm{card}(\mathcal{M})$, then the κ-homogeneity can be used iteratively to extend the map $a_i \mapsto a'_i$ ($i < \alpha$) to an automorphism of $\mathcal{M}$. This suggests the following definition.

Definition 6.49. Let κ be an infinite cardinal. A structure $\mathcal{M}$ is *strongly* κ-*homogeneous* if whenever α is an ordinal with $\alpha < \kappa$ and $(a_i)_{i<\alpha}$, $(a'_i)_{i<\alpha}$ are families in the universe of $\mathcal{M}$ such that $(\mathcal{M}, a_i \mid i < \alpha) \equiv (\mathcal{M}, a'_i \mid i < \alpha)$, there exists an automorphism $\mathcal{I}$ of $\mathcal{M}$ such that $\mathcal{I}(a_i) = a'_i$ for all $i < \alpha$.

The following theorem shows that, for arbitrarily large κ, every structure has elementary ultrapowers that are κ-saturated and κ-homogeneous.

Theorem 6.50. *For every infinite cardinal κ there exists an ultrafilter $\mathcal{U}$ with the following property: Whenever $\mathcal{M}$ is a metric structure of cardinality at most 2^κ, the $\mathcal{U}$-ultrapower of $\mathcal{M}$ is both κ^+-saturated and κ^+-homogeneous.*

We omit the proof of Theorem 6.50. The construction of the ultrafilter is due to S. Shelah [She71], and it builds on ideas of H. J. Keisler and K. Kunen. Shelah's epochal proof is for traditional first-order (i.e., discrete structures), but his argument was adapted by C. W. Henson and J. Iovino for structures based on Banach spaces [HI02, Corollary 12.3]. The proof for Banach structures applies to general metric structures without significant changes.

XI Extending the language

Given a signature L, a sort index set $\mathbf{S}$, indices $s_0, s_1, \ldots, s_n \in \mathbf{S}$, and a new function symbol $f : s_1 \times \cdots \times s_1 \to s_0$, we will denote by $L[f]$ the signature that results from adding f to L. If $\mathcal{M}$ is an L structure based on $(M^{(s)} \mid s \in \mathbf{S})$ and F is a function from $M^{(s_1)} \times \cdots \times M^{(s_n)}$ into $M^{(s_0)}$, then will denote by $(\mathcal{M}, F)$ the expansion of $\mathcal{M}$ to $L[f]$ that results from defining $f^{\mathcal{M}}$ as F.

Let L be a signature and consider the signature $L[f]$, where f is a function symbol not in L. Fix also an $L[f]$-theory T (which could be empty). In standard mathematical practice there are cases where, in models of T, the new symbol can be dispensed with because it is already definable through L-formulas. This leads to the notion of *definability*, which is the main concern of this subsection. There are several ways to formalize the concept of f being T-*definable*; for instance, one could say that

> If $(\mathcal{M}, F) \models T$, where $\mathcal{M}$ is an L-structure, then F is determined uniquely by T, i.e., if $(\mathcal{M}, F), (\mathcal{M}, F') \models T$, then $F = F'$.

or, alternatively,

> In all models of T, formulas involving f can be approximated by L-formulas.

Below, we prove that these two conditions are equivalent (see Theorem 6.53); but first we must formalize what we mean by "approximated" in the preceding statement.

Recall that if (M, d, a) is a pointed metric space, the open ball of radius r around the anchor point a is denoted or $B_M(r)$, or $B(r)$ if M is given by the context. If $\mathcal{M}$ is a metric structure and a is in $B_{M^{(s)}}(r)$, where $M^{(s)}$ is one of the sorts of $\mathcal{M}$, we may informally say that a is an element of $B_{\mathcal{M}}(r)$.

Definition 6.51. Let L be a signature, let f be a real-valued n-ary function symbol, and let T be a uniform $L[f]$-theory. We will say that f is *explicitly defined by T in L* if the following condition holds for every $r \in \mathbb{R}$ and every pair of nonempty intervals $I \subset J \subseteq \mathbb{R}$ with I closed and J open: There exists an L-formula $\varphi_{r,I,J} = \varphi(x_1, \ldots, x_n)$ such that, whenever $(\mathcal{M}, F) \models T$ and $a_1, \ldots, a_n$ are elements of $B_{\mathcal{M}}(r)$, we have

$$F(a_1, \ldots, a_n) \in I \quad \Rightarrow \quad \mathcal{M} \models \varphi[a_1, \ldots, a_n],$$
$$\mathcal{M} \models \varphi[a_1, \ldots, a_n] \quad \Rightarrow \quad F(a_1, \ldots, a_n) \in J.$$

The collection $(\varphi_{r,I,J} \mid r \in \mathbb{R}, \emptyset \neq I \subset J)$ is a *definition scheme* for F (modulo T).

Remark 6.52. A definition scheme $\Sigma = (\varphi_{r,I,J} \mid r \in \mathbb{R}, \emptyset \neq I \subset J)$ (I closed, J open) characterizes F uniquely in any structure $(\mathcal{M}, F) \models T$. Namely, if $a_1, \ldots, a_n \in B_{\mathcal{M}}(r)$, then $F(a_1, \ldots, a_n)$ is the unique real number t with the following property: $\mathcal{M} \models \varphi_{r,I,J}(a_1, \ldots, a_n)$ whenever $J \supset I \ni t$. Certainly, $F(a_1, \ldots, a_n)$ is one such number t (since Σ is a definition scheme for F). Conversely, if t has the stated property, we have $F(a_1, \ldots, a_n) \in J$ whenever $J \supset I \ni t$, hence $t = F(a_1, \ldots, a_n)$.

The theorem below is the most fundamental result about first-order definability. The topological proof we give here is not commonly known. We thank Xavier Caicedo for pointing out an error in an earlier draft of this manuscript and suggesting a correction.

Theorem 6.53. *Let L be a signature, let f be a real-valued function symbol, and let T be a uniform $L[f]$-theory. The following conditions are equivalent.*

(1) *f is explicitly defined by T in L.*

(2) *If $\mathcal{M}$ is an L-structure, then*

$$(\mathcal{M}, F), (\mathcal{M}, F') \models T \quad \Rightarrow \quad F = F'.$$

Before proving Theorem 6.53, let us make the following observation about general topological spaces. If Z is a topological space, two points $x, y \in Z$ are said to be *topologically indistinguishable*, denoted $x \equiv y$, if every neighborhood of x contains y and every neighborhood of y contains x. Now, we observe that if X, Y are regular topological spaces with X compact and $g : X \to Y$ is a continuous bijection, then g is a homeomorphism if and only if

$$g(x) \equiv g(y) \quad \Rightarrow \quad x \equiv y. \tag{5.2}$$

Indeed, any such g maps indistinguishable points to indistinguishable points (by continuity), so g induces a continuous map $\overline{g} : \overline{X} \to \overline{Y}$ between the spaces $\overline{X} = X/\equiv$ and $\overline{Y} = Y/\equiv$. Since topologically distinguishable points of a normal space have disjoint neighborhoods, $\overline{X}$ and $\overline{Y}$ are Hausdorff spaces, with $\overline{X}$ compact; moreover, $\overline{g}$ is a bijection, by (5.2), and hence a homeomorphism. Clearly, g is a homeomorphism also.

Proof. (1) $\Rightarrow$ (2): This is an immediate consequence of the remark following Definition 6.51.

As a preliminary step to showing (2) $\Rightarrow$ (1), we prove the following:

Claim 6.54. Assume that (2) holds. If L' is a signature extending L, and $\mathcal{M}$, $\mathcal{N}$ are L'-structures admitting $L'[f]$-expansions $(\mathcal{M}, F)$, $(\mathcal{N}, G)$ that are models of T, then

$$\mathcal{M} \equiv_{L'} \mathcal{N} \quad \Rightarrow \quad (\mathcal{M}, F) \equiv_{L'[f]} (\mathcal{N}, G).$$

To prove the claim, let L' extend L and let $\mathcal{M}$, $\mathcal{N}$ be elementarily equivalent L'-structures admitting expansions $(\mathcal{M}, F)$, $(\mathcal{N}, G)$ that are models of T. Using Theorem 6.50, fix an ultrafilter $\mathcal{U}$ such that the $\mathcal{U}$-ultrapowers $(\mathcal{M})_{\mathcal{U}}$, $(\mathcal{N})_{\mathcal{U}}$ are isomorphic L'-structures. By Łoś's Theorem 6.29, the $\mathcal{U}$-ultrapower $(\mathcal{M}, F)_{\mathcal{U}} = ((\mathcal{M})_{\mathcal{U}}, (F)_{\mathcal{U}})$ is an $L'[f]$-structure elementarily equivalent to $(\mathcal{M}, F)$; similarly, $(\mathcal{N}, G)_{\mathcal{U}} = ((\mathcal{N})_{\mathcal{U}}, (G)_{\mathcal{U}}) \equiv_{L'[f]} (\mathcal{N}, G)$. Let $\mathcal{I}$ be an L'-isomorphism from $(\mathcal{M})_{\mathcal{U}}$ into $(\mathcal{N})_{\mathcal{U}}$. *A fortiori*, $\mathcal{I}$ is an L-isomorphism between (the L-reducts of) $(\mathcal{M})_{\mathcal{U}}$ and $(\mathcal{N})_{\mathcal{U}}$. By (2), we have $\mathcal{I}((F)_{\mathcal{U}}) = (G)_{\mathcal{U}}$; thus, $(\mathcal{M}, F)_{\mathcal{U}}$ and $(\mathcal{N}, G)_{\mathcal{U}}$ are isomorphic $L'[f]$-structures, so $(\mathcal{M}, F) \equiv_{L'[f]} (\mathcal{N}, G)$. This proves the claim.

(2) $\Rightarrow$ (1): Assume that (2) holds. The signature of $L[f]$ specifies that $f : s_1 \times \cdots \times s_n \to s_{\mathbb{R}}$ for some sorts $s_1, \ldots, s_n$. Fix a new tuple $\bar{c} = c_1, \ldots, c_n$ of constant symbols such that c_i is of sort s_i for $i = 1, \ldots, n$ and let $L' = L[\bar{c}]$. For any fixed rational $r > 0$, let T'_r be the $L'[f]$-theory obtained by adding to T the sentences $\mathrm{d}(c_i, a_{s_i}) \leq r$, for $i = 1, \ldots, n$, where a_{s_i} is (the symbol for) the anchor of sort s_i. Clearly, T'_r is a uniform $L'[f]$-theory. Let $\mathscr{C}$ be the class of all models of T'_r, i.e., of $L'[f]$-structures of the form $(\mathcal{M}, b_1, \ldots, b_n, F)$, where $(\mathcal{M}, F)$ is a model of T, and b_i is an element of sort s_i of $\mathcal{M}$ that satisfies $\mathrm{d}(b_i, a_{s_i}) \leq r$ for $i = 1, \ldots, n$. Let $\mathscr{D}$ be the class of all L'-structures $\mathcal{N}$ that are L'-reducts of some $L'[f]$-structure $(\mathcal{N}, F) \in \mathscr{C}$ and let $g : \mathscr{C} \to \mathscr{D}$ be the map $(\mathcal{N}, F) \mapsto \mathcal{N}$. Note that g is surjective by definition.

We regard the classes $\mathscr{C}$ and $\mathscr{D}$ as topological spaces as follows. A basis for the closed classes of $\mathscr{C}$ is given by all subclasses of the form $\mathrm{Mod}_{\mathscr{C}}(\varphi)$ where φ is an $L'[f]$-sentence (the union of finitely many such basic closed subclasses is of the same form, since the logic is closed under disjunction); in other words, the closed subclasses of $\mathscr{C}$ are those of the form $\mathrm{Mod}_{\mathscr{C}}(\Phi)$, where Φ is an $L'[f]$-theory. Similarly, the closed subclasses of $\mathscr{D}$ are those of the form $\mathrm{Mod}_{\mathscr{D}}(\Phi)$, where Φ is an L'-theory.[9] Both $\mathscr{C}$ and $\mathscr{D}$ are regular, with $\mathscr{C}$ compact, by Remark 6.33, and the surjection g is clearly continuous.

Let $\mathcal{N} \in \mathscr{D}$, so $\mathcal{N}$ is the L'-reduct of some model $(\mathcal{N}, F) \models T'_r$. Let $\mathcal{M} = \mathcal{N} \restriction L$ be the L-reduct of $\mathcal{N}$. Clearly, $(\mathcal{M}, F)$ is a model of $T'_r \restriction L[f] = T$; moreover, by (2), $(\mathcal{M}, F)$ is the unique expansion of $\mathcal{M}$ to a model of T. Since T'_r extends T, $(\mathcal{N}, F)$ is necessarily the unique preimage of $\mathcal{N}$ under g, so g is injective; thus, g is a continuous bijection. By the claim and the topological observation immediately following the statement of the theorem, we conclude that g is a homeomorphism.

Fix intervals $\emptyset \neq I = [p, q] \subset J = (u, v)$ with p, q, u, v rational, and let

$$K = \mathrm{Mod}_{\mathscr{C}}(p \leq f(\bar{c}) \leq q), \qquad K' = \mathrm{Mod}_{\mathscr{C}}(f(\bar{c}) \leq u \vee f(\bar{c}) \geq v).$$

---

[9]Later, in Subsection XII, we shall refer to this as the *logic topology*.

Since K and K' are closed and disjoint, so are the subsets $g(K)$ and $g(K')$ of the homeomorphic image $\mathscr{D} = g(\mathscr{C})$, which is compact since $\mathscr{C}$ is. Since any compact regular space is normal, there exists a closed neighborhood Q of $g(K)$ disjoint from $g(K')$. By compactness of $g(K')$, Q may be taken of the form $\mathrm{Mod}_{\mathscr{D}}(\varphi(\bar{c}))$ for some L-formula $\varphi_{I,J,r} = \varphi(\bar{x})$. Since the interpretation of $\bar{c}$ is arbitrary in $B_{\mathcal{M}}(r)$, the scheme $(\varphi_{I,J,r} : r \in \mathbb{R}, \emptyset \neq I \subset J)$ defines f explicitly by T. □

Definition 6.55. If L is a signature, f is a real-valued function symbol, and T is a uniform $L[f]$-theory, we will say that f is T-*definable* in L if it satisfies the equivalent conditions of Theorem 6.53. When L is given by the context, we may simply say that f is T-*definable*; furthermore, if L and T are given by the context, we may say that f is "definable".

If T is a uniform theory, Theorem 6.53 allows us to see the real-valued functions that are T-definable as those that are left fixed by automorphisms of sufficiently saturated models of T. This observation yields Corollary 6.56 below.

For notational convenience, if f is a function symbol for a given function $F : \mathbb{R}^n \to \mathbb{R}$, we will liberally identify f with its interpretation F.

Corollary 6.56. *Let T be a uniform theory.*

(1) *A composition of functions that are T-definable is T-definable.*

(2) *Every continuous function $F : \mathbb{R}^n \to \mathbb{R}$ is T-definable.*

(3) *If $F : \mathbb{R}^{n+1} \to \mathbb{R}$ is T-definable by T, so are the functions $G_r : \mathbb{R}^n \to \mathbb{R}$ and $H_r : \mathbb{R}^n \to \mathbb{R}$ defined by*

$$G_r(x_1, \ldots, x_n) = \sup_{y \in B(r)} F(x_1, \ldots, x_n, y),$$

$$H_r(x_1, \ldots, x_n) = \inf_{y \in B(r)} F(x_1, \ldots, x_n, y).$$

Proof. Clauses (1) and (3) follow from the definitions. To prove (2), note that, since every signature comes equipped with the ordered field and lattice structure for $\mathbb{R}$ plus a constant for each rational, all polynomials functions with rational coefficients are definable in any theory. Therefore, (2) follows from the Stone-Weierstrass theorem. □

Definition 6.57. If L is a signature, f is an n-ary function symbol (not necessarily real-valued), and T is a uniform $L[f]$-theory, we will say that f is T-*definable in L* if the real-valued function $d(f(x_1, \ldots, x_n), y)$ is T-definable in L in the sense of Definition 6.55. As in Definition 6.55, if L or T are given by the context, we may omit them from the nomenclature.

Remark 6.58. It is not difficult to verify that, if f is a real-valued function symbol, there is no conflict between the notions of definability for f given by Definitions 6.55 and 6.57. (This is because if f is n-ary and F is an interpretation of f, then for any real number r, $d(F(a_1, \ldots, a_n), b) = r$ if and only if $b = \pm F(a_1, \ldots, a_n)$.)

XII   Spaces of types and the monster model

In this section, T will denote a fixed complete L-theory and we will denote by $\mathscr{T}$ the set of types that are realized in a model of T. Note that, if $(t_i)_{i \in I}$ is a family of types in $\mathscr{T}$ and t_i is realized in $\mathcal{M}_i$ then, by Corollary 6.38, there exists a model $\mathcal{N}$ of T such that $\mathcal{M}_i \prec \mathcal{N}$ for every $i \in I$. Thus, each t_i is realized in $\mathcal{N}$.

By Remark 6.42-(2), if $t = t(x_1, \ldots, x_n)$ is a type in $\mathscr{T}$, then there exists a positive rational r such that

$$\exists_r x_1 \ldots \exists_r x_n \bigwedge_{\varphi \in \varphi} \varphi(x_1, \ldots, x_n) \in t, \qquad \text{for each finite } \varphi \subseteq t.$$

For each choice of n and r, we will denote by $\mathscr{T}_n^{(r)}$ be the set of all $t(x_1, \ldots, x_n) \in \mathscr{T}$ that satisfy this condition.

If $\varphi(x_1, \ldots, x_n)$ is an L-formula, let

$$[\varphi] = \{ t \in \mathscr{T} \mid \varphi \in t \}.$$

The collection of sets of this form is closed under finite unions and intersections. We define a topology on $\mathscr{T}$ by letting these be the basic closed sets. We will refer to this topology as the *logic topology*. We shall always regard $\mathscr{T}$ as a topological space via the logic topology.

Remark 6.59. The Compactness Theorem 6.31 says exactly that $\mathscr{T}_n^{(r)}$ is compact for each n and r.

The complement of a set of the form $[\varphi]$ is not necessarily of the same form; however, if $t \in \mathscr{T}$, by Proposition 6.43-(3b), we have $t \notin [\varphi]$ if and only if there exists $\psi \in t$ and $\psi' > \psi$ such that $t \in [\psi'] \subseteq [\varphi]^c$. Thus, for $t \in \mathscr{T}$, the sets of the form $[\psi']$ where ψ' is an approximation of a formula $\psi \in t$ form a local neighborhood base around t.

Let $\varphi(x_1, \ldots, x_n)$ be the formula $\exists_r x_1 \ldots \exists_r x_n \bigwedge_{1 \le i \le n} x_i = x_i$. Then, $[\varphi] = \mathscr{T}_n^{(r)}$. Every type contains a formula φ of this kind, for some n and some r. Therefore, the space $\mathscr{T}$ is locally compact.

Let A be a set of parameters, i.e., A is a set of elements of some fixed model $\mathcal{M}$ of T. If $(t_i)_{i \in I}$ is a family in $\mathscr{T}_n^{(r)}$ such that t_i is a type over A for each $i \in I$ and $\mathcal{U}$ is an ultrafilter of I, then, by the compactness of $\mathscr{T}_n^{(r)}$, the limit $\mathcal{U} \lim_i t_i$ is a type over A. Conversely, if $t(\bar{x})$ is a type over A, since t is finitely satisfiable in $\mathcal{M}$ (see Definition 6.40), there exists a family $(\bar{a}_i)_{i \in I}$ in $\mathcal{M}$ and an ultrafilter $\mathcal{U}$ on I such that $t = \mathcal{U} \lim_i \text{tp}_A(\bar{a}_i)$. Thus, types over model $\mathcal{M}$ can be viewed as ultrafilter limits of types realized within $\mathcal{M}$.

Let κ be a cardinal that is both larger than the cardinality of every model of T needed in our proofs and larger than the cardinality of any indexed family of types mentioned. Let now $\mathfrak{C}$ be a κ-saturated, κ-homogeneous model of T (see Theorem 6.50). Then we can assume that:

1. Every model of T occurs as an elementary submodel of $\mathfrak{C}$ (see Proposition 6.46).

2. Every type (over a set of parameters is a model of T) is realized in $\mathfrak{C}$.

3. If $(a_1, \ldots, a_n)$ and $(b_1, \ldots, b_n)$ are n-tuples of elements of $\mathfrak{C}$ such that $\text{tp}_A(a_1, \ldots, a_n) = \text{tp}_A(b_1, \ldots, b_n)$, then there exists an automorphism f of $\mathfrak{C}$ such that $f(a_i) = b_i$ for $i = 1, \ldots, n$.

4. A real-valued function f is definable in T over a set of parameters A if and only if f is fixed by every automorphism of $\mathfrak{C}$ that fixes A pointwise fixes the graph of f. (See the remarks preceding Corollary 6.56.)

We shall henceforth refer to $\mathfrak{C}$ as a *big model* or a *monster model* for T. Monster models are a time-saving device, and what makes them convenient are the properties listed above. For each complete theory considered, we will always work within a fixed monster model. The particular choice of monster model will be irrelevant for our discussions since any two such models can be embedded elementarily in a common one. Thus, we may informally refer to "the" monster model of T.

There is another topology on types that plays an important role in the model theory of metric spaces. Let $t(x_1, \ldots, x_n)$ and $t'(x_1, \ldots, x_n)$ are types consistent with T (i.e., t and t' are realized in the monster model) over the same set of parameters (which can be thought of as a subset of the monster model). We define $d(t, t')$ as the infimum of all distances $d(\bar{a}, \bar{a}')$ where $\bar{a}$ realizes t and $\bar{a}'$ realizes t'. Note that this infimum is realized by some pair $\bar{a}, \bar{a}'$ in the monster model. It can be readily verified, using the saturation of the monster model, that d is a metric and that $(\mathscr{T}_n^{(r)}, d)$ is compact and complete.

Appendix: On the notion of "finitary" properties of metric structures

We conclude this manuscript with some general remarks on the meaning of the informal term "finitary" in our context. Describing a certain mathematical property as finitary presupposes, in our view, the existence of a formal language $\mathcal{L}$ in which the property can be formulated (here we use the term "language" in an abstract sense not restricted to first-order languages or Henson's language of positive bounded formulas). There is an inherent tradeoff between the expressive power of logical languages and the strength of their model-theoretic properties. On one hand, if the language $\mathcal{L}$ is rich enough (for example, if $\mathcal{L}$ admits infinite conjunctions and disjunctions), it may capture complex properties with, say, a single formula; however, a powerful model-theoretic property like the Compactness Theorem (Theorem 6.31) is bound to fail for such $\mathcal{L}$. On the other hand, compactness of first-order-like logics (including Henson's logic) is, in essence, a reflection of the limited expressive power of the language. In fact, for metric structures, there is no logic strictly more expressive than Henson logic satisfying both the Compactness Theorem and the elementary chain property (Proposition 6.28) [Iov01].[10]

A strong feature of the notion of approximate satisfaction of positive bounded formulas is that it inherently captures "asymptotically approximable" properties of L-structures: By definition, the approximate satisfaction of an L-formula φ amounts to the discrete satisfaction of the full set φ_+ of formulas ψ approximating φ. Now, if Φ is a set of positive bounded of L-formulas such that $\Phi \approx \varphi$ (i.e., every model of Φ satisfies φ approximately), then every approximation of φ admits a finite-length proof from Φ plus the axioms for the real numbers and metric spaces.[11] but φ itself may not admit a single such proof. A property P captured by the approximate satisfaction of a single formula φ should by all rights be called finitary, although P may not admit a finite-length proof. Still, it is enough for each rational $\epsilon > 0$ to prove φ_ϵ, the formula obtained from φ by ϵ-relaxing every inequality and quantifier bound in φ. This approach, though *sensu stricto* infinitary, is much the one used when a proof in analysis starts with the incantation: *"Let $\epsilon > 0$ be given."* (Of course, it is always desirable that "the same" proof works for all $\epsilon > 0$ in the sense that ϵ only appears in formulas used in the proof as a parameter, so that a uniform scheme proves all φ_ϵ.)

More generally, any property P equivalent to the simultaneous satisfaction of a collection Ψ of positive bounded formulas is finitary in the following sense: If P holds for all structures in a class $\mathscr{C}$ axiomatized by a theory T, then P admits, in principle, a proof scheme (say, a syntactic proof for each $\psi \in \Psi_+$). Equivalently, the *failure* of P in a structure $\mathcal{M}$ is equivalent to the discrete satisfaction $\mathcal{M} \models {}^{\underline{w}}\psi$ of the weak negation of a single formula $\psi \in \Psi_+$. Thus,

---

[10]Chapter 4 of this volume is devoted to the maximality of continuous logic.

[11]This was first observed by C. Ward Henson in the 1980s. He gave lectures on this material, but his lecture notes were not formally published. The idea of finite provability via approximations can be traced back to Mostowski, who proved [Mos62] that for any rational $r < 1$, the set of sentences of Łukasiewicz logic that take truth value $> r$ in all structures is recursively enumerable. Simultaneously, E. Specker's student B. Scarpellini had proved in his 1961 dissertation that the valid sentences of Łukasiewicz logic are not axiomatizable.

a finitary property is witnessed by the absence of counterexamples having finite-length proofs.[12]

The property that a sequence (a_n) in a metric L-structure be convergent cannot, in general, be captured by the simultaneous satisfaction of any collection (whether finite or infinite) of positive bounded L-formulas. However, the property "(a_n) *is convergent*" is equivalent to the disjunction of the properties "$E_\bullet$ *is a rate of metastability for* (a_n)" over all possible metastability rates $E_\bullet$. For a specific $E_\bullet$, the latter property is equivalent to the conjunction of the (infinitely many) properties "$E_{\epsilon,\eta}$ *is a rate of* $[\epsilon, \eta]$-*metastability for* (a_n)" for all $\epsilon > 0$ and samplings η. Thus, from the perspective discussed above, metastable convergence with rate $E_\bullet$ is a finitary property of metric structures, while convergence with no specified rate is not.

---

[12]As pointed out by Avigad *et al.* [ADR12], from the perspective of constructive analysis, this is an instance of Kreisel's no-counterexample interpretation [Kre51, Kre52], which is in turn a particular case of Gödel's Dialectica interpretation [Göd58].

Bibliography

[ADR12]   Jeremy Avigad, Edward T. Dean, and Jason Rute. A metastable dominated convergence theorem. *J. Log. Anal.*, 4:Paper 3, 19, 2012.

[AI13]   Jeremy Avigad and José Iovino. Ultraproducts and metastability. *New York J. Math.*, 19:713–727, 2013.

[BYBHU08] Itaï Ben Yaacov, Alexander Berenstein, C. Ward Henson, and Alexander Usvyatsov. Model theory for metric structures. In *Model theory with applications to algebra and analysis. Vol. 2*, volume 350 of *London Math. Soc. Lecture Note Ser.*, pages 315–427. Cambridge Univ. Press, Cambridge, 2008.

[BYU10]   Itaï Ben Yaacov and Alexander Usvyatsov. Continuous first order logic and local stability. *Trans. Amer. Math. Soc.*, 362(10):5213–5259, 2010.

[Cai]   Xavier Caicedo. Maximality of continuous logic. Chapter 4 of this volume.

[CI14]   Xavier Caicedo and José N. Iovino. Omitting uncountable types and the strength of $[0, 1]$-valued logics. *Ann. Pure Appl. Logic*, 165(6):1169–1200, 2014.

[CK62]   Chen-Chung Chang and H. Jerome Keisler. Model theories with truth values in a uniform space. *Bull. Amer. Math. Soc.*, 68:107–109, 1962.

[CK66]   Chen-Chung Chang and H. Jerome Keisler. *Continuous model theory*. Annals of Mathematics Studies, No. 58. Princeton Univ. Press, Princeton, N.J., 1966.

[Cut83]   Nigel J. Cutland. Nonstandard measure theory and its applications. *Bull. London Math. Soc.*, 15(6):529–589, 1983.

[Cut00]   Nigel J. Cutland. *Loeb measures in practice: recent advances*, volume 1751 of *Lecture Notes in Mathematics*. Springer-Verlag, Berlin, 2000.

[Göd58]   Kurt Gödel. Über eine bisher noch nicht benützte Erweiterung des finiten Standpunktes. *Dialectica*, 12:280–287, 1958.

[Hen75]   C. Ward. Henson. When do two Banach spaces have isometrically isomorphic nonstandard hulls? *Israel J. Math.*, 22(1):57–67, 1975.

[Hen76]   C. Ward Henson. Nonstandard hulls of Banach spaces. *Israel J. Math.*, 25(1-2):108–144, 1976.

[HI02]   C. Ward Henson and José Iovino. Ultraproducts in analysis. In *Analysis and logic (Mons, 1997)*, volume 262 of *London Math. Soc. Lecture Note Ser.*, pages 1–110. Cambridge Univ. Press, Cambridge, 2002.

[Iov01]   José Iovino. On the maximality of logics with approximations. *J. Symbolic Logic*, 66(4):1909–1918, 2001.

[Iov09]   José Iovino. Analytic structures and model theoretic compactness. In *The many sides of logic*, volume 21 of *Stud. Log. (Lond.)*, pages 187–199. Coll. Publ., London, 2009.

[Iov14]   José Iovino. *Applications of model theory to functional analysis.* Dover Publications, Inc., Mineola, NY, 2014.

[Kre51]   G. Kreisel. On the interpretation of non-finitist proofs. I. *J. Symbolic Logic*, 16:241–267, 1951.

[Kre52]   G. Kreisel. On the interpretation of non-finitist proofs. II. Interpretation of number theory. Applications. *J. Symbolic Logic*, 17:43–58, 1952.

[Loe75]   Peter A. Loeb. Conversion from nonstandard to standard measure spaces and applications in probability theory. *Trans. Amer. Math. Soc.*, 211:113–122, 1975.

[Łoś55]   Jerzy Łoś. Quelques remarques, théorèmes et problèmes sur les classes définissables d'algèbres. In *Mathematical interpretation of formal systems*, pages 98–113. North-Holland Publishing Co., Amsterdam, 1955.

[Mos62]   A. Mostowski. Axiomatizability of some many valued predicate calculi. *Fund. Math.*, 50:165–190, 1961/1962.

[Ros97]   David A. Ross. Loeb measure and probability. In *Nonstandard analysis (Edinburgh, 1996)*, volume 493 of *NATO Adv. Sci. Inst. Ser. C Math. Phys. Sci.*, pages 91–120. Kluwer Acad. Publ., Dordrecht, 1997.

[She71]   Saharon Shelah. Every two elementarily equivalent models have isomorphic ultrapowers. *Israel J. Math.*, 10:224–233, 1971.

[Tao]   Terence Tao. Walsh's ergodic theorem, metastability, and external Cauchy convergence. `http://terrytao.wordpress.com`.

[Tao08]   Terence Tao. Norm convergence of multiple ergodic averages for commuting transformations. *Ergodic Theory Dynam. Systems*, 28(2):657–688, 2008.

[Wal12]   Miguel N. Walsh. Norm convergence of nilpotent ergodic averages. *Ann. of Math. (2)*, 175(3):1667–1688, 2012.

Chapter 6

Randomizations of scattered sentences

H. Jerome Keisler
University of Wisconsin-Madison
Madison, Wisconsin, USA

1 Introduction

The notion of a scattered sentence φ of the infinitary logic $L_{\omega_1\omega}$ was introduced by Michael Morley [13] in connection with Vaught's conjecture. The notion of a randomization was introduced by the author in [10] and developed in the setting of continuous model theory by Itaï Ben Yaacov and the author in [6]. The **pure randomization theory** is a continuous theory with a sort $\mathbb{K}$ for random elements and a sort $\mathbb{E}$ for events, and a set of axioms that say that there is an event corresponding to each first order formula with random elements in its argument places, and there is an atomless probability measure on the events. By a **separable randomization** of a first order theory T we mean a separable model of the pure randomization theory in which each axiom of T has probability one.

In [1], Uri Andrews and the author showed that if T is a complete theory with at most countably many countable models up to isomorphism, then T has few separable randomizations, which means that all of its separable randomizations are very simple in a sense explained below. In this chapter we generalize that result by replacing the theory T with an infinitary sentence φ, and establish relationships between sentences with countably many countable models, scattered sentences, sentences with few separable randomizations, and Vaught's conjecture.

Let φ be a sentence of $L_{\omega_1\omega}$ whose models have at least two elements, and let $I(\varphi)$ be the class of isomorphism types of countable models of φ. In [13], Morley showed that if φ is scattered, then $I(\varphi)$ has cardinality at most $\aleph_1$, and if φ is not scattered, then $I(\varphi)$ has cardinality continuum. The absolute form of Vaught's conjecture for φ says that if φ is scattered, then $I(\varphi)$ is at most countable.

In the version of continuous model theory developed in [5], the universe of a structure is a complete metric space with distance playing the role of equality, and formulas take values in the unit interval $[0, 1]$ with 0 interpreted as true. A model is separable if its universe has a countable dense subset. The **randomization signature** L^R has two sorts, $\mathbb{K}$ for random elements and $\mathbb{E}$ for events. L^R has a function symbol $[\![\psi(\cdot)]\!]$ of sort $\mathbb{K}^n \to \mathbb{E}$ for each first order formula $\psi(\vec{v})$ with n free variables. The continuous term $[\![\psi(\vec{\mathbf{f}})]\!]$ is interpreted as the event that the formula $\psi(\vec{v})$ is satisfied by the n-tuple $\vec{\mathbf{f}}$ of random elements. In the event sort $\mathbb{E}$, L^R has the Boolean operations and a predicate μ. The continuous formula $\mu(\mathsf{E})$ takes values in $[0, 1]$ and is interpreted as the probability of the event E.

In Theorem 5.1 we show that in any separable model of the pure randomization theory, the function $[\![\psi(\cdot)]\!]$ can be extended in a natural way from the case that $\psi(\vec{v})$ is a first order formula to the case that $\psi(\vec{v})$ is a formula of $L_{\omega_1\omega}$. We can then define a **separable randomization** of an infinitary sentence φ to be a separable model of the pure randomization theory in which $[\![\varphi]\!]$ has probability one.

A **basic randomization of** φ is a very simple kind of separable randomization of φ that is determined up to isomorphism by taking a countable subset $J \subseteq I(\varphi)$ and assigning a probability $\rho(j)$ to each $j \in J$. A basic randomization of φ has a model $\mathcal{M}_j$ of isomorphism type j for each $j \in J$, and a partition of $[0, 1)$ into Borel sets B_j of measure $\rho(j)$. The events are the Borel subsets of $[0, 1)$ with the usual measure, and the random elements are the Borel functions that send B_j into $\mathcal{M}_j$ for each $j \in J$.

We say that φ **has few separable randomizations** if every separable randomization of φ is isomorphic to a basic randomization of φ.

In Theorem 9.6, we show that if $I(\varphi)$ is countable, then φ has few separable randomizations. In Theorem 10.1 we show that if φ has few separable randomizations, then φ is scattered. Therefore, if the absolute form of Vaught's conjecture holds for φ, then φ has few separable randomizations if and only if $I(\varphi)$ is countable, and also if and only if φ is scattered. In Theorem 10.3 we show that if Martin's axiom for $\aleph_1$ holds and φ is scattered, then φ has few separable randomizations.

Section 2 reviews some results we need in the literature about scattered sentences and Vaught's conjecture. Section 3 contains a review of some previous results about randomizations. In Section 4 we introduce the basic randomizations of φ. In Section 5 we introduce the separable randomizations of φ. In Section 6 we develop a key tool for constructing separable randomizations, called a countable generator, and in Section 7 we show that every separable randomization of φ is isomorphic to one that can be constructed in that way. In Section 8 we show that every separable randomization of φ can be elementarily embedded in some basic randomization if and only if only countably many first order types are realized in countable models of φ. The methods developed in Sections 6 through 8 are used to prove our main results are in Sections 9 and 10. In Section 11 we list some open questions that are related to our results.

2 Scattered sentences

We fix a countable[1] first order signature L, and all first order structures mentioned are understood to have signature L. We refer to [9] for the infinitary logic $L_{\omega_1\omega}$. Note in particular that every formula of $L_{\omega_1\omega}$ has at most finitely many free variables. By a **countable fragment** L_A of $L_{\omega_1\omega}$ we mean a countable set of formulas of $L_{\omega_1\omega}$ that contains the first order formulas

---

[1]In this chapter, "countable" means "of cardinality at most $\aleph_0$".

and is closed under subformulas, finite Boolean combinations, quantifiers, and change of free variables.

In general, the class of countable first order structures is a proper class. To avoid this problem, let $\mathbb{M}(L)$ be the class of countable structures with signature L, whose universe is $\mathbb{N}$ or an initial segment of $\mathbb{N}$. Then $\mathbb{M}(L)$ is a set, and every countable structure is isomorphic to some element of $\mathbb{M}(L)$. We define the **isomorphism type** of a countable structure $\mathcal{M}$ to be the set of all $\mathcal{H} \in \mathbb{M}(L)$ such that $\mathcal{H}$ is isomorphic to $\mathcal{M}$.

Consider a sentence φ of $L_{\omega_1\omega}$ that has at least one model. By the Löwenheim-Skolem Theorem, φ has at least one countable model. We let $I(\varphi)$ be the set of all isomorphism types of countable models of φ. By a **Scott sentence** for a countable structure $\mathcal{M}$ we mean an $L_{\omega_1\omega}$ sentence θ such that $\mathcal{M} \models \theta$, and every countable model of θ is isomorphic to $\mathcal{M}$.

Result 2.1. *(Scott's Theorem, [15]) Every countable structure has a Scott sentence.*

We let I be the set of all isomorphism types of countable structures of cardinality ≥ 2. Thus $I = I((\exists x)(\exists y)x \neq y)$. For each $i \in I$, we choose once and for all a Scott sentence θ_i for the countable models of isomorphism type i. We say that two countable L-structures $\mathcal{M}, \mathcal{H}$ are α-equivalent if they satisfy the same $L_{\omega_1\omega}$-sentences of quantifier rank at most α. By Scott's theorem, $\mathcal{M}$ and $\mathcal{H}$ are isomorphic if and only if they are α-equivalent for all countable α.

Several equivalent characterizations of scattered sentences were given in [4]. We will take one of these as our definition.

Definition 2.2. An $L_{\omega_1\omega}$ sentence φ is **scattered** if for each countable ordinal α, there are at most countably many α-equivalence classes of countable models of φ. A first order theory T is scattered if the sentence $\bigwedge T$ is scattered.

Result 2.3. *(Morley [13]) If φ is scattered, then $I(\varphi)$ has cardinality at most $\aleph_1$, and if φ is not scattered, then $I(\varphi)$ has cardinality $2^{\aleph_0}$.*

The **Vaught conjecture for** φ ([18]) says that $I(\varphi)$ is either countable or has cardinality $2^{\aleph_0}$. The **absolute Vaught conjecture for** φ (see Steel [17]) says that if φ is scattered, then $I(\varphi)$ is countable. It is called absolute because its truth does not depend on the underlying model of ZFC. In $ZFC + GCH$ the Vaught conjecture trivially holds for all φ. In $ZFC + \neg CH$, the absolute Vaught conjecture for φ is equivalent to the Vaught conjecture for φ.

Definition 2.4. (Morley [13]) An **enumerated structure** $(\mathcal{M}, a)$ is a countable structure $\mathcal{M}$ with signature L together with a mapping a from $\mathbb{N}$ onto the universe M of $\mathcal{M}$.

Consider a countable fragment L_A and an enumerated structure $(\mathcal{M}, a)$. We take 2^{L_A} to be the Polish space whose elements are the functions from L_A into $\{0, 1\}$. We say that a point $t \in 2^{L_A}$ **codes** $(\mathcal{M}, a)$ if for each formula $\psi \in L_A$ with at most the free variables $v_0, \ldots, v_{n-1}$, $t(\psi) = 0$ if and only if $\mathcal{M} \models \psi(a_0, \ldots, a_{n-1})$. Note that each enumerated structure is coded by a unique $t \in 2^{L_A}$.

The lemma below is a variant of Theorem 3.3 in [4], and follows from its proof.

Lemma 2.5. *Let φ be a sentence of $L_{\omega_1\omega}$. The following are equivalent:*

  (i) φ is not scattered.

  (ii) There is a countable fragment L_A of $L_{\omega_1\omega}$ and a perfect set $P \subseteq 2^{L_A}$ such that each $t \in P$ codes an enumerated model $(\mathcal{M}(t), a(t))$ of φ, and if $s \neq t$ in P, then $\mathcal{M}(s)$ and $\mathcal{M}(t)$ do not satisfy the same L_A-sentences.

---

3 Randomizations of theories

3.1 Continuous structures

We assume familiarity with the basic notions about continuous model theory as developed in [5]. We give some brief reminders here.

In continuous model theory, the universe of a structure is a complete metric space, and the universe of a pre-structure is a pseudo-metric space. A structure (or pre-structure) is said to be **separable** if its universe is a separable metric space (or pseudo-metric space). Formulas take truth values in $[0, 1]$, and are built from atomic formulas using continuous connectives on $[0, 1]$ and the quantifiers $\sup, \inf$. The value 0 in interpreted as truth, and a model of a set U of sentences is a continuous structure in which each $\Phi \in U$ has truth value 0.

We extend the notions of embedding and elementary embedding to pre-structures in the natural way. Given pre-structures $\mathcal{P}, \mathcal{N}$, we write $h : \mathcal{P} \prec \mathcal{N}$ (h is an **elementary embedding**) if h preserves the truth values of all formulas. If $h \colon \mathcal{P} \prec \mathcal{N}$ where h is the inclusion mapping, we write $\mathcal{P} \prec \mathcal{N}$ and say that $\mathcal{P}$ is an **elementary submodel** of $\mathcal{N}$ (leaving off the 'pre-' for brevity). If $h \colon \mathcal{P} \prec \mathcal{N}$, h preserves distance but is not necessarily one-to-one. Note that compositions of elementary embeddings are elementary embeddings. We write $h \colon \mathcal{P} \cong \mathcal{N}$ if $h : \mathcal{P} \prec \mathcal{N}$ and every element of $\mathcal{N}$ is at distance zero from some element of $h(\mathcal{P})$. We say that $\mathcal{P}$ and $\mathcal{N}$ are **isomorphic**, and write $\mathcal{P} \cong \mathcal{N}$, if $h \colon \mathcal{P} \cong \mathcal{N}$ for some h. By Remark 2.4 of [1], $\cong$ is an equivalence relation on pre-structures.

We call $\mathcal{N}$ a **reduction of** $\mathcal{P}$ if $\mathcal{N}$ is obtained from $\mathcal{P}$ by identifying elements at distance zero, and call $\mathcal{N}$ a **completion of** $\mathcal{P}$ if $\mathcal{N}$ is a structure obtained from a reduction of $\mathcal{P}$ by completing the metrics. Every pre-structure has a reduction, that is unique up to isomorphism. The mapping that identifies elements at distance zero is called the **reduction mapping**, and is an isomorphism from a pre-structure onto its reduction. Similarly, every pre-structure $\mathcal{P}$ has a completion, that is unique up to isomorphism, and the reduction map is an elementary embedding of $\mathcal{P}$ into its completion.

Following [6], we say that $\mathcal{P}$ is **pre-complete** if the metrics in a reduction of $\mathcal{P}$ are already complete. Thus if $\mathcal{P}$ is pre-complete, the reductions and completions of $\mathcal{P}$ are the same, and $\mathcal{P}$ is isomorphic to its completion.

3.2   Randomizations

We assume that:

- L is a countable first order signature.

- T_2 is the theory with the single axiom $(\exists x)(\exists y)x \neq y$.

- T is a theory with signature L that contains T_2.

- φ is a sentence of $L_{\omega_1 \omega}$ that implies T_2.

Note that T_2 is just the theory whose models have at least two elements, and $I(\varphi) \subseteq I(T_2) = I$. The randomization theory of T is a continuous theory T^R whose signature L^R has two sorts, a sort $\mathbb{K}$ for random elements of models of T, and a sort $\mathbb{E}$ for events in an underlying probability space. The probability of the event that a first order formula holds for a tuple of random elements will be expressible by a formula of continuous logic. The signature L^R has an n-ary function symbol $[\![\theta(\cdot)]\!]$ of sort $\mathbb{K}^n \to \mathbb{E}$ for each first order formula θ of L with n free variables, a $[0,1]$-valued unary predicate symbol μ of sort $\mathbb{E}$ for probability, and the Boolean operations $\top, \bot, \sqcap, \sqcup, \neg$ of sort $\mathbb{E}$. The signature L^R also has distance predicates $d_{\mathbb{E}}$ of sort $\mathbb{E}$ and $d_{\mathbb{K}}$ of sort $\mathbb{K}$. In L^R, we use $\mathsf{B}, \mathscr{C}, \ldots$ for variables or parameters of sort $\mathbb{E}$, and $\mathsf{B} \doteq \mathscr{C}$ means $d_{\mathbb{E}}(\mathsf{B}, \mathscr{C}) = 0$. For readability we write $\forall, \exists$ for $\sup, \inf$.

The axioms of T^R, which are taken from [6], are as follows:

Validity Axioms
$$\forall \vec{x}([\![\psi(\vec{x})]\!] \doteq \top)$$

where $\forall \vec{x}\, \psi(\vec{x})$ is logically valid in first order logic.

Boolean Axioms The usual Boolean algebra axioms in sort $\mathbb{E}$, and the statements
$$\forall \vec{x}([\![(\neg\theta)(\vec{x})]\!] \doteq \neg[\![\theta(\vec{x})]\!])$$
$$\forall \vec{x}([\![(\varphi \vee \psi)(\vec{x})]\!] \doteq [\![\theta(\vec{x})]\!] \sqcup [\![\psi(\vec{x})]\!])$$
$$\forall \vec{x}([\![(\theta \wedge \psi)(\vec{x})]\!] \doteq [\![\theta(\vec{x})]\!] \sqcap [\![\psi(\vec{x})]\!])$$

Distance Axioms
$$\forall x \forall y\, d_{\mathbb{K}}(x,y) = 1 - \mu[\![x = y]\!], \qquad \forall \mathsf{B} \forall \mathscr{C}\, d_{\mathbb{E}}(\mathsf{B}, \mathscr{C}) = \mu(\mathsf{B} \triangle \mathscr{C})$$

Fullness Axioms (or Maximal Principle)

$$\forall \vec{y} \exists x ([\![\theta(x, \vec{y})]\!] \doteq [\![(\exists x \theta)(\vec{y})]\!])$$

Event Axiom

$$\forall \mathsf{B} \exists x \exists y (\mathsf{B} \doteq [\![x = y]\!])$$

Measure Axioms

$$\mu[\top] = 1 \wedge \mu[\bot] = 0$$

$$\forall \mathsf{B} \forall \mathscr{C} (\mu[\mathsf{B}] + \mu[\mathscr{C}] = \mu[\mathsf{B} \sqcup \mathscr{C}] + \mu[\mathsf{B} \sqcap \mathscr{C}])$$

Atomless Axiom

$$\forall \mathsf{B} \exists \mathscr{C} (\mu[\mathsf{B} \sqcap \mathscr{C}] = \mu[\mathsf{B}]/2)$$

Transfer Axioms

$$[\![\theta]\!] \doteq \top$$

where $\theta \in T$.

By a **separable randomization of** T we mean a separable pre-model of T^R. In this chapter we will focus on the **pure randomization theory** T_2^R. T_2^R has the single transfer axiom $[\![(\exists x)(\exists y)x \neq y]\!] \doteq \top$. Note that for any theory $T \supseteq T_2$, any model of T^R is a model of the pure randomization theory. By a **separable randomization** we mean a separable randomization of T_2^R. A separable randomization is called **complete** if it is a model of T_2^R, and **pre-complete** if it is a pre-complete model of T_2^R.

We will use $\mathcal{M}, \mathcal{H}$ to denote models of T_2 with signature L, and use $\mathcal{N}$ and $\mathcal{P}$ to denote models or pre-models of T_2^R with signature L^R. The universe of $\mathcal{M}$ will be denoted by M. A pre-model of T_2^R will be a pair $\mathcal{N} = (\mathcal{K}, \mathcal{E})$ where $\mathcal{K}$ is the part of sort $\mathbb{K}$ and $\mathcal{E}$ is the part of sort $\mathbb{E}$. We write $[\![\theta(\vec{f})]\!]^{\mathcal{N}}$ for the interpretation of $[\![\theta(\vec{v})]\!]$ in a pre-structure $\mathcal{N}$ at a tuple $\vec{f}$, and write $[\![\theta(\vec{f})]\!]$ for $[\![\theta(\vec{f})]\!]^{\mathcal{N}}$ when $\mathcal{N}$ is clear from the context.

Result 3.1. *([6], Theorem 2.7) Every model or pre-complete model $\mathcal{N} = (\mathcal{K}, \mathcal{E})$ of T_2^R has perfect witnesses, i.e.,*

(i) for each first order formula $\theta(x, \vec{y})$ and each $\vec{g}$ in $\mathcal{K}^n$ there exists $\mathbf{f} \in \mathcal{K}$ such that

$$[\![\theta(\mathbf{f}, \vec{g})]\!] \doteq [\![(\exists x \theta)(\vec{g})]\!];$$

(ii) for each $\mathsf{B} \in \mathcal{E}$ there exist $\mathbf{f}, \mathbf{g} \in \mathcal{K}$ such that $\mathsf{B} \doteq [\![\mathbf{f} = \mathbf{g}]\!]$.

We let $\mathcal{L}$ be the family of Borel subsets of $[0, 1)$, and let $([0, 1), \mathcal{L}, \lambda)$ be the usual probability space, where λ is the restriction of Lebesgue measure to $\mathcal{L}$. We let $\mathcal{M}^{[0,1)}$ be the set of functions with countable range from $[0, 1)$ into M such that the inverse image of any element of M belongs to $\mathcal{L}$. The elements of $\mathcal{M}^{[0,1)}$ are called **random elements of** $\mathcal{M}$.

Definition 3.2. The **Borel randomization of** $\mathcal{M}$ is the pre-structure $(\mathcal{M}^{[0,1)}, \mathcal{L})$ for L^R whose universe of sort $\mathbb{K}$ is $\mathcal{M}^{[0,1)}$, whose universe of sort $\mathbb{E}$ is $\mathcal{L}$, whose measure μ is given by $\mu(\mathsf{B}) = \lambda(\mathsf{B})$ for each $\mathsf{B} \in \mathcal{L}$, and whose $[\![\psi(\cdot)]\!]$ functions are

$$[\![\psi(\vec{\mathbf{f}})]\!] = \{t \in [0,1) : \mathcal{M} \models \psi(\vec{\mathbf{f}}(t))\}.$$

(So $[\![\psi(\vec{\mathbf{f}})]\!] \in \mathcal{L}$ for each first order formula $\psi(\vec{v})$ and tuple $\vec{\mathbf{f}}$ in $\mathcal{M}^{[0,1)}$). Its distance predicates are defined by

$$d_{\mathbb{E}}(\mathsf{B}, \mathscr{C}) = \mu(\mathsf{B} \triangle \mathscr{C}), \quad d_{\mathbb{K}}(\mathbf{f}, \mathbf{g}) = \mu([\![\mathbf{f} \neq \mathbf{g}]\!]),$$

where $\triangle$ is the symmetric difference operation.

Result 3.3. *([6], Corollary 3.6) Every Borel randomization of a countable model of T_2 is a pre-complete separable randomization (in other words, a pre-complete separable model of T_2^R).*

Result 3.4. *([1], Theorem 4.5) Suppose $\mathcal{N}$ is pre-complete and elementarily embeddable in the Borel randomization $(\mathcal{M}^{[0,1)}, \mathcal{L})$ of a countable model of T_2. Then $\mathcal{N}$ is isomorphic to an elementary submodel of $(\mathcal{M}^{[0,1)}, \mathcal{L})$ whose event sort is all of $\mathcal{L}$.*

---

4   Basic randomizations

Basic randomizations are generalizations of Borel randomizations. They are very simple continuous pre-structures of sort L^R. Intuitively, a basic randomization is a combination of countably many Borel randomizations of first order structures. Andrews and Keisler [1] considered basic randomizations that are combinations of Borel randomizations of models of a single complete theory T, and called them called *product randomizations*.

Definition 4.1. Suppose that

- J is a countable subset of I;

- $[0,1) = \bigcup_{j \in J} \mathsf{B}_j$ is a partition of $[0,1)$ into Borel sets of positive measure;

- for each $j \in J$, $\mathcal{M}_j$ has isomorphism type j;

- $\prod_{j \in J} \mathcal{M}_j^{\mathsf{B}_j}$ is the set of all functions $\mathbf{f} \colon [0,1) \to \bigcup_{j \in J} M_j$ such that for all $j \in J$,

$$(\forall t \in \mathsf{B}_j)\mathbf{f}(t) \in M_j \text{ and } (\forall a \in M_j)\{t \in \mathsf{B}_j : \mathbf{f}(t) = a\} \in \mathcal{L};$$

- $(\prod_{j \in J} \mathcal{M}_j^{\mathsf{B}_j}, \mathcal{L})$ is the pre-structure for L^R whose measure and distance functions are as in Definition 3.2. and $[\![\psi(\cdot)]\!]$ functions are

$$[\![\psi(\vec{\mathbf{f}})]\!] = \bigcup_{j \in J} \{t \in \mathsf{B}_j \colon \mathcal{M}_j \models \psi(\vec{\mathbf{f}}(t))\},$$

$(\prod_{i \in J} \mathcal{M}_i^{\mathsf{B}_i}, \mathcal{L})$ is called a **basic randomization**. Given a basic randomization, we let $\mathcal{M}_t = \mathcal{M}_j$ whenever $j \in J$ and $t \in \mathsf{B}_j$. By a **basic randomization of** φ we mean a basic randomization such that $\mathcal{M}_j \models \varphi$ for each $j \in J$.

Remark 4.2.

1. In a basic randomization, the set $\bigcup_{j \in J} M_j$ is countable, so each $\mathbf{f} \in \prod_{j \in J} \mathcal{M}_j^{\mathsf{B}_j}$ has countable range.

2. If $\mathcal{M}_j \cong \mathcal{H}_j$ for each $j \in J$, then $(\prod_{j \in J} \mathcal{M}_j^{\mathsf{B}_j}, \mathcal{L}) \cong (\prod_{j \in J} \mathcal{H}_j^{\mathsf{B}_j}, \mathcal{L})$.

3. Every basic randomization $(\prod_{j \in J} \mathcal{M}_j^{\mathsf{B}_j}, \mathcal{L})$ is isomorphic to a basic randomization $(\prod_{j \in J} \mathcal{H}_j^{\mathsf{B}_j}, \mathcal{L})$ such that for each $j \in J$, $\mathcal{H}_j \in \mathbb{M}(L)$ (so the universe of $\mathcal{H}_j$ is $\mathbb{N}$ or an initial segment of $\mathbb{N}$).

4. If $\mathcal{M}_j \prec \mathcal{H}_j$ for each $j \in J$, then $(\prod_{j \in J} \mathcal{M}_j^{\mathsf{B}_j}, \mathcal{L}) \prec (\prod_{j \in J} \mathcal{H}_j^{\mathsf{B}_j}, \mathcal{L})$. (In this part we do not require that $\mathcal{H}_j$ has isomorphism type j).

Lemma 4.3. *Every basic randomization* $\mathcal{P} = (\prod_{j \in J} \mathcal{M}_j^{\mathsf{B}_j}, \mathcal{L})$ *is a pre-model of the pure randomization theory.*

Proof. All of the axioms for T_2^R, except the Fullness Axioms, hold trivially. Therefore $\mathcal{P}$ is a pseudo-metric space in both sorts. By Result 3.3, $(\mathcal{M}_j^{[0,1)}, \mathcal{L})$ satisfies the Fullness Axioms for each $j \in J$, and it follows easily that $\mathcal{P}$ also satisfies the Fullness Axioms, and thus is a pre-model of T_2^R. ∎ 4.3

We next introduce useful mappings from a basic randomization $(\prod_{j \in J} \mathcal{M}_j^{\mathsf{B}_j}, \mathcal{L})$ to the Borel randomizations $(\mathcal{M}_j^{[0,1)}, \mathcal{L})$.

Definition 4.4. Suppose $\mathsf{B} \in \mathcal{L}$ and $\lambda(\mathsf{B}) > 0$. We say that a mapping ℓ **stretches** B to $[0,1)$ if ℓ is a Borel bijection from B to $[0,1)$, ℓ^{-1} is also Borel, and for each Borel set $\mathsf{A} \subseteq \mathsf{B}$, $\lambda(\ell(\mathsf{A})) = \lambda(\mathsf{A})/\lambda(\mathsf{B})$.

Let $\mathcal{P} = (\prod_{j \in J} \mathcal{M}_j^{\mathsf{B}_j}, \mathcal{L})$ be a basic randomization, and for each $j \in J$, choose an ℓ_j that stretches B_j to $[0,1)$. Define the mapping $\ell_j \colon \mathcal{P} \to (\mathcal{M}_j^{[0,1)}, \mathcal{L})$ by

$$(\ell_j(\mathbf{f}))(t) = \mathbf{f}(\ell_j^{-1}(t)), \quad \ell_j(\mathsf{E}) = \ell_j(\mathsf{B}_j \cap \mathsf{E}).$$

Remark 4.5. Let $\mathcal{P} = (\prod_{j \in J} \mathcal{M}_j^{\mathsf{B}_j}, \mathcal{L})$ be a basic randomization.

1. For each $j \in J$, there exists a mapping ℓ_j that stretches B_j to $[0,1)$.

2. ℓ_j maps $\mathcal{P}$ onto $\mathcal{P}_j := (\mathcal{M}_j^{[0,1)}, \mathcal{L})$.

3. For each first order formula $\psi(\vec{v})$ and tuple $\vec{\mathbf{f}}$ of elements of $\mathcal{P}$ of sort $\mathbb{K}$.

$$\lambda([\![\psi(\vec{\mathbf{f}})]\!]^{\mathcal{P}}) = \sum_{j \in J} \lambda(\mathsf{B}_j)\lambda([\![\psi(\ell_j\vec{\mathbf{f}})]\!]^{\mathcal{P}_j}).$$

4. $d_{\mathbb{K}}^{\mathcal{P}}(\mathbf{f}, \mathbf{g}) = \sum_{j \in J} \lambda(\mathsf{B}_j)d_{\mathbb{K}}^{\mathcal{P}_j}(\ell_j(\mathbf{f}), \ell_j(\mathbf{g}))$.

Proof. Since $\nu(\mathsf{A}) = \lambda(\mathsf{A})/\lambda(\mathsf{B}_j)$ is a probability measure on B_j, (1) follows from Theorem 17.41 in [8]. (2)–(4) are clear.   ∎$_{4.5}$

The following result is a generalization of Theorem 7.3 of [1], but the proof we give here is different.

Theorem 4.6. *Every basic randomization is pre-complete and separable.*

Proof. Let $\mathcal{P} = (\prod_{j \in J} \mathcal{M}_j^{\mathsf{B}_j}, \mathcal{L})$ be a basic randomization. By Result 3.3, $\mathcal{P}$ is separable and pre-complete in the event sort. For each $j \in J$, pick a mapping ℓ_j that stretches B_j to $[0, 1)$. Pick an element $\mathbf{a} \in \prod_{j \in J} \mathcal{M}_j^{\mathsf{B}_j}$.

Separability in sort $\mathbb{K}$: By 3.3, for each $j \in J$, there is a countable set C_j that is dense in $\mathcal{M}_j^{[0,1)}$. For each finite $F \subseteq J$, let D_F be the set of all $\mathbf{f}$ such that for all $j \in F$, $\mathbf{f}$ agrees with some element of $\ell_j^{-1}C_j$ on B_j, and $\mathbf{f}$ agrees with $\mathbf{a}$ on $[0, 1) \setminus \bigcup_{i \in F} \mathsf{B}_j$. Then $D = \bigcup_F D_F$ is a countable subset of $\prod_{j \in J} \mathcal{M}_j^{\mathsf{B}_j}$. For each $\varepsilon > 0$, there is a finite $F \subseteq J$ such that $\sum_{j \in F} \mu(\mathsf{B}_j) \geq 1 - \varepsilon$. It follows that for each $\mathbf{g} \in \prod_{j \in J} \mathcal{M}_j^{\mathsf{B}_j}$, there exists $\mathbf{f} \in D_F$ such that for each $j \in F$, $d_{\mathbb{K}}(\ell_j(\mathbf{f}), \ell_j(\mathbf{g})) < \varepsilon/(|F| + 1)$, and therefore by Remark 4.5, $d_{\mathbb{K}}(\mathbf{f}, \mathbf{g}) < 2\varepsilon$. Hence D is dense in $\prod_{j \in J} \mathcal{M}_j^{\mathsf{B}_j}$.

Pre-completeness in sort $\mathbb{K}$: Suppose that $\langle \mathbf{f}_n \rangle_{n \in \mathbb{N}}$ is a Cauchy sequence of sort $\mathbb{K}$. By Remark 4.5, for each $j \in J$, $\langle \ell_j(\mathbf{f}_n) \rangle_{n \in \mathbb{N}}$ is a Cauchy sequence in $\mathcal{M}_j^{[0,1)}$. By Result 3.3, $\mathcal{M}_j^{[0,1)}$ is pre-complete, so there exists $\mathbf{g}_j$ in $\mathcal{M}_j^{[0,1)}$ such that $\lim_{n \to \infty} d_{\mathbb{K}}(\ell_j(\mathbf{f}_n), \mathbf{g}_j) = 0$. Let $\mathbf{g}$ be the function that agrees with $\ell_j^{-1}\mathbf{g}_j$ on B_j for each $j \in J$. Then $\mathbf{g}_j = \ell_j(\mathbf{g}))$ for each $j \in J$, so $\lim_{n \to \infty} d_{\mathbb{K}}(\ell_j(\mathbf{f}_n), \ell_j(\mathbf{g})) = 0$. By Remark 4.5, $\lim_{n \to \infty} d_{\mathbb{K}}(\mathbf{f}_n, \mathbf{g}) = 0$ in $\mathcal{P}$.   ∎$_{4.6}$

Definition 4.7. By a **probability density function** on I we mean a function $\rho \colon I \to [0, 1]$ such that $\rho(i) = 0$ for all but countably many $i \in I$, and $\sum_i \rho(i) = 1$.

For each basic randomization $\mathcal{P} = (\prod_{j \in J} \mathcal{M}_j^{\mathsf{B}_j}, \mathcal{L})$, the function $\rho(i) = \lambda(\mathsf{B}_i)$ for $i \in J$, and $\rho(i) = 0$ for $i \in I \setminus J$, is called the **density function of** $\mathcal{P}$.

Remark 4.8. It is easily seen that ρ is a probability density function on I if and only if ρ is the density function of some basic randomization.

The following result is a generalization of Theorem 7.5 of [1], and is proved in the same way.

Theorem 4.9. *Two basic randomizations are isomorphic if and only if they have the same density function.*

If a continuous structure $\mathcal{N}$ is isomorphic to a basic randomization $\mathcal{P}$, the density function of $\mathcal{P}$ is also called a density function of $\mathcal{N}$. Thus such an $\mathcal{N}$ has a unique density function, which characterizes $\mathcal{N}$ up to isomorphism.

---

5 Events defined by infinitary formulas

In this section we consider arbitrary complete separable randomizations. By definition, each complete separable randomization has an event function $[\![\psi(\cdot)]\!]^{\mathcal{N}}$ of sort $\mathbb{K}^n \to \mathbb{E}$ for each first order formula $\psi(\vec{v})$ with n free variables. The following theorem extends this to the case where $\psi(\vec{v})$ is a formula of the infinitary logic $L_{\omega_1\omega}$.

Theorem 5.1. *Let $\mathcal{N} = (\mathcal{K}, \mathcal{E})$ be a complete separable randomization, and let Ψ_n be the class of $L_{\omega_1\omega}$ formulas with n free variables. There is a unique family of functions $[\![\psi(\cdot)]\!]^{\mathcal{N}}$, $\psi \in \bigcup_n \Psi_n$, such that:*

(i) When $\psi \in \Psi_n$, $[\![\psi(\cdot)]\!]^{\mathcal{N}} : \mathcal{K}^n \to \mathcal{E}$.

(ii) When ψ is a first order formula, $[\![\psi(\cdot)]\!]^{\mathcal{N}}$ is the usual event function for the structure $\mathcal{N}$.

(iii) $[\![\neg\psi(\vec{\mathbf{f}})]\!]^{\mathcal{N}} = \neg[\![\psi(\vec{\mathbf{f}})]\!]^{\mathcal{N}}$.

(iv) $[\![(\psi_1 \vee \psi_2)(\vec{\mathbf{f}})]\!]^{\mathcal{N}} = [\![\psi_1(\vec{\mathbf{f}})]\!]^{\mathcal{N}} \sqcup [\![\psi_2(\vec{\mathbf{f}})]\!]^{\mathcal{N}}$.

(v) $[\![\bigvee_k \psi_k(\vec{\mathbf{f}})]\!]^{\mathcal{N}} = \sup_k [\![\psi_k(\vec{\mathbf{f}})]\!]^{\mathcal{N}}$.

(vi) $[\![(\exists u)\theta(u, \vec{\mathbf{f}})]\!]^{\mathcal{N}} = \sup_{\mathbf{g} \in \mathcal{K}} [\![\theta(\mathbf{g}, \vec{\mathbf{f}})]\!]^{\mathcal{N}}$.

Moreover, for each $\psi \in \Psi_n$, the function $[\![\psi(\cdot)]\!]^{\mathcal{N}}$ is Lipschitz continuous with bound one, that is, for any pair of n-tuples $\vec{\mathbf{f}}, \vec{\mathbf{h}} \in \mathcal{K}^n$ we have

$$d_{\mathbb{E}}([\![\psi(\vec{\mathbf{f}})]\!]^{\mathcal{N}}, [\![\psi(\vec{\mathbf{h}})]\!]^{\mathcal{N}}) \leq \sum_{m<n} d_{\mathbb{K}}(\mathbf{f}_m, \mathbf{h}_m).$$

Proof. We argue by induction on the complexity of formulas. Assume that the result holds for all subformulas of ψ. If ψ is a first order formula or a negation or finite disjunction, it is clear that the result holds for ψ.

Suppose $\psi = \bigvee_k \psi_k$. We show that the supremum exists. For each $m \in \mathbb{N}$ we have

$$[\![\bigvee_{k=0}^{m} \psi_k(\vec{\mathbf{f}})]\!]^{\mathcal{N}} = \bigsqcup_{k=0}^{m} [\![\psi_k(\vec{\mathbf{f}})]\!]^{\mathcal{N}}.$$

This is increasing in k, so by the completeness of the metric $d_{\mathbb{E}}$ on $\mathcal{E}$, $\lim_{k \to \infty} [\![\bigvee_{j=0}^{k} \psi_j(\vec{\mathbf{f}})]\!]^{\mathcal{N}}$ exists and is equal to $\sup_k [\![\psi_k(\vec{\mathbf{f}})]\!]^{\mathcal{N}}$. By hypothesis, the Lipschitz condition holds for each ψ_k. It follows that the Lipschitz condition also holds for ψ.

Now suppose $\psi(\vec{v}) = (\exists u)\theta(u, \vec{v})$. We again show first that the supremum exists. By separability, there is a countable dense subset $D = \{\mathbf{d}_k : k \in \mathbb{N}\}$ of $\mathcal{K}$. It follows from the axioms of T_2^R that there is a sequence $\langle \mathbf{g}_k \rangle_{k \in \mathbb{N}}$ of elements of $\mathcal{K}$ such that $\mathbf{g}_0 = \mathbf{d}_0$ and for each k, $\mathbf{g}_{k+1}$ agrees with $\mathbf{g}_k$ on the event $[\![\theta(\mathbf{g}_k, \vec{\mathbf{f}})]\!]^{\mathcal{N}}$ and agrees with $\mathbf{d}_k$ elsewhere. Then for each $m \in \mathbb{N}$ we have

$$[\![\theta(\mathbf{g}_m, \vec{\mathbf{f}})]\!]^{\mathcal{N}} = \bigsqcup_{k=0}^{m} [\![\theta(\mathbf{d}_k, \vec{\mathbf{f}})]\!]^{\mathcal{N}}.$$

So whenever $k \leq m$, we have

$$[\![\theta(\mathbf{g}_k, \vec{\mathbf{f}})]\!]^{\mathcal{N}} \sqsubseteq [\![\theta(\mathbf{g}_m, \vec{\mathbf{f}})]\!]^{\mathcal{N}},$$

and hence

$$\mathsf{E} := \lim_{k \to \infty} [\![\theta(\mathbf{g}_k, \vec{\mathbf{f}})]\!]^{\mathcal{N}} = \sup_{k \in \mathbb{N}} [\![\theta(\mathbf{g}_k, \vec{\mathbf{f}})]\!]^{\mathcal{N}}$$

exists in $\mathcal{E}$.

Consider any $\mathbf{h} \in \mathcal{K}$. To show that the supremum $\sup_{\mathbf{h} \in \mathcal{K}} [\![\theta(\mathbf{h}, \vec{\mathbf{f}})]\!]^{\mathcal{N}}$ exists in $\mathcal{E}$, it suffices to show that $[\![\theta(\mathbf{h}, \vec{\mathbf{f}})]\!]^{\mathcal{N}} \sqsubseteq \mathsf{E}$, because it will then follow that E is the desired supremum. Let $\varepsilon > 0$. For some $k \in \mathbb{N}$ we have $d_{\mathbb{K}}(\mathbf{d}_k, \mathbf{h}) < \varepsilon$. Moreover,

$$[\![\mathbf{d}_k = \mathbf{h} \wedge \theta(\mathbf{h}, \vec{\mathbf{f}})]\!]^{\mathcal{N}} = [\![\mathbf{d}_k = \mathbf{h} \wedge \theta(\mathbf{d}_k, \vec{\mathbf{f}})]\!]^{\mathcal{N}} \sqsubseteq [\![\theta(\mathbf{g}_k, \vec{\mathbf{f}})]\!]^{\mathcal{N}} \sqsubseteq \mathsf{E}.$$

Then

$$[\![\theta(\mathbf{h}, \vec{\mathbf{f}})]\!]^{\mathcal{N}} \sqcap \neg\mathsf{E} \sqsubseteq [\![\mathbf{d}_k \neq \mathbf{h}]\!]^{\mathcal{N}},$$

so

$$\mu([\![\theta(\mathbf{h}, \vec{\mathbf{f}})]\!]^{\mathcal{N}} \sqcap \neg\mathsf{E}) \leq \mu([\![\mathbf{d}_k \neq \mathbf{h}]\!]^{\mathcal{N}}) = d_{\mathbb{K}}(\mathbf{d}_k, \mathbf{h}) < \varepsilon.$$

Since this holds for all $\varepsilon > 0$, we have $[\![\theta(\mathbf{h}, \vec{\mathbf{f}})]\!]^{\mathcal{N}} \sqsubseteq \mathsf{E}$.

To prove the Lipschitz condition for ψ, we consider a pair of n-tuples $\vec{\mathbf{f}}, \vec{\mathbf{h}} \in \mathcal{K}^n$. By the preceding paragraph we have

$$[\![\psi(\vec{\mathbf{f}})]\!]^{\mathcal{N}} = \lim_{k \to \infty} [\![\theta(\mathbf{g}_k, \vec{\mathbf{f}})]\!]^{\mathcal{N}}, \quad [\![\psi(\vec{\mathbf{h}})]\!]^{\mathcal{N}} = \lim_{k \to \infty} [\![\theta(\mathbf{g}_k, \vec{\mathbf{h}})]\!]^{\mathcal{N}}.$$

Therefore, for each $\varepsilon > 0$ there exists $k \in \mathbb{N}$ such that

$$d_{\mathbb{E}}([\![\theta(\mathbf{g}_k, \vec{\mathbf{f}})]\!]^{\mathcal{N}}, [\![\psi(\vec{\mathbf{f}})]\!]^{\mathcal{N}}) < \varepsilon, \quad d_{\mathbb{E}}([\![\theta(\mathbf{g}_k, \vec{\mathbf{h}})]\!]^{\mathcal{N}}, [\![\psi(\vec{\mathbf{h}})]\!]^{\mathcal{N}}) < \varepsilon.$$

By the Lipschitz condition for $\theta(u, \vec{v})$, we have

$$d_{\mathbb{E}}(\llbracket \theta(\mathbf{g}_k, \vec{\mathbf{f}}) \rrbracket^{\mathcal{N}}, \llbracket \theta(\mathbf{g}_k, \vec{\mathbf{h}}) \rrbracket^{\mathcal{N}}) \leq \sum_{i < n} d_{\mathbb{K}}(\mathbf{f}_j, \mathbf{h}_j).$$

Then by the triangle inequality, for every $\varepsilon > 0$ we have

$$d_{\mathbb{E}}((\llbracket \psi(\vec{\mathbf{f}}) \rrbracket^{\mathcal{N}}, d_{\mathbb{E}}(\llbracket \psi(\vec{\mathbf{h}}) \rrbracket^{\mathcal{N}}) < \sum_{i < n} d_{\mathbb{K}}(\mathbf{f}_j, \mathbf{h}_j) + 2\varepsilon,$$

so

$$d_{\mathbb{E}}((\llbracket \psi(\vec{\mathbf{f}}) \rrbracket^{\mathcal{N}}, d_{\mathbb{E}}(\llbracket \psi(\vec{\mathbf{h}}) \rrbracket^{\mathcal{N}}) \leq \sum_{i < n} d_{\mathbb{K}}(\mathbf{f}_j, \mathbf{h}_j).$$

$\blacksquare_{5.1}$

Remark 5.2. The proof of Theorem 5.1 only used the metric completeness of the sort $\mathbb{E}$ part of $\mathcal{N}$. Hence the result also holds in the case that $\mathcal{N}$ is a separable randomization that has a metric in sort $\mathbb{K}$ and a complete metric in sort $\mathbb{E}$.

Corollary 5.3. *Suppose that $\mathcal{N}, \mathcal{P}$ are complete separable randomizations and $h : \mathcal{N} \cong \mathcal{P}$. Then for every $L_{\omega_1 \omega}$ formula $\psi(\vec{v})$ and every tuple $\vec{\mathbf{f}}$ of sort $\mathbb{K}$ in $\mathcal{N}$, we have $h(\llbracket \psi(\vec{\mathbf{f}}) \rrbracket^{\mathcal{N}}) = \llbracket \psi(h\vec{\mathbf{f}}) \rrbracket^{\mathcal{P}}$.*

Proof. By Theorem 5.1 and an easy induction on the complexity of ψ. $\blacksquare_{5.3}$

When $\mathcal{P}$ is a pre-complete separable randomization, h is the reduction map from $\mathcal{P}$ onto its completion $\mathcal{N}$, and $\psi(\vec{v})$ is a formula of $L_{\omega_1 \omega}$, then $\llbracket \psi(h\vec{\mathbf{f}}) \rrbracket^{\mathcal{N}}$ is uniquely defined by Theorem 5.1. In that case, we will sometimes abuse notation and write $\mu(\llbracket \psi(\vec{\mathbf{f}}) \rrbracket^{\mathcal{P}})$ for $\mu(\llbracket \psi(h\vec{\mathbf{f}}) \rrbracket^{\mathcal{N}})$.

We can now define the notion of a separable randomization of φ.

Definition 5.4. We say that $\mathcal{N}$ is a **complete separable randomization of** φ if $\mathcal{N}$ is a complete separable randomization such that $\llbracket \varphi \rrbracket^{\mathcal{N}}$ is the true event $\top$. We call $\mathcal{P}$ a **separable randomization of** φ if the completion of $\mathcal{P}$ is a complete separable randomization of φ. We say that φ **has few separable randomizations** if every complete separable randomization of φ is isomorphic to a basic randomization.

Thus when φ has few separable randomizations, each complete separable randomization $\mathcal{N}$ of φ has a unique density function ρ, and ρ characterizes $\mathcal{N}$ up to isomorphism.

Corollary 5.5. *Let $\mathcal{P} = (\prod_{j \in J} \mathcal{M}_j^{\mathsf{B}_j}, \mathcal{L})$ be a basic randomization with completion $\mathcal{N}$, and let $h \colon \mathcal{P} \cong \mathcal{N}$ be the reduction map. For each $L_{\omega_1 \omega}$ formula $\psi(\vec{v})$ and tuple $\vec{\mathbf{f}}$ in $\prod_{j \in J} \mathcal{M}_j^{\mathsf{B}_j}$, $\llbracket \psi(h\vec{\mathbf{f}}) \rrbracket^{\mathcal{N}}$ is the reduction of the event*

$$\bigcup_{j \in J} \{ t \in \mathsf{B}_j \colon \mathcal{M}_j \models \psi(\vec{\mathbf{f}}(t)) \}.$$

Hence $\mathcal{P}$ is a basic randomization of φ if and only if $\mathcal{P}$ is a basic randomization and $\mathcal{P}$ is a separable randomization of φ.

Proof. In the case that $\psi(\vec{v})$ is an atomic formula, the result holds by definition. A routine induction on the complexity of formulas gives the result for arbitrary $L_{\omega_1\omega}$ formulas. ∎5.5

Note that the complete separable randomizations of the sentence $\bigwedge T$ are exactly the separable models of the continuous theory T^R. With more overhead, we could have taken an alternative approach in which the complete separable randomizations of an $L_{\omega_1\omega}$ sentence φ are exactly the separable models of a theory φ^R in an infinitary continuous logic such as the logic in [7]. The idea would be to consider a countable fragment L_A of $L_{\omega_1\omega}$, and have the randomization signature $(L_A)^R$ contain a function symbol $[\![\psi(\cdot)]\!]$ for each formula $\psi(\vec{v})$ of L_A. Then Theorem 5.1 shows that every separable randomization can be expanded in a unique way to a model with the signature $(L_A)^R$ that satisfies the infinitary sentences corresponding to the conditions (i)–(v). In this approach, φ^R would be the theory in infinitary continuous logic with the axioms of the pure randomization theory plus the above infinitary sentences and an axiom stating that $[\![\varphi]\!] \doteq \top$.

---

6   Countable generators of randomizations

In this section we give a general method of constructing pre-complete separable randomizations. In the next section we will show that every pre-complete separable randomization is isomorphic to one that can be constructed in that way.

Definition 6.1. Assume that $(\Omega, \mathcal{E}, \nu)$ is an atomless probability space such that the metric space $(\mathcal{E}, d_{\mathbb{E}})$ is separable, and for each $t \in \Omega$, $\mathcal{M}_t$ is a countable model of T_2.

A **countable generator** (in $\langle \mathcal{M}_t \rangle_{t \in \Omega}$ over $(\Omega, \mathcal{E}, \nu)$) is a countable set C of elements $\mathbf{c} \in \prod_{t \in \Omega} \mathcal{M}_t$ such that:

(a) $M_t = \{\mathbf{c}(t) \colon \mathbf{c} \in C\}$ for each $t \in \Omega$, and

(b) For every first order atomic formula $\psi(\vec{v})$ and tuple $\vec{\mathbf{b}}$ in C,

$$\{t \in \Omega : \mathcal{M}_t \models \psi(\vec{\mathbf{b}}(t))\} \in \mathcal{E}.$$

Theorem 6.2. *Let C be a countable generator in $\langle \mathcal{M}_t \rangle_{t \in \Omega}$ over $(\Omega, \mathcal{E}, \nu)$. There is a unique pre-structure $\mathcal{P}(C) = (\mathcal{K}, \mathcal{E})$ such that:*

(c) *$\mathcal{K}$ is the set of all $\mathbf{f} \in \prod_{t \in \Omega} M_t$ such that $\{t \in \Omega \colon \mathbf{f}(t) = \mathbf{c}(t)\} \in \mathcal{E}$ for each $\mathbf{c} \in C$;*

(d) *$\top, \bot, \sqcup, \sqcap, \neg$ are the usual Boolean operations on $\mathcal{E}$, and μ is the measure ν;*

(e) for each first order formula $\psi(\vec{x})$ and tuple $\vec{\mathbf{f}}$ in $\mathcal{K}$, we have

$$[\![\psi(\vec{\mathbf{f}})]\!] = \{t \in \Omega : \mathcal{M}_t \models \psi(\mathbf{f}(t))\};$$

(f) $d_{\mathbb{E}}(\mathsf{B}, \mathscr{C}) = \nu(\mathsf{B} \triangle \mathscr{C}), \quad d_{\mathbb{K}}(\mathbf{f}, \mathbf{g}) = \mu([\![\mathbf{f} \neq \mathbf{g}]\!]).$

Moreover, $\mathcal{P}(C)$ is a pre-complete separable randomization.

Proof of Theorem 6.2. It is clear that $\mathcal{P}(C)$ is unique. We first show by induction on the complexity of formulas that condition (b) holds for all first order formulas ψ. The steps for logical connectives are trivial. For the quantifier step, suppose (b) holds for $\psi(u, \vec{v})$. Then by (a) and (c)–(f),

$$[\![(\exists u)\psi(u, \vec{\mathbf{b}})]\!] = \{t : \mathcal{M}_t \models (\exists u)\psi(u, \vec{\mathbf{b}}(t))\} = \{t : (\exists c \in M_t)\mathcal{M}_t \models \psi(c, \vec{\mathbf{b}}(t))\} =$$

$$= \{t : (\exists \mathbf{c} \in C)\mathcal{M}_t \models \psi(\mathbf{c}(t), \vec{\mathbf{b}}(t))\} = \bigcup_{\mathbf{c} \in C} [\![\psi(\mathbf{c}, \vec{\mathbf{b}})]\!] \in \mathcal{E},$$

so (b) holds for $(\exists u)\psi(u, \vec{v})$. By the definition of $\mathcal{K}$, for each tuple $\vec{\mathbf{g}}$ in $\mathcal{K}$ and $\vec{\mathbf{b}}$ in C, we have $[\![\vec{\mathbf{g}} = \vec{\mathbf{b}}]\!] \in \mathcal{E}$. Then for every first order formula $\psi(\vec{v})$ and tuple $\vec{\mathbf{g}}$ in $\mathcal{K}$,

$$[\![\psi(\vec{\mathbf{g}})]\!] = \bigcup \{[\![\psi(\vec{\mathbf{b}}) \wedge \vec{\mathbf{g}} = \vec{\mathbf{b}}]\!] : \vec{\mathbf{b}} \text{ is a tuple in } C\}.$$

We therefore have

(b') For each first order formula $\psi(\vec{v})$ and tuple $\vec{\mathbf{g}}$ in $\mathcal{K}$, $[\![\psi(\vec{\mathbf{g}})]\!] \in \mathcal{E}$.

This shows that $\mathcal{P}(C)$ is a pre-structure with signature L^R.

It is easily seen that $\mathcal{P}(C)$ satisfies all the axioms of T_2^R except possibly the Fullness and Event Axioms. We next show that $\mathcal{P}(C)$ has perfect witnesses. Once this is done, it follows at once that $\mathcal{P}(C)$ also satisfies the Fullness and Event Axioms, and hence is a pre-model of T_2^R.

Consider a first order formula $\psi(u, \vec{v})$ and a tuple $\vec{\mathbf{g}}$ in $\mathcal{K}$. For each $t \in \Omega$, there is a least $n(t) \in \mathbb{N}$ such that $\mathcal{M}_t \models (\exists u)\psi(u, \vec{\mathbf{g}}(t)) \rightarrow \psi(\mathbf{c}_{n(t)}(t), \vec{\mathbf{g}}(t))$. Since (b') holds and $C \subseteq \mathcal{K}$, the function $\mathbf{f}$ such that $\mathbf{f}(t) := \mathbf{c}_{n(t)}(t)$ belongs to $\mathcal{K}$. Therefore

$$[\![\psi(\mathbf{f}, \vec{\mathbf{g}})]\!] \doteq [\![(\exists u)\psi(u, \vec{\mathbf{g}})]\!].$$

Now consider an event $\mathsf{E} \in \mathcal{E}$. Since each $\mathcal{M}_t \models T_2$, we have $[\![(\exists u)u \neq \mathbf{c}_0]\!] \doteq \top$. Therefore there exists $\mathbf{f} \in \mathcal{K}$ such that $[\![\mathbf{f} \neq \mathbf{c}_0]\!] \doteq \top$. Then the function $\mathbf{g}$ such that $\mathbf{g}(t) = \mathbf{f}(t)$ for $t \in \mathsf{E}$ and $\mathbf{g}(t) = \mathbf{c}_0(t)$ for $t \notin \mathsf{E}$ belongs to $\mathcal{K}$, and $[\![\mathbf{f} = \mathbf{g}]\!] \doteq \mathsf{E}$. This shows that $\mathcal{P}(C)$ has perfect witnesses, so $\mathcal{P}(C)$ is a pre-model of T_2^R.

We now show that $\mathcal{P}(C)$ is pre-complete. This means that when d is either $d_{\mathbb{K}}$ or $d_{\mathbb{E}}$, for every Cauchy sequence $\langle x_n \rangle_{n \in \mathbb{N}}$ with respect to d, there exists x such that $d(x_n, x) \rightarrow 0$ as $n \rightarrow \infty$. This is clear for $d_{\mathbb{E}}$ because $(\Omega, \mathcal{E}, \nu)$ is countably additive. Suppose $\langle \mathbf{f}_n \rangle_{n \in \mathbb{N}}$ is a Cauchy sequence for $d_{\mathbb{K}}$. Let $C =$

$\{c_k : k \in \mathbb{N}\}$, and $C_m = \{c_0, \ldots, c_m\}$. For each $k \in \mathbb{N}$, $\langle [\![f_n = c_k]\!] \rangle_{n \in \mathbb{N}}$ is a Cauchy sequence with respect to $d_{\mathbb{E}}$. Therefore there exists $\mathsf{B}_k \in \mathcal{E}$ such that $\lim_{n \to \infty} d_{\mathbb{E}}([\![f_n = c_k]\!], \mathsf{B}_k) = 0$. Then $\mu(\mathsf{B}_k) = \lim_{n \to \infty} \mu([\![f_n = c_k]\!])$. We now cut the sets B_k down to disjoint sets with the same unions. Let $\mathsf{A}_0 = \mathsf{B}_0$, and for each m, let $\mathsf{A}_{m+1} = \mathsf{B}_{m+1} \setminus \bigcup_{k=0}^{m} \mathsf{B}_k$. Note that for all m,

$$\bigcup_{k=0}^{m} \mathsf{A}_k = \bigcup_{k=0}^{m} \mathsf{B}_k, \quad \mathsf{A}_k \subseteq \mathsf{B}_k, \quad (\forall k < m)\mathsf{A}_k \cap \mathsf{A}_m = \emptyset.$$

Claim. $\mu(\bigcup_{k=0}^{\infty} \mathsf{A}_k) = 1$.

Proof of Claim: Fix an $\varepsilon > 0$. We show that there exists m such that $\mu(\bigcup_{k=0}^{m} \mathsf{B}_k) > 1 - \varepsilon$. Note that for each m,

$$\mu(\bigcup_{k=0}^{m} \mathsf{B}_k) = \lim_{n \to \infty} \mu([\![f_n \in C_m]\!]).$$

Therefore it suffices to show that

$$(\exists m)(\forall n)\mu([\![f_n \in C_m]\!]) > 1 - \varepsilon.$$

Suppose this is not true. Then

$$(\forall m)(\exists n)\, \mu([\![f_n \notin C_m]\!]) \geq \varepsilon.$$

Since $C = \bigcup_m C_m$,

$$(\forall n)(\exists h)\, \mu([\![f_n \in C_h]\!]) \geq 1 - \varepsilon/2,$$

so

$$(\forall m)(\exists n)(\exists h)\mu([\![f_n \in (C_h \setminus C_m)]\!]) \geq \varepsilon/2.$$

It follows that there are sequences $n_0 < n_1 < \ldots$ and $m_0 < m_1 < \ldots$ such that

$$(\forall k)\mu([\![f_{n_k} \in (C_{m_{k+1}} \setminus C_{m_k})]\!]) \geq \varepsilon/2.$$

Therefore

$$(\forall k)(\forall h > k)d_{\mathbb{K}}(f_{n_k}, f_{n_h}) \geq \varepsilon/2.$$

This contradicts the fact that $\langle f_n \rangle_{n \in \mathbb{N}}$ is a Cauchy sequence, and the Claim is proved.

By Condition (c), there is an f in $\mathcal{P}(C)$ such that f agrees with c_k on A_k for each $k \in \mathbb{N}$. For each n and h we have

$$d_{\mathbb{K}}(f_n, f) = \mu([\![f_n \neq f]\!]) = \sum_{k=0}^{\infty} \mu([\![f_n \neq f]\!] \cap \mathsf{A}_k) = \sum_{k=0}^{\infty} \mu([\![f_n \neq c_k]\!] \cap \mathsf{A}_k) \leq$$

$$\leq \sum_{k=0}^{h} \mu([\![\mathbf{f}_n \neq \mathbf{c}_k]\!] \cap \mathsf{A}_k) + \mu(\bigcup_{k>h} \mathsf{A}_k) \leq \sum_{k=0}^{h} \mu([\![\mathbf{f}_n \neq \mathbf{c}_k]\!] \cap \mathsf{B}_k) + \mu(\bigcup_{k>h} \mathsf{A}_k)$$

$$\leq \sum_{k=0}^{h} d_{\mathbb{E}}([\![\mathbf{f}_n = \mathbf{c}_k]\!], \mathsf{B}_k) + \mu(\bigcup_{k>h} \mathsf{A}_k).$$

By the Claim, for each $\varepsilon > 0$ we may take h such that $\mu(\bigcup_{k>h} \mathsf{A}_k) < \varepsilon/2$. For all sufficiently large n we have

$$\sum_{k=0}^{h} d_{\mathbb{E}}([\![\mathbf{f}_n = \mathbf{c}_k]\!], \mathsf{B}_k) < \varepsilon/2,$$

and hence $d_{\mathbb{K}}(\mathbf{f}_n, \mathbf{f}) < \varepsilon$. It follows that $\lim_{n\to\infty} d_{\mathbb{K}}(\mathbf{f}_n, \mathbf{f}) = 0$, so $\mathcal{P}(C)$ is pre-complete.

We have not yet used the hypothesis that $(\mathcal{E}, d_{\mathbb{E}})$ is separable. We use it now to show that $\mathcal{P}(C)$ is separable. The Boolean algebra $\mathcal{E}$ has a countable subalgebra $\mathcal{E}_0$ such that $\mathcal{E}_0$ is dense with respect to $d_{\mathbb{E}}$, and $[\![\psi(\vec{\mathbf{b}})]\!] \in \mathcal{E}_0$ for each first order formula $\psi(\vec{v})$ and tuple $\vec{\mathbf{b}}$ in C. Let D be the set of all $\mathbf{f} \in \mathcal{K}$ such that for some $k \in \mathbb{N}$, $[\![\mathbf{f} \in C_k]\!] = \top$ and $[\![\mathbf{f} = \mathbf{c}_n]\!] \in \mathcal{E}_0$ for all $n \leq k$. Then D is countable and D is dense in $\mathcal{K}$ with respect to $d_{\mathbb{K}}$, so $\mathcal{P}(C)$ is separable. $\blacksquare_{6.2}$

Remark 6.3. Suppose C is a countable generator in $\langle \mathcal{M}_t \rangle_{t\in\Omega}$ over $(\Omega, \mathcal{E}, \nu)$, and let $\mathcal{P}(C) = (\mathcal{K}, \mathcal{E})$. Then:

1. $C \subseteq \mathcal{K}$.

2. If $C \subseteq D \subseteq \mathcal{K}$ and D is countable, then D is a countable generator in $\langle \mathcal{M}_t \rangle_{t\in\Omega}$.

3. For each $t \in \Omega$, $M_t = \{\mathbf{f}(t) : \mathbf{f} \in \mathcal{K}\}$.

4. If $\mathcal{M}_t \cong \mathcal{H}_t$ for all t, then there is a countable generator D in $\langle \mathcal{H}_t \rangle_{t\in\Omega}$ such that $\mathcal{P}(D) \cong \mathcal{P}(C)$.

Proof. We prove (4). For each t, choose an isomorphism $h_t \colon \mathcal{M}_t \cong \mathcal{H}_t$. For each $\mathbf{c} \in C$, define $h\mathbf{c}$ by $(h\mathbf{c})(t) = h_t(\mathbf{c}(t))$ and let $D = \{h\mathbf{c} : \mathbf{c} \in C\}$. Then D is a countable generator in $\langle \mathcal{H}_t \rangle_{t\in\Omega}$ and $\mathcal{P}(D) \cong \mathcal{P}(C)$. $\blacksquare_{6.3}$

The next corollary connects countable generators to basic randomizations.

Corollary 6.4. *Let* $\mathcal{N} = (\prod_{j\in J} \mathcal{M}_j^{\mathsf{B}_j}, \mathcal{L})$ *be a basic randomization.*

(i) *There is a countable generator* C *in* $\langle \mathcal{M}_t \rangle_{t\in[0,1)}$ *over* $([0,1), \mathcal{L}, \lambda)$ *such that* $C \subseteq \prod_{j\in J} \mathcal{M}_j^{\mathsf{B}_j}$.

(ii) *If* C *is as in* (i), *then* $\mathcal{P}(C) = \mathcal{N}$.

(iii) If C is a countable generator in $\langle \mathcal{H}_t \rangle_{t \in [0,1)}$ over $([0,1), \mathcal{L}, \lambda)$, $C \subseteq \prod_{j \in J} \mathcal{M}_j^{B_j}$, and $\mathcal{H}_t \prec \mathcal{M}_t$ for all t, then $\mathcal{P}(C) \prec \mathcal{N}$.

Proof. (i) For each $j \in J$, choose an enumerated structure $(\mathcal{M}_j, a_{j,0}, a_{j,1}, \ldots)$. Let $C = \{\mathbf{c}_n : n \in \mathbb{N}\}$ where $\mathbf{c}_n(t) = a_{j,n}$ whenever $j \in J$ and $t \in B_j$. C has the required properties.

(ii) Let $\mathcal{P}(C) = (\mathcal{K}, \mathcal{L})$. Since $C \subseteq \prod_{j \in J} \mathcal{M}_j^{B_j}$, for all $j \in J, a \in M_j$, and $\mathbf{c} \in C$ we have $\{t \in B_j : \mathbf{c}(t) = a\} \in \mathcal{L}$. It follows that for each $j \in J$ and $\mathbf{f}$,

$$(\forall a \in M_j)\{t \in B_j : \mathbf{f}(t) = a\} \in \mathcal{L} \Leftrightarrow (\forall \mathbf{c} \in C)\{t \in B_j : \mathbf{f}(t) = \mathbf{c}(t)\} \in \mathcal{L}.$$

Therefore $\mathcal{K} = \prod_{j \in J} \mathcal{M}_j^{B_j}$, and (ii) holds.

(iii) Let $\mathcal{P}(C) = (\mathcal{K}, \mathcal{L})$. For each $\mathbf{f} \in \mathcal{K}$ we have $[0,1) = \bigcup_{\mathbf{c} \in C} \{t : \mathbf{f}(t) = \mathbf{c}(t)\}$, and $\{t : \mathbf{f}(t) = \mathbf{c}(t)\} \in \mathcal{L}$ for all $\mathbf{c} \in C$. Therefore $\mathcal{K} \subseteq \prod_{j \in J} \mathcal{M}_j^{B_j}$. Since $\mathcal{H}_t \prec \mathcal{M}_t$, $[\![\psi(\cdot)]\!]$ has the same interpretation in $\mathcal{P}(C)$ as in $\mathcal{N}$ for every first order formula $\psi(\vec{v})$. Therefore $(\mathcal{K}, \mathcal{L})$ is a pre-substructure of $\mathcal{N}$. By quantifier elimination (Theorem 2.9 of [6]) we have $\mathcal{P}(C) \prec \mathcal{N}$. ∎6.4

The next result gives a very useful "pointwise" characterization of the event corresponding to an infinitary formula in a complete separable randomization that is isomorphic to $\mathcal{P}(C)$.

Proposition 6.5. *Suppose $\mathcal{N}$ is a complete separable randomization, C is a countable generator in $\langle \mathcal{M}_t \rangle_{t \in \Omega}$ over $(\Omega, \mathcal{E}, \nu)$, and $h : \mathcal{P}(C) \cong \mathcal{N}$. Then for every $L_{\omega_1 \omega}$ formula $\psi(\vec{v})$ and tuple $\vec{\mathbf{f}}$ of sort $\mathbb{K}$ in $\mathcal{P}(C)$, we have*

$$\{t : \mathcal{M}_t \models \psi(\vec{\mathbf{f}}(t))\} \in \mathcal{E}, \qquad [\![\psi(h\vec{\mathbf{f}})]\!]^{\mathcal{N}} = h(\{t : \mathcal{M}_t \models \psi(\vec{\mathbf{f}}(t))\}).$$

Moreover, $\mathcal{N}$ is a separable randomization of φ if and only if $\mu(\{t : \mathcal{M}_t \models \varphi\}) = 1$.

Proof. This is proved by a straightforward induction on the complexity of $\psi(\vec{v})$ using Theorems 5.1 and 6.2. ∎6.5

---

7   A representation theorem

In this section we show that every complete separable randomization of φ is isomorphic to $\mathcal{P}(C)$ for some countable generator C in countable models of φ.

We will use the following result, which is a consequence of Theorem 3.11 of [3], and generalizes Proposition 2.1.10 of [2].

Proposition 7.1. *For every pre-complete model $\mathcal{N}'$ of T^R, there is an atomless probability space $(\Omega, \mathcal{E}, \nu)$ and a family of models $\langle \mathcal{M}_t \rangle_{t \in \Omega}$ of T such that $\mathcal{N}'$ is isomorphic to a pre-complete model $\mathcal{N} = (\mathcal{K}, \mathcal{E})$ of T^R such that $\mathcal{K} \subseteq \prod_{t \in \Omega} \mathcal{M}_t$ and $\mathcal{N}$ satisfies Conditions (d), (e), and (f) of Theorem 6.2.*

Proof. Proposition 2.1.10 of [2] gives this result in the case that T is a complete theory, with the additional conclusion that there is a single model $\mathcal{M}$ of T such that $\mathcal{M}_t \prec \mathcal{M}$ for all $t \in \Omega$ [2]. The same argument works in the general case, but without the model $\mathcal{M}$. ■7.1

Proposition 7.2. *Suppose $\mathcal{N}'$ is pre-complete and elementarily embeddable in a basic randomization. Then $\mathcal{N}'$ is isomorphic to a pre-complete elementary submodel $\mathcal{N}$ of a basic randomization $(\prod_{j \in J} \mathcal{M}_j^{\mathsf{B}_j}, \mathcal{L})$ such that the event sort of $\mathcal{N}$ is all of $\mathcal{L}$. Moreover, Conditions (d), (e), and (f) of Theorem 6.2 hold for $\mathcal{N} = (\mathcal{K}, \mathcal{L})$ and $(\prod_{j \in J} \mathcal{M}_j^{\mathsf{B}_j}, \mathcal{L})$.*

Proof. Suppose $\mathcal{N}' \cong \mathcal{N}'' \prec (\prod_{j \in J} \mathcal{M}_j^{\mathsf{B}_j}, \mathcal{L})$. For each $j \in J$, let ℓ_j be a mapping that stretches B_j to $[0, 1)$. Then ℓ_j maps $\mathcal{N}''$ onto a pre-complete elementary submodel $\mathcal{N}_j$ of $(\mathcal{M}_j^{[0,1)}, \mathcal{L})$. By Result 3.4, $\mathcal{N}_j$ is isomorphic to a pre-complete elementary submodel of $(\mathcal{M}_j^{[0,1)}, \mathcal{L})$ with event sort $\mathcal{L}$. Using the inverse mappings ℓ_j^{-1}, it follows that $\mathcal{N}''$ is isomorphic to a pre-complete elementary submodel $\mathcal{N} = (\mathcal{K}, \mathcal{L}) \prec (\prod_{j \in J} \mathcal{M}_j^{\mathsf{B}_j}, \mathcal{L})$ with event sort $\mathcal{L}$. It is easily checked that $\mathcal{N}$ satisfies Conditions (d), (e), and (f) of Theorem 6.2. ■7.2

Theorem 7.3. *(Representation Theorem) Every pre-complete separable randomization $\mathcal{N}$ of φ is isomorphic to $\mathcal{P}(C)$ for some countable generator C in a family of countable models of φ. Moreover, if $\mathcal{N}$ is elementarily embeddable in some basic randomization, then C can be taken to be over the probability space $([0, 1), \mathcal{L}, \lambda)$.*

Proof. Let $\mathcal{N}'$ be a pre-complete separable randomization of φ. By Proposition 7.1, there is an atomless probability space $(\Omega, \mathcal{E}, \nu)$ and a family of models $\langle \mathcal{M}_t \rangle_{t \in \Omega}$ such that $\mathcal{N}'$ is isomorphic to a pre-complete model $\mathcal{N} = (\mathcal{K}, \mathcal{E})$ of T_2^R where $\mathcal{K} \subseteq \prod_{t \in \Omega} \mathcal{M}_t$ and $\mathcal{N}$ satisfies Conditions (d), (e), and (f) of Theorem 6.2. If $\mathcal{N}'$ is elementarily embeddable in a basic randomization, then by Proposition 7.2, we may take $(\Omega, \mathcal{E}, \nu) = ([0, 1), \mathcal{L}, \lambda)$.

Since $\mathcal{N}$ is separable, there is a countable pre-structure $(\mathcal{J}_0, \mathcal{A}_0) \prec \mathcal{N}$ that is dense in $\mathcal{N}$. We will use an argument similar to the proofs of Lemmas 4.7 and 4.8 of [1]. By Result 3.1, $\mathcal{N}$ has perfect witnesses. Hence, by listing the first order formulas, we can construct a chain of countable pre-structures $(\mathcal{J}_n, \mathcal{A}_n), n \in \mathbb{N}$ such that for each n:

- $(\mathcal{J}_n, \mathcal{A}_n) \subseteq (\mathcal{J}_{n+1}, \mathcal{A}_{n+1}) \subseteq \mathcal{N}$;

---

[2] In [2], $\mathcal{P}$ is called a neat randomization of $\mathcal{M}$.

- for each first order formula $\theta(u, \vec{v})$ and tuple $\vec{\mathbf{g}}$ in $\mathcal{J}_n$ there exists $\mathbf{f} \in \mathcal{J}_{n+1}$ such that
$$[\![\theta(\mathbf{f}, \vec{\mathbf{g}})]\!] \doteq [\![(\exists u \theta)(\vec{\mathbf{g}})]\!];$$

- For each $\mathsf{B} \in \mathcal{A}_n$ there exist $\mathbf{f}, \mathbf{g} \in \mathcal{J}_{n+1}$ such that $\mathsf{B} \doteq [\![\mathbf{f} = \mathbf{g}]\!]$.

The union
$$\mathcal{P} = (\mathcal{J}, \mathcal{A}) = \bigcup_n (\mathcal{J}_n, \mathcal{A}_n)$$

is a countable dense elementary submodel of $\mathcal{N}$ that has perfect witnesses. Therefore for each first order formula $\theta(u, \vec{v})$ and each tuple $\vec{\mathbf{g}}$ in $\mathcal{J}$, there exists $\mathbf{f} \in \mathcal{J}$ such that

$$[\![(\exists u)\theta(u, \vec{\mathbf{g}})]\!]^{\mathcal{N}} = [\![(\exists u)\theta(u, \vec{\mathbf{g}})]\!]^{\mathcal{P}} \doteq [\![\theta(\mathbf{f}, \vec{\mathbf{g}})]\!]^{\mathcal{P}} = [\![\theta(\mathbf{f}, \vec{\mathbf{g}})]\!]^{\mathcal{N}}.$$

Since $\mathcal{J}$ is countable, there is an event $\mathsf{E} \in \mathcal{E}$ such that $\nu(\mathsf{E}) = 1$ and for every tuple $\vec{\mathbf{g}}$ in $\mathcal{J}$ there exists $\mathbf{f} \in \mathcal{J}$ so that

$$(\forall t \in \mathsf{E})\mathcal{M}_t \models [(\exists u)\theta(u, \vec{\mathbf{g}}(t)) \leftrightarrow \theta(\mathbf{f}(t), \vec{\mathbf{g}}(t))].$$

For each $t \in \Omega$ let $\mathcal{H}_t = \{\mathbf{f}(t) \colon \mathbf{f} \in \mathcal{J}\}$. By the Tarski-Vaught test, we have $\mathcal{H}_t \prec \mathcal{M}_t$, and hence $\mathcal{H}_t \models T_2$, for each $t \in \mathsf{E}$.

Pick a countable model $\mathcal{H}$ of φ. For any set $\mathsf{D} \subseteq \mathsf{E}$ such that $\mathsf{D} \in \mathcal{E}$ and $\nu(\mathsf{D}) = 1$, let C^{D} be the set of all functions that agree with an element of $\mathcal{J}$ on D and take a constant value in $\mathcal{H}$ on $\Omega \setminus \mathsf{D}$. Let $\mathcal{H}_t^{\mathsf{D}} = \mathcal{H}_t$ for $t \in \mathsf{D}$, and $\mathcal{H}_t^{\mathsf{D}} = \mathcal{H}$ for $t \in \Omega \setminus \mathsf{D}$. Then $\mathcal{H}_t^{\mathsf{D}}$ is a model of T_2 for each $t \in \Omega$, and C^{D} is a countable generator in $\langle \mathcal{H}_t^{\mathsf{D}} \rangle_{t \in \Omega}$. By Theorem 6.2, $\mathcal{P}(C^{\mathsf{D}})$ is a pre-complete separable randomization. The reduction of $(\mathcal{J}, \mathcal{A})$ is dense in the reductions of $\mathcal{N}$ and of $\mathcal{P}(C^{\mathsf{D}})$, and both $\mathcal{N}$ and $\mathcal{P}(C^{\mathsf{D}})$ are pre-complete. Therefore $\mathcal{N} \cong \mathcal{P}(C^{\mathsf{D}})$.

In particular, C^{E} is a countable generator in $\langle \mathcal{H}_t^{\mathsf{E}} \rangle_{t \in \Omega}$, and $\mathcal{N} \cong \mathcal{P}(C^{\mathsf{E}})$. Now let $\mathsf{D} = \{t \in \mathsf{E} \colon \mathcal{H}_t^{\mathsf{E}} \models \varphi\}$. Since $\mathcal{N}$ is a pre-complete randomization of φ, we see from Proposition 6.5 that $\mu(\mathsf{D}) = 1$. Then $\mathcal{H}_t^{\mathsf{D}} \models \varphi$ for all $t \in \Omega$, $\mathcal{P}(C^{\mathsf{D}}) \cong \mathcal{N}$, and C^{D} is a countable generator in a family of countable models of φ.   ∎$_{7.3}$

---

8 Elementary embeddability in a basic randomization

Let $S_n(T)$ be the set of first order n-types realized in countable models of T, and $S_n(\varphi)$ be the set of first order types realized in countable models of φ. Note that $S_0(\varphi) = \{Th(\mathcal{M}) \colon \mathcal{M} \models \varphi\}$.

Theorem 3.12 in [6] and Proposition 5.7 in [1] show that:

Result 8.1. *Let T be complete. The following are equivalent:*

(i) $\bigcup_n S_n(T)$ *is countable.*

(ii) *Every complete separable randomization of T is elementarily embeddable in the Borel randomization of a countable model of T.*

(iii) *For every complete separable randomization $\mathcal{N}$ of T, $n \in \mathbb{N}$, and n-tuple $\vec{\mathbf{f}}$ of sort $\mathbb{K}$ in $\mathcal{N}$, there is a type $p \in S_n(T)$ such that $\mu([\![\bigwedge p(\vec{\mathbf{f}})]\!]^{\mathcal{N}}) > 0$.*

In Theorem 8.3 below, we generalize this result by replacing a complete theory T and a Borel randomization by an arbitrary $L_{\omega_1\omega}$ sentence φ and a basic randomization.

We will use Proposition 6.2 of [1], which can be formulated as follows.

Result 8.2. *Let T be complete. The following are equivalent:*

(i) *$\mathcal{N}$ is a complete separable randomization of T and for each n and each n-tuple $\vec{\mathbf{f}}$ in $\mathcal{K}$, $\sum_{q \in S_n(T)} \mu([\![\bigwedge q(\vec{\mathbf{f}})]\!]^{\mathcal{N}}) = 1$.*

(ii) *$\mathcal{N}$ is elementarily embeddable in the Borel randomization of a countable model of T.*

Theorem 8.3. *The following are equivalent:*

(i) $\bigcup_n S_n(\varphi)$ *is countable.*

(ii) *Every complete separable randomization of φ is elementarily embeddable in a basic randomization.*

(iii) *For every complete separable randomization $\mathcal{N}$ of φ, $n \in \mathbb{N}$, and n-tuple $\vec{\mathbf{f}}$ in $\mathcal{K}$, there is a type $p \in S_n(\varphi)$ such that $\mu([\![\bigwedge p(\vec{\mathbf{f}})]\!]^{\mathcal{N}}) > 0$.*

In (ii), we do not know whether the basic randomization can be taken to be a basic randomization of φ.

Proof of Theorem 8.3. We first assume (i) and prove (ii). Let $\mathcal{N}$ be a complete separable randomization of φ. By Theorem 7.3, there is a countable generator C in a family of countable models $\langle \mathcal{M}_t \rangle_{t \in \Omega}$ of φ over an atomless probability space $(\Omega, \mathcal{E}, \nu)$, such that $\mathcal{N} \cong \mathcal{P}(C) = (\mathcal{K}, \mathcal{E})$. For each $t \in \Omega$, $\mathcal{M}_t$ is a countable model of φ, so $Th(\mathcal{M}_t) \in S_0(\varphi)$. By (i), $S_0(\varphi)$ is countable. Let $\mathsf{B}_T = \{t \in \Omega\colon \mathcal{M}_t \models T\}$. By Proposition 6.5, $\mathsf{B}_T \in \mathcal{E}$. Let $G = \{T \in S_0(\varphi)\colon \nu(\mathsf{B}_T) > 0\}$, and consider any $T \in G$. Let ν_T be the atomless probability measure on $(\Omega, \mathcal{E})$ such that $\nu_T(\mathsf{E}) = \nu(\mathsf{E} \cap \mathsf{B}_T)/\nu(\mathsf{B}_T)$. (Note that ν_T is the conditional probability of ν with respect to B_T.) Let $\mathcal{N}_T$ be the structure $(\mathcal{K}, \mathcal{E})$ with the probability measure ν_T instead of ν. Then $\mathcal{N}_T$ is a pre-complete separable randomization of both φ and T. Let $S_n = S_n(T) \cap S_n(\varphi)$. Since $S_n(\varphi)$ is countable, $(\forall \vec{v}) \bigvee_{q \in S_n} \bigwedge q(\vec{v})$ is a sentence of $L_{\omega_1\omega}$ and is a consequence of φ. Therefore

$$\nu_T([\![(\forall \vec{v}) \bigvee_{q \in S_n} \bigwedge q(\vec{v})]\!]) = 1.$$

Then for every n-tuple $\vec{\mathbf{f}}$ in $\mathcal{K}$, $\sum_{q \in S_n(T)} \nu_T([\bigwedge q(\vec{\mathbf{f}})]) = 1$. Hence by Result 8.2, there is a countable model $\mathcal{H}_T$ of T and an elementary embedding

$$h_T \colon \mathcal{N}_T \prec (\mathcal{H}_T^{[0,1)}, \mathcal{L}).$$

Let $\{\mathsf{A}_T \colon T \in G\}$ be a Borel partition of $[0,1)$ such that $\lambda(\mathsf{A}_T) = \nu(\mathsf{B}_T)$ for each T. Let J be the set of isomorphism types of the models $\{\mathcal{H}_T \colon T \in G\}$. For each $T \in G$ let $\mathcal{H}_j = \mathcal{H}_T$, $h_j = h_T$, and $\mathsf{A}_j = \mathsf{A}_T$ where j is the isomorphism type of $\mathcal{H}_T$. Then $\mathcal{P} = (\prod_{j \in J} \mathcal{H}_j^{\mathsf{A}_j}, \mathcal{L})$ is a basic randomization. For each $j \in J$, let ℓ_j be a mapping that stretches A_j to $[0,1)$, and let $\ell_j \colon \mathcal{P} \to (\mathcal{H}_j^{[0,1)}, \mathcal{L})$ be the mapping defined in Definition 4.4. We then get an elementary embedding of $\mathcal{N}$ into $\mathcal{P}$ by sending each $\mathsf{E} \in \mathcal{E}$ to the set $\bigcup_{j \in J} \ell_j^{-1}(h_j(\mathsf{E}))$, and sending each $\mathbf{f} \in \mathcal{K}$ to the function that agrees with $\ell_j^{-1}(h_j(\mathbf{f}))$ on A_j for each $j \in J$.

We next assume (ii) and prove (iii). Let $\mathcal{N} = (\mathcal{K}, \mathcal{E})$ be a complete separable randomization of φ, and let $\vec{\mathbf{f}}$ be an n-tuple in $\mathcal{K}$. By (ii), there is an elementary embedding h from $\mathcal{N}$ into a basic randomization $\mathcal{P} = (\prod_{j \in J} \mathcal{H}_j^{\mathsf{A}_j}, \mathcal{L})$. Then there is a $j \in J$ and a set $\mathsf{B} \subseteq \mathsf{A}_j$ such that $\lambda(\mathsf{B}) > 0$ and $(h\vec{\mathbf{f}})$ is constant on B. Let $r = \lambda(\mathsf{B})$. Let p be the type of $h(\vec{\mathbf{f}})$ in $\mathcal{H}_j$. Then for each $\theta(\vec{v}) \in p$ we have $\mathcal{P} \models \mu([\theta(h\vec{\mathbf{f}})]) \geq r$. Since h is an elementary embedding, for each $\theta \in p$ we have $\mathcal{N} \models \mu([\theta(\vec{\mathbf{f}})]) \geq r$. Therefore

$$\mu([\bigwedge p(\vec{\mathbf{f}})]^{\mathcal{N}}) = \inf_{\theta \in p} \mu([\theta(\vec{\mathbf{f}})]^{\mathcal{N}}) \geq r > 0,$$

and (iii) is proved.

Finally, we assume that (i) fails and prove that (iii) fails. Since (i) fails, there exists n such that $S_n(\varphi)$ is uncountable. We introduce some notation. Let L_0 be the set of all atomic first order formulas. Let 2^{L_0} be the Polish space whose elements are the functions $s \colon L_0 \to \{0,1\}$. As in Section 2, we say that a point $t \in 2^{L_0}$ **codes** an enumerated structure $(\mathcal{M}, a)$ if for each formula $\theta(v_0, \ldots, v_{n-1}) \in L_0$, $t(\theta) = 0$ if and only if $\mathcal{M} \models \theta[a_0, \ldots, a_{n-1}]$. We note for each $t \in 2^{L_0}$, any two enumerated structures that are coded by t are isomorphic. When t codes an enumerated structure, we choose one and denote it by $(\mathcal{M}(t), a(t))$. For each $L_{\omega_1 \omega}$ formula $\psi(v_0, \ldots, v_{n-1})$, let $[\psi]$ be the set of all $t \in 2^{L_0}$ such that $(\mathcal{M}(t), a(t))$ exists and $\mathcal{M}(t) \models \psi[a_0(t), \ldots, a_{n-1}(t)]$.

Claim. There is a perfect set $P \subseteq [\varphi]$ such that for all s, t in P, we have

$$(\mathcal{M}(s), a_0(s), \ldots, a_{n-1}(s)) \equiv (\mathcal{M}(t), a_0(t), \ldots, a_{n-1}(t))$$

if and only if $s = t$.

Proof of Claim: By Proposition 16.7 in [8], for each $L_{\omega_1 \omega}$ formula ψ, $[\psi(\vec{v})]$ is a Borel subset of 2^{L_0}. In particular, $[\varphi]$ is Borel. Let E be the set of pairs $(s, t) \in [\varphi] \times [\varphi]$ such that

$$(\mathcal{M}(s), a_0(s), \ldots, a_{n-1}(s)) \equiv (\mathcal{M}(t), a_0(t), \ldots, a_{n-1}(t)).$$

E is obviously an equivalence relation on $[\varphi]$. Since $S_n(\varphi)$ is uncountable, E has uncountably many equivalence classes. We show that E is Borel. Let F be the set of all first order formulas $\theta(v_0, \ldots, v_{n-1})$. For each $\theta \in F$, let

$$E_\theta = \{(s,t) \in [\varphi] \times [\varphi]: s \in [\theta] \leftrightarrow t \in [\theta]\}.$$

Since $[\varphi]$ and $[\theta]$ are Borel, E_θ is Borel. Moreover, F is countable, and $E = \bigcap_{\theta \in F} E_\theta$. Therefore E is a Borel equivalence relation. By Silver's theorem in [14], there is a perfect set $P \subseteq [\varphi]$ such that whenever $s, t \in [\varphi]$, we have $(s,t) \in E$ if and only if $s = t$, as required in the Claim.

By Theorem 6.2 in [8], P has cardinality $2^{\aleph_0}$. By the Borel Isomorphism Theorem (15.6 in [8]), there is a Borel bijection β from $[0,1)$ onto P whose inverse is also Borel. Each $s \in P$ codes an enumerated model $(\mathcal{M}(s), a(s))$ of φ. For each $t \in [0,1)$ and $n \in \mathbb{N}$, $a_n(\beta(t)) \in M(\beta(t))$, so for each n the composition $\mathbf{c}_n = a_n \circ \beta$ is a function such that $\mathbf{c}_n(t) \in M(\beta(t))$. Let $C = \{\mathbf{c}_n: n \in \mathbb{N}\}$. Then for each t, we have

$$\{\mathbf{c}(t): \mathbf{c} \in C\} = \{a_n(\beta(t)): n \in \mathbb{N}\} = M(\beta(t)),$$

so C satisfies Condition (a) of Definition 6.1.

We next show that C is a countable generator. We will then show that the completion of $\mathcal{P}(C)$ is a separable randomization of φ that is not elementarily embeddable in a basic randomization.

For each $\theta \in L_0$, the set

$$P \cap [\theta] = \{s \in P: \mathcal{M}(s) \models \theta[a_0(s), \ldots, a_{n-1}(s)]\}$$

is Borel. Since β and its inverse are Borel functions, it follows that

$$\{t \in [0,1): \mathcal{M}(\beta(t)) \models \theta(\mathbf{c}_0(t), \ldots, \mathbf{c}_{n-1}(t))\} \in \mathcal{L}.$$

Thus C satisfies condition (b) of Definition 6.1, and hence is a countable generator in the family $\langle \mathcal{M}(\beta(t)) \rangle_{t \in [0,1)}$ of countable models of φ over the probability space $([0,1), \mathcal{L}, \lambda)$.

By Theorem 6.2 and Proposition 6.5, $\mathcal{P}(C)$ is a pre-complete separable randomization of φ. Then the completion $\mathcal{N}$ of $\mathcal{P}(C)$ is a complete separable randomization of φ. By the properties of P, for each first-order n-type p, there is at most one $t \in [0,1)$ such that $(\mathbf{c}_0(t), \ldots, \mathbf{c}_{n-1}(t))$ realizes p in $\mathcal{M}(\beta(t))$. Then

$$\mu([\![\bigwedge p(\mathbf{c}_0, \ldots, \mathbf{c}_{n-1})]\!]^{\mathcal{N}}) = 0.$$

Therefore $\mathcal{N}$ cannot be elementarily embeddable in a basic randomization. This shows that (iii) fails, and completes the proof. $\blacksquare_{8.3}$

9 Sentences with few separable randomizations

In this section we show that any infinitary sentence that has only countably many countable models has few separable randomizations (Theorem 9.6 below). We begin by stating a result from [1].

Result 9.1. *([1], Theorem 6.3). If T is complete and $I(T)$ is countable, then T has few separable randomizations.*

Theorem 9.6 below will generalize this result by replacing the complete theory T by an arbitrary $L_{\omega_1\omega}$ sentence φ.

The following lemma is a consequence of Theorem 7.6 in [1]. The underlying definitions are somewhat different in [1], so for completeness we give a direct proof here.

Lemma 9.2. *Let $\mathcal{N} = (\mathcal{H}^{[0,1)}, \mathcal{L})$ be the Borel randomization of a countable model $\mathcal{H}$ of T_2. Suppose $\mathcal{M}_t \cong \mathcal{H}$ for each $t \in [0,1)$, and C is a countable generator in $\langle \mathcal{M}_t \rangle_{t \in [0,1)}$ over $([0,1), \mathcal{L}, \lambda)$. Then $\mathcal{P}(C) \cong \mathcal{N}$.*

Remark 9.3. In the special case that $\mathcal{M}_t = \mathcal{M}$ for all $t \in [0,1)$ and $C \subseteq \mathcal{M}^{[0,1)}$, Corollary 6.4 and Remark 4.2 (ii) immediately give

$$\mathcal{P}(C) = (\mathcal{M}^{[0,1)}, \mathcal{L}) \cong \mathcal{N}.$$

This argument does not work in the general case, where the structures $\mathcal{M}_t$ may vary with t and there is no measurability requirement on the elements of C.

Proof of Lemma 9.2. Let $\mathcal{P}(C) = (\mathcal{J}, \mathcal{L})$. Let H denote the universe of $\mathcal{H}$. Let $\{\mathbf{f}_1, \mathbf{f}_2, \ldots\}$ and $\{\mathbf{g}'_1, \mathbf{g}'_2, \ldots\}$ be countable dense subsets of $\mathcal{J}$ and $\mathcal{H}^{[0,1)}$, respectively.

Claim. There is a sequence $\langle \mathbf{g}_1, \mathbf{g}_2, \ldots \rangle$ in $\mathcal{J}$, and a sequence $\langle \mathbf{f}'_1, \mathbf{f}'_2, \ldots \rangle$ in $\mathcal{H}^{[0,1)}$, such that the following statement $S(n)$ holds for each $n \in \mathbb{N}$:

For all $t \in [0,1)$,

$$(\mathcal{M}_t, (\mathbf{f}_1, \ldots, \mathbf{f}_n, \mathbf{g}_1, \ldots, \mathbf{g}_n)(t)) \cong (\mathcal{H}, (\mathbf{f}'_1, \ldots, \mathbf{f}'_n, \mathbf{g}'_1, \ldots, \mathbf{g}'_n)(t)).$$

Once the Claim is proved, it follows that for each first order formula $\psi(\vec{u}, \vec{v})$,

$$[\![\psi(\vec{\mathbf{f}}, \vec{\mathbf{g}})]\!]^{\mathcal{P}(C)} = [\![\psi(\vec{\mathbf{f}}', \vec{\mathbf{g}}')]\!]^{\mathcal{N}},$$

and hence there is an isomorphism $h: \mathcal{P}(C) \cong \mathcal{N}$ such that $h(\mathsf{E}) = \mathsf{E}$ for all $\mathsf{E} \in \mathcal{L}$, and $h(\mathbf{f}_n) = \mathbf{f}'_n$ and $h(\mathbf{g}_n) = \mathbf{g}'_n$ for all n.

Proof of Claim: Note that the statement $S(0)$ just says that $\mathcal{M}_t \cong \mathcal{H}$ for all $t \in [0,1)$, and is true by hypothesis. Let $n \in \mathbb{N}$ and assume that we already have functions $\mathbf{g}_1, \ldots, \mathbf{g}_{n-1}$ in $\mathcal{J}$ and $\mathbf{f}'_1, \ldots, \mathbf{f}'_{n-1}$ in $\mathcal{H}^{[0,1)}$ such that the statement $S(n-1)$ holds. Thus for each $t \in [0,1)$, there is an isomorphism

$$h_t : (\mathcal{M}_t, (\mathbf{f}_1, \ldots, \mathbf{f}_{n-1}, \mathbf{g}_1, \ldots, \mathbf{g}_{n-1})(t)) \cong (\mathcal{H}, (\mathbf{f}'_1, \ldots, \mathbf{f}'_{n-1}, \mathbf{g}'_1, \ldots, \mathbf{g}'_{n-1})(t)).$$

We will find functions $\mathbf{g}_n \in \mathcal{J}, \mathbf{f}'_n \in \mathcal{H}^{[0,1)}$ such that $S(n)$ holds.

Let Z be the set of all isomorphism types of structures

$$(\mathcal{H}, a_1, \ldots, a_{n-1}, b_1, \ldots, b_{n-1}, a, b),$$

and for each $z \in Z$ let θ_z be a Scott sentence for structures of isomorphism type z. Since H is countable, Z is countable. For each $a \in H$ and $t \in [0,1)$ let $z(a,t)$ be the isomorphism type of

$$(\mathcal{H}, (\mathbf{f}'_1, \ldots, \mathbf{f}'_{n-1}, \mathbf{g}'_1, \ldots, \mathbf{g}'_{n-1})(t), a, \mathbf{g}'_n(t)).$$

Then $z(a,t) \in Z$.

For each $a \in H$ and $c \in C$, let $\mathsf{B}(a,c)$ be the set of all $t \in [0,1)$ such that

$$(\mathcal{M}_t, (\mathbf{f}_1, \ldots, \mathbf{f}_{n-1}, \mathbf{g}_1, \ldots, \mathbf{g}_{n-1})(t), \mathbf{f}_n(t), c(t)) \models \theta_{z(a,t)}.$$

By Proposition 6.5, each of the sets $\mathsf{B}(a,c)$ is Borel. By taking $a \in H$ such that $a = h_t(\mathbf{f}_n(t))$, and $c \in C$ such that $c(t) = h_t^{-1}(\mathbf{g}'_n(t))$, we see that for every $t \in [0,1)$ there exist $a \in H$ and $c \in C$ with $t \in \mathsf{B}(a,c)$. Thus

$$[0,1) = \bigcup \{\mathsf{B}(a,c) : a \in H, c \in C\}.$$

Every countable family of Borel sets with union $[0,1)$ can be cut down to a countable partition of $[0,1)$ into Borel sets. Thus there is a partition

$$\langle \mathsf{D}(a,c) : a \in H, c \in C \rangle$$

of $[0,1)$ into Borel sets $\mathsf{D}(a,c) \subseteq \mathsf{B}(a,c)$.

Let $\mathbf{f}'_n$ be the function that has the constant value a on each set $\mathsf{D}(a,c)$, and let $\mathbf{g}_n$ be the function that agrees with c on each set $\mathsf{D}(a,c)$. Then $\mathbf{f}'_n$ is Borel and thus belong to $\mathcal{H}^{[0,1)}$, and $\mathbf{g}_n$ belongs to $\mathcal{J}$. Moreover, whenever $t \in \mathsf{D}(a,c)$ we have $t \in \mathsf{B}(a,c)$ and hence

$$(\mathcal{M}_t, (\mathbf{f}_1, \ldots, \mathbf{f}_n, \mathbf{g}_1, \ldots, \mathbf{g}_n)(t)) \cong (\mathcal{H}, (\mathbf{f}'_1, \ldots, \mathbf{f}'_n, \mathbf{g}'_1, \ldots, \mathbf{g}'_n)(t)).$$

So the functions $\mathbf{f}'_n$ and $\mathbf{g}_n$ satisfy the condition $S(n)$. This completes the proof of the Claim and of Lemma 9.2. $\blacksquare$ 9.3

Recall that for each $i \in I$, θ_i is a Scott sentence for structures of isomorphism type i.

Lemma 9.4. *Let* $\mathcal{P} = (\prod_{j \in J} (\mathcal{H}_j)^{\mathsf{A}_j}, \mathcal{L})$ *be a basic randomization. Then for each complete separable randomization* $\mathcal{N}$, *the following are equivalent:*

(i) $\mathcal{N}$ *is isomorphic to* $\mathcal{P}$.

(ii) $\mu([\![\theta_j]\!]^{\mathcal{N}}) = \lambda(\mathsf{A}_j)$ *for each* $j \in J$.

Proof. Assume (i) and let $h : \mathcal{P} \cong \mathcal{N}$. By Corollary 6.4, $\mathcal{P} = \mathcal{P}(C)$ for some countable generator C in $\langle \mathcal{H}_t \rangle_{t \in [0,1)}$ over $([0,1), \mathcal{L}, \lambda)$. By Proposition 6.5, for each $j \in J$ we have

$$[\![\theta_j]\!]^{\mathcal{N}} = h(\{t \in [0,1): \mathcal{H}_t \models \theta_j\}) = h(\mathsf{A}_j),$$

so (ii) holds.

We now assume (ii) and prove (i). Since the events $\mathsf{A}_j, j \in J$ form a partition of $[0,1)$, $\sum_{j \in J} \lambda(\mathsf{A}_j) = 1$, so by (ii) we have $\sum_{j \in J} \mu([\![\theta_j]\!]^{\mathcal{N}}) = 1$. Therefore $[\![\bigvee_{j \in J} \theta_j]\!]^{\mathcal{N}} = \top$, so $\mathcal{N}$ is a randomization of the sentence $\varphi = \bigvee_{j \in J} \theta_j$. Since $I(\varphi)$ is countable, $\bigcup_n S_n(\varphi)$ is countable. Then by Theorem 8.3, $\mathcal{N}$ is elementarily embeddable in a basic randomization. By Theorem 7.3, $\mathcal{N}$ is isomorphic to $\mathcal{P}(C)$ for some countable generator C in a family $\langle \mathcal{M}_t \rangle_{t \in [0,1)}$ of countable models of φ over the probability space $([0,1), \mathcal{L}, \lambda)$. By Proposition 6.5, for each $j \in J$ the set $\mathsf{B}_j = \{t \in [0,1): \mathcal{M}_t \models \theta_j\} \in \mathcal{L}$ and $\lambda(\mathsf{B}_j) = \mu([\![\theta_j]\!]^{\mathcal{N}}) = \lambda(\mathsf{A}_j)$. By Theorem 4.9, $\mathcal{P} \cong \mathcal{P}' = (\prod_{j \in J}(\mathcal{H}_j)^{\mathsf{B}_j}, \mathcal{L})$. For each $j \in J$, let ℓ_j be a mapping that stretches B_j to $[0,1)$.

Our plan is to use Lemma 9.2 to show that the images of $\mathcal{P}(C)$ and $\mathcal{P}'$ under ℓ_j are isomorphic for each j. Intuitively, this shows that for each j, the part of $\mathcal{P}(C)$ on B_j is isomorphic to the part of $\mathcal{P}'$ on A_j. The isomorphisms on these parts can then be combined to get an isomorphism from $\mathcal{P}(C)$ to $\mathcal{P}'$.

Here are the details. For each j, $\mathcal{P}_j = (\mathcal{H}_j^{[0,1)}, \mathcal{L})$ is the Borel randomization of $\mathcal{H}_j$, and ℓ_j maps $\mathcal{P}'$ to $\mathcal{P}_j$ and maps C to a countable generator $\ell_j(C)$ in $\langle \mathcal{M}'_t \rangle_{t \in [0,1)}$ over $([0,1), \mathcal{L}, \lambda)$, where $\mathcal{M}'_t = \mathcal{M}_{\ell_j^{-1}(t)}$. Note that for each $j \in J$ and $t \in \ell_j(\mathsf{B}_j)$, we have $\mathcal{M}'_t \cong \mathcal{H}_j$. Therefore by Lemma 9.2, we have an isomorphism $h_j \colon \mathcal{P}(\ell_j(C)) \cong \mathcal{P}_j$ for each $j \in J$. By pulling these isomorphisms back we get an isomorphism $h \colon \mathcal{P}(C) \cong \mathcal{P}'$ as follows. For an element $\mathbf{f}$ of $\mathcal{P}(C)$ of sort $\mathbb{K}$, $h(\mathbf{f})$ is the element of $\mathcal{P}'$ that agrees with $\ell_j^{-1}(h_j(\ell_j(\mathbf{f})))$ on the set B_j for each j. Since $\mathcal{N} \cong \mathcal{P}(C)$ and $\mathcal{P}' \cong \mathcal{P}$, (i) holds. ∎$_{9.4}$

Lemma 9.5. *The following are equivalent.*

(i) φ *has few separable randomizations.*

(ii) *For every complete separable randomization* $\mathcal{N}$ *of* φ, *there is a countable set* $J \subseteq I$ *such that* $[\![\bigvee_{j \in J} \theta_j]\!]^{\mathcal{N}} = \top$.

(iii) *For every complete separable randomization* $\mathcal{N}$ *of* φ, $\mu([\![\theta_i]\!]^{\mathcal{N}}) > 0$ *for some* $i \in I$.

Proof. It follows from Lemma 9.4 that (i) implies (ii). It is trivial that (ii) implies (iii).

We now assume (ii) and prove (i). Let $\mathcal{N}$ be a complete separable randomization of φ and let J be as in (ii). By removing j from J when $[\![\theta_j]\!]^{\mathcal{N}} = \bot$, we may assume that $\mu([\![\theta_j]\!]^{\mathcal{N}}) > 0$ for each $j \in J$. We also have

$$\sum_{j \in J} \mu([\![\theta_j]\!]^{\mathcal{N}}) = \mu([\![\bigvee_{j \in J} \theta_j]\!]^{\mathcal{N}}) = 1.$$

For each $j \in J$, choose $\mathcal{H}_j \models j$. Choose a partition $\{\mathsf{A}_j : j \in J\}$ of $[0, 1)$ such that $\mathsf{A}_j \in \mathcal{L}$ and $\lambda(\mathsf{A}_j) = \mu([\![\theta_j]\!]^{\mathcal{N}})$ for each $j \in J$. Then by Lemma 9.4, $\mathcal{N}$ is isomorphic to the basic randomization $(\prod_{j \in J} \mathcal{H}_j^{\mathsf{A}_j}, \mathcal{L})$. Therefore (i) holds.

We assume that (ii) fails and prove that (iii) fails. Since (ii) fails, there is a complete separable randomization $\mathcal{N}$ of φ such that for every countable set $J \subseteq I$, $\mu([\![\bigvee_{i \in I} \theta_j]\!]^{\mathcal{N}}) < 1$. The set $J = \{i \in I : \mu([\![\theta_j]\!]^{\mathcal{N}}) > 0\}$ is countable. By Theorem 7.3, $\mathcal{N}$ is isomorphic to $\mathcal{P}(C)$ for some countable generator C in a family $\langle \mathcal{M}_t \rangle_{t \in \Omega}$ of countable models of φ over a probability space $(\Omega, \mathcal{E}, \nu)$. By Proposition 6.5, the set $\mathsf{E} = \{t : \mathcal{M}_t \models \bigvee_{j \in J} \theta_j\}$ belongs to $\mathcal{E}$, and $\nu(\mathsf{E}) = \mu([\![\bigvee_{j \in J} \theta_j]\!]^{\mathcal{N}}) < 1$. Let $\mathcal{P}'$ be the pre-structure $\mathcal{P}(C)$ but with the measure ν replaced by the measure υ defined by $\upsilon(\mathsf{D}) = \nu(\mathsf{D} \setminus \mathsf{E})/\nu(\Omega \setminus \mathsf{E})$. This is the conditional probability of D given $\Omega \setminus \mathsf{E}$. Then the completion $\mathcal{N}'$ of $\mathcal{P}'$ is a separable randomization of φ such that $\mu([\![\theta_i]\!]^{\mathcal{N}'}) = 0$ for every $i \in I$, so (iii) fails. ∎9.5

Here is our generalization of Result 9.1.

Theorem 9.6. *If $I(\varphi)$ is countable, then φ has few separable randomizations.*

Proof. Suppose $J = I(\varphi)$ is countable. Then φ has the same countable models as the sentence $\bigvee_{j \in J} \theta_j$. Let $\mathcal{N}$ be a complete separable randomization of φ. By Theorem 7.3, $\mathcal{N} \cong \mathcal{P}(C)$ for some countable generator C in a family of $\langle \mathcal{M}_t \rangle_{t \in \Omega}$ countable models of φ. By Proposition 6.5,

$$\mu([\![\bigvee_{j \in J} \theta_j]\!]^{\mathcal{N}}) = \mu([\![\bigvee_{j \in J} \theta_j]\!]^{\mathcal{P}(C)}) = \mu(\{t : \mathcal{M}_t \models \bigvee_{j \in J} \theta_j\}) = \mu(\{t : \mathcal{M}_t \models \varphi\}) = 1.$$

Therefore $[\![\bigvee_{j \in J} \theta_j]\!]^{\mathcal{N}} = \top$, so φ satisfies Condition (ii) of Lemma 9.5. By Lemma 9.5, φ has few separable randomizations. ∎9.6

---

10   Few separable randomizations versus scattered

In this section we prove two main results. First, any infinitary sentence with few separable randomizations is scattered. Second, Martin's axiom for $\aleph_1$ implies that every scattered infinitary sentence has few separable randomizations. We also discuss the connection between these results and the absolute Vaught conjecture.

Theorem 10.1. *If φ has few separable randomizations, then φ is scattered.*

Proof. Suppose φ is not scattered. By Lemma 2.5, there is a countable fragment L_A of $L_{\omega_1\omega}$ and a perfect set $P \subseteq 2^{L_A}$ such that:

- Each $s \in P$ codes an enumerated model $(\mathcal{M}(s), a(s))$ of φ, and

- If $s \neq t$ in P then $\mathcal{M}(s)$ and $\mathcal{M}(t)$ do not satisfy the same L_A-sentences.

By Theorem 6.2 in [8], P has cardinality $2^{\aleph_0}$. By the Borel Isomorphism Theorem (15.6 in [8]), there is a Borel bijection β from $[0, 1)$ onto P whose inverse is also Borel. For each $s \in P$, $(\mathcal{M}(s), a(s))$ can be written as $(\mathcal{M}(s), a_0(s), a_1(s), \ldots)$. For each $t \in [0, 1)$, let $\mathcal{M}_t = \mathcal{M}(\beta(t))$. It follows that:

(i) $\mathcal{M}_t \models \varphi$ for each $t \in [0, 1)$, and

(ii) If $s \neq t$ in P then $\mathcal{M}_s$ and $\mathcal{M}_t$ do not satisfy the same L_A-sentences.

For each $n \in \mathbb{N}$, the composition $\mathbf{c}_n = a_n \circ \beta$ belongs to the Cartesian product $\prod_{t \in [0,1)} M_t$. For each $t \in [0, 1)$, we have

$$\{\mathbf{c}_n(t) \colon n \in \mathbb{N}\} = \{a_n(\beta(t)) \colon n \in \mathbb{N}\} = M(\beta(t)) = M_t.$$

Consider an atomic formula $\psi(\vec{v})$ and a tuple $(\mathbf{c}_{i_1}, \ldots, \mathbf{c}_{i_n}) \in C$. ψ belongs to the fragment L_A. The set

$$\{s \in P \colon \mathcal{M}(s) \models \psi(a_{i_1}(s), \ldots, a_{i_n}(s))\} = \{s \in P \colon s(\psi(v_{i_1}, \ldots, v_{i_n})) = 0\}$$

is Borel in P. Since β and its inverse are Borel functions, it follows that

$$\{t \in [0, 1) \colon \mathcal{M}_t \models \psi(\mathbf{c}_{i_1}(t), \ldots, \mathbf{c}_{i_n}(t))\} \in \mathcal{L}.$$

Thus C satisfies conditions (a) and (b) of Definition 6.1, and hence is a countable generator in $\langle \mathcal{M}_t \rangle_{t \in [0,1)}$ over $([0, 1), \mathcal{L}, \lambda)$.

By (ii), for each $i \in I$, there is at most one $t \in [0, 1)$ such that $\mathcal{M}_t \models \theta_i$. By Theorem 6.2 and Proposition 6.5, the randomization $\mathcal{N} = \mathcal{P}(C)$ generated by C is a separable pre-complete randomization of φ. The event sort of $\mathcal{N}$ is $([0, 1), \mathcal{L}, \lambda)$. Therefore, for each $i \in I$, the event $[\![\theta_i]\!]^{\mathcal{N}}$ is either a singleton or empty, and thus has measure zero. So by Lemma 9.5, φ does not have few separable randomizations. $\blacksquare_{10.1}$

Corollary 10.2. *Assume that the absolute Vaught conjecture holds for the $L_{\omega_1\omega}$ sentence φ. Then the following are equivalent:*

(i) $I(\varphi)$ is countable;

(ii) φ has few separable randomizations;

(iii) φ is scattered.

Proof. (i) implies (ii) by Result 9.1. (ii) implies (iii) by Theorem 10.1. The absolute Vaught conjecture for φ says that (iii) implies (i). ■$_{10.2}$

Our next theorem will show that if ZFC is consistent, then the converse of Theorem 10.1 is consistent with ZFC.

The Lebesgue measure is said to be $\aleph_1$-additive if the union of $\aleph_1$ sets of Lebesgue measure zero has Lebesgue measure zero. Note that the continuum hypothesis implies that Lebesgue measure is not $\aleph_1$-additive. Solovay and Tennenbaum [16] proved the relative consistency of Martin's axiom $MA(\aleph_1)$, and Martin and Solovay [12] proved that $MA(\aleph_1)$ implies that the Lebesgue measure is $\aleph_1$-additive. Hence if ZFC is consistent, then so is ZFC plus the Lebesgue measure is $\aleph_1$-additive. See [11] for an exposition.

Theorem 10.3. *Assume that the Lebesgue measure is $\aleph_1$-additive. If φ is scattered, then φ has few separable randomizations.*

Proof. Suppose φ is scattered. Then there are at most countably many ω-equivalence classes of countable models of φ, so there are at most countably many first order types that are realized in countable models of φ. Thus $\bigcup_n S_n(\varphi)$ is countable.

Let $\mathcal{N}$ be a complete separable randomization of φ. By Theorem 8.3, $\mathcal{N}$ is elementarily embeddable in some basic randomization. By Theorem 7.3, there is a countable generator C in a family $\langle \mathcal{M}_t \rangle_{t \in [0,1)}$ of countable models of φ over $([0,1), \mathcal{L}, \lambda)$ such that $\mathcal{N} \cong \mathcal{P}(C)$. By Proposition 6.5, for each $i \in I(\varphi)$ we have $\mathsf{B}_i := \{t \colon \mathcal{M}_t \models \theta_i\} \in \mathcal{L}$. Moreover, the events B_i are pairwise disjoint and their union is $[0,1)$. By Result 2.3, $I(\varphi)$ has cardinality at most $\aleph_1$.

Let $J := \{i \in I(\varphi) \colon \lambda(\mathsf{B}_i) > 0\}$. Then J is countable. The set $I(\varphi) \setminus J$ has cardinality at most $\aleph_1$, so by hypothesis we have $\lambda(\bigcup_{j \in J} \mathsf{B}_j) = 1$. Pick an element $j_0 \in J$. For $j \neq j_0$ let $\mathsf{A}_j = \mathsf{B}_j$. Let A_{j_0} contain the other elements of $[0,1)$, so $\mathsf{A}_{j_0} = \mathsf{B}_{j_0} \cup ([0,1) \setminus \bigcup_{j \in J} \mathsf{B}_j)$. Then $\langle \mathsf{A}_j \rangle_{j \in J}$ is a partition of $[0,1)$. For each $j \in J$, choose a model $\mathcal{H}_j$ of isomorphism type j. Then $\mathcal{P} = (\prod_{j \in J} \mathcal{H}_j^{\mathsf{A}_j}, \mathcal{L})$ is a basic randomization of φ. For each $j \in J$ we have $\lambda(\llbracket \theta_j \rrbracket^{\mathcal{N}}) = \lambda(\mathsf{A}_j)$, so by Lemma 9.4, $\mathcal{N}$ is isomorphic to $\mathcal{P}$. This shows that φ has few separable randomizations. ■$_{10.3}$

Corollary 10.4. *Assume that the Lebesgue measure is $\aleph_1$-additive. Then the following are equivalent.*

(i) *For every φ, the absolute Vaught conjecture holds.*

(ii) *For every φ, if φ has few separabable randomiztions then $I(\varphi)$ is countable.*

Proof. Corollary 10.2 shows that (i) implies (ii).

Assume that (i) fails. Then there is a scattered sentence φ such that $|I(\varphi)| = \aleph_1$. By Theorem 10.3, φ has few separable randomizations. Therefore (ii) fails. ■$_{10.4}$

11   Some open questions

Question 11.1. Suppose $\mathcal{N}$ and $\mathcal{P}$ are complete separable randomizations. If

$$\mu([\![\varphi]\!]^{\mathcal{N}}) = \mu([\![\varphi]\!]^{\mathcal{P}})$$

for every $L_{\omega_1\omega}$ sentence φ, must $\mathcal{N}$ be isomorphic to $\mathcal{P}$?

Question 11.2. Suppose C and D are countable generators in $\langle \mathcal{M}_t \rangle_{t\in\Omega}$, $\langle \mathcal{H}_t \rangle_{t\in\Omega}$ over the same probability space $(\Omega, \mathcal{E}, \nu)$. If $\mathcal{M}_t \cong \mathcal{H}_t$ for ν-almost all $t \in \Omega$, must $\mathcal{P}(C)$ be isomorphic to $\mathcal{P}(D)$?

Question 11.3. (Possible improvement of Theorem 8.3.) If $\bigcup_n S_n(\varphi)$ is countable, must every complete separable randomization of φ be elementarily embeddable in a basic randomization of φ?

Question 11.4. Can Theorem 10.3 be proved in ZFC (without the hypothesis that the Lebesgue measure is $\aleph_1$-additive)?

Added in December 2016: The above question was answered affirmatively in a forthcoming paper, "Scattered Sentences Have Few Separable Randomizations", by Uri Andrews, Isaac Goldbring, Sherwood Hachtman, H. Jerome Keisler, and David Marker.

Bibliography

[1] Uri Andrews and H. Jerome Keisler. Separable Models of Randomizations. Journal of Symbolic Logic **80** (2015), 1149–1181.

[2] Uri Andrews, Isaac Goldbring, and H. Jerome Keisler. Definable Closure in Randomizations. Annals of Pure and Applied Logic **166** (2015), pp. 325–341.

[3] Itaï Ben Yaacov. On Theories of Random Variables. Israel J. Math **194** (2013), 957–1012.

[4] John Baldwin, Sy Friedman, Martin Koerwein, and Michael Laskowski. Three Red Herrings around Vaught's Conjecture. Transactions of the American Math Society, **368**:22, 2016.

[5] Itaï Ben Yaacov, Alexander Berenstein, C. Ward Henson and Alexander Usvyatsov. Model Theory for Metric Structures. In Model Theory with Applications to Algebra and Analysis, vol. 2, London Math. Society Lecture Note Series, **350** (2008), 315–427.

[6] Itaï Ben Yaacov and H. Jerome Keisler. Randomizations of Models as Metric Structures. Confluentes Mathematici **1** (2009), 197–223.

[7] C.J. Eagle. Omitting Types in Infinitary [0,1]-valued Logic. Annals of Pure and Applied Logic **165** (2014), 913–932.

[8] Alexander Kechris. Classical Descriptive Set Theory. Springer-Verlag, 1995.

[9] H. Jerome Keisler. Model Theory for Infinitary Logic. North-Holland, 1971.

[10] H. Jerome Keisler. Randomizing a Model. Advances in Mathematics **143** (1999), 124–158.

[11] Kenneth Kunen. Set Theory. Studies in logic 34, London: College Publications (2011).

[12] Donald Martin and Robert Solovay. Internal Cohen Extensions. Annals of Mathematical Logic **2** (1970), 143–178.

[13] Michael Morley. The Number of Countable Models. Journal of Symbolic logic 35 (1970), 14-18.

[14] Jack H. Silver. Counting the Number of Equivalence Classes of Borel and Coanalytic Equivalence Relations. Ann. Math. Logic **18** (1980), 1–28.

[15] Dana Scott. Logic with Denumerably Long Formulas and Finite Strings of Quantifiers. In *The Theory of Models*, ed. by J. Addison et al., North-Holland, 1965, 329–341.

[16] Robert Solovay and Stanley Tennenbaum. Iterated Cohen Extensions and Souslin's Problem. Annals of Mathematics **94** (1971), 201–245.

[17] John R. Steel. On Vaught's conjecture. In Cabal Seminar 76-77, Lecture Notes in Math., 689, Springer (1978) pp. 193–208.

[18] Robert L. Vaught. Denumerable Models of Complete Theories, in Infinitistic Methods, Warsaw 1961, pp. 303–321

Chapter 7

Existentially closed locally finite groups (Sh312)

Saharon Shelah

The Hebrew University of Jerusalem
Jerusalem, Israel
Rutgers, The State University of New Jersey
Piscataway, New Jersey, USA

Partially supported by the ISF, Israel Science Foundation.
Publication 312 on Shelah's list. I would like to thank Alice Leonhardt for the beautiful typing.

0   Introduction

0(A)   Background

On lf (locally finite) groups and exlf (existentially closed locally finite) groups, see the book by Kegel-Wehrfritz [KW73]; exlf groups were originally called ulf (universal locally finite) groups, which we change as the word "universal" has been used in this context with a different meaning, see Definition 0.21 and Claim 0.14.
Recall:

Definition 0.1. 1) G is a lf (locally finite) group <u>if</u> G is a group and every finitely generated subgroup is finite.
2) G is an exlf (existentially closed lf) group (in [KW73] it is called ulf, universal locally finite group) <u>when</u> G is a locally finite group and for any finite groups $K \subseteq L$ and embedding of K into G, the embedding can be extended to an embedding of L into G.
3) Let $\mathbf{K}_{\mathrm{lf}}$ be the class of lf (locally finite) groups (partially ordered by $\subseteq$, being a subgroup) and let $\mathbf{K}_{\mathrm{exlf}}$ be the class of existentially closed $G \in \mathbf{K}_{\mathrm{lf}}$.

In particular there is one and only one exlf group of cardinality $\aleph_0$. Hall proved that every lf group can be extended to an exlf group, as follows. It suffices for a given lf group G to find $H \supseteq G$ such that if $K \subseteq L$ are finite and f embeds K into G, then some $g \supseteq f$ embed L into H. To get such H, for finite $K \subseteq G$ let $E_{G,K} = \{(a,b) : a,b \in G$ and $aK = bK\}$ and let $G^{\oplus}$ be the group of permutations f of G such that for some finite $K \subseteq G$ we have $a \in G \Rightarrow aE_{G,K}f(a)$; now $b \in G$ should be identified with $f_b \in G^{\oplus}$ where f_b is defined by $f_b(x) = xb$ hence $f_b \in G^{\oplus}$ because if $b \in K \subseteq G$, then $a \in G \Rightarrow f_b(a) = ab \in abK = aK$ and $f_{b_2} \circ f_{b_1}(x) = (xb_1)b_2 = x(b_1b_2) = f_{b_1b_2}(x)$. Now $H = G^{\oplus}$ is essentially as required.

The proof gives a canonical extension. This means for example that every automorphism of G can be extended to an automorphism of $G^{\oplus}$ and, moreover, we can do it uniformly so as to preserve isomorphisms. Still we may like to have more; (for a given lf infinite group G) the extension $G^{\oplus}$ defined above is of cardinality $2^{|G|}$ rather than the minimal value - $|G|+\aleph_0$ (not to mention having to repeat this ω times in order to get an exlf extension). Also if $G_1 \subseteq G_2$, then the connection between $G_1^{\oplus}$ and $G_2^{\oplus}$ is not clear, i.e., failure of "naturality". A major point of the present work is a construction of a canonical existentially closed extension of G which has those two additional desirable properties, see e.g., 3.15.

Note that in model theoretic terminology the exlf groups are the $(\mathbf{D}, \aleph_0)$-homogeneous groups, with $\mathbf{D}$ the set of isomorphism types of finite groups or more exactly complete qf (quantifier free) types of finite tuples generating a finite group, see, e.g., [Sh:88r, §2]. We use quantifier free types as we use embeddings (rather than, e.g., elementary embeddings). Let $\mathbf{D}(G)$ be the set of qf-complete types of finite sequences from the group G.

Let $\mathbf{K}_{\mathrm{exlf}}$ be the class of exlf groups. By Grossberg and Shelah [GrSh:174], if $\lambda = \lambda^{\aleph_0}$, then no $G \in \mathbf{K}_{\lambda}^{\mathrm{exlf}} := \{H \in \mathbf{K}_{\mathrm{exf}} : |H| = \lambda\}$ is universal in it, i.e., such that every other member is embeddable into it. But if κ is a compact cardinal and $\lambda > \kappa$ is strong limit of cofinality $\aleph_0$, then there is a universal exlf in cardinality λ, (this is a special case of a general theorem).

Wehrfritz asked about the categoricity of the class of exlf groups in any $\lambda > \aleph_0$. This was answered by Macintyre-Shelah [McSh:55] which proved that in every $\lambda > \aleph_0$ there are 2^λ nonisomorphic members of $\mathbf{K}_{\lambda}^{\mathrm{exlf}}$. This was disappointing in some sense: in $\aleph_0$ the class is categorical, so the question was perhaps motivated by the hope that also general structures in the class can be understood to some extent.

A natural and frequent question on a class of structures is the existence of rigid members, i.e., ones with no nontrivial automorphism. Now any exlf group $G \in \mathbf{K}_{\mathrm{lf}}$ has nontrivial automorphisms - the inner automorphisms (recalling it has a trivial center). So the natural question is about complete members where a group is called complete **iff** it has no non-inner automorphism.

Concerning the existence of a complete, locally finite group of cardinality λ: Hickin [Hic78] proved one exists in $\aleph_1$ (and more, e.g., he finds a family of $2^{\aleph_1}$ such groups pairwise far apart, i.e., no uncountable group is embeddable in two of them). Thomas [Tho86] assumed G.C.H. and built one in every successor cardinal (and more, e.g., it has no Abelian or just solvable subgroup of the same cardinality). Related are Shelah and Ziegler [ShZi:96], which investigate $\mathbf{K}_{G_*}$ where G_* is an existentially closed countable group and

(*) $\mathbf{K}_{G_*}$ is the class of groups G such that every finitely generated subgroup of G is embeddable into G_*

(**) $\mathbf{K}_{G_*}^{\mathrm{excl}}$ is the class of groups G which are $\mathbb{L}_{\infty,\aleph_0}$-equivalent to G_* (excl stands for existentially closed); equivalently $G \in \mathbf{K}_{G_*}$, every finitely generated subgroup of G_* is embeddable into G and if $\bar{a}, \bar{b} \in {}^nG$ realize the same qf type in G, then some inner automorphism of G maps $\bar{a}$ to $\bar{b}$

(***) we can replace "group G_*" by any other structure.

Giorgetta and Shelah [GgSh:83] build in cardinality continuum $G \in \mathbf{K}_{\mathrm{exlf}}$ with no uncountable Abelian subgroup and similarly for $\mathbf{K}_{G_*}^{\mathrm{excl}}$ and also for the similarly defined $\mathbf{K}_{F'_*}^{\mathrm{excl}}$, F_* an existentially closed countable fixed division ring.

In 1985 the author wrote notes (in Hebrew) for proving that there are anti-prime constructions and complete exlf groups when, e.g., $\lambda = \mu^+, \mu^{\aleph_0} = \mu$;

using black boxes and "anti-prime" construction, i.e., using definable types as below; here we exclusively use qf (quantifier free) types; this was announced in [Sh:300, pg. 418], but the work was not properly finished. To do so is our aim here.

Meanwhile Dugas and Göbel [DG93, Th.2] prove that for $\lambda = \lambda^{\aleph_0}$ and $G_0 \in \mathbf{K}^{\mathrm{lf}}_{\leq \lambda}$ there is a complete $G \in \mathbf{K}^{\mathrm{exlf}}_{\lambda^+}$ extending G_0; moreover 2^{λ^+} pairwise nonisomorphic ones. Then Braun and Göbel [BG03] got better results for complete locally finite p-groups. Those constructions build an increasing continuous chain $\langle G_\alpha : \alpha < \lambda^+ \rangle$, each G_α of cardinality λ, such that $G_{\alpha+1}$ is the wreath product of G_α and suitable Abelian locally finite groups, $G = \bigcup_\alpha G_\alpha$ is the desired group. This gives a tight control over the group and implies, e.g., that only few (i.e., $\leq \lambda$) members commute with G_0. Here we are interested in groups G' which are "more existentially closed", e.g., "for every $G' \subseteq G$ of cardinality $< |G|$, there are $|G|$ elements commuting with it" such properties are called "being full"; note that fullness implies that a restriction on the cardinal is necessary and not so without it, see 5.5.

We show that though the class $\mathbf{K}_{\mathrm{exlf}}$ is very "unstable"; there is a large enough set of definable types so we can imitate stability theory and have reasonable control in building exlf groups, using quantifier free types. This may be considered a "correction" to the nonstructure results discussed above.

Before we turn to explaining our results, we deal with the so-called schemes needed for explaining them.

0(B) Schemes

We deal with a class $\mathbf{K}$ of structures; usually it is the class of locally finite groups, but some of the results hold for suitable universal classes, see §6.

In §1 we present somewhat abstractly our results relying on the existence of a dense and closed so-called $\mathfrak{S}$, a set of schemes of definitions of the relevant types.

Central here are so-called schemes. For <u>models theorists</u> they are for a given $G \in \mathbf{K}_{\mathrm{lf}}$ and finite sequence $\bar{a} \subseteq G$ (realizing a suitable quantifier free type) a definition of a complete (quantifier free) type over G so realized in some extensions of G from $\mathbf{K}_{\mathrm{lf}}$, which does not split over $\bar{a}$; alternatively, you may say that they are definitions of a complete type over G which does not split over $\bar{a}$ and its restriction.

For <u>algebraists</u>, they are our replacement of free products $G_1 *_{G_0} G_2$, but $\mathbf{K}_{\mathrm{lf}}$ is not closed under free product, in fact, fail amalgamation. So we are interested in replacements in the cases G_0 is finite, so we waive symmetry.

Convention 0.2. 1) $\mathbf{K}$ is a universal class of structures (i.e., all of the same

vocabulary, closed under isomorphisms and $M \in \mathbf{K}$ iff every finite generated substructure belongs to $\mathbf{K}$; usually $\mathbf{K} = \mathbf{K}_{\mathrm{lf}}$).
2) $G, H, \ldots \in \mathbf{K}$ and use Convention 0.17.

Definition 0.3. For $H \in \mathbf{K}, n < \omega$, a set $A \subseteq H$ and $\bar{a} \in {}^n H$ let $\mathrm{tp}(\bar{a}, A, H) = \mathrm{tp}_{\mathrm{bs}}(\bar{a}, A, H)$ be the basic type of $\bar{a}$ in H over A, that is:

$\{\varphi(\bar{x}, \bar{b}) : \quad \varphi$ is a basic (atomic or negation of atomic) formula in the variables $\bar{x} = \langle x_\ell : \ell < n \rangle$ and the parameters $\bar{b}$, a finite sequence from A, which is satisfied by $\bar{a}$ in $H\}$.

So if $\mathbf{K}$ is a class of groups, without loss of generality φ is $\sigma(\bar{x}, \bar{b}) = e$ or $\sigma(\bar{x}, \bar{b}) \neq e$ for some group-term σ, a so-called "word", (for $\mathbf{K}_{\mathrm{olf}}$ we also have $\sigma_1(\bar{x}, \bar{b}) < \sigma_2(\bar{x}, \bar{b})$) <u>but</u> we may write $p(\bar{y}) = \mathrm{tp}_{\mathrm{bs}}(\bar{b}, A, H)$ or $p(\bar{z}) = \mathrm{tp}_{\mathrm{bs}}(\bar{c}, A, H)$ or just p when the sequence of variables is clear from the context.
2) We say $p(\bar{x})$ is an $n - bs$-type over G <u>when</u> it is a set of basic formulas in the variables $\bar{x} = \langle x_\ell : \ell < n \rangle$ and parameters from G, such that $p(\bar{x})$ is consistent, which means: if $K \subseteq G$ is f.g. and $q(\bar{x})$ is a finite subset of $p(\bar{x})$ and $q(\bar{x})$ is over K (i.e., all the parameters appearing in $q(\bar{x})$ are from K) <u>then</u> $q(\bar{x})$ is realized in some $L \in \mathbf{K}$ extending K. We say $\bar{a}$ realizes p in H if $G \subseteq H$ and $\varphi(\bar{x}, \bar{b}) \in p \Rightarrow H \models \varphi[\bar{a}, \bar{b}]$.
3) $\mathbf{S}_{\mathrm{bs}}^n(G) = \{\mathrm{tp}_{\mathrm{bs}}(\bar{a}, G, H) : G \subseteq H, H$ is from $\mathbf{K}$ and $\bar{a} \in {}^n H\}$ and $\mathbf{S}_{\mathrm{bs}}(G) = \bigcup_n \mathbf{S}_{\mathrm{bs}}^n(G)$; if $\mathbf{K}$ is not clear from the context we should write $\mathbf{S}_{\mathrm{bs}}^n(G, \mathbf{K}), \mathbf{S}_{\mathrm{bs}}(G, \mathbf{K})$.

Observation 0.4. *For every $p \in \mathbf{S}_{\mathrm{bs}}^n(M)$ and $M \in \mathbf{K}$ there are $N, \bar{a}$ such that $M \subseteq N \in \mathbf{K}, \bar{a} \in {}^n N$ realizes $p, G_N = c\ell(G_M + \bar{a}, N)$ <u>and</u> if $M \subseteq N' \in \mathbf{K}$ and $\bar{a}'$ realizes p in N', then there is $N'' \subseteq N'$ and an isomorphism f from N onto N'' extending id_M such that $f(\bar{a}) = \bar{a}'$.*

Remark 0.5. 0) In 0.4 we later use the convention of 0.15(1),(3).
1) We are particularly interested in types which are definable in some sense over small sets.
2) We can define "$p \in \mathbf{S}_{\mathrm{bs}}^n(M)$" syntactically, because for a set p of basic formulas $\varphi(\bar{x}, \bar{a}), \bar{a}$ from M which is complete (i.e., if $\varphi(\bar{x}, \bar{a})$ is an atomic formula over M, then $\varphi(\bar{x}, \bar{a}) \in p$ or $\neg\varphi(\bar{x}, \bar{a}) \in p$), we have $p \in \mathbf{S}_{\mathrm{bs}}^n(M)$ iff for every f.g. $N \subseteq M$ we have $p \upharpoonright N := \{\varphi(\bar{x}, \bar{a}) \in p : \bar{a} \subseteq N\} \in \mathbf{S}_{\mathrm{bs}}^n(N)$.
3) Why do we use types below which do not split over a finite subgroup and the related set of schemes? As we like to get a canonical extension of $M \in \mathbf{K}$ it is natural to use a set of types closed under automorphisms of M, and as their number is preferably $\leq \|M\|$, it is natural to demand that any such type is, in some sense, definable over some finite subset of M.

As in [Sh:3]:

Definition 0.6. We say that $p = \text{tp}_{\text{bs}}(\bar{a}, G, H) \in \mathbf{S}^n_{\text{bs}}(G)$ does not split over $K \subseteq G$ when for every $m < \omega$ and $\bar{b}_1, \bar{b}_2 \in {}^mG$ satisfying $\text{tp}_{\text{bs}}(\bar{b}_1, K, G) = \text{tp}_{\text{bs}}(\bar{b}_2, K, G)$ we have $\text{tp}_{\text{bs}}(\bar{b}_1{}^\smallfrown\bar{a}, K, H) = \text{tp}_{\text{bs}}(\bar{b}_2{}^\smallfrown\bar{a}, K, H)$.

Definition 0.7. 1) Let $\mathbf{D}(\mathbf{K}) = \bigcup_n \mathbf{D}_n(\mathbf{K})$, where $\mathbf{D}_n(\mathbf{K}) = \{\text{tp}_{\text{bs}}(\bar{a}, \emptyset, M) :$ $\bar{a} \in {}^nM$ and $M \in \mathbf{K}\}$.

2) Assume[1] $p(\bar{x})$ is a k-type, that is, $\bar{x} = \langle x_\ell : \ell < k \rangle$ and for some $p'(\bar{x})$ we have $p(\bar{x}) \subseteq p'(\bar{x}) \in \mathbf{D}_k(\mathbf{K})$ and $m < \omega$. We let $\mathbf{D}_{p(\bar{x}),m}(\mathbf{K}) = \mathbf{D}_m(p(\bar{x}), \mathbf{K})$ be the set of $q(\bar{x}, \bar{y}) \in \mathbf{D}_{k+m}(\mathbf{K})$ such that $q(\bar{x}, \bar{y}) \supseteq p(\bar{x})$, which means that there is $M \in \mathbf{K}$ and $\bar{a} \in {}^kM$ realizing $p(\bar{x})$ and $(\bar{a}, \bar{b})$ realizing $q(\bar{x}, \bar{y})$ in M, i.e., $\ell g(\bar{a}) = k, \ell g(\bar{b}) = m$ and $\bar{a}{}^\smallfrown\bar{b}$ realizes $q(\bar{x}, \bar{y})$.

3) In part (2) let $\mathbf{D}_{p(\bar{x})}(\mathbf{K}) = \cup\{\mathbf{D}_m(p(\bar{x}), \mathbf{K}) : m < \omega\}$.

Remark 0.8. Below $\mathfrak{s} \in \Omega_{n,k}[\mathbf{K}]$ is a scheme to fully define a type $q(\bar{z}) \in \mathbf{S}^n_{\text{bs}}(M)$ for a given parameter $\bar{a} \in {}^kM$ such that $q(\bar{z})$ does not split over $\bar{a}$. Sometimes $\mathfrak{s}$ is not unique but if, e.g., $M \in \mathbf{K}_{\text{exlf}}$ it is.

Definition 0.9. 1) Let $\Omega[\mathbf{K}]$ be the set of schemes, i.e., $\cup\{\Omega_{n,k}[\mathbf{K}] : k, n < \omega\}$ where $\Omega_{n,k}[\mathbf{K}]$ is the set of (k, n)-schemes $\mathfrak{s}$ which means, see below.

1A) We say $\mathfrak{s}$ is a (k, n)-scheme when for some $p(\bar{x}) = p_\mathfrak{s}(\bar{x}_\mathfrak{s})$ with $\ell g(\bar{x}_\mathfrak{s}) = k$, (and $k_\mathfrak{s} = k(\mathfrak{s}) = k, n_\mathfrak{s} = n(\mathfrak{s}) = n$) we have:

(a) $\mathfrak{s}$ is a function with domain $\mathbf{D}_{p(\bar{x})}(\mathbf{K})$ such that for each m it maps $\mathbf{D}_{p(\bar{x}),m}(\mathbf{K})$ into $\mathbf{D}_{k+m+n}(\mathbf{K})$

(b) if $s(\bar{x}, \bar{y}) \in \mathbf{D}_{p(\bar{x}),m}(\mathbf{K})$ and $r(\bar{x}, \bar{y}, \bar{z}) = \mathfrak{s}(s(\bar{x}, \bar{y}))$ then $r(\bar{x}, \bar{y}, \bar{z}){\restriction}(k + m) = s(\bar{x}, \bar{y})$; that is, if $(\bar{a}, \bar{b}, \bar{c})$, i.e., $\bar{a}{}^\smallfrown\bar{b}{}^\smallfrown\bar{c}$, realizes $r(\bar{x}, \bar{y}, \bar{z})$ in $M \in \mathbf{K}$ so $k = \ell g(\bar{a}), m = \ell g(\bar{b}), n = \ell g(\bar{c})$, then $\bar{a}{}^\smallfrown\bar{b}$ realizes $s(\bar{x}, \bar{y})$ in M; see 1.2(1)

(c) in clause (b), moreover, if $\bar{b}' \in {}^{\omega>}M$, $\text{Rang}(\bar{b}') \subseteq \text{Rang}(\bar{a}{}^\smallfrown\bar{b})$ then $\bar{a}{}^\smallfrown\bar{b}'{}^\smallfrown\bar{c}$ realizes the type $\mathfrak{s}(\text{tp}_{\text{bs}}(\bar{a}{}^\smallfrown\bar{b}', \emptyset, M))$; this is to avoid $\mathfrak{s}$'s which define contradictory types[2].

2) Assume $\mathfrak{s} \in \Omega_{n,k}[\mathbf{K}]$ and $M \in \mathbf{K}$ and $\bar{a} \in {}^kM$ realizes $p_\mathfrak{s}(\bar{x}_\mathfrak{s})$; we let $q_\mathfrak{s}(\bar{a}, M)$ be the unique $r(\bar{z}) = r(z_\mathfrak{s}) \in \mathbf{S}^n_{\text{bs}}(M)$ such that for any $\bar{b} \in {}^{\omega>}M$ letting $r_{\bar{b}}(\bar{x}, \bar{y}, \bar{z}) := \mathfrak{s}(\text{tp}_{\text{bs}}(\bar{a}{}^\smallfrown\bar{b}, \emptyset, M))$ we have $r_{\bar{b}}(\bar{a}, \bar{b}, \bar{z}) \subseteq r(\bar{z})$.

3) We call $\mathfrak{s}$ full when $p_\mathfrak{s}(\bar{x}) \in \mathbf{D}_{k(\mathfrak{s})}(\mathbf{K})$.

4) For technical reasons we allow $\bar{x}_\mathfrak{s} = \langle x_{\mathfrak{s},\ell} : \ell \in u \rangle, u \subseteq \mathbb{N}, |u| = k_\mathfrak{s}$ and in this case ${}^{k(\mathfrak{s})}M$ will mean ${}^uM = \{\langle a_\ell : \ell \in u \rangle : a_\ell \in M$ for $\ell \in u\}$ and we do not pedantically distinguish between u and k. Similarly for $n_\mathfrak{s}$ and $\bar{z}$, the reason is 1.1, 1.6(4).

---

[1]This is used to define the set $\mathfrak{S}$ of schemes; for this section the case $p(\bar{x}) = p'(\bar{x})$ is enough as we can consider all the completions but the general version is more natural in counting a set $\mathfrak{S}$ of schemes and in considering actual examples.

[2]But some $\mathfrak{s}$'s satisfying (a),(b) but failing this may give a consistent type in an interesting class of cases.

Convention 0.10. $\mathfrak{S}$ will denote a subset of $\Omega[\mathbf{K}]$.

0(C)   The Results

In particular (in the so-called first avenue, see below):

Theorem 0.11. *Let λ be any cardinal $\geq |\mathfrak{S}|$.*
1) For every $G \in \mathbf{K}^{\mathrm{lf}}_{\leq \lambda}$ there is $H_G \in K^{\mathrm{exlf}}_{\lambda}$ which is λ-full over G (hence over any $G' \subseteq G$; see Definition 1.15) and $\mathfrak{S}$-constructible over it (see 1.19).
2) If $H \in \mathbf{K}^{\mathrm{lf}}_{\leq \lambda}$ is λ-full over $G(\in \mathbf{K}^{\mathrm{lf}}_{\leq \lambda})$ then H_G from above can be embedded into H over G, see 1.23(4).

This is proved by 1.23 and §2. So in some sense H_G is prime over G, that is, it is prime but not among the members of $\mathbf{K}^{\mathrm{exlf}}_{\lambda}$, i.e., for a different class. Still we would like to have canonicity, hence uniqueness. There are some additional avenues helpful toward this.

The second avenue tries to get results which are nicer by assuming $\mathfrak{S}$ is so-called symmetric which is the parallel of being stable in this context. Under this assumption we prove the existence of a canonical closure of a locally finite group to an exlf one. This is done in 1.12 and 1.13.

The third avenue is without assuming "$\mathfrak{S}$ is symmetric" but using a more complicated construction, for which we have similar, somewhat weaker results using special linear orders. The failure of symmetry seems to draw you to order the relevant pairs $(\mathfrak{s}, \bar{a})$ for G. That is, trying to repeat the construction in 1.12(2), without symmetry we have to well order or at least linearly order $\mathrm{def}(G) = \mathrm{def}_{\mathfrak{S}}(G)$ which is essentially the set of relevant complete quantifier types over G over a finite set of parameters, see Definition 1.1; this suffices by 1.8(9). At first glance we have to linearly order $\mathrm{def}(G)$, but we take a list of $\mathrm{def}(G)$, with each appearing λ times and linearly order it such that it does not induce a linear order of $\mathrm{def}(G)$. See below.
So we prove (in 1.30, 1.31, 1.33)

Theorem 0.12. *1) We can for every* lf *group G, define $G^{c\ell}$ such that:*

(a) *if $G \in K^{\mathrm{lf}}_{\leq \lambda}$, then $G \subseteq G^{c\ell} \in K^{\mathrm{exlf}}_{\lambda}$*

(b) *$G^{c\ell}$ is unique up to isomorphism over G.*

2) Also[3] essentially it commutes with extensions, i.e., $G_1 \subseteq G_2 \Rightarrow G^{c\ell}_1 \subseteq G^{c\ell}_1$, pedantically

---

[3]See on this in 3.14.

(c) *if $G_1 \subseteq G_2$ and G_ℓ^{cl} is as above* <u>*then*</u> *there is an embedding h of G_1^{cl} into G_2^{cl} such that $h(G_1^{cl}) \cap G_2 = G_1$*

(c)′ *restricting ourselves to $\{G \in \mathbf{K}_{\mathrm{lf}}\colon$ every $x \in G$ is a singleton$\}$ we have:*

(b)″ *G^{cl} is really unique*

(c)″ *$G_1 \subseteq G_2 \Rightarrow G_1^{cl} \subseteq G_2^{cl}$.*

To stress the generality in addition to the class $\mathbf{K}_{\mathrm{lf}}$ of lf-groups we use $\mathbf{K}_{\mathrm{olf}}$, the class of ordered locally finite groups (see 0.15); for them the proof of the existence of a suitable $\mathfrak{S}$ is easier. Naturally for $\mathbf{K}_{\mathrm{olf}}$ we certainly do not have a symmetric $\mathfrak{S}$.

In §2, we show that $\mathfrak{S}$ as needed in §1 exists, but not necessarily symmetric and define and investigate some specific schemes used later; also we define and investigate NF, a relative of free amalgamation. In §3, we find a fourth avenue which is more specific to the class of lf groups. We show that we can induce symmetry, i.e., define symmetric constructions even for nonsymmetric $\mathfrak{S}$ hence get somewhat better results, see 3.15. In particular we construct reasonable closures.

In §4(A), we show that we can find amalgamation preserving commuting and so can get a new relative NF^3 of NF. In §4(B) we deal with some related schemes (of types). In §4(C) we deal with types with infinitely many variables.

In §5, we prove the existence of a complete group $G_* \in \mathbf{K}_\lambda^{\mathrm{exlf}}$ when $\lambda = \mu^+, \mu = \mu^{\aleph_0}$. Moreover, we prove the existence of a complete extension $G_* \in \mathbf{K}_\lambda^{\mathrm{exlf}}$ of an arbitrary $G \in \mathbf{K}_{\leq\mu}^{\mathrm{lf}}$.

Some of the definitions and claims work also in quite a general framework, but it is not clear at present how interesting this is. Still we consider some expansions of $\mathbf{K}_{\mathrm{lf}}$, and comment on them in §6.

We here also consider the partial order $\leq_{\mathfrak{S}}$ on $\mathbf{K}$, where $G_1 \leq_{\mathfrak{S}} G_2$ means that every finite $\bar{a} \subseteq G_2$ realizes over G_2 a type from $\mathrm{def}_{\mathfrak{S}}(G_1)$. Note that on $(\mathbf{K}, \leq_{\mathfrak{S}})$ we may generalize stability theory, in particular when $\mathfrak{S}$ is symmetric (see §1) or when we use the symmetrized version (see §3). In particular, we can investigate orthogonality, parallelism, super-stability, and indiscernible sets which Δ-converge ([Sh:300] or [Sh:300a]). A class somewhat similar to $\mathbf{K}_{\mathrm{lf}}$, for an existentially closed countable group L is $\mathbf{K}_L$, the class of groups G such that every f.g. subgroup is embeddable into L. We further investigate $\mathbf{K}_{\mathrm{lf}}$ in [Sh:1098] and in more general direction in a work in preparation with Gianluca Paolini.

We thank Omer Zilberboim and Gianluca Paolini for some help in the proofs of this chapter and a referee for many useful comments to clarify this chapter.

0(D) Preliminaries

Definition 0.13. 1) Let $\mathbf{K}_\lambda^{\mathrm{lf}}$ be the class of $G \in \mathbf{K}_{\mathrm{lf}}$ of cardinality λ, let $\mathbf{K}_\lambda^{\mathrm{exlf}}$ be the class of $G \in \mathbf{K}_{\mathrm{exlf}}$ of cardinality λ; see Definition 0.1.
2) Let $\mathrm{fsb}(M)$ be the set of f.g. (finitely generated) substructures of M.

Note that $\mathbf{K}_{\mathrm{exlf}}$ is the same $\mathbf{K}_{\mathrm{ulf}}$ as defined by Hall as proved in Macintyre-Shelah [McSh:55], Wood [Woo72].

Claim 0.14. The following conditions on a locally finite group G are equivalent:

(A) G is ulf which means:

  (a) every finite group is embeddable into G

  (b) if H_1, H_2 are isomorphic finite subgroups of G, <u>then</u> for some $x \in G$, conjugation by x maps H_2 onto (here equivalently into) H_2, i.e., $x^{-1} H_1 x = H_2$

(B) $G \in \mathbf{K}_{\mathrm{exlf}}$.

Proof. $(B) \Rightarrow (A)$
    Clause <u>(A)(a)</u>: let H be a finite group, let $H_1 = \{e_H\} \subseteq H$ so a subgroup of $\overline{H}$ and let $H_2 = H$ and let $h_1 : H_1 \to G$ be defined by $h_2(e_H) = e_G$. So by Clause (B) there is an extension h_2 of h_1 embedding $H_2 = H$ into G, so $h_2(H)$ is as required.

Clause <u>(A)(b):</u>: let $H_1, H_2 \subseteq G$ be finite subgroups and let $H_3 \subseteq G$ be the finite subgroup which $H_1 \cup H_2$ generates. There is a finite group H_4 extending H_3 such that: any partial automorphism of H_3 is included in some conjugation in H_4. Let $h_3 : H_3 \to H_3 \subseteq G$ be the identity, hence by Clause (B) recalling $G \in \mathbf{K}_{\mathrm{exlf}}$, there is an embedding h_4 of H_4 into G extending h_3.
    So in $h_4(H_4) \subseteq G$ there is a conjugation as required.

<u>$(A) \Rightarrow (B)$</u>:
    Let $H_1 \subseteq H_2$ be finite groups and h_1 be an embedding of H_1 into G. Let $H_4 \supseteq H_2$ be a finite group such that any automorphism of H_1 is included in an inner automorphism of H_1. By Clause (A)(a) there is an embedding h_4 of H_4 into G. By Clause (A)(b) there is $x \in G$ such that $H_4' := x^{-1} h_1(H_1) x \subseteq G$ is equal to $h_4(H_1)$.
    Now $h_4' = (\square_x \restriction h_4(H_4)) \circ h_4$ embeds H_4 into G and maps H_1 onto $h_1(H_1)$; but the embedding h_4' does not necessarily extend h_1. However, by clause (A)(b), for some $y \in h_4'(H_4), \square_y h_4'$ embeds H_4 (hence H_2) and extends h_1 as required. $\qquad\square_{0.14}$

We may use the class $\mathbf{K}_{\mathrm{olf}}$ of linearly ordered lf groups; it is closely related and some issues are more transparent for it. $\mathbf{K}_{\mathrm{olf}}$ is defined as follows.

Definition 0.15. 1) Let $\mathbf{K}_{\mathrm{olf}}$ be the class of structures M which are an expansion of a lf group $G = G_M$ by a linear order $<_M$, also this class is partially ordered by $M_1 \subseteq M_2$, M_1 a substructure of M_2.
2) We say that $M \in \mathbf{K}_{\mathrm{olf}}$ is _existentially closed_ as in 0.13(2) and define $\mathbf{K}_\lambda^{\mathrm{olf}}$ as in 0.1(3).
3) If $M \in \mathbf{K}_{\mathrm{lf}}$, then we let $G_M = M$.

Remark 0.16. For $\mathbf{K}_{\mathrm{lf}}$ conceivably there is a symmetric dense $\mathfrak{S}$, hence a very natural canonical exlf-closure. Without it we can either use a somewhat less natural one (using linear orders, see end of §1) or "make it symmetric by brute force" (see §3). But for the class $\mathbf{K}_{\mathrm{olf}}$ we can use only the linear orders, so every M has a canonical existentially closed extension, but it is more difficult to make it unique up to isomorphism. We shall in 6.2 introduce another class, $\mathbf{K}_{\mathrm{clf}}$, locally finite groups with choice.

Convention 0.17. 1) Except in §6, $\mathbf{K}$ is the class $\mathbf{K}_{\mathrm{lf}}$ of locally finite groups or $\mathbf{K}_{\mathrm{olf}}$ of ordered locally finite groups (we may use $\leq_{\mathbf{K}}$ but here $\mathbf{K}$ is partially ordered by $\subseteq$, being a substructure) and see 0.16.
2) Let xlf-group mean a member of $\mathbf{K}$. Let $\mathbf{K}_{\mathrm{ec}}$ be the class of existentially closed members of $\mathbf{K}$.
3) In §2, §3, §4, §5 we use only $\mathbf{K}_{\mathrm{lf}}$; in §1 you can restrict yourself to $\mathbf{K} = \mathbf{K}_{\mathrm{lf}}$ but in §6 we have further cases on which we comment.

The following definition is for the more general framework.

Definition 0.18. 1) For $M, N \in \mathbf{K}$ let $M \leq_{\mathrm{fsb}} N$ mean that if $K \subseteq L$ are f.g., $K \subseteq M, L \subseteq N$, <u>then</u> there is an embedding of L into M over K.
2) For $M, N \in \mathbf{K}$ let $M \leq_{\Sigma_1} N$ <u>means</u> that $M \subseteq N$ and if $\bar{a} \in {}^{\ell g(\bar{y})}M, \bar{b} \in {}^{\ell g(\bar{x})}N$ and $\varphi(\bar{x}, \bar{y}) \in \mathbb{L}(\tau_{\mathbf{K}})$ is quantifier free and $N \models \varphi[\bar{b}, \bar{a}]$ then for some $\bar{b}' \in {}^{\ell g(\bar{x})}M$ we have $M \models \varphi[\bar{b}', \bar{a}]$.
3) Let $M_\ell \in \mathbf{K}, \bar{a}_\ell \in {}^{n(\ell)}(M_\ell)$ for $\ell = 1, 2$. We say that a relation on $M_1 \times M_2$ is quantifier-free definable in $(M_1, \bar{a}_1, M_2, \bar{a}_2)$ <u>when</u> it is a Boolean combination of finitely many simple ones, where R is a simple n-place relation on $M_1 \times M_2$ <u>when</u> R is the set of n-tuples $((b_0, c_0), \dots, (b_{n-1}, c_{n-1}))$ such that $b_i \in M_1, c_i \in M_2$ for $i < n$ and

$$M_1 \models \varphi_1[b_0, \dots, b_{n-1}, \bar{a}_1]$$

$$M_2 \models \varphi_2[c_0, \dots, c_{n-1}, \bar{a}_2]$$

for some quantifier-free formulas φ_1, φ_2 in $\mathbb{L}(\tau_{\mathbf{K}})$ and finite sequences $\bar{a}_1, \bar{a}_2$ from M_1, M_2 respectively.

Remark 0.19. 1) Note that 0.18(3) is not actually used, but just indicates the form of definability used.

2) Note that $\leq_{\Sigma_1}$ for $\mathbf{K}_{\mathrm{lf}}$ and $\mathbf{K}_{\mathrm{olf}}$ is the same as $\leq_{\mathrm{fsb}}$. For other classes, see §6, if the vocabulary is finite and we deal with locally finite structures they are still the same. Otherwise, by our choice of "does not split" we have to use $\leq_{\mathrm{fsb}}$. But if we prefer to use $\leq_{\Sigma_1}$ we have to strengthen the definition of "does not split" to make the proof of 1.10(1) work.

Convention 0.20. Let $M_1, M_2 \in \mathbf{K}, M_1 \subseteq M_2$ and $\bar{a} \in {}^n(M_2)$, so $\bar{a} = (a_0, a_1, a_2, \ldots, a_{n-1})$.

1) Denote by $cl(M_1 + \bar{a}, M_2)$ the substructure generated by $M_1 \cup \bar{a} = M_1 \cup \{a_0, a_1, \ldots, a_{n-1}\}$ in M_2.

2) For a group G and $A \subseteq G$ let

- $\mathbf{C}_G(A) = \{g \in G : G \models \text{"}ag = ga\text{"} \text{ for every } a \in A\}$

- $\mathbf{Z}(G) = \mathbf{C}_G(G)$

- $\mathbf{N}_G(A) = \{c \in G : c^{-1}Ac = A\}$.

3) For a group G, $\mathrm{aut}(G)$ is the group of automorphisms of G and $\mathrm{inner}(G)$ is the normal subgroup of $\mathrm{aut}(G)$ consisting of the inner automorphisms of G.

A side issue here is:

Definition 0.21. 1) For a class $\mathbf{K}$ of structures (of a fixed vocabulary) we say $M \in \mathbf{K}$ is λ-<u>universal</u> in $\mathbf{K}$ <u>when</u> every $N \in \mathbf{K}$ of cardinality λ can be embedded into it.

2) We say $M \in \mathbf{K}$ is $(\leq \lambda)$-<u>universal</u> in $\mathbf{K}$ <u>when</u> every $N \in \mathbf{K}$ of cardinality $\leq \lambda$ can be embedded into M.

3) We say $M \in \mathbf{K}$ is <u>universal</u> <u>when</u> it is λ-universal for $\lambda = \|M\|$.

4) Assume $\mathfrak{k} = (K_{\mathfrak{k}}, \leq_{\mathfrak{k}}), K_{\mathfrak{k}}$ as a class of τ-structures (for some vocabulary $\tau = \tau_{\mathfrak{k}}$), closed under isomorphism, $\leq_{\mathfrak{k}}$ a partial order on $K_{\mathfrak{k}}$ preserved under isomorphisms. Above "$M \in K_{\mathfrak{k}}$ is λ-universal in $\mathfrak{k}$" means that if $N \in K_{\mathfrak{k}}$ has cardinality λ, then there is a $\leq_{\mathfrak{k}}$-embedding f of N into M, i.e., f is an isomorphism from N onto some $N' \leq_{\mathfrak{k}} M$; similarly in the other variants.

The problem of the existence of universal members of $\mathbf{K}_\lambda^{\mathrm{lf}}$ is connected to

Question 0.22. Fixing κ and an ideal J on κ, what is $\lambda_{\mu,\kappa}(J, \mathbf{K})$, which is the minimal cardinal (or ∞) λ which is $> \mu$ and there is no sequence $\langle (G_\alpha, \bar{a}_\alpha) : \alpha < \lambda \rangle$ such that $G_\alpha \in \mathbf{K}_{\leq \mu}, \bar{a}_\alpha \in {}^\kappa(G_\alpha)$ and there are no $H \in \mathbf{K}$ and $\alpha < \beta < \lambda$ and embeddings f_1, f_2 of G_α, G_β respectively into H such that $\{i < \kappa : f(a_{\alpha,i}) \neq a_{\beta,i}\} \in J$?

Notation 0.23. 1) Let G, H, K denote members of $\mathbf{K}$.

2) Let p, q, r and s denote types.

3) $\mathfrak{s}$ denotes a scheme of defining types, here qf.

4) t denotes a member of some $\mathrm{def}(G)$, i.e., a pair $(\mathfrak{s}, \bar{a})$ which defines a type in $\mathbf{S}_{\mathrm{bs}}^{n(\mathfrak{s})}(G)$.

5) For $A \subseteq M$ let $c\ell(A, M) = \langle A \rangle_M$ be the closure of the set A under the functions of M, i.e., the subgroup of M which A generates when M is, as usual, a group.

6) We may write, e.g., $A + B, A + \bar{a}, \sum_{i < \alpha} \bar{a}_i$ instead of $A \cup B, A \cup \mathrm{Rang}(\bar{a}), \bigcup_{i < \alpha} \mathrm{Rang}(\bar{a}_i)$.

1   Definable types

What is accomplished in §1 and under what assumptions? We have to assume that there are dense $\mathfrak{S} \in \Omega[\mathbf{K}]$ to get existentially closed H (see §2). Still there are $\mathfrak{S}$'s and any $\mathfrak{S}$ can be extended to a closed one, preserving density. For any $\mathfrak{S}$, the partial order $\leq_{\mathfrak{S}}$ on $\mathbf{K}$ is quite reasonable: not fully so-called a.e.c., but still close enough. In $(\mathbf{K}, \leq_{\mathfrak{S}})$ for regular λ we can find over any $G \in K_{<\lambda}$ a prime H among the $H \in \mathbf{K}_{\lambda}$ extending G which are so-called $(\lambda, \mathfrak{S})$-full over it, see 1.23. Also we can find such H quite definable in three ways. First avenue is to allow order. Second avenue is to assume $\mathfrak{S}$ is symmetric, then H is canonical and commutes with extensions (1.13, 1.16, 1.23, 1.17). Third avenue relies on linear order. We still get uniqueness, but rely on linear ordering of $\mathrm{def}(G)$ and the commutation with extension is problematic. However, we may use pair (I, E), I a linear order, E an equivalence relation on I and "dedicate" each equivalence class to some $t \in \mathrm{def}(G)$, so we can avoid linearly ordering $\mathrm{def}(G)$, see 1.30, 1.33; see more in §3.

1(A)   The Framework

Definition 1.1. 1) For $G \in \mathbf{K}$ let $\mathrm{def}(G)$ be the set of pairs $t = (\mathfrak{s}, \bar{a}) = (\mathfrak{s}_t, \bar{a}_t)$ such that $\mathfrak{s} \in \Omega[\mathbf{K}]$ and $\bar{a} \in {}^{\omega>}G$ realizes $p_{\mathfrak{s}}(\bar{x}_{\mathfrak{s}})$ and let $q_t(G) = q_{\mathfrak{s}_t}(\bar{a}_t, G)$ and $p_t(\bar{x}_t) = p_{\mathfrak{s}}(\bar{x}_{\mathfrak{s}}), k(t) = k(\mathfrak{s}), n(t) = n(\mathfrak{s})$.
2) We say $\mathfrak{s}_1, \mathfrak{s}_2$ are disjoint when $\bar{x}_{\mathfrak{s}_1}, \bar{x}_{\mathfrak{s}_2}$ are <u>disjoint</u> as well as $\bar{z}_{\mathfrak{s}_1}, \bar{z}_{\mathfrak{s}_2}$. Similarly for $t_1, t_2 \in \mathrm{def}(G)$.
3) We say $\mathfrak{s}_1, \mathfrak{s}_2$ are <u>congruent</u>, written $\mathfrak{s}_1 \equiv \mathfrak{s}_2$ <u>when</u> we get $\mathfrak{s}_2$ from $\mathfrak{s}_1$ by replacing $\bar{x}_{\mathfrak{s}_1}, \bar{z}_{\mathfrak{s}_1}$ by other sequences of variables, $\bar{x}_{\mathfrak{s}_2}, \bar{z}_{\mathfrak{s}_2}$ (again with no repetitions, of the same length respectively, of course). Similarly for $t_1, t_2 \in \mathrm{def}(G)$ (the aim is to be able to get <u>disjoint</u> congruent copies; we do not always remember to replace a scheme by some congruent copy).
4) We say $\mathfrak{S}$ is <u>invariant when</u>: if $\mathfrak{s}_1, \mathfrak{s}_2 \in \Omega[\mathbf{K}]$ are congruent, then $\mathfrak{s}_1 \in \mathfrak{S} \Leftrightarrow \mathfrak{s}_2 \in \mathfrak{S}$.
5) The invariant closure of $\mathfrak{S}$ is defined naturally. Let $|\mathfrak{S}|$ mean its cardinality up to congruency, that is, $|\mathfrak{S}/ \equiv|$; if not said otherwise we use invariant $\mathfrak{S}$.
6) We define the (equivalence) relation $\approx_G$ on $\mathrm{def}(G)$ by $t_1 \approx_G t_2$ <u>iff</u> $t_1, t_2 \in \mathrm{def}(G)$ and $q_{t_1}(G) = q_{t_2}(G)$.

Claim 1.2. 1) If $\mathfrak{s} \in \Omega_{n,k}[\mathbf{K}]$ and $G \in \mathbf{K}, \bar{a} \in {}^{k}M$ <u>then</u> indeed $q_{\mathfrak{s}}(\bar{a}, G) \in \mathbf{S}^n_{\mathrm{bs}}(G)$ so exist and is unique and does not split over $\bar{a}$, see Definition 0.9(2); if $\bar{a}$ is empty, i.e., $k_{\mathfrak{s}} = 0$ we may write $q_{\mathfrak{s}}(G)$.
1A) If $G_1 \subseteq G_2 \subseteq \mathbf{K}$ and $t \in \mathrm{def}(G_1)$ <u>then</u> $t \in \mathrm{def}(G_2)$ and $q_t(G_1) \subseteq q_t(G_2)$.

2) Assume $G \subseteq H \in \mathbf{K}$ and G is existentially closed or just $G \leq_{\Sigma_1} H \in \mathbf{K}$. If $t_1, t_2 \in \mathrm{def}(G)$ then $q_{t_1}(G) = q_{t_2}(G)$ iff $q_{t_1}(H) = q_{t_2}(H)$.

3) Let $K \subseteq G \in \mathbf{K}, G$ be existentially closed or just every $r \in \mathbf{S}_{\mathrm{bs}}^{<\omega}(K)$ is realized in G, K is finite, and $p \in \mathbf{S}_{\mathrm{bs}}^n(G)$.

The type p does not split over K iff there are $\mathfrak{s} \in \Omega[\mathbf{K}]$ and a finite sequence $\bar{a}$ from K (even listing K) realizing $p_{\mathfrak{s}}(\bar{x})$ such that $p = q_{\mathfrak{s}}(\bar{a}, M)$.

4) If $G \subseteq H, \mathfrak{s} \in \Omega[\mathbf{K}], \bar{a} \in {}^{k(\mathfrak{s})}G$ realizes $p_{\mathfrak{s}}(\bar{x}_{\mathfrak{s}})$ and $\bar{c} \in {}^{n(\mathfrak{s})}H$ realizes $q_{\mathfrak{s}}(\bar{a}, G)$ in H and $\sigma(\bar{z}_{\mathfrak{s}}, \bar{x}_{\mathfrak{s}})$ is a group-term then $\sigma^H(\bar{c}, \bar{a}) \in G \Rightarrow \sigma^H(\bar{c}, \bar{a}) \in cl(\bar{a}, G)$.

4A) In (4), if $\bar{a}' = \bar{a}^{\frown}\bar{a}''$, then $\sigma^H(\bar{c}, \bar{a}'') \in G \Rightarrow \sigma^H(\bar{c}, \bar{a}'') \in cl(\bar{a}'', G)$ because p also does not split over $\bar{a}^*$ if $\bar{a}^* \subseteq G, \bar{a} \subseteq cl(\bar{a}^*, H)$.

Proof. 1) Let $K_* \subseteq G$ be the subgroup of G generated by $\bar{a}$.

First, there are H and $\bar{c}$ such that:

$$(*)_{H,\bar{c}}^1 \quad G \subseteq H \in \mathbf{K} \text{ and } \bar{c} \in {}^nH \text{ such that } \mathrm{tp}(\bar{c}, G, H) = q_{\mathfrak{s}}(\bar{a}, G).$$

Why? For every $K \in \mathbf{K}^* := \{K \subseteq G : K \text{ finite extending } K_*\}$ we can choose a pair $(H_K, \bar{c}_K)$ such that: $K \subseteq H_K \in \mathbf{K}, H_K$ is finite, $\bar{c}_K \in {}^n(H_K), H_K$ is generated by $K \cup \bar{c}_K$ and for some $\bar{b}$ listing $K, \mathrm{tp}_{\mathrm{bs}}(\bar{a}^{\frown}\bar{b}^{\frown}\bar{c}_K, \emptyset, H_K) = \mathfrak{s}(\mathrm{tp}_{\mathrm{bs}}(\bar{a}^{\frown}\bar{b}, \emptyset, G))$.

[Why? By Definition 0.9(1A)(b).] Now for every $K_1 \subseteq K_2$ from $\mathbf{K}_*$ we can choose an embedding f_{K_2,K_1} from H_{K_1} into H_{K_2} extending id_{K_1} and mapping $\bar{c}_{K_1}$ to $\bar{c}_{K_2}$. [Why? By Definition 0.9(1A)(c).]

As H_{K_1} is generated by $K_1 \cup \bar{c}$, this mapping is unique. Now if $K_1 \subseteq K_2 \subseteq K_3$ are from $\mathbf{K}_*$, then $f_{K_3,K_2} \circ f_{K_2,K_1}$ is an embedding of H_{K_1} into H_{K_3} extending id_{K_1} and mapping $\bar{c}_{K_1}$ to $\bar{c}_{K_3}$; hence by the previous sentence $f_{K_3,K_2} \circ f_{K_2,K_1} = f_{K_3,K_1}$. Hence $\langle H_{K_1}, f_{K_2,K_1} : K_1 \subseteq K_2 \text{ are from } \mathbf{K}_* \rangle$ has a direct limit, i.e., we can find a group H and $\bar{f} = \langle f_K : K \in \mathbf{K}_* \rangle$ such that f_K embed H_K into H and for every $K_1 \subseteq K_2$ from $\mathbf{K}_*$ we have $f_{K_1} = f_{K_2} \circ f_{K_2,K_1}$. Without loss of generality $H = \cup\{f_K(H_K) : K \in \mathbf{K}_*\}$ hence H is a locally finite group and $\{f_K : K \in \mathbf{K}_*\}$ embeds G into H, so without loss of generality $G \subseteq H$. Letting $\bar{c} = f_K(\bar{c}_K)$ for any $K \in \mathbf{K}_*$, clearly $(H, \bar{c})$ is as required in $(*)_{H,\bar{c}}^1$.

$$(*)_2 \quad \mathrm{tp}_{\mathrm{bs}}(\bar{c}, G, H) \text{ belongs to } \mathbf{S}_{\mathrm{bs}}^n(G).$$

[Why? By the definitions of $\mathbf{S}_{\mathrm{bs}}^n(G)$ because $G \subseteq H \in \mathbf{K}$ and $\bar{c} \in {}^nH$.]

$$(*)_3 \quad q_{\mathfrak{s}}(\bar{a}, G) \text{ is unique and does not split over } \bar{a}.$$

[Why? See Definition 0.9(1A)(c).]

1A) See Definition 0.9(2).

2) For $\ell = 1, 2$ we have $q_{t_\ell}(G) \subseteq q_{t_\ell}(H)$, moreover, $q_{t_\ell}(G) = \{\varphi(\bar{z}_{n(t)}, \bar{b}) \in q_{t_\ell}(H) : \bar{b} \subseteq G\}$. For the other direction, note that $\bar{a}_{t_1}, \bar{a}_{t_2} \subseteq G$ and assume $q_{t_1}(H) \neq q_{t_2}(H)$, hence there are m and $\bar{b} \in {}^mH$ and a basic formula $\varphi(\bar{y}_m, \bar{z}_n)$

such that $\varphi(\bar{b}, \bar{z}_n) \in q_{t_1}(H), \neg\varphi(\bar{b}, \bar{z}_n) \in q_{t_2}(H)$. Now there is $\bar{b}' \in {}^m G$ such that $\mathrm{tp}_{\mathrm{bs}}(\bar{b}', \bar{a}_{t_1} {}^\frown \bar{a}_{t_2}, G) = \mathrm{tp}_{\mathrm{bs}}(\bar{b}, \bar{a}_{t_1} {}^\frown \bar{a}_{t_2}, H)$ because $G \leq_{\Sigma_1} H$ and our choice of $\mathbf{K}$. As $q_{t_\ell}(H)$ does not split over $\bar{a}_{t_\ell}$, clearly $\varphi(\bar{b}', \bar{z}_n) \in q_{t_\ell}(H) \Leftrightarrow \varphi(\bar{b}, \bar{z}_n) \in q_{t_\ell}(H)$ for $\ell = 1, 2$.

Together with an earlier sentence, $\varphi(\bar{b}', \bar{z}_n) \in q_{t_1}(H), \neg\varphi(\bar{b}', \bar{z}_n) \in q_{t_2}(H)$ hence by the first sentence in the proof of 1.2(2) we have $\varphi(\bar{b}', \bar{z}_n) \in q_{t_1}(G)$ and $\neg\varphi(\bar{b}', \bar{z}_n) \in q_{t_2}(G)$ hence $q_{t_2}(G) \neq q_{t_2}(G)$ so we are also done with the "other" direction.

3) The implication "if" holds by 1.2(1). For the other direction assume p does not split over K. As K is finite, let $k = |K|$ and let $\bar{a} \in {}^n K \subseteq {}^n G$ list K.

We now define $\mathfrak{s}$ by:

(a) $p_{\mathfrak{s}} = \mathrm{tp}_{\mathrm{bs}}(\bar{a}, \emptyset, K)$ so $k_{\mathfrak{s}} = k$

(b) $q = \mathfrak{s}(s(\bar{x}, \bar{y}))$ iff for some $\bar{b} \in {}^m G$ we have:

- $s(\bar{x}, \bar{y}) = \mathrm{tp}(\bar{a} {}^\frown \bar{b}, \emptyset, G)$
- $q = \mathrm{tp}_{\mathrm{bs}}(\bar{a} {}^\frown \bar{b} {}^\frown \bar{c}, \emptyset, G)$ for some $\bar{c} \in {}^n G$ realizing $p{\restriction}(\bar{a} {}^\frown \bar{b})$.

Now $\mathfrak{s}$ is well defined because on one hand p does not split over $\bar{a}$, and on the other hand G is existentially closed or just every $r \in \mathbf{S}_{\mathrm{bs}}^{<\omega}(K)$ is realized in G.
4) By 1.2(1A) without loss of generality G is existentially closed, assume $\sigma^H(\bar{c}, \bar{a}) \in G$ and let $b = \sigma^H(\bar{c}, \bar{a})$. If $b \notin cl(\bar{a}, G)$ there is $b' \in G \backslash \{b\}$ realizing $\mathrm{tp}_{\mathrm{bs}}(b, K, G)$ because $\mathbf{K}$ has disjoint amalgamation for finite members. As $q_{\mathfrak{s}}(\bar{a}, G)$ does not split over $\bar{a}$ and $b', b \in G$ realize the same type over $\bar{a}$ it follows that $H \models$ "$(\sigma(\bar{c}, \bar{a}) = b) \equiv (\sigma(\bar{c}, \bar{a}) = b')$", an obvious contradiction.
4A) Should be clear. □$_{1.2}$

Example 1.3. There is $\mathfrak{s} \in \Omega[\mathbf{K}]$ such that:

(a) $k_{\mathfrak{s}} = 0$ and $n_{\mathfrak{s}} = 1$;

(b) if $G \subseteq H \in \mathbf{K}$ and $a \in H$, then: a realizes $q_{\mathfrak{s}}(<>, G)$ iff $a \in H \backslash G$ has order 2 and commute with every member of G.

Definition 1.4. 1) For $\mathfrak{S} \subseteq \Omega[\mathbf{K}]$ we define the two place relation $\leq_{\mathfrak{S}}$ on $\mathbf{K}$ as follows: $M \leq_{\mathfrak{S}} N$ iff $M \subseteq N$ (and they belong to $\mathbf{K}$) and for every $n < \omega$ and $\bar{c} \in {}^n N$ we can find $k < \omega$ and $\bar{a} \in {}^k M$ and $\mathfrak{s} \in \mathfrak{S}$ such that $p_{\mathfrak{s}}(\bar{x}) \subseteq \mathrm{tp}_{\mathrm{bs}}(\bar{a}, \emptyset, M) \in \mathbf{D}_k(\mathbf{K})$ and $\mathrm{tp}_{\mathrm{bs}}(\bar{c}, M, N) = q_{\mathfrak{s}}(\bar{a}, M)$ recalling $q_{\mathfrak{s}}(\bar{a}, M) \in \mathbf{S}_{\mathrm{bs}}^n(M)$.
2) For $M \in \mathbf{K}$ and $\mathfrak{S} \subseteq \mathfrak{S}[\mathbf{K}]$ let

(a) $\mathbf{S}_{\mathfrak{S}}^n(M) = \{q_{\mathfrak{s}}(\bar{a}, M) : \mathfrak{s} \in \mathfrak{S}$ satisfies $n_{\mathfrak{s}} = n$ and $\bar{a} \in {}^{k(\mathfrak{s})} M$ realizes $p_{\mathfrak{s}}(\bar{x}_{\mathfrak{s}})\}$

(b) $\mathrm{def}_{\mathfrak{S}}(M) = \{t \in \mathrm{def}(M) : \mathfrak{s}_t \in \mathfrak{S}\}$

(c) $\mathbf{S}_{\mathfrak{S}}(M) = \cup\{\mathbf{S}_{\mathfrak{S}}^n(M) : n < \omega\}$.

3) We say $M \in \mathbf{K}$ is $\mathfrak{S}$-existentially closed <u>when</u> for every $\mathfrak{s} \in \mathfrak{S}$, finite[4] $G \subseteq M$ and $\bar{a} \in {}^{\omega >}G$ realizing $p_{\mathfrak{s}}(\bar{x})$ the type $q_{\mathfrak{s}}(\bar{a}, G)$ is realized in M; (this is equivalent to being existentially closed if $\mathfrak{S}$ is dense, see Definition 1.6(2) below).

Definition 1.5. We say $\mathfrak{S} \subseteq \Omega[\mathbf{K}]$ is <u>symmetric</u> <u>when</u> : if $\mathfrak{s}_1, \mathfrak{s}_2 \in \mathfrak{S}, M \subseteq N$ are from $\mathbf{K}$ and $\bar{c}_\ell \in {}^{n(\mathfrak{s}_\ell)}N$ realizes $q_{\mathfrak{s}_\ell}(\bar{a}_\ell, M)$ in N (so $\bar{a}_\ell \in {}^{k(\mathfrak{s}_\ell)}M$ realizes $p_{\mathfrak{s}_\ell}(\bar{x}_{\mathfrak{s}_\ell})$) and $M_\ell = cl(M+\bar{c}_\ell, N) \subseteq N$ for $\ell = 1, 2$ <u>then</u> $\bar{c}_1$ realizes $q_{\mathfrak{s}_1}(\bar{a}_1, M_2)$ in N iff $\bar{c}_2$ realizes $q_{\mathfrak{s}_2}(\bar{a}_2, M_1)$ in N.

Definition 1.6. 1) We say $\mathfrak{S}$ is closed <u>when</u> it is dominating-<u>closed</u> and composition-closed, see below and invariant of course.
1A) $\mathfrak{S}$ is <u>composition-closed</u> <u>when</u> if $H_0 \subseteq H_1 \subseteq H_2 \in \mathbf{K}, \bar{a}_\ell \in {}^{n(\ell)}(H_\ell)$ for $\ell = 0, 1, 2$ and $\mathrm{tp}_{\mathrm{bs}}(\bar{a}_{\ell+1}, H_\ell, H_{\ell+1}) = q_{\mathfrak{s}_\ell}(\bar{a}_\ell, H_\ell) \in \mathbf{S}_{\mathfrak{S}}^{n(\ell+1)}(H_\ell)$ and $H_{\ell+1} = cl(H_\ell + \bar{a}_\ell, H_{\ell+1}), \mathfrak{s}_\ell \in \mathfrak{S}$ for $\ell = 0, 1$ <u>then</u> $\mathrm{tp}_{\mathrm{bs}}(\bar{a}_1\hat{\,}\bar{a}_2, H_0, H_2) = q_{\mathfrak{s}}(\bar{a}_0, H_0)$ for some $\mathfrak{s} \in \mathfrak{S} \cap \Omega_{n(1)+n(2),n(0)}[\mathbf{K}]$.
1B) $\mathfrak{S}$ is <u>dominating-closed</u> <u>when</u>: if $H_0 \subseteq H_1 \in \mathbf{K}, \bar{a}_1 \in {}^{k(1)}(H_0), \bar{c}_1 \in {}^{n(1)}(H_1)$, $\mathrm{tp}_{\mathrm{bs}}(\bar{c}_1, H_0, H_1) = q_{\mathfrak{s}}(\bar{a}_1, H_0) \in \mathbf{S}_{\mathfrak{S}}^{n(1)}(H_0)$ and $\bar{c}_2 \in {}^{n(2)}(H_1)$ and $\bar{a}_2 \in {}^{k(2)}(H_0)$, $\mathrm{Rang}(\bar{a}_2) \supseteq \mathrm{Rang}(\bar{a}_1)$ and $\bar{c}_2 \subseteq cl(\bar{a}_2 + \bar{c}_1, H_1)$ <u>then</u> $\mathrm{tp}(\bar{c}_2, H_0, H_1) = q_{\mathfrak{s}}(\bar{a}_2, H_0)$ for some $\mathfrak{s} \in \mathfrak{S}$.
2) We say $\mathfrak{S}$ is <u>weakly dense</u> <u>when</u>: every $\mathfrak{S}$-existentially closed $G \in K$ is existentially closed.
3) We say $\mathfrak{S}$ is <u>dense</u> when: for every $G_0 \subseteq H \in \mathbf{K}, G_0 \subseteq G_1 \in \mathbf{K}, G_0, G_1$ are finite and $\bar{c} \in {}^n(G_1)$ <u>there is</u> $p(\bar{z}) \in \mathbf{S}_{\mathfrak{S}}^n(H)$ which extends $\mathrm{tp}_{\mathrm{bs}}(\bar{c}, G_0, G_1)$. Moreover, $p(\bar{z}) = q_{\mathfrak{s}}(\bar{a}, H)$ for some $\mathfrak{s} \in \mathfrak{S}$ and $\bar{a}$ from G_0.
4) For disjoint $\mathfrak{s}_1, \mathfrak{s}_2 \in \mathfrak{S}$ define $\mathfrak{s} = \mathfrak{s}_1 \oplus \mathfrak{s}_2$ with $p_{\mathfrak{s}}(\bar{x}_{\mathfrak{s}}) = p_{\mathfrak{s}_1}(\bar{x}_{\mathfrak{s}_1}) \cup p_{\mathfrak{s}_2}(\bar{x}_{\mathfrak{s}_2})$, recalling $\bar{x}_{\mathfrak{s}_1}, \bar{x}_{\mathfrak{s}_2}$ are disjoint, as follow: if $G_0 \subseteq G_1 \subseteq G_2$ are from $\mathbf{K}$ and $\bar{a}_1 \in {}^{k(\mathfrak{s}_1)}G_0, \bar{a}_2 \in {}^{k(\mathfrak{s}_2)}G_0, \bar{a}_\ell$ realizes $p_{\mathfrak{s}_\ell}(\bar{x}_{\mathfrak{s}_\ell})$ in $G_0 \in \mathbf{K}$ and $\bar{c}_\ell \in {}^{n(\mathfrak{s}_\ell)}(G_{\ell+1})$ realizes $q_{\mathfrak{s}_\ell}(\bar{a}_\ell, G_\ell)$ for $\ell = 1, 2$ <u>then</u> $\bar{c}_1\hat{\,}\bar{c}_2$ realizes $q_{\mathfrak{s}}(\bar{a}_1\hat{\,}\bar{a}_2, G_0)$ in G_2.
4A) For (disjoint) $t_1, t_2 \in \mathrm{def}(G)$ we define $t_1 \oplus t_2 = t_1 \oplus_G t_2$ similarly.
5) We define $\bigoplus_{k<m} \mathfrak{s}_k, \bigoplus_{k<m} t_k$ similarly using associativity, see 1.8(5).
6) Let $\mathfrak{s}_1 \le \mathfrak{s}_2$ means: if $G \in \mathbf{K}, \bar{a}_2 \in {}^{u(2)}G$ realizes $p_{\mathfrak{s}_2}(\bar{x}_{\mathfrak{s}_2}), G \subseteq H, \bar{c}_2 \in {}^{n(\mathfrak{s}_2)}H$ realizes $q_{\mathfrak{s}_2}(\bar{a}_2, G)$ <u>then</u> $\mathrm{dom}(\bar{x}_{\mathfrak{s}_1}) \subseteq u(2)$ and $\bar{c}_2 {\restriction} \mathrm{dom}(\bar{z}_{\mathfrak{s}_2})$ realizes $q_{\mathfrak{s}_1}(\bar{a}_2 {\restriction} k(\mathfrak{s}_1), G)$ and $p_{\mathfrak{s}_2}(\bar{x}_{\mathfrak{s}_2}) {\restriction} \bar{x}_{\mathfrak{s}_1} = p_{\mathfrak{s}_1}(\bar{x}_{\mathfrak{s}_1})$.
7) Let $\mathfrak{s}_1 \le_{\bar{h}} \mathfrak{s}_2$ means that $\bar{h} = (h', h'')$, h' is a one-to-one function from $\mathrm{dom}(\bar{x}_{\mathfrak{s}_1})$ into $\mathrm{dom}(\bar{x}_{\mathfrak{s}_2})$ and h'' is a one-to-one function from $\mathrm{dom}(\bar{z}_{\mathfrak{s}_1})$ into $\mathrm{dom}(\bar{z}_{\mathfrak{s}_2})$ such that: if $\mathrm{tp}_{\mathrm{bs}}(\bar{c}_2, G, H) = q_{\mathfrak{s}_2}(\bar{a}_2, G)$ and $\bar{a}_1 = \langle a_{2,h''(\ell)} : \ell \in \mathrm{dom}(\bar{a}_1)\rangle$ and $\bar{c}_1 = \langle c_{2,h(\ell)} : \ell \in \mathrm{dom}(\bar{c}_2)\rangle$, then $\mathrm{tp}_{\mathrm{bs}}(\bar{c}_1, G, H) =$

---

[4] For general $\mathbf{K}$: we use finitely generated $G \subseteq M$; generally this change is needed.

$q_{\mathfrak{s}_1}(\bar{a}_1, G, H)$. Similarly $t_1 \leq_{\bar{h}} t_2$ for $t_1, t_2 \in \text{def}(G)$. If $h' \cup h''$ is well defined, we may write $h' \cup h''$ instead of $\bar{h}$.

Remark 1.7. 0) Concerning 1.6(7) the point of disjoint $\mathfrak{s}_1, \mathfrak{s}_2$ and congruency is to avoid using this. So we may ignore it as well as 1.9(2),(3), 3.4(3), 3.5(4), 3.6(5).
1) Note that the operation $\mathfrak{s}_1 \oplus \mathfrak{s}_2$ is not necessarily commutative, e.g., for $\mathbf{K}_{\text{olf}}$ it cannot be.
2) In, e.g., Definition 1.6(1A), in general $\mathfrak{s}$ is not uniquely determined by the relevant information $\text{tp}_{\text{bs}}(\bar{a}_1\hat{\,}\bar{a}_2\hat{\,}\bar{c}_1\hat{\,}\bar{c}_2, H_0, H_2)$ and the lengths of $\bar{a}_1, \bar{a}_2, \bar{c}_1, \bar{c}_2$ <u>but</u> if H_1 is existentially closed, it is. We could have written the definition in a computational form.
3) So $\mathfrak{s}_1 \leq \mathfrak{s}_1$ means $\mathfrak{s}_1 \leq_{\bar{h}} \mathfrak{s}_2$ with h_ℓ the identity for $\ell = 1, 2$.

1.8 Definition/Claim. 1) For any $\mathfrak{S} \subseteq \Omega[\mathbf{K}]$ we can define its closure as the minimal closed (and <u>invariant</u>, of course) $\mathfrak{S}_1 \subseteq \Omega[\mathbf{K}]$ which includes it, see 1.6(1); we denote it by $cl(\mathfrak{S}) = cl(\mathfrak{S}; \mathbf{K})$.
2) Similarly for <u>dominating-closure</u> $\text{docl}(\mathfrak{S})$ and <u>composition-closure</u> $\text{cocl}(\mathfrak{S})$.
3) Those closures preserves density and countability (and being invariant), and have the obvious closure properties.
4) Also dominating-closure preserve being composition closed.
5) The operation $\oplus$ on $\Omega[\mathbf{K}]$ is well defined and associative. If $\mathfrak{S} \subseteq \Omega[\mathbf{K}]$ is closed under $\oplus$, for transparency, <u>then</u> $\mathfrak{S}$ is symmetric (see 1.5) <u>iff</u> the operation $\oplus$ on $\mathfrak{S}$ is commutative (when defined). Similarly for $\text{def}_{\mathfrak{S}}(G)$.
6) $\Omega[\mathbf{K}]$ has cardinality $\leq 2^{\aleph_0}$; generally $\leq 2^{|\tau(\mathbf{K})| + \aleph_0}$.
7) $\leq_{\mathfrak{S}}$ is a transitive relation on $\mathbf{K}$, if $\mathfrak{S} \subseteq \Omega[\mathbf{K}]$ is closed.
8) If $H_0 \subseteq H_1 \subseteq H_2, \mathfrak{s} \in \Omega[\mathbf{K}]$ and $\text{tp}_{\text{bs}}(\bar{c}, H_1, H_2) = q_{\mathfrak{s}}(\bar{a}, H_1)$ and $\bar{a} \in {}^{k(\mathfrak{s})}H_0$ <u>then</u> $\text{Rang}(\bar{c}) \cap H_1 = \text{Rang}(\bar{c}) \cap H_0$.
9) Assume $\mathfrak{S}$ is dense and closed. If $G \subseteq H \in \mathbf{K}$ and G is finite, then $G \leq_{\mathfrak{S}} H$.
10) If $\mathfrak{s} = \mathfrak{s}_0 \oplus \ldots \oplus \mathfrak{s}_{n-1}$ and $i(0) < \ldots i(k-1) < n$ and $\mathfrak{s}' = \mathfrak{s}_{i(0)} \oplus \ldots \oplus \mathfrak{s}_{i(k-1)}$ <u>then</u> $\mathfrak{s}' \leq \mathfrak{s}$.

Proof. This is natural, noting that (8) is specific for our present $\mathbf{K}$, see 1.2(4).
$\square_{1.8}$

Claim 1.9. 0) The operation $\oplus$ is well defined, that is:

(a) if $\mathfrak{s}_1, \mathfrak{s}_2 \in \Omega[\mathbf{K}]$ are disjoint <u>then</u> $\mathfrak{s}_1 \oplus \mathfrak{s} \in \Omega[\mathbf{K}]$ is well defined;

(b) if $t_1, t_2 \in \text{def}(G)$ are disjoint <u>then</u> $t_1 \oplus t_2 \in \text{def}(G)$.

1) The operation $\oplus$ on disjoint pairs from $\text{def}(G)$ respects congruency, see Definition 1.1(3). If $\mathfrak{s}_1, \mathfrak{s}_2 \in \Omega[\mathbf{K}]$ <u>then</u> $(\mathfrak{s}_1 / \equiv) \oplus (\mathfrak{s}_2 / \equiv)$ is well defined, i.e., if $\mathfrak{s}'_\ell, \mathfrak{s}''_\ell$ are congruent to $\mathfrak{s}_\ell$ for $\ell = 1, 2$ and $\mathfrak{s}' = \mathfrak{s}'_1 \oplus \mathfrak{s}'_2, \mathfrak{s}'' = \mathfrak{s}''_1 \oplus \mathfrak{s}''_2$ are well defined (equivalently for $\ell = 1, 2$ the two schemes $\mathfrak{s}'_\ell, \mathfrak{s}''_\ell$ are disjoint), then $\mathfrak{s}', \mathfrak{s}''$ are congruent. (So we may forget to be pedantic about this.)

2) If $(\mathfrak{s}, \bar{a}) = (\mathfrak{s}_1, \bar{a}_1) \oplus_G (\mathfrak{s}_2, \bar{a}_2)$ <u>then</u> $(\mathfrak{s}_\ell, \bar{a}_\ell) \leq (\mathfrak{s}, \bar{a})$.

3) If in $\mathrm{def}(G)$ we have $(\mathfrak{s}_\ell, \bar{a}_\ell) \leq_{h_\ell} (\mathfrak{s}'_\ell, \bar{a}'_\ell)$ for $\ell = 1, 2$ and $\mathrm{Dom}(h_1) \cap \mathrm{Dom}(h_2) = \emptyset$, $\mathrm{Rang}(h_1) \cap \mathrm{Rang}(h_1) = \emptyset$ <u>then</u> $(\mathfrak{s}_1, \bar{a}_1) \oplus (\mathfrak{s}_2, \bar{a}_2) \leq_{h_1 \cup h_2} (\mathfrak{s}'_1, \bar{a}_1) \oplus (\mathfrak{s}'_2, \bar{a}_2)$. Similarly for $\bar{h}_1, \bar{h}_2$.

Proof. Straightforward. $\qquad\qquad \square_{1.9}$

Claim 1.10. Assume $\mathfrak{S} \subseteq \Omega[\mathbf{K}]$ is dominating-closed and $G_0 \subseteq G_1 \in \mathbf{K}$ and $G_0 \leq_\mathfrak{S} G_2$ and, for transparency, $G_1 \cap G_2 = G_0$ and[5] $G_0 \leq_{\Sigma_1} G_2$ (holds if G_0 is existentially closed in $\mathbf{K}$).

1) There is $G_3 \in \mathbf{K}$ such that $G_1 \leq_\mathfrak{S} G_3$ and $G_2 \subseteq G_3$ and $G_3 = \langle G_1 \cup G_2 \rangle_{G_3}$ and $G_1 \leq_{\Sigma_1} G_3$.

2) G_3 above is unique up to isomorphism over $G_1 \cup G_2$.

3) If $\mathfrak{S}$ is symmetric and $G_0 \leq_\mathfrak{S} G_1$ in part (1) <u>then</u> also $G_2 \leq_\mathfrak{S} G_3$.

Proof. Straightforward, e.g.,

1) Let $\bar{c} = \langle c_\alpha : \alpha < \alpha(*) \rangle$ list the elements of G_2, and for every finite $u \subseteq \alpha(*)$ let $\bar{x}_u = \langle x_\alpha : \alpha \in u \rangle$ and $p_u^0(\bar{x}_u) = \mathrm{tp}_{\mathrm{bs}}(\bar{c} \upharpoonright u, G_0, G_2)$ hence by assumption, there is $\mathfrak{s}_u \in \mathfrak{S}$ (up to congruency) and $\bar{a}_u \in {}^{k(\mathfrak{s}_u)}(G_0)$ such that $p_u^0(\bar{x}) = q_{\mathfrak{s}_u}(\bar{a}_u, G_0)$ so $\mathrm{dom}(\bar{x}_{\mathfrak{s}_u}) = u$. We define $p_u^1(\bar{x}_u) \in \mathbf{S}(G_1)$ as $q_{\mathfrak{s}_u}(\bar{a}_u, G_1)$. We define G_3 as a group extending G_1 generated by $G_1 \cup \{c_\alpha : \alpha < \alpha(*)\}$ such that $\bar{c} \upharpoonright u$ realizes $p_u^1(\bar{x}_u)$ for every finite $u \subseteq \alpha(*)$. But for this to work we have to prove that for finite $u \subseteq v \subseteq \alpha(*)$ we have $p_u^1(\bar{x}_u) \subseteq p_v^1(\bar{x}_v)$. This is straightforward recalling 1.2(1A).

Lastly, $G_1 \leq_{\Sigma_1} G_3$ is easy, too. $\qquad\qquad \square_{1.10}$

Remark 1.11. 1) We may consider an alternative definition of $\leq_\mathfrak{S}$:

- $\bullet_1$ $G \leq_\mathfrak{S} H$ iff for every finite $A \subseteq H$ there are $\bar{c} \in {}^{\omega >}H, \bar{a} \in {}^{\omega >}G$ and $\mathfrak{s} \in \mathfrak{S}$ such that: $\bar{a}$ realizes $p_\mathfrak{s}(\bar{x}_\mathfrak{s})$, $\bar{c}$ realizes $q_\mathfrak{s}(\bar{a}, G)$ in H and $A \subseteq \mathrm{Rang}(\bar{c})$.

An even weaker version is:

- $\bullet_2$ as in $\bullet_1$ but "$A \subseteq \mathrm{Rang}(\bar{c})$" is replaced by $A \subseteq c\ell(G \cup \bar{c}, H)$.

2) But, e.g., for $\bullet_1$, to prove $\leq_\mathfrak{S}$ is transitive we need a stronger version of composition-closed: if $G_0 \subseteq G_1 \subseteq G_2$ and for $\ell = 0, 1, \bar{c}_\ell \in {}^{n(\ell)}(G_{\ell+1})$ realizes $q_{\mathfrak{s}_\ell}(\bar{a}_\ell, G_\ell)$ and $\mathrm{Rang}(\bar{b}_0) \subseteq \mathrm{Rang}(\bar{a}_1)$ <u>then</u> for some $\mathfrak{s} \in \mathfrak{S}, p_\mathfrak{s}(\bar{x}_\mathfrak{s}) = p_{\mathfrak{s}_0}(\bar{x}_\mathfrak{s})$ and $\bar{a}_1 {}^{\frown} \bar{a}_2$ realizes $q_\mathfrak{s}(\bar{a}_0, G_0)$.

3) In any case for closed $\mathfrak{S}$ the three definitions are equivalent, i.e., those in $\bullet_1$, in $\bullet_2$ and in 1.4(1).

4) Does the operation $\oplus_G$ respect $\approx_G$, see Definition 1.1, i.e., if $t_1 \approx_G t'_1$ and $t_2 \approx_G t'_2$ <u>then</u> $t_1 \oplus_G t_2 \approx_G t'_1 \oplus_G t'_2$?; all this assuming the operations are well defined, i.e., the disjointness demands from 1.6(4) are satisfied. We do not see a reason for this to hold.

---

[5]If $G_2 = \langle G_0 \cup A \rangle$, A finite then for part (1) this is not necessary.

1(B)   Constructions

Before we present the more systematic construction from [Sh:c, Ch.IV], we give a self-contained direct definition and proof for the existence of a canonical existentially closed extension of $G \in \mathbf{K}$ when $\mathfrak{S}$ is symmetric, i.e., the "second avenue" in §0B. We shall deal with the nonsymmetric case later.

Definition 1.12. Assume $\mathfrak{S} \subseteq \Omega[\mathbf{K}]$ is symmetric.
1) We say H is a $\mathfrak{S}$-<u>closure</u> of G <u>when</u> there is a sequence $\langle G_n : n < \omega \rangle$ such that $G_0 = G, H = \cup \{G_n : n < \omega\}$ and G_{n+1} is a one-step $\mathfrak{S}$-closure of G_n, see below.
2) We say that H is a <u>one-step</u> $\mathfrak{S}$-<u>closure</u> of G <u>when</u>:

(a) $G \subseteq H$ are from $\mathbf{K}$;

(b) $S := \operatorname{def}(G) = \{(\mathfrak{s}, \bar{a}) : \mathfrak{s} \in \mathfrak{S}$ and $\bar{a} \in {}^{\omega >}G$ realizes $p_\mathfrak{s}(\bar{x}_\mathfrak{s})\}$ and let $t = (\mathfrak{s}_t, \bar{a}_t) = (\mathfrak{s}(t), \bar{a}(t))$ for $t \in S$;

(c) $\bar{c}_t \in {}^{n(\mathfrak{s}(t))}H$ realizes $q_{\mathfrak{s}_t}(\bar{a}_t, G)$ for $t \in S$;

(d) H is generated by $\{\bar{c}_t : t \in S\} \cup G$;

(e) $\bar{c}_t$ realizes $q_{\mathfrak{s}_t}(\bar{c}_t, c\ell(\cup\{\bar{c}_s : s \in S \setminus \{t\}\} \cup G, H)$ inside H for every $t \in S$.

Claim 1.13. Let $\mathfrak{S} \subseteq \Omega[\mathbf{K}]$ be symmetric.
1) For every $G \in \mathbf{K}$ there is a one-step $\mathfrak{S}$-closure H of G.
2) For every $G \in \mathbf{K}$ there is an $\mathfrak{S}$-closure H of G.
3) In both parts (1) and (2) we have $|G| \leq |H| \leq |G| + |\mathfrak{S}| + \aleph_0$.
4) In both parts (1) and (2), H is unique up to isomorphism over G.
5) If the pair (G_ℓ, H_ℓ) is as in part (1), or as in part (2) for $\ell = 1, 2$ and $G_1 \subseteq G_2$ <u>then</u> H_1 can be embedded into H_2 over G_1.
6) In both parts (1) and (2) there is a set theoretic class function $\mathbf{F}$ computing H from G, pedantically for every $G \in \mathbf{K}$ and ordinal α not in the transitive closure $\operatorname{tr} - c\ell(G)$ of $G, \mathbf{F}_\alpha(G)$ is well defined such that:

(A) (a) $\mathbf{F}_\alpha(G) \in \mathbf{K}_{\mathrm{lf}}$ is of cardinality $\leq |G| + \aleph_0 + |\alpha|$

(b) if $\alpha = 0$, then $\mathbf{F}_\alpha(G) = G$

(c) the sequence $\langle \mathbf{F}_\beta(G) : \beta \leq \alpha \rangle$ is increasing continuous

(d) $\mathbf{F}_{\alpha+1}(G)$ is a one-step closure of $\mathbf{F}_\alpha(G)$

(B) if $G_1 \subseteq G_2 \wedge G_2 \cap \mathbf{F}_\alpha(G_1) = G_1 \wedge \bigwedge\limits_{\ell=1}^{2} \emptyset = (\alpha + 1) \cap \operatorname{tr} - c\ell(G_\ell)$, then $\mathbf{F}_\alpha(G_1) \subseteq \mathbf{F}_\alpha(G_2)$; this is "naturality"; an alternative is 0.12(2).

7) In fact we do not have to use the axiom of choice.

Proof. Should be clear (alternatively, below we do more). $\qquad \square_{1.13}$

Remark 1.14. Similarly in §3.

Definition 1.15. 1) We say N is $(\lambda, \mathfrak{S})$-<u>full</u> over M when: $M \subseteq N$ and <u>if</u> $M \subseteq M_1 \subseteq N$ and $M_1 = c\ell(M + A, N)$ for some $A \subseteq M_1$ of cardinality $< \lambda$ and $\mathfrak{s} \in \mathfrak{S}$ and $\bar{a} \in {}^{k(\mathfrak{s})}M_1$ realizes $p_\mathfrak{s}(\bar{x}_\mathfrak{s})$ in M_1, then $q_\mathfrak{s}(\bar{a}, M_1)$ is realized in N.
2) We may write "N is $\mathfrak{S}$-full over M" when $\lambda = \|N\|$ is regular <u>or</u>, in general, when there is a list $\langle a_\alpha : \alpha < \|N\| \rangle$ of N such that for every $\alpha < \|N\|$ and $\mathfrak{s} \in \mathfrak{S}$ we have: if $M_\alpha = c\ell(M + \{a_\beta : \beta < \alpha\}, N)$ and $\bar{a} \in {}^{k(\mathfrak{s})}M_\alpha$ realizes $p_\mathfrak{s}(\bar{x}_\mathfrak{s})$, then the type $q_\mathfrak{s}(\bar{a}, M_\alpha)$ is realized in N by $\|N\|$ elements.
3) We may omit $\mathfrak{S}$ when $\mathfrak{S} = \Omega[\mathbf{K}]$.

Claim 1.16. Let $\mathfrak{S}$ be symmetric.
1) If $\mathfrak{S} \subseteq \mathfrak{S}(\mathbf{K})$ is closed (see 1.6(1)) <u>then</u> $(\mathbf{K}, \leq_\mathfrak{S})$ is a weak a.e.c. with amalgamation[6] (even canonical), see [Sh:88r, 1.2] or [Sh:h, Ch.I], i.e., in the Definition of a.e.c. we have Ax 0,(I),(II),(III),(V) but LST$(\mathbf{K}, \leq_\mathfrak{S})$ may be ∞ and we omit Ax(IV), see 1.18 below.
2) If $\mathfrak{S} \subseteq \Omega[\mathbf{K}]$ is dense and closed (see 1.6) <u>then</u> for every $M \in \mathbf{K}_\lambda$ there is an existentially closed $N \in \mathbf{K}_\lambda$ which $\leq_\mathfrak{S}$-extends it, in fact any $\mathfrak{S}$-closure of M can serve.
3) If N is $(\lambda, \mathfrak{S})$-full over M_1 and $M_0 \subseteq M_1$, <u>then</u> N is $(\lambda, \mathfrak{S})$-full over M_0; also in Definition 1.15 without loss of generality $\bar{a}$ is from $M \cup A$, i.e., $\bar{a} \in {}^{k(\mathfrak{s})}(M \cup A)$.
4) If $M \in \mathbf{K}_{\leq \lambda}$ <u>then</u> there is a model $N, (\lambda, \mathfrak{S})$-full over M of cardinality $\leq \lambda + \|M\| = \lambda$; moreover, if $\mathfrak{S}$ is dense, then M is existentially closed.
5) In (4), we can add: if $N' \in \mathbf{K}$ is $(\lambda, \mathfrak{S})$-full over M <u>then</u> we can find an embedding of N into N' over M.

Proof. 1) Easy.
2) Easy by 1.13 and see more below.
3) Easy.
4) We choose $G_n \in \mathbf{K}$ by induction on n such that:

(a) $G_0 = M$;

(b) $G_{n+1} \supseteq G_n$ is as in Definition 1.12 but each t appears λ times, i.e.,

- $G_{n+1} = c\ell(\cup\{\bar{c}^n_{t,\alpha} : t \in \mathrm{def}_\mathfrak{S}(G_n) \text{ and } \alpha < \lambda\} \cup G_n, G_{n+1})$ where
- $\mathrm{tp}_{\mathrm{bs}}(\bar{c}^n_{t,\alpha}, G_{n,t,\alpha}, G_{n+1}) = q_t(\bar{a}_t, G_{n,t,\alpha})$ where
- $G_{n,t,\alpha} = c\ell(\cup\{\bar{c}^n_{t_1,\alpha_1} : t_1 \in \mathrm{def}_\mathfrak{S}(G_n), \alpha_1 < \lambda \text{ but } (t_1, \alpha_1) \neq (t, \alpha)\} \cup G_n, G_{n+1})$.

---

[6]Not enough for quoting results.

Let $\hat{G} = \bigcup_n G_n$ and we shall show that $\hat{G}$ is $(\lambda, \mathfrak{S})$-full over G. We can ignore the case $\lambda = \aleph_0$. Assume $A \subseteq \hat{G}, |A| < \lambda$ and $t_* \in \text{def}_{\mathfrak{S}}(\hat{G})$ and $\bar{a}_{t_*} \subseteq cl(G_0 + A, \hat{G})$, hence we can find $\bar{S}$ such that:

$(*)$ (a) $\bar{S} = \langle S_n : n < \omega \rangle$;

(b) $S_n \subseteq \text{def}_{\mathfrak{S}}(G_n) \times \lambda$ and $\bigcup_m S_m$ has cardinality $< |A|^+ + \aleph_0$;

(c) if $(t, \alpha) \in S_n$, then $\bar{a}_t \subseteq cl(\cup\{\bar{c}^m_{t_1,\alpha_1} : m < n \text{ and } (t_1, \alpha_1) \in S_m\} \cup G_0, G_n)$;

(d) $A \subseteq \bigcup_n A_n \cup G_0$ where $A_n = \cup\{\bar{c}^m_{t,\alpha} : (t,\alpha) \in S_m \text{ and } m < n\}$;

(e) for some $n_*, (t_*, 0) \in S_{n_*}$.

We have to prove that $q_{t_*}(cl(A \cup G, \hat{G}))$ is realized in M. Choose α_* such that $(t_*, \alpha_*) \notin S_{n_*}$ and prove by induction on $n \geq n_*$ that $\bar{c}^{n_*}_{t_*,\alpha_*}$ realizes $q_{t_*}(cl(A_n \cup G_0, \hat{G}))$.

For $n = n_*$ this is obvious, so assume this holds for n and we shall prove for $n + 1$.

For this it suffices to prove, for every finite $u \subseteq S_n$ that $\bar{c}^{n(*)}_{t_*,\alpha_*}$ realizes $q_{t_*}(cl(A_n \cup G_0 \cup \{\bar{c}^n_{t,\alpha} : (t,\alpha) \in u\}, \hat{G})$; we prove this by induction on $|u|$. Now if $|u| = 0$ this holds by the induction hypothesis on n and if $|u| > 0$, let $\beta \in u$ and use the induction for $u' = u\backslash\{\beta\}$ and $\mathfrak{S}$ being symmetric.

5) We can find a list $\langle (n_\zeta, t_\zeta, \alpha_\zeta) : \zeta < \lambda \rangle$ of $\{(n, t, \alpha) : n < \omega \text{ and } (t, \alpha) \in S_n\}$ such that $\bar{a}_{t_\zeta} \subseteq cl(\cup\{(\bar{c}^{n_\xi}_{t_\xi,\alpha_\xi} : \xi < \zeta\} \cup M, N)$.

Now choose $f(\bar{c}^{n_\zeta}_{t_\zeta,\alpha_\zeta}) \subseteq N'$ by induction on ζ. $\square_{1.16}$

1.17 Discussion. 1) So by 1.13(2), 1.16(2) if there is a symmetric closed dense $\mathfrak{S}$ <u>then</u> for every lf group G there is a "nice" extension of G to an existentially closed one $\hat{G}$, that is we have:

(a) uniqueness (by 1.13(4))

(b) cardinality $\leq |\theta| + |\mathfrak{S}|$ (by 1.13(3))

(c) extending G (see 1.12(1))

(d) being existentially closed (see 1.16(2)).

2) Fixing λ and demanding $G \in \mathbf{K}_{\leq\lambda}$ we can add

(e) $\hat{G}$ is $(\lambda, \mathfrak{S})$-full over M

(f) if $H \supseteq G$ is $(\lambda, \mathfrak{S})$-full <u>then</u> there is an embedding of $\hat{G}$ into H over G.

1.18 Discussion. Concerning 1.16(1), if we assume $\langle G_\alpha : \alpha \leq \delta + 1 \rangle$ is $\subseteq$-increasing continuous and $\alpha < \delta \Rightarrow G_\alpha \leq_{\mathfrak{S}} G_{\delta+1}$, does it follow that $G_\delta \leq_{\mathfrak{S}} G_{\delta+1}$? This is Ax(IV) of the definition of a.e.c. Well, if δ has uncountable cofinality and each G_α is existentially closed, then yes. The point is that the relevant types do not split over <u>finite</u> sets. If we deal with "not split over countable sets" we need $\mathrm{cf}(\delta) \geq \aleph_2$, etc.

So $(\mathbf{K}, \leq_{\mathfrak{S}})$ is not an a.e.c. in general failing Ax(IV); in fact, e.g., we may prove for the maximal $\mathfrak{S}$ that this axiom fails, see the proof of 5.1.

Now we turn to constructions not necessarily assuming "$\mathfrak{S}$ is symmetric" presenting the "first avenue" in §0(B).

Definition 1.19. 1) We say that $\mathscr{A} = \langle G_i, \bar{a}_j, w_j, K_j : i \leq \alpha, j < \alpha \rangle$ is an $\mathbf{F}^{\mathrm{sch}}_{\aleph_0} - \mathfrak{S}$-<u>construction</u> (for $\mathbf{K}$) <u>when</u> :

(a) G_i for $i \leq \alpha$ is an $\leq_{\mathfrak{S}}$-increasing continuous sequence of members of $\mathbf{K}$;

(b) G_{i+1} is generated by $G_i \cup \bar{a}_i, \bar{a}_i$ a finite sequence;

(c) w_i is a finite subset of i;

(d) $K_i \subseteq G_i$ is finite;

(d)$^+$ moreover, $K_i \subseteq \langle G_0 + \sum\limits_{j \in w_i} \bar{a}_j \rangle_{G_i}$, may add K_j generated by $\cup \{ \bar{a}_j : j \in w_i \} \cup (K_i \cap G_i)$;

(e) $\mathrm{tp}_{\mathrm{bs}}(\bar{a}_i, G_i, G_{i+1}) \in \mathbf{S}^{\ell g(\bar{a}_i)}_{\mathfrak{S}}(G_i)$ as witnessed by K_i, i.e., it is $q_{\mathfrak{s}}(\bar{a}, G_i)$ for some $\bar{a} \in {}^{\omega >}K_i$ realizing $p_{\mathfrak{s}}$ for some $\mathfrak{s} \in \mathfrak{S}$.

2) We may say above that G_α is $\mathbf{F}^{\mathrm{sch}}_{\aleph_0} - \mathfrak{S}$-constructible over G_0; and may also say that $\mathscr{A}$ is an $\mathfrak{S}$-construction over G_0. We let $\alpha = \ell g(\mathscr{A}), G_i = G^{\mathscr{A}}_i, \bar{a}_i = \bar{a}^{\mathscr{A}}_i, w_j = w^{\mathscr{A}}_j, K_j = K^{\mathscr{A}}_j$.

3) We say above that $\mathscr{A}$ is a <u>definite</u> $\mathbf{F}^{\mathrm{sch}}_{\aleph_0} - \mathfrak{S}$-<u>construction</u> <u>when</u> for every $j < \alpha$ we have also $t_j = t^{\mathscr{A}}_j \in \mathrm{def}(G^{\mathscr{A}}_j)$ such that $\bar{a}_{t_j} \in {}^{\omega >}(K_j)$ and $\bar{a}^{\mathscr{A}}_j$ realizes $q_{t_j}(G_j)$ (note that in 1.19(1)(e) we have "for some $\mathfrak{s}_j$", so $\mathscr{A}$ does not determine the $\mathfrak{s}$'s (or here the t_j; so every $\mathbf{F}^{\mathrm{sch}}_{\aleph_0} - \mathfrak{S}$-construction can be expanded to a definite one, but not necessarily uniquely).

4) We say $\mathscr{A}$ is a λ-<u>full definite</u> $\mathbf{F}^{\mathrm{sch}}_{\aleph_0} - \mathfrak{S}$-<u>construction</u> <u>when</u> α is divisible by λ and for every $i < \alpha$ and $t \in \mathrm{def}(G_i)$, the set $\{j : j \in (i, \alpha)$ and $t^{\mathscr{A}}_j = t\}$ is an unbounded subset of $\alpha(*)$ of order type divisible by λ.

1.20 Discussion. We may replace 1.19(1)(e) by "$\mathrm{tp}_{\mathrm{bs}}(\bar{a}_i, G_i, G_{i+1})$ does not split over K_i", this is like the case $\mathbf{F}^p_{\aleph_0}$ in [Sh:c, Ch.IV, Def.2.6, pg. 168] and [Sh:c, Ch.IV, Lemma 2.20, pg. 168] and is equal to $\mathbf{F}^{\mathrm{nsp}}_{\aleph_0}$ in [Sh:900, §1,1.1-1.12], both for first order theories, but we seemingly lose the following:

Observation 1.21. *1) If $\mathscr{A}$ is a $\mathbf{F}^{\mathrm{sch}}_{\aleph_0} - \mathfrak{S}$-construction and $G^{\mathscr{A}}_0 \subseteq G$ and $G \cap G^{\mathscr{A}}_{\ell g(\mathscr{A})} = G^{\mathscr{A}}_0$ then there is an $\mathbf{F}^{\mathrm{sch}}_{\aleph_0} - \mathfrak{S}$-construction $\mathscr{B}$ with $G^{\mathscr{B}}_0 = G, \ell g(\mathscr{B}) = \ell g(\mathscr{A})$ and $G^{\mathscr{B}}_{\ell g(\mathscr{B})} = \langle G^{\mathscr{A}}_{\ell g(\mathscr{A})} \cup G \rangle_{G^{\mathscr{A}}_{\ell g(\mathscr{A})}}$.*
2) Like (1) but with definite $\mathbf{F}^{\mathrm{sch}}_{\aleph_0} - \mathfrak{S}$-constructions and then add in the end $t^{\mathscr{B}}_j = t^{\mathscr{A}}_j$ for $j < \ell g(\mathscr{A})$.
3) For the definite version, see 1.19(3), we get even uniqueness in (2).

1.22 Discussion. In 1.24 below, we may consider (see [Sh:f, Ch.IV,§1]):

Ax(V.1): If $(q, G, L) \in \mathbf{F}, G \subseteq H \in \mathbf{K}; \bar{a}, \bar{b} \in {}^{\omega >}H; q = \mathrm{tp}_{\mathrm{bs}}(\bar{a}^\frown \bar{b}, G, H)$ and $p = \mathrm{tp}_{\mathrm{bs}}(\bar{a}, \langle G + \bar{b} \rangle_H, H)$ then $(p, \langle G + \bar{b} \rangle_H, L) \in \mathbf{F}$.

Ax(V.2): A notational variant of (V1), so ignore.

The following claim (together with §2, the existence of countable dense $\mathfrak{S}$) proves Theorem 0.11.

Claim 1.23. 1) If $G \in \mathbf{K}$ is of cardinality $\leq \lambda$ and $\mathfrak{S} \subseteq \Omega[\mathbf{K}]$ is closed and dense and of cardinality $\leq \lambda$ (if $\lambda \geq 2^{\aleph_0}$ this follows) then there is an $\mathbf{F}^{\mathrm{sch}}_{\aleph_0} - \mathfrak{S}$-construction $\mathscr{A}$ such that:

(a) $\alpha^{\mathscr{A}} = \lambda$;

(b) $G^{\mathscr{A}}_0 = G$;

(c) $G^{\mathscr{A}}_\lambda \in \mathbf{K}$ is existentially closed of cardinality λ;

(d) $\mathscr{A}$ is λ-full, that is for every $\mathfrak{s} \in \mathfrak{S}$ and $\bar{a} \in {}^{k(\mathfrak{s})}(G^{\mathscr{A}}_\lambda)$ realizing $p_{\mathfrak{s}}(\bar{x})$, for λ ordinals $\alpha < \lambda$ we have: $\mathrm{tp}_{\mathrm{bs}}(\bar{a}_\alpha, G^{\mathscr{A}}_\alpha, G^{\mathscr{A}}_{\alpha+1}) = q_{\mathfrak{s}}(\bar{a}, G^{\mathscr{A}}_\alpha)$.

2) Assume $\lambda \geq \|G\| + |\mathfrak{S}|$ is regular. Then we can find $H \in \mathbf{K}_\lambda$ which is $\mathbf{F}^{\mathrm{sch}}_{\aleph_0} - \mathfrak{S}$-constructible over G, is $(\lambda, \mathfrak{S})$-full over M and is embeddable over M into any N' which is $(\lambda, \mathfrak{S})$-full over G, in fact G_λ from part (1) is as required.
3) If $\mathfrak{S}$ is symmetric and is closed and H_1, H_2 are $\mathbf{F}^{\mathrm{sch}}_{\aleph_0} - \mathfrak{S}$-constructible over G and $(\lambda, \mathfrak{S})$-full over G and of cardinality λ then H_1, H_2 are isomorphic over G.
4) If $\lambda \geq \|G\|$ and $\mathscr{A}$ is an $\mathbf{F}^{\mathrm{sch}}_{\aleph_0} - \mathfrak{S}$-construction of H over G and $\ell g(\mathscr{A}) = \lambda$ then for every $H' \in \mathbf{K}$ which is $(\lambda, \mathfrak{S})$-full over G, we have H is embeddable into H' over G.

Proof. By [Sh:c, Ch.VI, §3] as all the relevant axioms there apply (see below or [Sh:c, Ch.IV, §1, pg.153]) or just check directly. Of course, we can use a monster $\mathfrak{C}$ for groups, but use only sets A such that $c\ell(A, \mathfrak{C}) = \langle A \rangle_{\mathfrak{C}}$ is locally finite, and we use quantifier-free types. $\square_{1.23}$

Now we make the connection to [Sh:c, Ch.IV].

1.24 Definition/Claim. 1) Let $\mathfrak{S} \subseteq \Omega[\mathbf{K}]$ be closed and below let $\lambda = \lambda(\mathbf{F}_{\mathfrak{S}})$ be $\aleph_0$. Then $\mathbf{F} = \mathbf{F}_{\mathfrak{S}}$ is defined as the set of triples (p, G, A) such that: A is finite, for some $B \subseteq G \in \mathbf{K}$ we have $A \subseteq B, c\ell(B) = c\ell(B, G) = G \in \mathbf{K}$, $p \in \mathbf{S}_{\mathfrak{S}}^{<\omega}(c\ell(B))$ is $q_{\mathfrak{s}}(\bar{b}, c\ell(B))$ for some $\mathfrak{s} \in \mathfrak{S}, \bar{b} \subseteq c\ell(A)$ over A; we may restrict ourselves to the case $B = c\ell(B, G)$. Note that: as here we do not have a monster model $\mathfrak{C}$ we can either demand $B \in \mathbf{K}$ <u>or</u> demand $B \subseteq G \in \mathbf{K}$, but then, it is more natural to write (p, G, A) instead of (p, A).
2) $\mathbf{F}$ satisfies the axioms (from [Sh:c, Ch.IV,§1] written below in the present notation) except possibly V, VI, VIII, X.1, X.2, XI.1, XI.2.
3) If $\mathfrak{S}$ is symmetric <u>then</u> $\mathbf{F}$ satisfies also Ax(VI).
4) If $\mathfrak{S}$ is dense <u>then</u> $\mathbf{F}$ satisfies also Ax(X.1).

Remark 1.25. If $\mathfrak{S}$ is compact (see 1.6(5)), <u>then</u> $\mathbf{F}$ satisfies Ax(VIII), i.e.,
   Ax(VIII) when $\mathfrak{S}$ is compact: If $\langle G_i : i \leq \delta + 1 \rangle$ is $\subseteq$-increasing contin-uous in $\mathbf{K}, L \subseteq G_0$ finite, $p \in \mathbf{S}_{\mathfrak{S}}(G_\lambda)$ and $i < \delta \Rightarrow (p{\restriction}G_i, G_i, L) \in \mathbf{F}$ <u>then</u> $(p, G_\delta, L) \in \mathbf{F}$.

[Why? By the Definition; also holds when $\mathrm{cf}(\delta) > \aleph_0$.]

Proof. Isomorphism - Ax(I): preservation under isomorphism.
   Obvious.

Concerning trivial $\mathbf{F}$-types:
Ax(II1): If $K \subseteq L \subseteq G \in \mathbf{K}, |L| < \lambda, K$ is finite, $\bar{a} \in {}^{\omega>}K$ and $p = \mathrm{tp}_{\mathrm{bs}}(\bar{a}, L, G)$, then $(p, G, K) \in \mathbf{F}$.
   [Why? Trivially; recall $\lambda = \aleph_0$.]

Axiom(II2)-(II3)-(II4): irrelevant here.

Concerning monotonicity:
Ax(III1): If $L \subseteq G_1 \subseteq G_2$ and $(p, G_2, L) \in \mathbf{F}$, then $(p{\restriction}G_1, G_1, L) \in \mathbf{F}$.

[Why? Because if $\bar{a} \in {}^{\omega>}L, L \subseteq G_1 \subseteq G_2 \in \mathbf{K}$ and $q_{\mathfrak{s}}(\bar{a}, G_2)$ is well defined and equal to p, <u>then</u> $q_{\mathfrak{s}}(\bar{a}, G_1) = q_{\mathfrak{s}}(\bar{a}, G_2){\restriction}G_1)$, see Claim 1.2(1A).]

Ax(III2): If $L \subseteq L_1 \subseteq G, |L_1| < \lambda$, i.e., L_1 is finite and $(p, G, L) \in \mathbf{F}$ <u>then</u> $(p, G, L_1) \in \mathbf{F}$.

[Why? By the definition.]

Ax(IV): If $\bar{a}, \bar{b} \in {}^{\omega>}H, L \subseteq G \subseteq H, (\mathrm{tp}_{\mathrm{bs}}(\bar{b}, G, H), G, L) \in \mathbf{F}$ and $\mathrm{Rang}(\bar{a}) \subseteq \mathrm{Rang}(\bar{b})$ <u>then</u> $(\mathrm{tp}_{\mathrm{bs}}(\bar{a}, G, H), G, L) \in \mathbf{F}$.

[Why? Straightforward as $\mathfrak{S}$ is domination closed, see Definition 1.6(1B).]

Concerning transitivity and symmetry:

Ax(VI): ($\mathfrak{S}$ is symmetric). If $G \subseteq H \in \mathbf{K}, \bar{a}, \bar{b} \in {}^{\omega>}H$ and $L_1, L_2 \subseteq G$ are finite and $(\mathrm{tp}_{\mathrm{bs}}(\bar{b}, \langle G + \bar{a} \rangle_H, H), \langle G + \bar{a} \rangle_H, L_1) \in \mathbf{F}$ and $(\mathrm{tp}_{\mathrm{bs}}(\bar{a}, G, H), G, L_2) \in \mathbf{F}$ <u>then</u> $(\mathrm{tp}_{\mathrm{bs}}(\bar{a}, \langle G + \bar{b} \rangle_H, H), \langle G + \bar{b} \rangle_H, L_1) \in \mathbf{F}$.

[Why? By $\mathfrak{S}$ being symmetric.]

Ax(VII): If $G \subseteq H \in \mathbf{K}, \bar{a}, \bar{b} \in {}^{\omega >}H, (\mathrm{tp}_{\mathrm{bs}}(\bar{a}, \langle G + \bar{b} \rangle_H, H), \langle G + \bar{b} \rangle_H, L) \in \mathbf{F}$ and $(\mathrm{tp}_{\mathrm{bs}}(\bar{b}, G, H), G, L) \in \mathbf{F}$ hence $L \subseteq G$ is finite, <u>then</u> $(\mathrm{tp}_{\mathrm{bs}}(\bar{a} \char`\^ \bar{b}, G, H), G, L) \in \mathbf{F}$.

[Why? By $\mathfrak{S}$ being composition-closed, see Definition 1.6(1A).]

Concerning continuity:

Ax(IX): irrelevant as $\lambda = \aleph_0$.

Concerning existence:

Ax(X.1): If $L_1 \subseteq G \in \mathbf{K}, L_1 \subseteq L_2$ finite, $\bar{a} \in {}^{\omega >}(L_2)$ <u>then</u> for some p extending $\overline{\mathrm{tp}_{\mathrm{bs}}}(\bar{a}, L_1, L_2)$ and finite $L \subseteq G$ we have $(p, G, L) \in \mathbf{F}$, moreover, without loss of generality $L = L_1$.

[Why? By $\mathfrak{S}$ being dense.]

Ax(X.2): irrelevant and follows by the moreover in Ax(X.1).

Ax(XI.1): If $p \in \mathbf{S}_{\mathrm{bs}}(G_1), (p, G_1, L) \in \mathbf{F}$ hence $p \in \mathbf{S}_{\mathfrak{S}}^n(G_1)$ for some n and $G_1 \subseteq G_2$ <u>then</u> there is $q \in \mathbf{S}_{\mathrm{bs}}^n(G_2)$ extending p such that $(q, G_1, L_2) \in \mathbf{F}$ for L_2, so $\bar{q} \in \mathbf{S}_{\mathfrak{S}}^n(G_2)$; moreover, in fact, $L_2 = L$ is O.K.

[Why? Use the same $\mathfrak{s} \in \mathfrak{S}$.]

Ax(XI.2): irrelevant and really follows by the moreover in (XI.1). $\qquad \square_{1.24}$

Definition 1.26. A sequence $\mathbf{I} = \langle \bar{a}_s : s \in I \rangle$ in $G \in \mathbf{K}$ is κ-convergent <u>when</u> for some $m, s \in I \Rightarrow \bar{a}_s \in {}^m G$ and for every finite $K \subseteq G$ and some $q \in \mathbf{S}^m(K)$ for all but $< \kappa$ members s of $\mathbf{I}, q = \mathrm{tp}_{\mathrm{bs}}(\bar{a}_s, K, G)$.

Remark 1.27. 1) So $\mathbf{F}_{\mathfrak{S}}$-constructions preserve "$\mathbf{I}$ is κ-convergent". Moreover, if $\mathbf{I}$ is κ-convergent in $G \in \mathbf{K}$ and $G \leq_{\mathfrak{S}} H$, where $\mathfrak{S} \subseteq \Omega[\mathbf{K}]$ <u>then</u> I is κ-convergent in H.

2) We can assume I is a linear order with no last member and of cofinality $\geq \kappa$ and replace "all but $< \kappa$ of the $s \in I$" by "every large enough $s \in I$". See more in [Sh:950, §(1C)].

1(C) Using Order

We now turn to the third avenue of §(0B) to deal with the general and not necessarily symmetric case. Can we get uniqueness for nonsymmetric $\mathfrak{S}$? Can we get every automorphism extendable, etc.? The answer is that at some price, yes. A major point in the construction was the use of linear well-ordered index set (λ in 1.23(1) or $\alpha^{\mathscr{A}}$ in general). But actually, we can use linear non-well-ordered index sets, so those index sets can have automorphisms which help us toward uniqueness. The solution here is not peculiar to locally finite groups.

Definition 1.28. We say (I, E) is λ-suitable <u>when</u> (we may omit λ when $\lambda = |I|$, we may write $(I, P_i)_{i<\lambda}$ with $\langle P_i : i < \lambda \rangle$ listing the E-equivalence classes (with no repetitions)):

(a) I is a linear order;

(b) E is an equivalence relation on I with λ equivalence classes;

(c) every permutation of I/E is induced by some automorphism of the linear order which preserves equivalence and nonequivalence by E;

(d) each E-equivalence class has cardinality $|I|$.

Claim 1.29. Let $T = \mathrm{Th}(\mathbb{R}, <, E)$ where Th stands for "the first order theory of", $E := \{(a, b) : a, b \in \mathbb{R} \text{ and } a - b \in \mathbb{Q}\}$; so $(A, <, E) \models T$ iff $(A, <)$ is a dense linear order with neither first nor last element, E an equivalence relation with each equivalence class a dense subset of A and with infinitely many equivalence classes.

1) If $\lambda = \lambda^{<\lambda}$ and (I, E) is a saturated model of T of cardinality λ, <u>then</u> (I, E) is suitable[7]

2) For every λ the $(I, P_i)_{i<\lambda}$ from [Sh:E62, §2] (see history there) is λ-suitable and $|I| = \lambda$.

3) There is a definable sequence $\langle (I_\lambda, P_i^\lambda)_{i<\lambda} : \lambda$ an infinite cardinal$\rangle$ such that $(I_\lambda, P_i^\lambda)_{i<\lambda}$ is λ-suitable and is increasing with λ and this definition is absolute.

Proof. 1) Obvious.

2),3) See there. $\square_{1.29}$

Claim 1.30. Assume

(A) $G \in \mathbf{K}$ is of cardinality λ;

(B) $\mathfrak{S} \subseteq \Omega[\mathbf{K}]$ is closed and dense;

(C) (a) $i(*) \leq \lambda$ and $\mathscr{S} = \{t_i = (\mathfrak{s}_i, \bar{a}_i) : i < i(*)\}$ lists $\mathrm{def}_{\mathfrak{S}}(G)$, i.e., the pairs $(\mathfrak{s}, \bar{a})$, as in clause (d) of 1.23(1) or 1.23(3);

(b) each such pair appears exactly once;

(c) let $t_i = (\mathfrak{s}_*, <>)$ for $i \in [i(*), \lambda)$ so $\mathfrak{s}_* \in \mathfrak{S}, k_{\mathfrak{s}_*} = 0, n_{\mathfrak{s}} = 1, i(*) = \|\mathrm{def}_{\mathfrak{S}}(G)\|, \mathfrak{s}_*$ is from 1.3; so $\bar{a}_i = \langle\rangle$;

(D) $(I, P_i)_{i<\lambda}$ is λ-suitable, see Definition 1.28.

---

[7]By similar arguments, if $\lambda = \lambda^\mu$ <u>then</u> there is a μ-suitable $(I, P_i)_{i<\mu}$ but $|I/E| = \mu < \lambda$. We can use any model of cardinality λ which is strongly μ^+-sequence homogeneous; this means that every partial automorphism of cardinality $\leq \mu$ can be extended to an automorphism.

Then we can find $H, \mathbf{c} = \langle \bar{c}_r : r \in I \rangle$ (the ordered one-step $(\lambda, \mathfrak{S})$-closure), such that:

(a) $H \in \mathbf{K}$ is a $\leq_{\mathfrak{S}}$-extension of G;

(b) $\bar{c}_r \in {}^{n(t_i)}H$ if $r \in P_i, i < \lambda$;

(c) H is generated by $G \cup \{\bar{c}_r : r \in I\}$;

(d) if $i < \lambda$ and $r \in P_i$, then $\bar{c}_r$ realizes in H over $c\ell(G \cup \{\bar{c}_s : s <_I r\}, H)$ the type defined by $(\mathfrak{s}_i, \bar{a}_i)$;

(e) every automorphism of G can be extended to an automorphism of H.

Proof. Straightforward; e.g., to define H we should choose $q_{r_0,\ldots,r_{n-1}}$ for every $r_0 <_I \ldots <_I r_{n-1}$ by induction on n such that in the end $q_{r_0,\ldots,r_{n-1}} = \text{tp}_{\text{bs}}(\bar{c}_{r_0} ^\frown \ldots ^\frown \bar{c}_{r_{n-1}}, G, H)$, by clause (d), and prove that:

(*) if $m \leq n$ and $h : \{0, \ldots, m-1\} \to \{0, \ldots, n-1\}$ is increasing, then $q_{r_{h(0)},\ldots,r_{h(m-1)}} \leq_h q_{r_0,\ldots,r_{n-1}}$.

Note that clause (e) follows by clauses (a)-(d) above recalling clause (c) of Definition 1.28.

Why? Let π be an automorphism of G, for each $i < \lambda$ we have $(\mathfrak{s}_i, \bar{a}_i) \in \mathscr{S}$ and also $(\mathfrak{s}_i, \pi(\bar{a}_i)) \in \mathscr{S}$, so by the choice of $\langle (\mathfrak{s}_i, \bar{a}_i) : i < \lambda \rangle$ there is a unique $j < \lambda$ such that $i \geq i(*) \Rightarrow j = i$ and $(\pi(\bar{a}_i), \mathfrak{s}_i) = (\bar{a}_j, \mathfrak{s}_j)$, so let $j = \hat{\pi}(i)$. So $\hat{\pi}$ is a permutation of λ. By "$(I, P_i)_{i<\lambda}$ is λ-suitable" there is an automorphism $\check{\pi}$ of the linear order I such that $i < \lambda \Rightarrow \check{\pi}(P_i) = P_{\hat{\pi}(j)}$. Clearly there is a unique automorphism $\dot{\pi}$ of H such that $\pi = \dot{\pi} {\restriction} G$ and $\dot{\pi}(\bar{c}_i) = \bar{c}_{\check{\pi}(i)}$. $\square_{1.30}$

Definition 1.31. 1) We say H is an ordered one-step $(\lambda, \mathfrak{S})$-closure of G, pedantically the ordered one-step $(I, E) - \mathfrak{S}$-closure of G, <u>when</u> $G, H, \mathbf{c}$ are as in 1.30.
2) We say H is an ordered $(\lambda, \mathfrak{S})$-closure of G, pedantically the ordered $(I, E) - \mathfrak{S}$-closure of G when:

(a) $H = \bigcup_n H_n$

(b) $H_0 = G$

(c) H_{n+1} is the one-step $(I, E) - \mathfrak{S}$-closure of H_n.

Remark 1.32. In what way is 1.30 weaker? We have to choose the listing of $\text{def}(G)$ in clause (C). Also for $G_1 \subseteq G_2$ it is not clear why $H_1 \subseteq H_2$, where (G_ℓ, H_ℓ) is as above. But see 1.29(3).

Conclusion 1.33. *The parallel of parts (2)-(6) of 1.13 holds.*

Proof. Straightforward, for part (6) of 1.13 use 1.29(3). $\square_{1.33}$

2 There are enough reasonable schemes

2(A) There is a Dense Set of Schemes

We like to find $\mathfrak{S}$'s as in §1 for $\mathbf{K}_{\mathrm{lf}}$, in particular to prove that there are dense $\mathfrak{S}$, so we have to look in details at amalgamations of lf-groups under special assumptions.

Recall the well known: for finite groups $G_0 \subseteq G_\ell \in \mathbf{K}$ for $\ell = 1, 2$ we can amalgamate G_1, G_2 over G_0 by embedding into suitable finite permutations group; see the proof of the theorem of Hall, explained in the second paragraph of §(0A).

Concerning the $\mathbf{K}_{\mathrm{olf}}$ versions of 2.2, see later in 6.7.

Convention 2.1. K is $\mathbf{K}_{\mathrm{lf}}$.

Definition 2.2. 1) Let $\mathbf{X_K} = \mathbf{X(K)}$, the set of amalgamation tries, be the set of $\mathbf{x}$ such that: $\mathbf{x}$ is a quintuple $(G_0, G_1, G_2, \mathbf{I}_1, \mathbf{I}_2) = (G_{\mathbf{x},0}, G_{\mathbf{x},1}, \ldots)$ satisfying:

(a) $G_0 \subseteq G_\ell \in \mathbf{K}$ for $\ell = 1, 2$;

(b) $\mathbf{I}_\ell$ is a set of representatives of the left G_0-cosets in G_ℓ, i.e., $\langle gG_0 : g \in \mathbf{I}_\ell \rangle$ is a partition of G_ℓ (so without repetitions) for $\ell = 1, 2$;

(c) $e_{G_{\mathbf{x},0}} \in \mathbf{I}_{\mathbf{x},1} \cap \mathbf{I}_{\mathbf{x},2}$.

2) For $\mathbf{x}$ as above let

(a) $\mathscr{U} = \mathscr{U}_{\mathbf{x}} = \{(g_0, g_1, g_2) : g_\ell \in G_\ell \text{ for } \ell = 0, 1, 2 \text{ and } g_1 \in \mathbf{I}_1, g_2 \in \mathbf{I}_2\}$;

(b) for $\ell = 1, 2$ and $g \in \mathbf{I}_\ell$ let $\mathscr{U}_g^\ell = \mathscr{U}_{\mathbf{x},g}^\ell := \{(g_0, g_1, g_2) \in \mathscr{U}_{\mathbf{x}} : g_\ell = g\}$;

(c) $\mathbf{j_x} = \mathbf{j}_{\mathbf{x},1} \cup \mathbf{j}_{\mathbf{x},2}$, see below;

(d) for $\ell = 0, 1, 2$ let $\mathbf{j}_\ell = \mathbf{j}_{\mathbf{x},\ell}$ be the following embedding of G_ℓ into $\mathrm{per}(\mathscr{U}_{\mathbf{x}})$, the group of permutations of $\mathscr{U}_{\mathbf{x}}$, so let $g \in G_\ell$ and we should define $\mathbf{j}_\ell(g)$, so let $(g_0, g_1, g_2) \in \mathscr{U}_{\mathbf{x}}$ and we define $(g_0', g_1', g_2') = (\mathbf{j}_\ell(g))(g_0, g_1, g_2)$ from $\mathscr{U}_{\mathbf{x}}$ as follows:

$\underline{\ell = 0}$: $g'_0 = g_0 g$ in G_0 and $g'_1 = g_1, g'_2 = g_2$;

$\underline{\ell = 1}$: $g'_1 g'_0 = g_1 g_0 g$ in G_1 and $g'_2 = g_2$;

$\underline{\ell = 2}$: $g'_2 g'_0 = g_2 g_0 g$ in G_2 and $g'_1 = g_1$.

3) Let $G_{\mathbf{x}} = G_{\mathbf{x},3}$ be the subgroup of $\mathrm{Sym}(\mathscr{U}_{\mathbf{x}})$ which $\mathrm{Rang}(\mathbf{j}_{\mathbf{x},1}) \cup \mathrm{Rang}(\mathbf{j}_{\mathbf{x},2})$ generates where $G_{\mathbf{x}} \models$ "$f_1 f_2 = f_3$" means that for every $u \in \mathscr{U}_{\mathbf{x}}, f_3(u) = f_2(f_1(u))$, i.e., we look at the permutation as acting from the right.

4) Let $\leq_{\mathbf{X}(K)}$ be the following partial order on $\mathbf{X_K} : \mathbf{x} \leq_{\mathbf{X}(K)} \mathbf{y}$ iff:

(a) $\mathbf{x}, \mathbf{y} \in \mathbf{X_K}$;

(b) $G_{\mathbf{x},0} = G_{\mathbf{y},0}$;

(c) $G_{\mathbf{x},\ell} \subseteq G_{\mathbf{y},\ell}$ for $\ell = 1, 2$;

(d) $\mathbf{I}_{\mathbf{x},\ell} = \mathbf{I}_{\mathbf{y},\ell} \cap G_{\mathbf{x},\ell}$ for $\ell = 1, 2$.

5) We say (f_1, f_2) embeds $\mathbf{x} \in \mathbf{X_K}$ into $\mathbf{y} \in \mathbf{X_K}$ when:

(a) f_ℓ embeds $G_{\mathbf{x},\ell}$ into $G_{\mathbf{y},\ell}$ for $\ell = 1, 2$;

(b) $f_1 \restriction G_{\mathbf{x},0} = f_2 \restriction G_{\mathbf{y},0}$ maps $G_{\mathbf{x},0}$ onto $G_{\mathbf{y},0}$.

6) We say (f_1, f_2) is an isomorphism from $\mathbf{x} \in \mathbf{X_K}$ onto $\mathbf{y} \in \mathbf{X_K}$ when above f_ℓ is onto $G_{\mathbf{y},\ell}$ for $\ell = 1, 2$.

Observation 2.3. *Let $\mathbf{x}$ be as in Definition 2.2, i.e., it is an amalgamation try.*

0) If $G_0 \subseteq G_\ell \in \mathbf{K}$ for $\ell = 1, 2$ then for some $\mathbf{x} \in \mathbf{X_K}$ we have $G_{\mathbf{x},\ell} = G_\ell$ for $\ell = 0, 1, 2$.

1) In Definition 2.2(2), for $\ell = 0, 1, 2$ if $g \in G_{\mathbf{x},\ell}$ then $\mathbf{j}_{\mathbf{x},\ell}(g)$ is a permutation of $\mathscr{U}_{\mathbf{x}}$, in fact, its restriction to $\mathscr{U}^{3-\ell}_{g_1}$ is a permutation for each $g_1 \in G_{3-\ell}$.

2) Moreover, in part (1) the mapping $\mathbf{j}_{\mathbf{x},\ell}$ embeds the group $G_{\mathbf{x},\ell}$ into the group of permutation of $\mathscr{U}_{\mathbf{x}}$ hence into $G_{\mathbf{x}}$.

3) The mapping $\mathbf{j}_{\mathbf{x},0}$ is equal to $\mathbf{j}_{\mathbf{x},1} \restriction G_{\mathbf{x},0}$ and also to $\mathbf{j}_{\mathbf{x},2} \restriction G_{\mathbf{x},0}$.

4) If $G_{\mathbf{x},\ell}$ is finite for $\ell = 0, 1, 2$ then $|G_{\mathbf{x}}| \leq (|G_{\mathbf{x},1}| \times |G_{\mathbf{x},2}|/|G_{\mathbf{x},0}|)!$.

5) If $\mathbf{x}$ is an amalgamation try and $G_{\mathbf{x},0} \subseteq G'_\ell \subseteq G_{\mathbf{x},\ell}$ so G'_ℓ is a subgroup of $G_{\mathbf{x},\ell}$, for $\ell = 1, 2$ then for one and only one amalgamation try $\mathbf{y}$ we have $G_{\mathbf{y},0} = G_{\mathbf{x},0}, G_{\mathbf{y},\ell} = G'_\ell$ for $\ell = 1, 2$ and $\mathbf{I}_{\mathbf{y},\ell} = \mathbf{I}_{\mathbf{x},\ell} \cap G'_\ell$ so $\mathbf{y} \leq_{\mathbf{X}(K)} \mathbf{x}$.

6) Moreover, in part (5), if $\mathbf{z}$ is an amalgamation try with $(G_{\mathbf{z},0}, G_{\mathbf{z},1}, G_{\mathbf{z},2}) = (G_{\mathbf{x},0}, G'_1, G'_2)$ then for some $\mathbf{x}'$, the pair $(\mathbf{x}', \mathbf{z})$ is like $(\mathbf{x}, \mathbf{y})$ in (5) and $(G_{\mathbf{x}',0}, G_{\mathbf{x}',1}, G_{\mathbf{x}',2}) = (G_{\mathbf{x},0}, G_{\mathbf{x},1}, G_{\mathbf{x},2})$.

7) In part (5) there is a unique homomorphism f from $G = \langle \mathbf{j}_{\mathbf{x},1}(G'_1) \cup \mathbf{j}_{\mathbf{x},2}(G'_2) \rangle_{\mathrm{Sym}(\mathscr{U}_{\mathbf{x}})}$ onto $G_{\mathbf{y}}$ such that $\ell \in \{1, 2\} \wedge g \in G'_\ell \Rightarrow \mathbf{j}_{\mathbf{y},\ell}(g) = f(\mathbf{j}_{\mathbf{x},\ell}(g))$.

8) In part (5), if G'_1, G'_2 are finite, then $\langle \mathbf{j}_{\mathbf{x},1}(G'_1) \cup \mathbf{j}_{\mathbf{x},2}(G'_2) \rangle_{G_{\mathbf{x}}}$ has at most $(n_*!)^{m_*}$ members where $n_* = |G'_1| \times |G'_2| \times |G_{\mathbf{x},0}|^3$ and $m_* = (n_*!)^{|G'_1|+|G'_2|}$.

Proof. Straightforward. For example:

1) For $\ell = 1$ and $f, h \in G_1$ and $(g_0, g_1, g_2) \in \mathscr{U}$ let $(\mathbf{j}_1(f))(g_0, g_1, g_2) = (g_0', g_1', g_2')$ and $(\mathbf{j}_1(h))(g_0', g_1', g_2') = (g_0'', g_1'', g_2'')$. Then $g_2 = g_2'$ and $g_2' = g_2''$ and in G_1 we have $g_1 g_0 f = g_1' g_0'$ and $g_1' g_0' h = g_1'' g_0''$, hence $g_2 = g_2''$ and $g_1'' g_0'' = g_1' g_0' h = (g_1 g_0 f) h = (g_1 g_0)(fh)$, so $\mathbf{j}_1(fh)(g_0, g_1, g_2) = (g_0'', g_1'', g_2'') = (\mathbf{j}_1(h))(\mathbf{j}_1(f))(g_0, g_1, g_2)$, i.e., $G_{\mathbf{x}} \models$ "$\mathbf{j}_1(fh) = \mathbf{j}_1(f)\mathbf{j}_1(h)$".

2) Clearly $|G_{\mathbf{x},\ell}| = |\mathbf{I}_{\mathbf{x},\ell}| \times |G_{\mathbf{x},0}|$ for $\ell = 1, 2$ hence $|\mathscr{U}_{\mathbf{x}}| = |\mathbf{I}_{\mathbf{x},1}| \times |\mathbf{I}_{\mathbf{x},2}| \times |G_{\mathbf{x},0}| = (|G_{\mathbf{x},1}|/|G_{\mathbf{x},0}|) \times (|G_{\mathbf{x},2}|/|G_{\mathbf{x},0}| \times |G_{\mathbf{x},0}| = |G_{\mathbf{x},1}| \times |G_{\mathbf{x},2}|/|G_{\mathbf{x},0}|$.

Hence $|G_{\mathbf{x}}| \leq |\mathrm{Sym}(\mathscr{U}_{\mathbf{x}})| = (|\mathscr{U}_{\mathbf{x}}|)! = (|G_{\mathbf{x},1}| \times |G_{\mathbf{x},2}|/|G_{\mathbf{x},0}|!)$ as stated.

3) Let $G_0 = G_{\mathbf{x},0}$. We define $E = \{((g_0', g_1', g_2'), (g_0'', g_1'', g_2''))) \in \mathscr{U}_{\mathbf{x}} \times \mathscr{U}_{\mathbf{x}} : G_0 g_1' G_1' = G_0 g_1'' G_1'$ and $G_0 g_2' G_2' = G_0 g_2'' G_2'\}$, this is an equivalence relation on $\mathscr{U}_{\mathbf{x}}$, each equivalence class has $\leq (|G_1'| \times |G_2'| \times |G_{\mathbf{x},0}|^3) = n_*$ members.

[Why? As if $(g_0', g_1', g_2') \in (g_0, g_1, g_2)/E$ then $g_0' \in G_0, g_1' \in G_0 g_1 G_1', g_2' \in G_0 g_2 G_2'$ and $|G_0 g_\ell G_\ell'| \leq |G_0| \times |G_\ell'|$.]

Also each of the permutations of $\mathscr{U}_{\mathbf{x}}$ from $\mathbf{j}_{\mathbf{x},1}(G_1') \cup \mathbf{j}_{\mathbf{x},2}(G_2')$ maps each E-equivalence class onto itself. Hence for $n \in [1, n_*]$ there are $\leq m_n^* := n!^{|G_1'|+|G_2'|-1}$ isomorphism types of structures of the form: $N = (|N|, F_f^N)_{f \in G_1' \cup G_2'}$, where $|N|$, the universe, has exactly n elements and is an E-equivalence class, and for each $f \in G_1' \cup G_2'$ we have: F_f^N is a permutation of this equivalence class and $F_{e(G_0)}^N$ is the identity. Clearly as $\sum_{n \leq n_*} (n!)^{|G_1'|+|G_2'|-1} \leq (n_*!)^{|G_1'|+|G_2'|} = m_*$, the subgroup $\langle \mathbf{j}_{1,\mathbf{x}}(G_1') \cup \mathbf{j}_{2,\mathbf{x}}(G_2') \rangle_{G_{\mathbf{x}}}$ of $G_{\mathbf{x}}$ has at most $(n_*!)^{m_*}$ members. Of course, the argument gives better bounds, e.g., the number of relevant N's is much smaller and using a finer E. $\square_{2.3}$

Claim 2.4. In Definition 2.2, $\mathbf{j}_{\mathbf{x},1}(G_1) \cap \mathbf{j}_{\mathbf{x},2}(G_2) = \mathbf{j}_{\mathbf{x},\ell}(G_0)$.

Proof. Assume that $a_\ell \in G_\ell$ and $b_\ell = \mathbf{j}_{\mathbf{x},\ell}(a_\ell)$ for $\ell = 1, 2$. It suffices to show that: if $b_1 = b_2$ <u>then</u> $a_1, a_2 \in G_{\mathbf{x},0}$ and $a_1 = a_2$. We check to what b_ℓ maps the triple $(e, e, e) \in \mathscr{U}_{\mathbf{x}}$: by the definition of $\mathbf{j}_{\mathbf{x},1}, \mathbf{j}_{\mathbf{x},2}$ we have:

- $b_1((e, e, e)) = (g_0', g_1, e) \in \mathscr{U}_{\mathbf{x}}$ where $G_1 \models g_1 g_0' = b_1$;

- $b_2((e, e, e)) = (g_0'', e, g_2) \in \mathscr{U}_{\mathbf{x}}$ where $G_2 \models g_2 g_0'' = b_2$.

So if $b_1 = b_2$, then $(g_0', g_1, e) = b_1((e, e, e) = b_2((e, e, e,)) = (g_0'', e, g_2)$, hence $g_0' = g_0'' \wedge g_1 = e \wedge e = g_2$; this implies that $g_0' = b_1, g_0'' = b_2$ hence $g_0' = g_0''$, also $g_0'' \in G_0$ together $a_1 = a_2$ so we are done. $\square_{2.4}$

Definition 2.5. 1) Let[8] $\mathrm{NF}_{\mathrm{rfin}}(G_0, G_1, G_2, G_3)$ means that $G_\ell \subseteq G_3 (\in \mathbf{K})$ for $\ell < 3$ and $\mathrm{NF}_{\mathrm{fin}}(G_0, G_1, G_2, \langle G_1 \cup G_2 \rangle_{G_3})$, see below.

2) Let $\mathrm{NF}_{\mathrm{fin}}(G_0, G_1, G_2, G_3)$ mean that:

---

[8]NF stands for nonforking.

(a) $G_0 \subseteq G_\ell \subseteq G_3 \in \mathbf{K}$ are finite groups for $\ell = 1, 2$;

(b) $G_3 = \langle G_1 \cup G_2 \rangle_{G_3}$;

(c) if $\mathbf{x} \in \mathbf{X_K}$ and $G_0 = G_{\mathbf{x},0}, G_1 \subseteq G_{\mathbf{x},1}, G_2 \subseteq G_{\mathbf{x},2}$ <u>then</u> there is a homomorphism $\mathbf{f}$ from G_3 into $G_{\mathbf{x}}$ such that $\mathbf{f} \restriction G_\ell = \mathbf{j}_{\mathbf{x},\ell} \restriction G_\ell$ for $\ell = 1, 2$;

(d) if $a \in G_3 \backslash \{e_{G_3}\}$ <u>then</u> for some $\mathbf{x}, \mathbf{f}$ as above we have $\mathbf{f}(a) \neq e_{G_3}$.

Remark 2.6. Note the choice "$G_\ell \subseteq G_{\mathbf{x},\ell}$" rather than $G_\ell = G_{\mathbf{x},\ell}$ in clause (c) of 2.5.

Now the amalgamation in Definition 2.5 is very nice but do we have existence, in $\mathbf{K}_{\mathrm{lf}}$ of course? The following Claim 2.7(3) answers positively.

Claim 2.7. 1) In clause (c) of Definition 2.5(2), the homomorphism $\mathbf{f}$ is unique.
1A) If $\mathrm{NF}_{\mathrm{fin}}(G_0^\iota, G_1^\iota, G_2^\iota, G_3^\iota)$ for $\iota = 1, 2$ and $\mathbf{f}_\ell$ is an isomorphism from G_ℓ^1 onto G_ℓ^2 such that $\mathbf{f}_0 \subseteq \mathbf{f}_\ell$ for $\ell = 0, 1, 2$ <u>then</u> there is one and only one isomorphism $\mathbf{f}_3$ from G_3^1 onto G_3^2 extending $\mathbf{f}_1 \cup \mathbf{f}_2$.
2) In Definition 2.5, necessarily $G_1 \cap G_2 = G_0$.
3) If $G_0 \subseteq G_\ell \in \mathbf{K}$ are finite for $\ell = 1, 2$ <u>then</u> we can find $\bar{f}, \bar{H}$ such that

(a) $\bar{f} = \langle f_0, f_1, f_2 \rangle$;

(b) $\bar{H} = \langle H_\ell : \ell \leq 3 \rangle$;

(c) $\mathrm{NF}_{\mathrm{fin}}(H_0, H_1, H_2, H_3)$;

(d) f_ℓ is an isomorphism from G_ℓ onto H_ℓ for $\ell = 0, 1, 2$;

(e) $f_0 \subseteq f_1$ and $f_0 \subseteq f_2$.

Proof. 1), 1A) Obvious.
2) By Claim 2.4 recalling clause (c) of 2.5(2).
3) Follows by 2.3(8) but we elaborate. Let $\bar{G} = \langle G_\ell : \ell = 0, 1, 2 \rangle$ and

$(*)_1$ let $\mathbf{X}_{\bar{G}} := \{ \mathbf{x} \in \mathbf{X_x} : G_{\mathbf{x},0} = G_0$ and $G_{\mathbf{x},\ell}$ is a lf group extending G_ℓ for $\ell = 1, 2 \}$;

$(*)_2$ for $\mathbf{x} \in \mathbf{X}_{\bar{G}}$ let: $\mathbf{n}_{\bar{g}}(\mathbf{x}) =$ the number of elements of $\langle \mathbf{j}_{\mathbf{x},1}(G_1) \cup \mathbf{j}_{\mathbf{x},2}(G_2) \rangle_{G_{\mathbf{x},3}}$.

We define $\mathbf{X}_{\bar{G}}^{\mathrm{mx}}$ as the set of $\mathbf{x}$ such that:

$(*)_3$ (a) $\mathbf{x} \in \mathbf{X}_{\bar{G}}$;

(b) if $\mathbf{y} \in \mathbf{X}_{\bar{G}}$ and $\mathbf{x} \leq \mathbf{y}$, then $n_{\bar{G}}(\mathbf{x}) = n_{\bar{G}}(\mathbf{y})$;

$(*)_4$ if $\mathbf{x}, \mathbf{z} \in \mathbf{X}_{\bar{G}}^{\mathrm{mx}}$ and $\mathbf{x} \leq_{\mathbf{X}(\mathbf{K})} \mathbf{z}$, <u>then</u> $n_{\bar{G}}(\mathbf{x}) \leq n_{\bar{G}}(\mathbf{z})$.

[Why? Because by 2.3(7) there is a homomorphism from $G_{\mathbf{z}} = \langle \mathbf{j}_{\mathbf{z},1}(G_1) \cup \mathbf{j}_{\mathbf{z},2}(G_2) \rangle$ onto $G_{\mathbf{x}} = \langle \mathbf{j}_{\mathbf{x},1}(G_1) \cup \mathbf{j}_{\mathbf{x},2}(G_2) \rangle$.]

$(*)_5$ for every $\mathbf{x} \in \mathbf{X}_{\bar{G}}$ there is $\mathbf{y} \in \mathbf{X}_{\bar{G}}^{\mathrm{mx}}$ such that $x \leq \mathbf{y}$; hence $\mathbf{X}_{\bar{G}}^{\mathrm{mx}} \neq \emptyset$.

[Why? By $(*)_4$ and 2.3(8).]

$(*)_6$ $(\mathbf{X}_{\bar{G}}, \leq_{\mathbf{X}[\mathbf{K}]})$ has amalgamation, that is

- if $\mathbf{x}_0 \leq_{\mathbf{X}[\mathbf{K}]} \mathbf{x}_\iota$ for $\iota = 1,2$ <u>then</u> we can find $\mathbf{x}_3$ and (f_1^ι, f_2^ι) for $\iota = 1,2$ such that:
  - (a) $\mathbf{x}_3 \in \mathbf{X}_{\mathbf{K}}$
  - (b) $\mathbf{x}_0 \leq_{\mathbf{X}[\mathbf{K}]} \mathbf{x}_3$
  - (c) (f_1^ι, f_2^ι) embeds $\mathbf{x}_\iota$ into $\mathbf{x}_3$ over $\mathbf{x}_0$
    (over $\mathbf{x}_0$ means: $f_1^\iota {\restriction} G_{\mathbf{x}_0,1} = \mathrm{id}_{G_{\mathbf{x}_0,1}}, f_2^\iota {\restriction} G_{\mathbf{x}_0,2} = \mathrm{id}_{G_{\mathbf{x}_0,2}}$).

[Why? For $\ell = 1,2$, we use the disjoint amalgamation for finite groups, i.e., find $(G_\ell, f_\ell^1, f_\ell^2)$ such that:

- $\bullet_1$ G_ℓ is a finite group extending $G_{\mathbf{x}_0,0}$

- $\bullet_2$ f_ℓ^1 embeds $G_{\mathbf{x}_1,\ell}$ into G_ℓ over $G_{\mathbf{x}_0,0}$

- $\bullet_3$ f_ℓ^2 embeds $G_{\mathbf{x}_2,\ell}$ into G_ℓ over $G_{\mathbf{x}_0,0}$

- $\bullet_4$ $f_\ell^1(G_{\mathbf{x}_1,\ell}) \cap f_\ell^2(G_{\mathbf{x},\ell}) = G_{\mathbf{x}_0,0}$.

Note that $f_\ell^1(\mathbf{I}_{\mathbf{x}_1,\ell}) \cap f_\ell^2(\mathbf{I}_{\mathbf{x}_2,\ell}) = \{e_{G_{\mathbf{x}_0,0}}\}$, moreover, $\langle gG_{\mathbf{x}_0,0} : g \in f_\ell^1(\mathbf{I}_{\mathbf{x}_1,\ell}) \cup f_\ell^2(\mathbf{I}_{\mathbf{x}_2,\ell}) \rangle$ is a sequence of pairwise disjoint sets. Hence there is $\mathbf{I}_\ell \subseteq G_\ell$ extending $f_\ell^1(\mathbf{I}_{\mathbf{x}_1,\ell}) \cup f_\ell^2(\mathbf{I}_{\mathbf{x}_2,\ell})$ such that $\langle gG_{\mathbf{x}_0,0} : g \in \mathbf{I}_\ell \rangle$ is a partition of G_ℓ.
Define $\mathbf{x}_3$ by:

- $\bullet_1'$ $G_{\mathbf{x}_3,0} = G_{\mathbf{x}_0,0}$

- $\bullet_2'$ $G_{\mathbf{x}_3,\ell} = G_\ell$ for $\ell = 1,2$

- $\bullet_3'$ $\mathbf{I}_{\mathbf{x}_3,\ell} = \mathbf{I}_\ell$ for $\ell = 1,2$.

Now check that $\mathbf{x}_3, (f_1^\iota, f_2^\iota)$ for $\iota = 1,2$ are as required.]

$(*)_7$ if $\mathbf{y} \in \mathbf{x}_{\bar{G}}^{\mathrm{mx}}$, then $\mathrm{NF}_{\mathrm{fin}}(G_{\mathbf{y},0}, G_{\mathbf{y},1}, G_{\mathbf{y},2}, G_{\mathbf{y},3})$.

[Should be clear now.]
Alternatively[9], use 2.15 below and 2.3(8). $\qquad\qquad\qquad$ $\square_{2.7}$

We give now further basic properties, mainly connecting it to nonsplitting (in 2.8(4)).

Claim 2.8. Assume $\mathrm{NF}_{\mathrm{rfin}}(G_0, G_1, G_2, G_3)$ hence $\mathrm{NF}_{\mathrm{fin}}(G_0, G_1, G_2, G_3) \Leftrightarrow$ $G_3 = \langle G_1 \cup G_2 \rangle_{G_3}$.
1) Symmetry: Also $\mathrm{NF}_{\mathrm{rfin}}(G_0, G_2, G_1, G_3)$ holds.
2) Monotonicity: If $G_0 \subseteq G'_\ell \subseteq G_\ell$ for $\ell = 1, 2$ and $G'_1 \cup G'_2 \subseteq G'_3 \subseteq G_3$ then $\mathrm{NF}_{\mathrm{rfin}}(G_0, G'_1, G'_2, G'_3)$.
3) Uniqueness: if $\mathrm{NF}_{\mathrm{fin}}(G'_0, G'_1, G'_2, G'_3)$ hence $G'_3 = \langle G'_1 \cup G'_2 \rangle_{G'_3}$, f_ℓ is an isomorphism from G'_ℓ into G_ℓ for $\ell = 0, 1, 2$ such that $f_1 {\restriction} G'_0 = f_0 = f_2 {\restriction} G'_0$ and f_0 is onto G_0, <u>then</u> there is an embedding f_3 of G'_3 into G_3 extending $f_1 \cup f_2$ (unique, of course; it is onto if and only if $G_3 = \langle G_1 \cup G_2 \rangle_{G_3}$ and f_ℓ is onto G_ℓ for $\ell = 1, 2$).
4) If $\bar{a} \in {}^{\omega >}(G_2)$ <u>then</u> $\mathrm{tp}_{\mathrm{bs}}(\bar{a}, G_1, G_3)$ does not split over G_0.

Proof. Straightforward but we elaborate.
1) Use the symmetry in the definition (recall that in §2 we have $\mathbf{K} = \mathbf{K}_{\mathrm{lf}}$ not $\mathbf{K}_{\mathrm{olf}}$!)
2) By 2.3(7) and use the uniqueness in 2.7(1). Alternatively, use 2.15 below and 2.3(8).
3) Easily, too.
4) Obvious by parts (2) and (3). $\qquad\qquad\qquad\qquad\qquad\qquad\qquad$ $\square_{2.8}$

Now above the restriction of G_1, G_2 to be finite is undesirable.

Definition 2.9. Let $\mathrm{NF}_f(G_0, G_1, G_2, G_3)$ or "G_1, G_2 are NF_f-stably amalgamated over G_0 inside G_3" mean that:

(a) $G_\ell \in \mathbf{K}$ for $\ell \leq 3$

(b) G_0 is finite

(c) $G_0 \subseteq G_\ell \subseteq G_3$ for $\ell = 1, 2$ and $G_1 \cap G_2 = G_0$

(d) if G'_1, G'_2 are finite groups and $G_0 \subseteq G'_\ell \subseteq G_\ell$ for $\ell = 1, 2$ and $G'_3 = \langle G'_1 \cup G'_2 \rangle_{G_3}$, then $\mathrm{NF}_{\mathrm{fin}}(G_0, G'_1, G'_2, G'_3)$.

Claim 2.10. <u>Stable Amalgamation over Finite Claim</u> 1) Existence: If $G_0 \in \mathbf{K}$ is finite and $\overline{G_0 \subseteq G_\ell} \in \mathbf{K}$ for $\ell = 1, 2$ and for transparency $G_1 \cap G_2 = G_0$ <u>then</u> for some G_3 we have $\mathrm{NF}_f(G_0, G_1, G_2, G_3)$ and $G_3 = \langle G_1 \cup G_2 \rangle_{G_3}$.
2) Uniqueness: In part (1), G_3 is unique up to isomorphism over $G_1 \cup G_2$.

---

[9]Or see [Sh:1098].

3) Monotonicity: If $G_0 \subseteq G'_\ell \subseteq G_\ell$ for $\ell = 1, 2$ and $\mathrm{NF}_\ell(G_0, G_1, G_1, G_2)$ then $\mathrm{NF}_f(G_0, G'_1, G'_2, G_3)$.

4) Symmetry: $\mathrm{NF}_f(G_0, G_1, G_2, G_3)$ holds iff $\mathrm{NF}_f(G_0, G_2, G_1, G_3)$ holds.

5) Definability: If $\mathrm{NF}_f(G_0, G_1, G_2, G_3)$, then $G_1 \leq_{\Omega[\mathbf{K}]} G_3$.

Proof. Straightforward by 2.7(3), 2.8(3), 2.8(2), 2.8(1), 2.8(4), i.e., existence, uniqueness, monotonicity, symmetry and definability respectively. $\square_{2.10}$

Now we go back to the major problem left in §1.

Claim 2.11. There is one and only one full $\mathfrak{s} \in \Omega[\mathbf{K}]$ such that $q = q_{\mathfrak{s}}(\bar{a}, G_1)$ when:

(a) $G_1 \in \mathbf{K}$ is existentially closed

(b) $q(\bar{x}) \in \mathbf{S}^n_{\mathfrak{S}_{\mathrm{atdf}}}(G)$ is $\mathrm{tp}_{\mathrm{bs}}(\bar{c}, G_1, G_3)$, see below

(c) $\mathrm{NF}_f(G_0, G_1, G_2, G_3)$

(d) $\bar{c} \in {}^{n(\mathfrak{s})}(G_2)$ and $\bar{a} \in {}^{k(\mathfrak{s})}(G_0)$ generate G_0.

Proof. By 2.10. $\square_{2.11}$

Definition 2.12. Let $\mathfrak{S}_{\mathrm{df}} \subseteq \Omega[\mathbf{K}]$ be the closure of $\mathfrak{S}_{\mathrm{atdf}}$, see Definition 1.6 where $\mathfrak{S}_{\mathrm{atdf}} \subseteq \mathfrak{S}(\mathbf{K})$ is the set of $\mathfrak{s} \in \Omega[\mathbf{K}]$ as in 2.11.

Claim 2.13. 1) $\mathfrak{S}_{\mathrm{df}}$ is well defined, see Definition 2.12, 2.8(3).
2) $\mathfrak{S}_{\mathrm{df}}$ is dense (see Definition 1.6(2)), closed and countable.

Proof. 1) Obvious.
2) $\mathfrak{S}_{\mathrm{df}}$ is dense: holds by 2.10 and 2.11 recalling Definition 2.9, 2.12.
$\mathfrak{S}_{\mathrm{df}}$ is closed: by its definition.
$\mathfrak{S}_{\mathrm{df}}$ is countable: as $\mathfrak{S}_{\mathrm{atdf}}$ is by 2.8(3) recalling 1.8(3). $\square_{2.13}$

2.14 Discussion. Is $\mathfrak{S}_{\mathrm{df}}$ symmetric? Not clear, however, in the end of §1 we have circumvented this and we shall in §3 circumvent this in another way.

Claim 2.15. 1) Assume $G_0 \subseteq G_\ell \in \mathbf{K}$ and G_ℓ is existentially closed for $\ell = 1, 2$ and G_0 finite.

Then we can find $\mathbf{x} \in \mathbf{X_K}$ such that $G_\ell = G_{\mathbf{x}, \ell}$ for $\ell = 0, 1, 2$ and $(\mathbf{j}_{\mathbf{x}, 0}(G_0) \subseteq \mathbf{j}_{\mathbf{x}, \ell}(G_\ell) \leq_{\mathfrak{S}_{\mathrm{df}}} G_{\mathbf{x}})$ and $\mathrm{NF}_f(\mathbf{j}_{\mathbf{x}, 0}(G_0), \mathbf{j}_{\mathbf{x}, 1}(G_1), \mathbf{j}_{\mathbf{x}, 2}(G_2), G_{\mathbf{x}})$.

2) Assume $G_0 \subseteq G'_\ell \subseteq G_\ell$ and G'_ℓ finite (or just $(G_\ell : G'_\ell) = |G_\ell|$) and $\mathbf{y} \in \mathbf{X_K}, G_{\mathbf{y}, 0} = G_0$ and $G_{\mathbf{y}, \ell} = G'_\ell$ for $\ell = 1, 2$. Then in part (1) we can demand that $\mathbf{x}$ extends $\mathbf{y}$.

Remark 2.16. If $G_0 \subseteq G_\ell \in \mathbf{K}$ for $\ell = 1, 2$, then we can find infinite $G'_1, G'_2 \in \mathbf{K}$ extending G_1, G_2 respectively as $\mathbf{K}$ is closed under (finite) product (for $\mathbf{K}_{\mathrm{olf}}$ use lexicographic order).

Proof. 1) By the definitions it is easy. That is, for $\ell = 1, 2$ we can choose $\mathbf{I}_\ell$ as in 2.2(1)(b) satisfying:

(*) if $G'_\ell \subseteq G_\ell$ is finite and extends G_0 and $\mathbf{I}' \subseteq G'_\ell$ is such that $e_{G_0} \in \mathbf{I}'$ and $\langle gG_0 : g \in \mathbf{I}' \rangle$ is a partition of G'_ℓ <u>then</u> we can find $g^* \in G_\ell$ such that $\{g^*g : g \in \mathbf{I}'\} \subseteq \mathbf{I}_\ell$.

Now think.
2) Similarly. $\qquad\qquad\qquad\qquad\qquad\qquad\qquad\qquad\qquad$ $\square_{2.15}$

2(B) Constructing Reasonable Schemes

We now give some examples of $\mathfrak{s} \in \Omega[\mathbf{K}]$.

Definition 2.17. 1) Let $\mathfrak{s}_{cg}$ be the $\mathfrak{s}$ from 2.18(2) below.
2) Let $\mathfrak{s}_{gl}$ be the $\mathfrak{s}$ from 2.18(3) below.

Claim 2.18. 1) For every $G \in \mathbf{K}_{lf}$ there are G^+ and a such that $G \subseteq G^+ \in \mathbf{K}_{lf}, G^+ = \langle G \cup \{a\}\rangle_{G^+}$; in G^+ the element a does not commute with any $b \in G \backslash \{e_G\}$, a has order 2 and the sets $G, a^{-1}Ga$ commute in G^+ and their intersection is $\{e_G\}$.
2) There is unique $\mathfrak{s} \subset \Omega[\mathbf{K}]$ such that $k_\mathfrak{s} = 0, n_\mathfrak{s} = 1, p_\mathfrak{s}$ is empty and in part (1) above $\mathrm{tp}_{\mathrm{bs}}(a, G, G^+)$ is $q_\mathfrak{s}(<>, G)$.
3) There is $\mathfrak{s} \in \Omega[\mathbf{K}]$ with $k_\mathfrak{s} = 1, n_\mathfrak{s} = 4, p_\mathfrak{s}(\bar{x}_\mathfrak{s}) = \{x_0 = x_0^{-1} \wedge x_0 \neq e\}$ such that: if $G \in \mathbf{K}_{lf}$ and $a \in G$ realizes $p_\mathfrak{s}(x_0)$ <u>then</u> there are $G^+, \bar{c}$ such that $G \subseteq G^+ = \langle G \cup \bar{c}\rangle_{G^+}, \mathrm{tp}_{\mathrm{bs}}(\bar{c}, G, G^+) = q_\mathfrak{s}(\langle a\rangle, G)$ and c_ℓ realizes $q_{\mathfrak{s}_{cg}}(<>, G)$ in G^+ for $\ell < n_\mathfrak{s}$ and $a \in \langle \bar{c}\rangle_{G^+}$.

Proof. We first make a less specific construction for any $G \in \mathbf{K}$.

For $n \geq 2$ let $\mathscr{U}_n = G \times n = \{(g, \iota) : g \in G, \iota < n\}$. For finite $K \subseteq G$ let $E_K := \{((g_1, \iota_1), (g_2, \iota_2)) : g_1, g_2 \in G$ and $\iota_1, \iota_2 < n$ and $g_1 K = g_2 K\}$, this is an equivalence relation on $\mathscr{U}_n$, each equivalence class has $\leq n \times |K|$ elements. For $\bar{a} \in {}^{\omega>}G$ let $E_{\bar{a}} = E_K$ when $K = \langle \mathrm{Rang}(\bar{a})\rangle_G$ which is finite.

For $\bar{a} \in {}^nG$ and π a permutation of $\{0, \ldots, n-1\}$ let $h_{\bar{a}, \pi}$ be the following function from $\mathscr{U}_n$ into $\mathscr{U}_n$:

$(*)_1$ $h_{\bar{a}, \pi}((g, \iota)) = (ga_\iota, \pi(\iota))$.

Clearly

$(*)_2$ $h_{\bar{a}, \pi}$ is a permutation of $\mathscr{U}_n$ which maps every $E_{\bar{a}}$-equivalence class onto itself.

Let H be the group of permutations of $\mathscr{U}_n$ generated by $\{h_{\bar{a},\pi} : \bar{a} \in {}^nG$ and π is a permutation of $\{0,\ldots,n-1\}\}$, now by $(*)_2$ it is easy to see that $H \in \mathbf{K}_{\mathrm{lf}}$ where, as in earlier cases,

- $H \models$ "$h = h_1 h_2$" iff $x \in \mathscr{U}_n \Rightarrow h(x) = h_2(h_1(x))$.

Now for $\iota < n$ let $\mathbf{j}_\iota$ be the following function from G into H:

$(*)_3$ $\mathbf{j}_\iota(a) = h_{\bar{b},\pi}$ when $\pi = $ the identity and b_k is a if $k = \iota$ and is e_G otherwise.

Now

$(*)_4$ for $\iota < n, \mathbf{j}_\iota$ is an embedding of G into H.

[Why? Check.]

We let $G^*, \mathbf{j}_*$ be such that $G^* \supseteq G$ and $\mathbf{j}_*$ is an isomorphism from G^* onto H extending $\mathbf{j}_0$.

For later use note:

$(*)_5$ for transparency we can use existentially closed G.

Also

$(*)_6$ (a) $G \leq_{\Omega[\mathbf{K}]} G^*$ equivalently $\mathbf{j}_0(G) = \mathbf{j}_*(G) \leq_{\Omega[\mathbf{K}]} H$;

(b) if $A \subseteq G$ and for $\ell < m$ we have $\bar{a}_\ell \in {}^nA$ and π_ℓ is a permutation of $\{0,\ldots,n-1\}$, then $p = \mathrm{tp}_{\mathrm{bs}}(\langle h_{\bar{a}_\ell,\pi_\ell} : \ell < m\rangle, \mathbf{j}_*(G), H)$ does not split over A;

(c) if above A is finite and $\bar{a}$ lists A <u>then</u> for some $\mathfrak{s} \in \Omega[\mathbf{K}], p = q_\mathfrak{s}(\bar{a}, \mathbf{j}_*(G))$.

Now we prove each part.

1) Let $n = 2$ and π be the permutation of $\{0,1\}$ such that $\pi(0) = 1, \pi(1) = 0$, and let $a = \mathbf{j}_*^{-1}(h_{<e_G,e_G>,\pi})$.

2) Should be clear.

3) First note that

$\oplus_1$ $\mathbf{j}_*^{-1}(h_{\bar{a},\pi})$ realizes $q_{\mathfrak{s}_{\mathrm{cg}}}(\langle\rangle, G)$ in G^* <u>when</u> for some $k \in \{1,\ldots,n-1\}$ we have

- $\bullet_1$ π is a permutation of $\{0,\ldots,n-1\}$ and has order two

- $\bullet_2$ $\pi(0) = k$

- $\bullet_3$ $\pi(k) = 0$

- $\bullet_4$ $\bar{a} \in {}^n G$ satisfies $a_{\pi(\iota)} = a_\iota^{-1}$ for $\iota < n$

- $\bullet_5$ if $\pi(\iota) \neq \iota, \iota < n$, then $a_\iota = e_G$, (or just a_0 belongs to the center of G).

[Why? By $(*)_2$ and the choice of H clearly $h_{\bar{a},\pi} \in H$ and inspecting $(*)_1$, easily $h_{\bar{a},\pi}$ has order two. By the choice of $\mathbf{j}_*, \pi$ as $\pi(0) = k, \pi(k) = 0$ and $a_k = e_G = a_0$, for $g \in G$ we get $H \models$ "$h_{\bar{a},\pi}^{-1} \mathbf{j}_0(g) h_{\bar{a},\pi} = \mathbf{j}_k(g)$". However, for every $g_1, g_2 \in G$ the elements $\mathbf{j}_0(g_1), \mathbf{j}_k(g_2)$ of H commute as $h_{\bar{a}_1,\pi_1}, h_{\bar{a}_0,\pi_2}$ commute in H, e.g., when $\pi_1 = \mathrm{id}_n = \pi_2$ and $\bigwedge_{\ell < n} (a_{1,\ell} = e \vee a_{2,\ell} = e)$. Lastly, $g_1, g_2 \in G \wedge \mathbf{j}_0(g_1) = \mathbf{j}_0(g_2) \Rightarrow g_1 = e_G = g_2$. Together we are done.]

Let $n = 3$ and for $\ell < 4$ let $g_\ell \in H$ be $h_{\bar{a}_\ell,\pi_\ell}$ where $\pi_\ell, \bar{a}_\ell$ are defined by (recall $a \in G$ is given and has order 2):

- $\oplus_2$ for $\ell < 4$ let π_ℓ be such that:

  $\ell = 0, 3$: the orbits are $\{0, 1\}, \{2\}$;

  $\ell = 1, 2$: the orbits are $\{0, 2\}, \{1\}$.

- $\oplus_3$ let $\bar{a}_\ell = \langle a_{\ell,i} : i < 3 \rangle$ be $\langle e, e, e \rangle, \langle e, a, e \rangle, \langle e, e, e \rangle, (e, e, e)$ for $\ell = 0, 1, 2, 3$.

Now

- $\oplus_4$ $c_\ell := \mathbf{j}_*^{-1}(h_{\bar{a}_\ell,\pi_\ell})$ realizes $q_{\mathfrak{s}_{\mathrm{cg}}}(\langle\rangle, G)$ for $\ell < 4$.

[Why? We apply $\oplus_1$ with k being 1 for $\ell = 0, 3$ and 2 for $\ell = 1, 2$. So we have to check $\bullet_1 - \bullet_4$ for each ℓ; now $\bullet_1 + \bullet_2 + \bullet_3$ holds by inspecting $\oplus_2$ and the choice of k and of π_ℓ.

Lastly, for $\bullet_4 + \bullet_5$ note that a, e has order 2 and $a_{\ell,0} = e_G = a_{\ell,k}$ by inspecting $\oplus_3$.]

- $\oplus_5$ $\mathrm{tp}_{\mathrm{bs}}(\langle c_0, c_1, c_2, c_3 \rangle, G, G^*)$ does not split over $\langle a \rangle$, moreover, is $q_t(\langle a \rangle, G)$ for some $t \in \Omega[\mathbf{K}]$.

[Why? Just think recalling $(*)_6$.]

Lastly,

- $\oplus_6$ $G^+ \models$ "$c_0 c_1 c_2 c_3 = a$".

[Why? This is equivalent to $H \models h_{\bar{a}_0,\pi_0} h_{\bar{a}_1,\pi_1} h_{\bar{a}_2,\pi_2} h_{\bar{a}_3,\pi_3} = \mathbf{j}_0(a)$. By the definition of the product we check how each $(g, \ell) \in \mathcal{U}_n$ is mapped (see above, so $h_{\bar{a}_0,\pi_0}$ is first) applying $h_{\bar{a}_\ell,\pi_\ell}$ in turn:

$$(g,0) \mapsto (ge,1) \mapsto (gea,1) \mapsto (geae,1) \mapsto (geaee,0) = (ga,0) = \mathbf{j}_0(a)(g,0)$$

and

$$(g,1) \mapsto (ge,0) \mapsto (gee,2) \mapsto (geee,0) \mapsto (geeee,1) = (g,1) = \mathbf{j}_0(a)(g,1)$$

$$(g,2) \mapsto (ge,2) \mapsto (gee,0) \mapsto (geee,2) \mapsto (geeee,2) = (g,2) = \mathbf{j}_0(a)(g,2).$$

So we are done.] $\qquad\qquad\qquad\qquad\qquad\qquad\qquad\qquad\qquad\square_{2.18}$

The following will be used in the proof of existence of complete existentially closed G.

Claim 2.19. 1) If (A), then (B) <u>where</u>:

(A)   (a) $G_n \subseteq G_{n+1} \in \mathbf{K}$ for $n < \omega$ and I a set;

     (b) $a_n^t \in G_{n+1}$ has order $k(t)$ and let $b_n^t = a_0^t \ldots a_n^t$ in G_{n+1} for $n < \omega, t \in I$;

     (c) $\bar{a}_n = \langle a_n^t : t \in I \rangle, \bar{b}_n = \langle b_n^t : t \in I \rangle$;

     (d) (α) $\mathrm{tp}_{\mathrm{bs}}(\bar{a}_n, G_n, G_{n+1})$ is increasing[10] with n;

            (β) $c\ell(\bar{a}_n, G_{n+1}) \cap G_n = \{e_{G_n}\}$; if $I = \{t\}$ this means that for every $i \in \{1, \ldots, k(t)\}$ we have:
$$G_{n+1} \models \text{``}(a_n^t)^i = e_{G_1}\text{''} \text{ iff } (a_n^t)^i \in G_n \text{ iff } i = k(t);$$

     (e) a_n^t commutes with every $c \in G_n$;

     (f) $G_\omega = \cup\{G_n : n < \omega\}$ hence $\in \mathbf{K}$;

(B) for some $\bar{b}_\omega, G_{\omega+1}$ we have:

     (a) $G_{\omega+1} \supseteq G_\omega$ belongs to $\mathbf{K}$;

     (b) $\bar{b}_\omega = \langle b_\omega^t : t \in I \rangle$ and $b_\omega^t \in G_{\omega+1}$;

     (c) $G_{\omega+1} = c\ell(G_\omega \cup \{b_\omega^t : t \in I\}, G_{\omega+1})$;

     (d) if $n < \omega$, then $p_n = \mathrm{tp}_{\mathrm{bs}}(\bar{b}_\omega, G_n, G_{\omega+1}) = \mathrm{tp}_{\mathrm{bs}}(\bar{b}_n, G_n, G_{n+1})$.

2) If we have (A) except omitting (A)(d)(β), still we have:

---

[10]So by (A)(e) this is equivalent to "$\mathrm{tp}(\bar{a}_n, \emptyset, G_{n+1})$ is constant".

(B)′ $(a) - (c)$ as above;

(d) $\bar{b}_\omega \!\restriction\! u$ realizes $\mathrm{tp}_{\mathrm{bs}}(\bar{b}_{n!}, G_{n!} \!\restriction\! u, G_{n+1})$ in $G_{\omega+1}$ when $u \subseteq I$ is finite and n is large enough.

Proof. 1) Letting $p_n(\bar{x}) = \mathrm{tp}_{\mathrm{bs}}(\bar{b}_n, G_n, G_{n+1})$, it is enough to prove:

$(*)_1$ $p_n \subseteq p_{n+1}$.

For this it is enough to prove, letting $\bar{y} = \langle y_t : t \in I \rangle$,

$(*)_2$ if $\sigma(\bar{y}, \bar{z})$ is a group-term and $\bar{c} \in {}^{(\ell g(\bar{z}))}(G_n)$ then $G_{n+2} \models$ "$\sigma(\bar{b}_{n+1}, \bar{c}) = e$" iff $G_{n+1} \models$ "$\sigma(\bar{b}_n, \bar{c}) = e$".

Towards proving $(*)_2$ note:

- $\bullet_{2.1}$ $\bar{c}$ and $\bar{b}_n$, and hence $\sigma(\bar{b}_n, \bar{c})$ are from G_{n+1},

- $\bullet_{2.2}$ a_{n+1}^t commutes with every $c_i (i < \ell g(\bar{c}))$ and with b_n^s for $s \in I$.

By clause (A)(b) of the assumption of the claim,

- $\bullet_{2.3}$ $b_{n+1}^t = b_n^t a_{n+1}^t$ and $b_n^t = b_{n-1}^t a_n^t$ stipulating $b_{-1}^t = e$.

Similarly,

- $\bullet_{2.4}$ $\bar{c}$ and $\bar{b}_{n-1}$ are from G_n;

- $\bullet_{2.5}$ a_n^t commute with every $c_i (i < \ell g(\bar{c}))$ and with b_{n-1}^s for $s \in I$.

Hence for some group term $\sigma_*(\bar{x})$:

- $\bullet_{2.6}$ $G_{n+2} \models$ "$\sigma(\bar{b}_{n+1}, \bar{c}) = \sigma(\bar{b}_n, \bar{c})\sigma_*(\bar{a}_{n+1})$";

- $\bullet_{2.7}$ $G_{n+1} \models$ "$\sigma(\bar{b}_n, \bar{c}) = \sigma(\bar{b}_{n-1}, \bar{c})\sigma_*(\bar{a}_n)$".

Hence by clauses (A)(d)$(\alpha), (\beta)$:

- $\bullet_{2.8}$ $\sigma_*(\bar{a}_n) \in G_n$ iff $\sigma_*(\bar{a}_n) = e_{G_n}$ iff $\sigma_*(\bar{a}_{n+1}) = e_{G_n}$ iff $\sigma_*(\bar{a}_{n+1}) \in G_{n+1}$;

- $\bullet_{2.9}$ if $\sigma_*(\bar{a}_n) \notin G_n$, hence $\sigma_*(\bar{a}_{n+1}) \notin G_{n+1}$, then both statements in $(*)_2$ fail because:

(α) $\sigma(\bar{b}_n, \bar{c})$ is from G_{n+1} and $\sigma_*(\bar{a}_{n+1}) \notin G_{n+1}$ so $\sigma(\bar{b}_{n+1}, \bar{c}) \notin G_{n+1}$ and thus $\sigma(\bar{b}_{n+1}, \bar{c}) \neq e_{G_n}$;

(β) similarly $\sigma(\bar{b}_n, \bar{c}) \notin G_n$ and thus $\sigma(\bar{b}_n, \bar{c}) \neq e_{G_n}$;

•2.10 if $\sigma_*(\bar{a}_n) \in G_n$ hence $\sigma_*(\bar{a}_n) = e = \sigma_*(\bar{a}_{n+1})$, then $\sigma(\bar{b}_{n+1}, \bar{c}) = \sigma(\bar{b}_n, \bar{c})$ and again we are done.

Together $(*)_2$ holds.

2) Similarly (and the same as part (1) when G_n is existentially closed for every n) but we elaborate. Without loss of generality I is finite; letting $p_n(\bar{y}) = \mathrm{tp}_{\mathrm{bs}}(\bar{a}_n, G_n)$, we need:

$(*)_1$ if $\bar{c}$ is a finite sequence from G_ω <u>then</u> the sequence $\langle \mathrm{tp}_{\mathrm{bs}}(\bar{b}_{n!}{}^{\frown}\bar{c}, \emptyset, G_{n!+1}) : n < \omega \rangle$ is eventually constant.

Let $K_n = c\ell(\bar{a}_n, G_{n+1})$, so by clause $(A)(d)(\alpha)$ of the assumption $|K_n|$ is constant, finite and $K_n \cap G_n$ is $\subseteq$-increasing with n. Hence for some $K_*, n(*)$ we have $n \geq n(*) \Rightarrow K_n \cap G_n = K_*$ and let $k(*) = |K_*|$. Without loss of generality $n(*) \geq k(*)$; so it is enough to prove

$(*)_2$ if $\bar{y} = \langle y_t : t \in I \rangle$ and $\sigma(\bar{y}, \bar{z})$ is a group term, $\bar{c} \in {}^{\ell g(\bar{z})}(G_n)$ and $n \geq n(*)$, <u>then</u> $G_{n+1} \models$ "$\sigma(\bar{b}_n, \bar{c}) = e$" <u>iff</u> $G_{n+k(*)+1} \models \sigma(\bar{b}_{n+k(*)}, \bar{c}) = e$".

As in part (1) we can prove that for some group term $\sigma_*(\bar{y})$ we have

⊞ if $n \geq n(*)$, then $G_{n+2} \models \sigma(\bar{b}_{n+1}, \bar{c}) = \sigma(\bar{b}_n, \bar{c})\sigma_*(\bar{a}_{n+1})$.

<u>Case 1</u>: "$\sigma_*(\bar{a}_n) \notin G_n$ for some, equivalently every, $n \geq n(*)$.
  In this case $G_{n+1} \models$ "$\sigma(\bar{b}_n, \bar{c}) \neq e$" for every $n \geq n(*)$.

<u>Case 2</u>: "$\sigma_*(\bar{a}_n) \in G_n$ for some, equivalently every, $n \geq n(*)$.
  In this case there is b such that $\sigma_*(\bar{a}_n, \bar{c}) = b$ for every $n \geq n(*)$. So for every $n \geq n(*)$ by induction on m we can prove $\sigma(\bar{b}_{n+m}, \bar{c}) = \sigma(\bar{b}_n, \bar{c}) \cdot b^m$. But necessarily $b \in K_*$ hence b has order dividing $|K_*| = k(*)$. Hence $n \geq n(*) \Rightarrow \sigma(\bar{b}_{n+k(*)}, \bar{c}) = \sigma(\bar{b}_n, \bar{c})$ and thus $n_2 > n_1 \geq n(*) \wedge k(*)|(n_2 - n_1) \Rightarrow \sigma(\bar{b}_{n_2}, \bar{c}) = \sigma(\bar{b}_{n_1}, \bar{c})$, and so we can finish easily. □$_{2.19}$

2.20 Definition/Claim. 1) For $k = 2, 3\ldots$ let $\mathfrak{s}_{\mathrm{ab}(k)}$ be the unique $\mathfrak{s} \in \Omega[\mathbf{K}_{\mathrm{lf}}]$ such that:

(a) $n(\mathfrak{s}) = 1, k(\mathfrak{s}) = 0$;

(b) if $G \subseteq H$ and $c \in H$ realizes $q_{\mathfrak{s}}(<>, G) = \mathrm{tp}_{\mathrm{bs}}(c, G, H)$ <u>then</u> c commutes with every $a \in G$;

(c) also for every $m < \omega, a^m = e_H$ iff $a^m \in G$ iff $k|m$.

2) Assume $K \in \mathbf{K}_{\mathrm{lf}}$ is finite and $\bar{c} \in {}^{|K|}K$ list it. <u>Then</u> let $\mathfrak{s} = \mathfrak{s}_{\mathrm{ab}}(\bar{c}, K)$ be the unique $\mathfrak{s} \in \Omega[\mathbf{K}_{\mathrm{lf}}]$ such that:

(a) $n(\mathfrak{s}) = \ell g(\bar{c}), k(\mathfrak{s}) = 0$ so $p_{\mathfrak{s}}(\bar{x}_{\mathfrak{s}}) = \emptyset$;

(b) if $G \subseteq H \in \mathbf{K}_{\mathrm{lf}}$ and $\bar{c}' \in {}^{n(\mathfrak{s})}H$ <u>then</u> the following are equivalent:

(α) $\mathrm{tp}(\bar{c}', G, H) = q_{\mathfrak{s}}(<>, G)$,

(β) $\bar{c}'$ commutes with G, realizes $\mathrm{tp}(\bar{c}, \emptyset, K)$ and $\langle \bar{c}' \rangle_H \cap G = \{e\}$.

Claim 2.21. Assume $\mathrm{NF}_f(G_0, G_1, G_2, G_3)$ and $a \in G_1 \backslash G_0, b \in G_2 \backslash G_0$. <u>Then</u> a, b commute in G_3 iff $a \in \mathbf{C}_{G_1}(G_0), b \in \mathbf{C}_{G_2}(G_0)$ and G_0 is commutative.

Remark 2.22. 1) NF_f is from Definition 2.9.
2) Recall $g^{[a]} = a^{-1}ga$.

Proof. First assume

$$\oplus \quad a \in \mathbf{N}_{G_1}(G_0) \text{ and } b \in \mathbf{N}_{G_2}(G_0).$$

Without loss of generality G_1, G_2 are existentially closed (by monotonicity of NF_f, see 2.10(3) and existence of existentially closed extensions). By 2.15, without loss of generality we can find $\mathbf{x} \in \mathbf{X}_\mathbf{K}$ such that $G_3 = G_\mathbf{x}, G_{\mathbf{x}, \ell} = G_\ell$ for $\ell < 3$ and let $f_a = \mathbf{j}_{\mathbf{x}, 1}(a), f_b = \mathbf{j}_{\mathbf{x}, 2}(b)$; we shall use the fact that we have some freedom in the choice of $\mathbf{x}$, see 2.15.

Let $(g_0, g_1, g_2) \in \mathscr{U}_\mathbf{x}$ and we should see whether $f_b \circ f_a((g_0, g_1, g_2)) = f_a \circ f_b((g_0, g_1, g_2))$; there are unique a', h_a, b', h_b such that:

$(*)_0$ (a) $g_1 a = a' h_a$ with $h_a \subset C_0, a' \in \mathbf{I}_{\mathbf{x}, 1}$;

(b) $g_2 b = b' h_b$ with $h_b \in G_0, b' \in \mathbf{I}_{\mathbf{x}, 2}$.

Now

$(*)_1$ $f_a((g_0, g_1, g_2)) = (h_a g_0^{[a]}, a', g_2)$.

[Why? As $g_1 g_0 a = g_1 a g_0^{[a]} = a'(h_a g_0^{[a]})$, noting that $g_0^{[a]} \in G_0$ because we are assuming that a normalize G_0 inside G_1.]

$(*)_2$ $f_b((h_a g_0^{[a]}, a', g_2)) = (h_b h_a^{[b]} g_0^{[a][b]}, a', b')$.

[Why? As $g_2(h_a g_0^{[a]})b = g_2 b(h_a^{[b]} g_0^{[a][b]}) = b'(h_b h_a^{[b]} g_0^{[a][b]})$.]
So,

$(*)_3$ $(f_b \circ f_a)((g_0, g_1, g_2)) = (h_b h_a^{[b]} g_0^{[a][b]}, a', b')$.

Now,

$(*)_4 \quad f_b((g_0, g_1, g_2)) = (h_b g_0^{[b]}, g_1, b')$.

[Why? As $g_2 g_0 b = g_2 b g_0^{[b]} = b' h_b g_0^{[b]}$).]

$(*)_5 \quad f_a((h_b g_0^{[b]}, g_1, b')) = (h_a h_b^{[a]} g_0^{[b][a]}, a', b')$.

[Why? As $g_1(h_b g_0^{[b]})a = g_1 a(h_b^{[a]} g_0^{[b][a]}) = a'(h_a h_b^{[a]} g_0^{[b][a]})$.]
    Hence,

$(*)_6 \quad (f_a \circ f_b)((g_0, g_1, g_2)) = (h_a h_b^{[a]} g_0^{[b][a]}, a', b')$.

Together we can deduce:

$(*)_7 \quad (f_b \circ f_a)(g_0, g_1, g_2) = (f_a \circ f_b)(g_0, g_1, g_2)$ iff $h_b h_a^{[b]} g_0^{[a][b]} = h_a h_b^{[a]} g_0^{[b][a]}$ in G_0.

Now, <u>not</u> assuming $\oplus$ we shall prove the claim by cases (using $(*)_7$ when $\oplus$ holds).

$\oplus_1$ a, b commute in G_3 <u>when</u>:

- $\bullet_1$ a commutes with G_0 in G_1,

- $\bullet_2$ b commutes with G_0 in G_2,

- $\bullet_3$ G_0 is commutative.

[Why? Note that the assumption $\oplus$ holds (by $\bullet_1 + \bullet_2$), and so let $\mathbf{x} \in \mathbf{X_K}$ be as above. For any $(g_0, g_1, g_2) \in \mathscr{U}_\mathbf{x}$, we can apply $(*)_7$ thus $h_a, h_b \in G_0$ are well defined, by $(*)_0$. Now as $h_b, h_a, g_0 \in G_0$ and $a \in \mathbf{C}_{G_1}(G_0), b \in \mathbf{C}_{G_1}(G_0)$ and G_0 is commutative, by the present assumptions, clearly $h_b h_a^{[b]} g_0^{[a][b]} = h_b h_a g_0 = h_a h_b g_0 = h_a h_b^{[a]} g_0^{[b][a]}$. As $G_3 = G_\mathbf{x}, G_\mathbf{x}$ is a group of permutations of $\mathscr{U}_\mathbf{x}$ and $(*)_7$ holds for any $(g_0, g_1, g_3) \in \mathscr{U}_\mathbf{x}$, clearly $f_a, f_b \in G_\mathbf{x}$ commute, so we are done.]

$\oplus_2$ a, b do not commute in G_3 <u>when</u>:

- a commutes with G_0,

- b commutes with G_0,

- G_0 is not commutative.

[Why? Choose $h_1, h_2 \in G_0$ which do not commute and let $(g_0, g_1, g_2) = (e_{G_0}, e_{G_1}, e_{G_2}) = (e, e, e)$; note that $ah_\ell^{-1} \notin G_0, bh_\ell^{-1} \notin G_0$ for $\ell = 1, 2$.

Above we could have chosen $\mathbf{x} \in \mathbf{X_K}$ such that $(G_{\mathbf{x}, \ell} = G_\ell$ for $\ell < 3$ and) $ah_1^{-1} \in \mathbf{I_{x,1}}, bh_2^{-1} \in \mathbf{I_{x,2}}$. Again $\oplus$ holds hence $(*)_7$ holds for any relevant $\mathbf{x}, g_0, g_1, g_2$. Recall $(g_0, g_1, g_2) := (e, e, e)$, so $g_1 a = ea = a = (ah_1^{-1})h_1$. So in $(*)_0(a)$, we get $a' = ah_1^{-1}$ and $h_a = h_1$. Similarly in $(*)_0(b)$ we get $b' = bh_2^{-1}, h_b = h_2$. So $f_a, f_b \in G_{\mathbf{x}}$ do not commute by $(*)_7$, because we get $h_b h_a^{[b]} g_0^{[a][b]} = h_b h_a g_0 = h_2 h_1 g_0 \neq h_1 h_2 g_0 = h_a h_b g_0 = h_a h_b^{[a]} g_0^{[b][a]}$, the inequality as $G_0 \models h_1 h_2 \neq h_2 h_1$ and we are done by $(*)_7$.]

$\oplus_3$ a, b does not commute in G_3 <u>when</u>:

- a normalizes G_0 in G_1,
- b normalizes G_0 in G_2,
- b does not commute with G_0 in G_2.

[Why? Again $\oplus$ holds hence we can apply $(*)_7$ for any relevant $\mathbf{x}, g_0, g_1, g_2$. Let $h_1 \in G_0$ be such that it does not commute with b in G_2 and let $h_2 = e_{G_0}$. Choose above $\mathbf{x} \in \mathbf{X_K}$ such that $ah_1^{-1} \in \mathbf{I_{x,1}}$ and $b = bh_2^{-1} \in \mathbf{I_{x,2}}$ and let $(g_0, g_1, g_2) = (e, e, e)$. Again in $(*)_0$ we get $a' = ah_1^{-1}, h_a = h_1$ and $b' = bh_2^{-1}, h_b = h_2 = e$. Now $h_b h_a^{[b]} g_0^{[a][b]} = e h_a^{[b]} e = h_a^{[b]} \neq h_a = h_a e e = h_a h_b^{[a]} g_0^{[b][a]}$, the inequality by the choice of $h_a = h_1$.]

$\oplus_4$ a, b do not commute in G_3 <u>when</u>:

- a normalizes G_0 in G_1,
- b normalizes G_0 in G_2,
- a does not commute with G_0 in G_2.

[Why? Like $\oplus_3$.]
    Next

$\oplus_5$ a, b does not commute in G_3 <u>when</u>:

- $a \in G_1 \backslash G_0$ does not normalize G_0.

Choose $h \in G_0$ such that $a^{-1}ha \notin G_0$ hence $ha \notin aG_0$ and, of course, $ha \notin G_0$ as $a \notin G_0, h \in G_0$ and similarly $bh^{-1} \in G_2 \backslash G_0$. Let $a' = ha$ so $a' \neq a$ because $h \neq e$.

Choose above $\mathbf{x} \in \mathbf{X_{K_{1f}}}$ such that $bh^{-1} \in \mathbf{I_{x,2}}$ and $a, a' \in \mathbf{I_{x,1}}$. Why can we choose such $\mathbf{I_{x,1}}$? Because $a' = ha \in G_1 \backslash G_0, a \in G_1 \backslash G_0$ and $aG_0 \neq a'G_0$, as otherwise for some $h_1 \in G_0$ we have $a' = ah_1$, and so $a^{-1}ha = a^{-1}a' = h_1 \in G_0$, contradicting the choice of h.

Let f_a, f_b be as above for this choice of $\mathbf{x}$.
Now consider $(e, e, e) \in \mathscr{U}_{\mathbf{x}}$ so

$(*)'_1$ $f_a((e, e, e)) = (e, a, e).$

[Why? As $a \in \mathbf{I}_{\mathbf{x},1}.$]

$(*)'_2$ $f_b((e, a, e)) = (h, a, bh^{-1}).$

[Why? Because $bh^{-1} \in \mathbf{I}_{\mathbf{x},2}, h \in G_0.$]

$(*)'_3$ $(f_b \circ f_a)(e, e, e) = (h, a, bh^{-1}).$

[Why? By $(*)'_1 + (*)'_2.$]

$(*)'_4$ $f_b((e, e, e)) = (h, e, bh^{-1}).$

[Why? Because $bh^{-1} \in \mathbf{I}_{\mathbf{x},2}$ and $h \in G_0.$]

$(*)'_5$ $f_a((h, e, bh^{-1})) = (e, a', bh^{-1}).$

[Why? As $eha = ha = a' = a'e$ and $a' \in \mathbf{I}_{\mathbf{x},1}.$]

$(*)'_6$ $(f_a \circ f_b)((e, e, e)) = (e, a', bh^{-1})).$

[Why? By $(*)'_4 + (*)'_5.$]
    By $(*)'_3 + (*)'_6$, as $a' \neq a$ the triple (e, e, e) exemplifies $\mathbf{j}_{\mathbf{x},1}(a), \mathbf{j}_{\mathbf{x},2}(b)$ do not commute in $G_{\mathbf{x}}$.
    Lastly,

$\oplus_6$ a, b do not commute in G_3 <u>when</u>:

  • $b \in G_2 \backslash G_0$ does not normalize $G_0.$

[Why? As in $\oplus_5.$]
    As we have covered all the cases we are done. $\square_{2.21}$

Claim 2.23. Assume $\mathfrak{S} \subseteq \Omega[\mathbf{K}_{\mathrm{lf}}]$ and $G_1 \leq_{\mathfrak{S}} G_2, G_1$ is existentially closed and $d \in G_2$. If conjugation by d (in G_2) maps G_1 onto itself <u>then</u> for some $c \in G_1$ we have $a \in G_1 \Rightarrow c^{-1}ac = d^{-1}ad$, i.e., $dc^{-1}a = adc^{-1}$, i.e., dc^{-1}, a commute in G_1.

Proof. Easy. Clearly there is $(\mathfrak{s}, \bar{a}) \in \mathrm{def}(G_1)$ such that $\mathrm{tp}_{\mathrm{bs}}(d, G_1, G_2) = q_{\mathfrak{s}}(\bar{a}, G_1))$, hence, if $b, b_1, c_1 \in G_1$ and $\mathrm{tp}_{\mathrm{bs}}(\langle b_1, c_1 \rangle, \bar{a}, G_1) = \mathrm{tp}_{\mathrm{bs}}(\langle b, d^{-1}bd \rangle, \bar{a}, G_1)$, then $d^{-1}b_1d = c_1$. Having disjoint amalgamation we have $x \in G_1 \Rightarrow d^{-1}xd \in c\ell(\bar{a}^\frown \langle x \rangle, G_1)$. We can continue or note that if there is no such $c \in G_1$, then every existentially closed G has a non-inner automorphism, contradiction. $\square_{2.23}$

3 Symmetryzing

Our intention is to start with $\mathfrak{S} \subseteq \Omega[\mathbf{K}]$ which may contain $\mathfrak{s}_1, \mathfrak{s}_2$ failing symmetry but have the nice conclusion as for symmetric $\mathfrak{S}$. Towards this we define the operation $\otimes$, related to $\oplus$ defined in Definition 1.6(4),(4A), and $\mathfrak{S} - \otimes$-construction (close but not the same as the constructions in Definition 1.12, 1.19, 1.24) and $\mathfrak{S} - \oplus$-constructions.

Note that $\mathfrak{S}_{\mathrm{atdf}}$ has "quasi symmetry", i.e., when the parameter (base of amalgamation) is the same, but when we allow increasing the base this is not clear. Now $\otimes$ is like $\oplus$ when we insist on it being symmetric. We use the construction here in §4,§5 where we sometimes give more details. Recall def(G) for $G \in \mathbf{K}$ is from Definition 1.1. Recall

Definition 3.1. For $t \in \mathrm{def}(G)$ let $q_t(G) = q_{\mathfrak{s}_t}(\bar{a}_t, G)$ and $n_t = n_{\mathfrak{s}_t}, k_t = k_{\mathfrak{s}_\ell}$ and see Definition 1.1(6).

Definition 3.2. 1) On def(G) we define a (partial) operation $\otimes$ by $t_1 \otimes t_2 = (\mathfrak{s}_{t_1} \otimes \mathfrak{s}_{t_2}, \bar{a}_{t_1} \hat{\ } \bar{a}_{t_2})$, see below.
2) $\mathfrak{s} = \mathfrak{s}_1 \otimes \mathfrak{s}_2$ means that $\mathfrak{s}_1, \mathfrak{s}_2$ are disjoint[11], $\bar{x}_{\mathfrak{s}} = \bar{x}_{\mathfrak{s}_1} \hat{\ } \bar{x}_{\mathfrak{s}_2}, \bar{z}_{\mathfrak{s}} = \bar{z}_{\mathfrak{s}_1} \hat{\ } \bar{z}_{\mathfrak{s}_2}$, so $k(\mathfrak{s}) = k(\mathfrak{s}_1) + k(\mathfrak{s}_n), n(\mathfrak{s}) = n(\mathfrak{s}_1) + n(\mathfrak{s}_2)$ and:

⊞ if $H \subseteq H^+ \in \mathbf{K}, \bar{a}_\ell \in {}^{k(\mathfrak{s}_\ell)}H$ realizes $p_{\mathfrak{s}_\ell}(\bar{x}_{\mathfrak{s}_\ell})$ in H and $\bar{c}_\ell \in {}^{n(\mathfrak{s}_\ell)}(H^+)$ for $\ell = 1, 2$, then $\bar{c}_1 \hat{\ } \bar{c}_2$ realizes $q_{\mathfrak{s}}(\bar{a}_1 \hat{\ } \bar{a}_2, H)$ __iff__:

    (a) $\bar{c}_\ell$ realizes $q_{\mathfrak{s}_\ell}(\bar{a}_\ell, H)$ in H^+, for $\ell = 1, 2$;

    (b) if $\sigma(\bar{z}_1, \bar{z}_2, \bar{y})$ is a group-term, $lg(\bar{z}_1) = n(\mathfrak{s}_1), lg(\bar{z}_2) = n(\mathfrak{s}_2)$ and $\bar{b} \in {}^{lg(\bar{y})}(H)$, then $(\alpha) \Leftrightarrow (\beta)$ where:

        (α) $H^+ \models$ "$\sigma(\bar{c}_1, \bar{c}_2, \bar{b}) = e_H$",

        (β) $(\sigma(\bar{z}_1, \bar{c}_2, \bar{b}) = e) \in q_{\mathfrak{s}_1}(\bar{a}_1, H^+)$ and
             $(\sigma(\bar{c}_1, \bar{z}_2, \bar{b}) = e) \in q_{\mathfrak{s}_2}(\bar{a}_2, H^+)$.

Claim 3.3. 1) If $\mathfrak{s}_1, \mathfrak{s}_2 \in \Omega[\mathbf{K}]$ __then__ $\mathfrak{s} = \mathfrak{s}_1 \otimes \mathfrak{s}_2$ belongs to $\Omega[\mathbf{K}]$.
2) If $G \in \mathbf{K}$ and $t_1, t_2 \in \mathrm{def}(G)$ __then__ $t = t_1 \otimes t_2 \in \mathrm{def}(G)$.

Proof. Straightforward. $\qquad\qquad\qquad\qquad\qquad\qquad\qquad\qquad$ □$_{3.3}$

Definition 3.4. 1) Let $\approx_G^*$ be the following two-place relation on def(G) : $(\mathfrak{s}_1, \bar{a}_1) \approx_G^* (\mathfrak{s}_2, \bar{a}_2)$ if both are in def(G) and $G \subseteq G^+ \in \mathbf{K} \Rightarrow q_{\mathfrak{s}_1}(\bar{a}_1, G^+) = q_{\mathfrak{s}_2}(\bar{a}_2, G^+)$, (compare with $\approx_G$ from 1.1(6)).
2) For $t_1, t_2 \in \mathrm{def}(G)$ let $t_1 \leq t_2$ means dom$(\bar{x}_{t_1}) \subseteq \mathrm{dom}(\bar{x}_{t_2}), \mathrm{dom}(\bar{z}_{t_1}) \subseteq \mathrm{dom}(z_{\bar{t}_2})$ and $\bar{a}_{t_1} = \bar{a}_{t_2}\lceil\mathrm{dom}(\bar{x}_{t_1})$, and if $G \subseteq G_1 \subseteq G_2$ and $\bar{c}_2$ realizes $q_{t_2}(G_1)$ in G_2, then $\bar{c}_2\lceil\mathrm{dom}(\bar{z}_{t_1})$ realizes $q_{t_1}(G)$ in G_2.
3) $t_1 \leq_{\bar{h}} t_2$ is defined similarly as in 1.6(7).

---

[11] As we use only invariant $\mathfrak{S}$, this is not a real restriction.

Claim 3.5. 0) $\approx_G^*$ is an equivalence relation on $\text{def}(G)$.

1) If $(\mathfrak{s}, \bar{a}) \in \text{def}(G_1)$ and $G_1 \subseteq G_2 \in \mathbf{K}$ <u>then</u> $q_{\mathfrak{s}}(\bar{a}, G_1) \subseteq q_{\mathfrak{s}}(\bar{a}, G_2)$ and $(\mathfrak{s}, \bar{a}) \in \text{def}(G_2)$.

2) If $G \in \mathbf{K}$ and $(\mathfrak{s}_\ell, \bar{a}_\ell) \in \text{def}(G)$ for $\ell = 1, 2$, <u>then</u> the satisfaction of $(\mathfrak{s}_1, \bar{a}_1) \approx_G^* (\mathfrak{s}_2, \bar{a}_2)$ depends just on $\mathfrak{s}_1, \mathfrak{s}_2$ and $\text{tp}_{\text{bs}}(\bar{a}_1 \hat{\,} \bar{a}_2, \emptyset, G)$.

3) Transitivity: in Definition 3.4(2), $\leq$ is indeed a partial order.

4) Moreover, if $(\mathfrak{s}_1, \bar{a}_1) \leq_{\bar{h}_1} (\mathfrak{s}_2, \bar{a}_2) \leq_{\bar{h}_2} (\mathfrak{s}_3, \bar{a}_3)$ <u>then</u> $(\mathfrak{s}_1, \bar{a}_1) \leq_{\bar{h}_2 \circ \bar{h}_1} (\mathfrak{s}_3, \bar{a}_3)$.

Proof. Easy. $\square_{3.5}$

Claim 3.6. 0) The operation $\otimes$ on disjoint pairs respects congruency (see Definition 1.1(3), Claim 1.9(1)).

1) If G is existentially closed, <u>then</u> the operation $\otimes$ respects $\approx_G^*$, i.e., if $t_1 \approx_G^* t_1'$ and $t_2 \approx_G^* t_2'$ <u>then</u> $t_1 \otimes t_2 \approx_G^* t_1' \otimes t_2'$ assuming the operations are well defined, of course.

2) If $(\mathfrak{s}, \bar{a}) = (\mathfrak{s}_1, \bar{a}_1) \otimes (\mathfrak{s}_2, \bar{a}_2)$, <u>then</u> $(\bar{\mathfrak{s}}_\ell, \bar{a}_\ell) \leq (\mathfrak{s}, \bar{a})$.

3) If in $\text{def}(G)$ we have $t_\ell \leq t_\ell'$ for $\ell = 1, 2$ and $t_1' \otimes t_2'$ is well defined (i.e., t_1', t_2' are disjoint) <u>then</u> $t_1 \otimes t_2 \leq t_1' \otimes t_2'$.

4) The operation $\otimes$ is associative and is symmetric, e.g., symmetry means: if $G \subseteq G^+$ and $(\mathfrak{s}_\ell, \bar{a}_\ell) \in \text{def}(G)$ and $\bar{c}_\ell^\ell \hat{\,} \bar{c}_{3-\ell}^\ell$ realizes $q_{t_\ell}(G)$ in G^+, where $t_\ell = (t_\ell, \bar{b}_\ell) = (\mathfrak{s}_\ell, \bar{a}_\ell) \otimes (\mathfrak{s}_{3-\ell}, \bar{a}_{3-\ell})$, (so assuming disjointness for transparency), for $\ell = 1, 2$, <u>then</u> $\text{tp}_{\text{bs}}(\bar{c}_1^1 \hat{\,} \bar{c}_2^1, G, G^+) = \text{tp}_{\text{bs}}(\bar{c}_1^2 \hat{\,} \bar{c}_2^2, G, G^+)$.

5) If in $\text{def}(G)$ we have $(\mathfrak{s}_\ell, \bar{a}_\ell) \leq_{h_\ell} (\mathfrak{s}_\ell', \bar{a}_\ell')$ for $\ell = 1, 2$ and $\text{Dom}(h_1) \cap \text{Dom}(h_2) = \emptyset, \text{Rang}(h_1) \cap \text{Rang}(h_2) = \emptyset$ <u>then</u> $(\mathfrak{s}_1, \bar{a}_1) \otimes (\mathfrak{s}_2, \bar{a}_2) \leq_{h_1 \cup h_2} (\mathfrak{s}_1', \bar{a}_1) \otimes (\mathfrak{s}_2', \bar{a}_2)$.

Proof. Straightforward. $\square_{3.6}$

Remark 3.7. 1) Also the operation $\oplus$ satisfies the parallels of 3.6(1),(2),(3) and the first demand in (4).

2) We may phrase 3.6(5) as in 3.6(3) and vice versa.

Definition 3.8. Assume $\mathfrak{S} \subseteq \Omega[\mathbf{K}]$ is closed.

1) We say $\mathfrak{S} \subseteq \Omega[\mathbf{K}]$ is $\otimes$-closed <u>when</u> (recalling it is invariant) if $\mathfrak{s}_\ell \in \mathfrak{S}$ for $\ell = 1, 2$ are disjoint <u>then</u> $\mathfrak{s} = \mathfrak{s}_1 \otimes \mathfrak{s}_2 \in \mathfrak{S}$.

2) The $\otimes$-closure of $\mathfrak{S}$ is the $\subseteq$-minimal $\otimes$-closed $\mathfrak{S}' \subseteq \Omega[\mathbf{K}]$ such that $\mathfrak{S} \subseteq \mathfrak{S}'$.

3) Let $G_3 = G_1 \overset{\mathfrak{S}}{\underset{G_0}{\bigotimes}} G_2$ or $G_3 = \otimes_{\mathfrak{S}}(G_0, G_1, G_2)$ mean:

$(*)$ (a) $G_0 \leq_{\mathfrak{S}} G_2 \subseteq G_3 \in \mathbf{K}$ and $G_0 \leq_{\mathfrak{S}} G_1 \subseteq G_3$ and $G_3 = \langle G_1 \cup G_2 \rangle_{G_3}$

(b) if $\text{tp}_{\text{bs}}(\bar{c}_\ell, G_0, G_\ell) = q_{\mathfrak{s}_\ell}(\bar{a}_\ell, G_0)$ so $\bar{c}_\ell \in {}^{\omega>}(G_\ell), \bar{a}_\ell \in {}^{\omega>}(G_0)$ for $\ell = 1, 2$, <u>then</u> $\text{tp}_{\text{bs}}(\bar{c}_1 \hat{\,} \bar{c}_2, G_0, G_3) = q_{\mathfrak{s}}(\bar{a}_1 \hat{\,} \bar{a}_2, G_0)$ when $(\mathfrak{s}, \bar{a}_1 \hat{\,} \bar{a}_2) = (\mathfrak{s}, \bar{a}_1) \otimes (\mathfrak{s}_2, \bar{a}_2)$; note that without loss of generality $\mathfrak{s}_1, \mathfrak{s}_2$ are disjoint (i.e., as in the proof of 1.10).

4) $\text{NF}_{\mathfrak{S}}^2(G_0, G_1, G_2, G_3)$ means that $G_0 \leq_{\mathfrak{S}} G_\ell \leq_{\mathfrak{S}} G_3$ for $\ell = 1, 2$ and the demands in (3) hold except that possibly $G_3 \neq \langle G_1 \cup G_2 \rangle_{G_3}$.

Claim 3.9. Assume $\mathfrak{S}$ is closed and moreover, $\otimes$-closed.
1) $G_3 = \otimes_{\mathfrak{S}}(G_0, G_1, G_2)$ iff $\mathrm{NF}^2_{\mathfrak{S}}(G_0, G_1, G_2, G_3)$ and $G_3 = \langle G_1 \cup G_2 \rangle_{G_3}$.
2) (disjointness): $\mathrm{NF}^2_{\mathfrak{S}}(G_0, G_1, G_2, G_3)$ implies $G_1 \cap G_2 = G_0$.
3) (uniqueness): If $G_3^\iota = \otimes_{\mathfrak{S}}(G_0^\iota, G_1^\iota, G_2^\iota)$ for $\iota = 1, 2$ and f_ℓ is an isomorphism from G_ℓ^1 onto G_ℓ^2 for $\ell = 1, 2$ and $G_0^1 = G_0^2$, $f_1 \upharpoonright G_0^1 = f_2 \upharpoonright G_0^2$ and G_0 is existentially closed[12] then there is one and only one isomorphism from G_3^1 onto G_3^2 extending $f_1 \cup f_2$ (which is well defined by (2)).
4) (symmetry): $\mathrm{NF}^2_{\mathfrak{S}}(G_0, G_1, G_2, G_3)$ iff $\mathrm{NF}^2_{\mathfrak{S}}(G_0, G_2, G_1, G_3)$.
5) (monotonicity): If $\mathrm{NF}^2_{\mathfrak{S}}(G_0, G_1, G_2, G_3)$ and $G_0 \subseteq G_\ell' \subseteq G_\ell$ for $\ell = 1, 2$ then $\mathrm{NF}^2_{\mathfrak{S}}(G_0, G_1', G_2', G_3)$.
6) (existence): If $G_0 \leq_{\mathfrak{S}} G_\ell$ for $\ell = 1, 2$ and G_0 is existentially closed and $G_1 \cap G_2 = G_0$ then for some $G_3 \in \mathbf{K}$ we have $\mathrm{NF}^2_{\mathfrak{S}}(G_0, G_1, G_2, G_3)$.

Remark 3.10. For parts (3) and (6) of 3.9 recall: for such G, if $t_1, t_2 \in \mathrm{def}(G), q_{t_1}(G) = q_{t_2}(G)$ and $G \subseteq G^+ \in \mathbf{K}$, then $q_{t_1}(G^+) = q_{t_2}(G^+)$.

Proof. Straightforward, e.g., for disjointness (part (2)) use Claim 1.2(4).
$\square_{3.9}$

An alternative to §1 from 1.12 on is: we repeat it with changes being that we use $\otimes$ instead of $\oplus$ and we incorporated the λ-fullness, also in 3.12(3) we choose another version. We have not sorted out whether we can generalize 1.16(5) based on 1.15 and 1.23(2).

Definition 3.11. 1) We say that $\mathscr{A}$ is a one-step $(\lambda, \mathfrak{S}) - \otimes$-construction when $\mathscr{A} = (G, H, \langle \bar{c}_\alpha, t_\alpha : \alpha < \alpha(\mathscr{A}) = \alpha_{\mathscr{A}} \rangle)$ satisfies:

(a) $G \subseteq H \in \mathbf{K}$;

(b) $t_\alpha \in \mathrm{def}_{\mathfrak{S}}(G)$ for $\alpha < \alpha(\mathscr{A})$;

(c) if $\alpha_0, \ldots, \alpha_{n-1} < \alpha(\mathscr{A})$ with no repetitions, then $\bar{c}_{\alpha_0} \hat{\ } \ldots \hat{\ } \bar{c}_{\alpha_{n-1}}$ realizes $t_{\alpha_0} \otimes \ldots \otimes t_{\alpha-1}$ over G in H;

(d) $H = \langle \cup \{\bar{c}_\alpha : \alpha < \alpha(\mathscr{A})\} \cup G \rangle_H$;

(e) $\langle t_\alpha : \alpha < \alpha(\mathscr{A}) \rangle$ lists $\mathrm{def}_{\mathfrak{S}}(G)$ each appearing exactly λ times.

---

[12]Why? The problem is that $G \leq_{\mathfrak{S}} H \in \mathbf{K}$ does not imply the existence of $\bar{t} = \langle t_{\bar{c}} : \bar{c} \in {}^{\omega >}H \rangle$ such that $t_{\bar{c}} \in \mathrm{def}(G), \mathrm{tp}_{\mathrm{bs}}(\bar{c}, G, H) = q_t(G)$ and if $\bar{c}^1, \bar{c}^2 \in {}^{\omega >}H, h : \ell g(\bar{c}^1) \to \ell g(\bar{c}^2)$ and $\bar{c}^2 = \langle c_{h(i)}^2 : i < \ell g(\bar{c}^1) \rangle$, then $t_{\bar{c}^1} \leq_h t_{\bar{c}^2}$. Moreover, even if there is such $\bar{t}$ we can "amalgamate for it" but this is not enough as $\bar{t}$ is not necessarily unique, which may give different results. Why is 3.9(3) O.K.? As in Definition 3.8(3), we ask "for every $\mathfrak{s}_1, \mathfrak{s}_2$". In other words if $G_0 \subseteq G_1, G_0 \subseteq G_2$ and $t_1, t_2 \in \mathrm{def}(G_0), \mathrm{tp}_{\mathrm{bs}}(\bar{c}_\ell, G_0, G_2) = q_{t_\ell}(G_0)$ for $\ell = 1, 2$ but $q_{t_1}(G_1) \neq q_{t_2}(G_1)$ we can amalgamate as in 3.8(3).

2) In (1) we may use any index set instead of $\alpha(\mathscr{A})$, e.g., $\text{def}_{\mathfrak{S}}(G)$ itself when $\lambda = 1, \text{def}_{\mathfrak{S}}(G) \times \lambda$ in general.

3) We say $\mathscr{A}$ is an $\alpha(\mathscr{A})$-$\underline{\text{step}}$-$(\lambda, \mathfrak{S}) - \otimes$-construction or $(\alpha(\mathscr{A}), \lambda, \mathfrak{S}) - \otimes$-construction $\underline{\text{when}}$

(a) $\mathscr{A} = \langle G_\alpha, \langle \bar{c}_{\beta,s}, t_{\beta,s} : s \in S_\beta \rangle : \alpha \leq \alpha(\mathscr{A}), \beta < \alpha(\mathscr{A}) \rangle$

(b) $(G_\alpha : \alpha \leq \alpha(\mathscr{A}))$ is increasing continuous (in $\mathbf{K}$)

(c) $(G_\alpha, G_{\alpha+1}, \langle \bar{c}_{\alpha,s}, t_{\alpha,s} : s \in S_\alpha \rangle)$ is a one-step $(\lambda, \mathfrak{S}) - \otimes$-construction.

4) In part (3), let $G_\alpha^{\mathscr{A}} = G_\alpha[\mathscr{A}]$ be G_α, etc., and in part (1) let $G^{\mathscr{A}} = G[\mathscr{A}]$ be G, etc.

5) In part (3) if $\alpha(\mathscr{A}) = \omega$, then we may omit it; also for every $\alpha < \alpha(\mathscr{A})$ the sequence $(G_\alpha^{\mathscr{A}}, G_{\alpha+1}^{\mathscr{A}}, \langle \bar{c}_{\alpha,s}, t_{\alpha,s} : s \in S_\alpha^{\mathscr{A}} \rangle)$ is called the α-th step of $\mathscr{A}$.

Definition 3.12. 1) We say H is a λ-full $\underline{\text{one-step}}$ $\mathfrak{S} - \otimes$-closure of G $\underline{\text{when}}$ there is a one-step $(\lambda, \mathfrak{S}) - \otimes$-construction $\mathscr{A}$ such that $G[\mathscr{A}] = G, H[\mathscr{A}] = H$. We may say H is λ-full one-step $\mathfrak{S} - \otimes$-constructible over G; similarly in part (2).

2) We say H is λ-full α-step $\mathfrak{S}$-closure over G or H is $(\alpha, \lambda, \mathfrak{S})$-closure of G $\underline{\text{when}}$ there is a $(\alpha, \lambda, \mathfrak{S}) - \otimes$-construction $\mathscr{A}$ with $G = G_0^{\mathscr{A}}, H = G_{\ell g(\mathscr{A})}^{\mathscr{A}}$.

3) We say G_* is $(\delta, \lambda, \mathfrak{S}) - \otimes$-full over G $\underline{\text{when}}$ for some $\bar{G} = \langle G_i : i \leq \delta \rangle$ increasing continuous sequence in $\mathbf{K}$, $G_0 = G, G_\delta = G_*$ and G_{i+1} is $(1, \lambda, \mathfrak{S}) - \otimes$-full over G_i which means some $G' \subseteq G_{i+1}$ is a one-step $(\lambda, \mathfrak{S}) - \otimes$-construction over G_i. If $\delta = \omega$ one may omit it writing $(\lambda, \mathfrak{S})$ instead of $(\delta, \lambda, \mathfrak{S})$.

4) We may in part (3) replace $\otimes$ by $\oplus$.

Claim 3.13. Assume $\mathfrak{S} \subseteq \Omega[\mathbf{K}]$ is $\otimes$-closed, α an ordinal, λ a cardinal.

1) If $G \in \mathbf{K}$ $\underline{\text{then}}$ there is a one-step $(\lambda, \mathfrak{S}) - \otimes$-construction $\mathscr{A}$ over G (i.e., $G_0^{\mathscr{A}} = G$) of cardinality $\leq \lambda + |G| + |\mathfrak{S}|$ and $\geq \lambda$.

2) If in part (1), $\mathscr{A}_1, \mathscr{A}_2$ are one-step-$(\lambda, \mathfrak{S}) - \otimes$-constructions over G $\underline{\text{then}}$ $H[\mathscr{A}_1], H[\mathscr{A}_2]$ are isomorphic over G.

3) For any $G \in \mathbf{K}$ there is an $(\alpha, \lambda, \mathfrak{S}) - \otimes$-construction $\mathscr{A}$ over G and $G_\alpha[\mathscr{A}]$ is unique up to isomorphism over G.

4) If $\mathfrak{S}$ is dense, H is an $(\alpha, \lambda, \mathfrak{S}) - \otimes$-closure of G and α is a limit ordinal $\underline{\text{then}}$ H is existentially closed and is $(\alpha, \lambda, \mathfrak{S})$-full over G.

Proof. Straightforward, as in 1.23(3). $\square_{3.13}$

3.14 Discussion. Essentially we know that if "$G_1 \subseteq G_2$" implies the $(\alpha, \lambda, \mathfrak{S})$-closure of G_1 is a subgroup of the $(\alpha, \lambda, \mathfrak{S})$-closure of G_2.

But we have a delicate problem: what if the $(\alpha, \lambda, \mathfrak{S})$-closure of G_1 is not disjoint to $G_2 \backslash G_1$?

We have similar problems with "the algebraic closure of a field" or "the field of quotients of a field", but there, if $G_1 \subseteq G_2$, then the closure G_1^+

of G_1 inside G_1 is definable (from G_2, G_1 and G_2^+). Here this is not true, but clearly this is not a serious problem. Ways to circumvent this appear in 0.12(2), 1.13(2) and below.

Claim 3.15. 1) We can choose $\hat{G} \in \mathbf{K}_{\text{exlf}}$ such that $\hat{G}$ extends $G \in \mathbf{K}_{\text{lf}}, G_1 \cong G_2 \Rightarrow \hat{G}_1 \cong \hat{G}_2$ and every embedding $f : G_1 \to G_2 \in \mathbf{K}_{\text{lf}}$ can be extended to $\hat{f} : \hat{G}_1 \to \hat{G}_2$ canonically.
1A) Moreover, $G_1 \subseteq G_2 \Rightarrow \hat{G}_1 \subseteq \hat{G}_2$ but pedantically see (2).
2) There is a set theoretic class function $\mathbf{F}$, that computes from $G \in \mathbf{K}, \alpha \in$ Ord, $\lambda \in$ Card, $\gamma \in$ Ord and $\mathfrak{S} \subseteq \Omega[\mathbf{K}]$ a group $H = \mathbf{F}(G, \alpha, \mathfrak{S}, \gamma)$ such that:

(a) $\mathbf{F}(G, \alpha, \mathfrak{S}, \gamma) \in \mathbf{K}$ extends G, moreover;

(b) $\mathbf{F}(G, \alpha, \gamma, \mathfrak{S})$ is an $(\alpha, \lambda, \mathfrak{S})$-closure of G;

(c) [uniqueness]: if $G_1, G_2 \in \mathbf{K}$ and g is an isomorphism from G_1 onto G_2 and $H_\ell = \mathbf{F}(G, \alpha, \mathfrak{S}, \gamma)$ for $\ell = 1, 2$ <u>then</u> there is an isomorphism g from H_1 onto H_2 extending g;

(d) we have $H_1 \subseteq H_2$ and $G_1 = H_1 \cap G_2$ <u>when</u> $G_1 \subseteq G_1 \in \mathbf{K}, \gamma > \alpha$ and[13] $\gamma > \sup(\text{Ord} \cap \text{tr} - cl(G_\ell))$ for $\ell = 1, 2$ and $H_\ell = \mathbf{F}(G_\ell, \alpha, \mathfrak{S}, \gamma)$;

(e) if we restrict ourselves to $G \in \mathbf{K}' = \{G \in \mathbf{K}:$ if $x \in G$, then x is a singleton$\}$ <u>then</u> $G_1 \subseteq G_2 \Rightarrow \mathbf{F}(G, \alpha, \mathfrak{S}) \subseteq \mathbf{F}(G, \alpha, S, 0)$.

*   *   *

In §4,§5 we intend to use also some relative of those constructions, including:

Definition 3.16. Assume $\bar{H} = \langle H_i : i < \delta \rangle$ is $\subseteq$-increasing in $\mathbf{K}$ and $H_\delta = \cup\{H_i : i < \delta\}$, (we shall use $\delta = \omega$). We say $\mathscr{A}$ is a one-step atomic $\mathfrak{S} - \otimes$-construction above $\bar{H}$, <u>when</u> (and we may say H is weakly atomically $\mathfrak{S} - \otimes$-constructible over $\bar{H}$, omitting $\bar{H}$ means for some $\bar{H}$ of length ω and we may replace $\alpha_{\mathscr{A}} = \alpha(\mathscr{A})$ by any index set) $\mathscr{A}$ has the following objects satisfying the following additional conditions:

(A) $(\bar{H}, H_\delta, H, \langle \bar{c}_\alpha, t_{\alpha,i}, \alpha < \alpha_{\mathscr{A}}, i < \delta \rangle)$;

(B) $H_\delta \subseteq H \in \mathbf{K}$;

(C) $t_{\alpha,i} \in \text{def}_{\mathfrak{S}}(H_i)$;

(D) $H = \langle \cup\{\bar{c}_\alpha : \alpha < \alpha_{\mathscr{A}}\} \cup H_\delta \rangle_H$;

---

[13]Recalling tr-cl is the (set-theoretic) transitive closure.

(E) $\bar{c}_\alpha$ realizes $q_{t_{\alpha,i}}(H_i)$ in H for $\alpha < \alpha_\mathscr{A}, i < \delta$;

(F) $\bar{c}_{\alpha,i} \subseteq H_{i+1}$ realizes $q_{t_{\alpha,i}}(H_i)$ for $i < \delta, \alpha < \alpha_\mathscr{A}$ and moreover;

(F)$^+$ assuming $\alpha(0) < \ldots < \alpha(n-1) < \alpha_\mathscr{A}$ and $\ell g(\bar{x}_\alpha) = \ell g(\bar{c}_\alpha)$ and
$\varphi = \varphi(\bar{x}_{\alpha(0)}, \ldots, x_{\alpha(n-1)}, \bar{y})$ we have[14]
$\varphi(\bar{x}_{\alpha(0)}, \ldots, \bar{x}_{\alpha(n-1)}, \bar{b}) \in \mathrm{tp}_{\mathrm{at}}(\bar{c}_{\alpha(0)}\,\hat{}\,\ldots\,\hat{}\,\bar{c}_{\alpha(n-1)}, G_\delta, H)$
iff $\bar{b} \subseteq {}^{\ell g(\bar{y})}G_\delta$ and for every permutation π of n,
$(\forall^\infty i(0) < \delta)(\forall^\infty i(1) < \delta), \ldots, (\forall^\infty i(n-1) < \delta)$
$\varphi[\bar{c}_{\alpha(0),i(\pi(0))}, \bar{c}_{\alpha(1),i(\pi(1))}, \ldots, \bar{c}_{d(n-1),\pi(n-1)}, \bar{b}]$
(used in the proof of $(*)_{5.2}$ stage C in the proof of 5.1); note that φ is not necessarily atomic.

Remark 3.17. 1) We may consider replacing clause $(F)^+$ by:

(F)* $\bar{c}_{\alpha(0)}\,\hat{}\,\ldots\,\hat{}\,\bar{c}_{\alpha(n-1)}$ realizes $q_{t_{\alpha(0)}\otimes\ldots\otimes t_{\alpha(n-1)}}$ for $\alpha(0) < \ldots < \alpha(n-1) < \alpha(\mathscr{A})$.

2) In this alternative version we do not need the existence of $\bar{c}_{\alpha,i} \subseteq H_{i+1}$, so it is easier to prove existence but the version above is the one we actually use. In particular the version in (1) would create problems in $(*)_{5.7}$ in the proof of 5.1; we may try to take care of this by changing the definition of L_β^* there.
3) A sufficient condition for having the assumptions of 3.16 appear in 2.19.

Observation 3.18. *Let* $\mathfrak{S}$ *be closed and* $\otimes$*-closed. Assume* $\langle G_i : i \le \alpha \rangle$ *is* $\subseteq$*-increasing continuous in* **K**.
1) In 3.11(1) we can prove $G^\mathscr{A} \le_\mathfrak{S} H^\mathscr{A}$ *and in 3.11(2), we can prove* $\langle G_\alpha^\mathscr{A} : \alpha \le \alpha_\mathscr{A} \rangle$ *is* $\le_\mathfrak{S}$*-increasing continuous.*
2) In 3.16, if $\bar{H}$ *is* $\le_\mathfrak{S}$*-increasing* then *we have* $i < \delta \Rightarrow H_i \subseteq_\mathfrak{S} H$.
3) Assume S *is a set of limit ordinals* $< \delta, \langle G_i : i \le \delta \rangle$ *is a* $\subseteq$*-increasing continuous sequence of members of* **K** *and* G_{i+1} *is a one-step* $\mathfrak{S} - \otimes$*-constructible over* G_i *for* $i \in \delta \setminus S$ *and* G_{i+1} *is weakly one-step* $\mathfrak{S}-\otimes$*-constructible over* $\bar{G}\!\upharpoonright\! C_i$ *for some unbounded* $C_i \subseteq i \setminus S$ *for each* $i \in S$, *(hence* i *is a limit ordinal). Then* $i < j \le \delta \wedge i \notin S \Rightarrow G_i \le_\mathfrak{S} G_j$.

Remark 3.19. The idea of $\mathfrak{s}_1 \otimes \mathfrak{s}_2$ can be applied to one $\mathfrak{s}$ (and is used in the end of the proof of $\boxplus_1$ in stage B of the proof of Theorem 5.1).

Toward this, in §4(B) we shall deal with finding such amalgamations and $\mathfrak{s}$'s.

3.20 Definition/Claim. Assume $\mathfrak{s} \in \Omega[\mathbf{K}_{1f}]$ and $H_1 \subseteq H_2 \in \mathbf{K}$ are finite, $\bar{a} \in {}^{k(\mathfrak{s})}(H_1), \bar{c} \in {}^{n(\mathfrak{s})}(H_2)$ and $\bar{a}, \bar{c}$ generate H_1, H_2 respectively, and $\bar{a}$ realizes $p_\mathfrak{s}(\bar{x}_\mathfrak{s})$ in H_1 and $\bar{c}$ realizes $q_\mathfrak{s}(\bar{a}, H_1)$ in H_2. Assume further K is a group of automorphisms of H_2 mapping H_1 onto itself. Then there is a one and only one t such that:

---

[14]Yes! $\mathrm{tp}_{\mathrm{at}}$ and not $\mathrm{tp}_{\mathrm{bs}}$.

(a) $\mathfrak{t} \in \Omega[\mathbf{K}_{\mathrm{lf}}]$

(b) $k(\mathfrak{t}) = k(\mathfrak{s})$ and $p_{\mathfrak{t}}(\bar{x}_{\mathfrak{t}}) = \mathrm{tp}_{\mathrm{qf}}(\bar{a}, \emptyset, H_1)$

(c) if $H_1 \subseteq G_1 \subseteq G_2, H_2 \subseteq G_2$ and $\bar{c}$ realizes $q_{\mathfrak{s}}(\bar{a}, G_1)$ in G_2 and $\bar{c}' \in {}^n(G_2)$ realizes $q_{\mathfrak{t}}(\bar{a}, G_2)$ <u>then</u> $\mathrm{tp}_{\mathrm{at}}(\bar{c}', G_1, G_2) = \cap\{\mathrm{tp}_{\mathrm{at}}(\pi(\bar{c}), G_1, G_2) : \pi \in K\}$.

Remark 3.21. Toward this, in §(4B) we deal with finding such amalgamations and $\mathfrak{s}$'s.

Proof. Straightforward. $\qquad\qquad\qquad\qquad\qquad\qquad\qquad\qquad$ $\square_{3.20}$

4 For fixing a distinguished subgroup

In the construction of complete members of $\mathbf{K}_{\mathrm{exlf}}$ (and related aims) we fix large enough $\mathfrak{S} \subseteq \Omega[\mathbf{K}]$ and build a $\subseteq$-increasing continuous sequence $\langle G_\alpha : \alpha < \lambda \rangle, |G_\alpha| < \lambda$; normally we demand for $\alpha < \beta < \lambda$ that "usually" $G_\alpha \leq_{\mathfrak{S}} G_\beta$ (i.e., except for $\delta \in S$, where $S \subseteq S_{\aleph_0}^\lambda$). But at some moment for $\alpha = \delta + n$, we like to use $p = \mathrm{tp}_{\mathrm{bs}}(c, G_\alpha, G_{\alpha+1})$ which extends some $r \in \mathbf{S}_{\mathrm{bs}}(K), K \subseteq G_\alpha$ finite but such that c commutes with G_δ. Toward this in §(4A) we deal with a relative NF^3 of NF_f, in which we demand $\mathbf{C}_{G_1}(G_3)$ is large, this continues §2 concentrating on the case G_0 is with trivial center. In §(4B) we use this to define some schemes from $\Omega[\mathbf{K}]$; see e.g., 4.10.

Another problem is that given G_1 instead of extending G_1 to G_2 such that $q_t(G_1)$ is realized by $\bar{c} \in {}^{\omega>}(G_2)$ for some $t \in \mathrm{def}_{\mathfrak{S}}(G_1)$, we like to have an infinite $\bar{c} = (\ldots {}^\frown \bar{c_i} {}^\frown \ldots)_{i \in I}$, with $\mathrm{tp}(\bar{c} \restriction u, G_1, G_2) \in q_{t_u}(G_1)$ for every finite $u \subseteq I$; used in stage D of the proof of Theorem 5.1. This is done in §4(C).

4(A)   Preserving Commutation

Claim 4.1. The subgroups H_1', H_2' of G_3 commute <u>when</u>:

($*$) (a) $\mathbf{x} \in \mathbf{X_K}$;

(b) $G_\ell = G_{\mathbf{x},\ell}, G_\ell' = \mathbf{j}_{\mathbf{x},\ell}(G_\ell)$ for $\ell = 0, 1, 2$;

(c) $G_3 = G_{\mathbf{x}}$;

(d) $H_1 \subseteq G_1$ and $H_1' = \mathbf{j}_{\mathbf{x},1}(H_\ell)$;

(e) $H_1 = \cup\{b(H_1 \cap G_0) : b \in \mathbf{I}_1\}$ where $\mathbf{I}_1 = \mathbf{I}_{\mathbf{x},1} \cap H_1$;

(f) if $g \in \mathbf{I}_{\mathbf{x},1}$ and[15] $b \in \mathbf{I}_1$, then $gb \in \mathbf{I}_{\mathbf{x},1}$;

(g) the subgroups G_0, H_1 of G_1 commute;

(h) $H_2 \subseteq G_2$ commutes with $G_0 \cap H_1$ and $H_2' = \mathbf{j}_{\mathbf{x},2}(H_2)$;

(i) $H_2 = \cup\{a(G_0 \cap H_2) : a \in \mathbf{I}_2\}$ where $\mathbf{I}_2 = \mathbf{I}_{\mathbf{x},2} \cap H_2$.

Remark 4.2. 1) Really here it suffices to deal with the case $G_0 \cap H_1 = \{e\}$.
2) A natural case is $\mathbf{Z}(G_0) = \{e_{G_0}\}, H_1 = \mathbf{C}_{G_1}(G_0), H_2 = G_2$.
3) See the proof of 5.1.

Notation 4.3. Let $\mathbf{X}_{\mathrm{lf}}^3 = \mathbf{X}_{\mathbf{K}_{\mathrm{lf}}}^3$ be the class of tuple $(\mathbf{x}, H_1, H_2)$ which satisfies ($*$) of Claim 4.1.

---

[15] As G_1 is locally finite, necessarily $\mathbf{I}_1$ is a subgroup of H_1.

Proof. Let $a \in H_2, b \in H_1, f_a = \mathbf{j}_{\mathbf{x},2}(a), f_b = \mathbf{j}_{\mathbf{x},1}(b)$, so by $(*)(d),(h)$ we just have to prove that $f_b f_a((g_0, g_1, g_2)) = f_a f_b((g_0, g_1, g_2))$ for any $(g_0, g_1, g_2) \in \mathscr{U}_{\mathbf{x}}$.

Clearly

- if $a \in G_0$ or $b \in G_0$ this holds.

[Why? First, if $a \in G_0$, then $f_a = \mathbf{j}_{\mathbf{x},2}(a) = \mathbf{j}_{\mathbf{x},0}(a) = \mathbf{j}_{\mathbf{x},1}(a) \in \mathbf{j}_{\mathbf{x},1}(G_1) = G'_1 \subseteq G_{\mathbf{x}}$ and as $b \in H_1 \subseteq G_{\mathbf{x}}$, by $(*)(g)$ we have $G_1 \models$ "a, b commute" hence $G_{\mathbf{x}} \models$ "$\mathbf{j}_{\mathbf{x},2}(a), \mathbf{j}_{\mathbf{x},1}(b)$ commute" and so $G_{\mathbf{x}} \models$ "f_a, f_b commute". Second, if $b \in G_0$, then $b \in G_0 \cap H_1 \subseteq G_0 \subseteq G_2$ and $a \in H_2 \subseteq G_2$, so by clause $(*)(h)$ clearly $G_2 \models$ "a, b commute" and we finish as above.]

Moreover, as $H_1 = \langle (G_0 \cap H_1) \cup \mathbf{I}_1 \rangle_{G_1}$ by clause $(*)(e)$, without loss of generality

$\boxplus_1$ $b \in \mathbf{I}_1 \subseteq \mathbf{I}_{\mathbf{x},1}$.

By clause $(*)(i)$, without loss of generality:

$\boxplus_2$ $a \in \mathbf{I}_2 \subseteq \mathbf{I}_{\mathbf{x},2}$.

Let[16] $f_{\mathbf{x}}((g_0, g_1, g_2)) = (g_0^x, g_1^x, g_2^x)$ and $f_y f_{\mathbf{x}}((g_0, g_1, g_2)) = (g_0^{x,y}, g_1^{x,y}, g_2^{x,y})$ for $x \in \{a, b\}$ and $y \in \{a, b\} \setminus \{x\}$.

We shall prove that $g_\ell^{a,b} = g_\ell^{b,a}$ for $\ell = 0, 1, 2$; this suffices.

Clearly,

- $\bullet_1$ $g_1^a = g_1$ and $g_2 g_0 a = g_2^a g_0^a$;

- $\bullet_2$ $g_2^{a,b} = g_2^a$ and $g_1^a g_0^a b = g_1^{a,b} g_0^{a,b}$;

- $\bullet_3$ $g_2^b = g_2$ and $g_1 g_0 b = g_1^b g_0^b$;

- $\bullet_4$ $g_1^{b,a} = g_1^b$ and $g_2^b g_0^b a = g_2^{b,a} g_0^{b,a}$.

Now

$\boxplus_3$ $g_1^{a,b} G_0 = g_1^{a,b} g_0^{a,b} G_0 = g_1^a g_0^a b G_0 = (g_1^a b)(g_0^a G_0) = (g_1^a b) G_0$.

[Why? As $g_0^{a,b} \in G_0$, by the second statement of $\bullet_2$, noting that b, g_0^a commute by $(*)(g)$, and as $g_0^a \in G_0$, respectively.]

But $g_1^a \in \mathbf{I}_{\mathbf{x},1}$ (as $(g_0^a, g_1^a, g_2^a) \in \mathscr{U}_{\mathbf{x}}$), and $b \in \mathbf{I}_1 \subseteq \mathbf{I}_{\mathbf{x},1}$ by $\boxplus_1$, hence by $(*)(f)$ we have $g_1^a b \in \mathbf{I}_{\mathbf{x},1}$ and also $g_1^{a,b} \in \mathbf{I}_{\mathbf{x},1}$ (as $(g_0^{a,b}, g_1^{a,b}, g_2^{a,b}) \in \mathscr{U}_{\mathbf{x}}$). So by $\boxplus_3$, $g_1^{a,b} G_0 = (g_1^a b) G_0$ and by the last sentence $g_1^{a,b}, g_1^a \in \mathbf{I}_{\mathbf{x},1}$ and thus

---

[16] Note that g_ℓ^x is <u>not</u> conjugation by x.

$\bullet_5$ $g_1^{a,b} = g_1^a b$.

So by $\bullet_5$ and the second equation in $\bullet_2$ we have $g_1^a b g_0^{a,b} = g_1^{a,b} g_0^{ab} = g_1^a g_0^a b = g_1^a b g_0^a$, the last equality as recalling b, g_0^a commute by $(*)(g)$, hence we have:

$\bullet_6$ $g_0^{a,b} = g_0^a$.

Similarly to $\boxplus_3$ we have

$\boxplus_4$ $g_1^b G_0 = g_1^b g_0^b G_0 = g_1 g_0 b G_0 = (g_1 b)(g_0 G_0) = (g_1 b) G_0$.

[Why? As $g_0^b \in G_0$, by $\bullet_3$ second statement, as b, g_0 commute by $(*)(g)$, and as $g_0 \in G_0$, respectively.]

Also $g_1 \in \mathbf{I}_{\mathbf{x},1}$ as $(g_0, g_1, g_2) \in \mathscr{U}_{\mathbf{x}}$ and $b \in \mathbf{I}_1$ by $\boxplus_1$ so recalling $(*)(f)$ we deduce $g_1, g_1 b \in \mathbf{I}_{\mathbf{x},1}$ thus from $\boxplus_4$ we deduce:

$\bullet_7$ $g_1^b = g_1 b$.

Hence by $\bullet_7$ and $\bullet_3$ second statement we have $g_1 b g_0^b = g_1^b g_0^b = g_1 g_0 b = g_1 b g_0$, the last equation recalling b, g_0 commute (by $(*)(g)$), hence we have:

$\bullet_8$ $g_0^b = g_0$.

So by $\bullet_4, \bullet_7, \bullet_1, \bullet_6$, b commuting with G_0 and $\bullet_2$ second statement respectively, we have

$\boxplus_5$ $g_1^{b,a} = g_1^b = (g_1 b) = (g_1^a b) = (g_1^a b)(g_0^a(g_0^{a,b})^{-1}) = (g_1^a g_0^a b)(g_0^{a,b})^{-1} = g_1^{a,b}$,

and thus

$\bullet_9$ $g_1^{b,a} = g_1^{a,b}$.

Also by $\bullet_4, \bullet_3, \bullet_8, \bullet_1, \bullet_6, \bullet_2$ we have

$\boxplus_6$ $g_2^{b,a} g_0^{b,a} = g_2^b g_0^b a = g_2 g_0^b a = g_2 g_0 a = g_2^a g_0^a = g_2^a g_0^{a,b} = g_2^{a,b} g_0^{a,b}$.

So

$\bullet_{10}$ $g_2^{b,a} g_0^{b,a} = g_2^{a,b} g_0^{a,b}$

but $g_0^{b,a}, g_0^{a,b} \in G_0$ and $g_2^{b,a}, g_2^{a,b} \in \mathbf{I}_{\mathbf{x},2}$ hence

$\bullet_{11}$ $g_2^{b,a} = g_2^{a,b}$ and $g_0^{b,a} = g_0^{a,b}$.

But $\bullet_{11} + \bullet_9$ imply that we are done. $\qquad\qquad\qquad\qquad\qquad\square_{4.1}$

The following claim is like Definition 2.5, but now we preserve a large $\mathbf{C}_{G_1}(G_0)$ using 4.1.

Definition 4.4. Let $\mathrm{NF}^3(\bar{G}, H_1, L, H_2)$ mean:

(A) (a) $\bar{G} = \langle G_\ell : \ell \leq 3 \rangle$ are from $\mathbf{K}_{\mathrm{lf}}$;

    (b) $G_0 \subseteq G_\ell$ for $\ell = 1, 2$;

    (c) G_0 is finite;

    (d) $H_1 \subseteq \mathbf{C}_{G_1}(G_0), L \subseteq H_1, L \cap G_0 = \{e_{G_0}\}, H_1 = \langle L, G_0 \cap H_1 \rangle_{G_1}$;

    (e) $G_1 \cap G_2 = G_0$;

    (f) $H_2 \subseteq \mathbf{C}_{G_2}(H_1 \cap G_0)$;

(B) (a) $G_\ell \subseteq G_3$ for $\ell = 1, 2$;

    (b) for $\sigma(\bar{x}, \bar{y})$ a group-term, $\bar{a} \in {}^{\ell g(\bar{x})}(G_1)$ and $\bar{b} \in {}^{\ell g(\bar{y})}(G_2)$ the following conditions are equivalent:

        • $G_3 \models \text{``}\sigma(\bar{a}, \bar{b}) = e_{G_3}\text{''}$,

        • if $(\mathbf{x}, H_1, H_2) \in \mathbf{X}^3_{\mathrm{lf}}, G_\ell = G_{\mathbf{x},\ell}$ for $\ell = 0, 1, 2$ and $\bar{a}' = \mathbf{j}_{\mathbf{x},1}(\bar{a})$ and[17] $\bar{b}' = \mathbf{j}_{\mathbf{x},2}(\bar{b})$ <u>then</u> $G_{\mathbf{x}} \models \text{``}\sigma(\bar{a}', \bar{b}') = e_{G_{\mathbf{x}}}\text{''}$.

Convention 4.5. In 4.4, if $H_1 = L$ we may in addition omit L. We may omit L, H_2 if $L = H_1, H_2 = \mathbf{C}_{G_2}(H_1 \cap G_0)$. Lastly, if $\mathbf{Z}(G_0) = \{e_{G_0}\}, L = \mathbf{C}_{G_1}(G_0)$ and $H_1 = L$ and $H_2 = G_2$, then we may omit H_1, L and H_2; see 4.6(3).

Claim 4.6. Assume $\bar{G} = \langle G_\ell : \ell < 3 \rangle, H_1, L, H_2$ are as in 4.4(A).
1) We can find $\mathbf{x}$ such that $(\mathbf{x}, H_1, H_2) \in \mathbf{X}^3_{\mathrm{lf}}$.
2) There is $G_3 \in \mathbf{K}$ such that $\mathrm{NF}^3(\langle G_0, G_1, G_2, G_3 \rangle, H_1, L, H_2)$ and G_3 is unique up to isomorphism over $G_1 \cup G_2$.
3) If $\bar{G}$ satisfies (A)(a),(b),(c) of Definition 4.4, $\mathbf{Z}(G_0) = \{e_{G_0}\}, H_1 = L = \mathbf{C}_{G_1}(G_0)$ and $H_2 = G_2$, <u>then</u> $(\bar{G}, H_1, L, H_2)$ satisfies 4.4(A).

Proof. 1) It suffices to prove we can choose $\mathbf{I}^*_1, \mathbf{I}^*_2$ satisfying the demands on $\mathbf{I}_{\mathbf{x},1}, \mathbf{I}_{\mathbf{x},2}$ in 4.1.
Why can we do it? For $\mathbf{I}^*_2$ the demands are just clauses (b),(c) from 2.2(1) and $(*)(i)$ of 4.1 so just choose $\mathbf{I}_2 \subseteq H_2$ such that $e_{G_0} \in \mathbf{I}_2$ and $\langle g(G_0 \cap H_2) : g \in \mathbf{I}_2 \rangle$ is a partition of H_2 and then let $\mathbf{I}^*_2$ be such that $\mathbf{I}_2 \subseteq \mathbf{I}^*_2 \subseteq G_2$ and $\langle gG_0 : g \in \mathbf{I}^*_2 \rangle$ is a partition of G_2. Clearly $\mathbf{I}^*_1$ is as required.

For $\mathbf{I}^*_1$ we have to take care of clauses (b),(c) of 2.2(1), of $(*)(e)$ (the parallel of $(*)(i)$) and of $(*)(f)$ from 4.1. For this let $H^+_1 := \langle G_0, H_1 \rangle_{G_1}$. First, choose $\mathbf{I}'_1 = L$ so that $e_{G_0} \in \mathbf{I}'_1$ and thus $\langle gG_0 : g \in \mathbf{I}'_1 \rangle$ is a partition of H^+_1. Why?

---

[17]We may add $\mathbf{I}_1 = L$.

Recalling that $L \subseteq H_1 \subseteq G_1, L \cap G_0 = \{e_{G_0}\}$ and $H_1 = \langle L, G_0 \cap H_1 \rangle_{G_1}$ and H_1 commute with G_0 in G_1; by clause (A)(d) we know that this is satisfied. Also let $\mathbf{J}_1 \subseteq G_1$ be such that $e_{G_0} = e_{G_1} \in \mathbf{J}_1$ and $\langle gH_1^+ : g \in \mathbf{J}_1 \rangle$ is a partition of G_1. Now let $\mathbf{I}_1^* = \{gb : g \in \mathbf{J}_1$ and $b \in \mathbf{I}_1'\}$.

Clearly $\langle gG_0 : g \in \mathbf{I}_1^* \rangle = \langle g(bG_0) : b \in \mathbf{I}_1', g \in \mathbf{J}_1 \rangle$ is a partition of G_1 (refining $\langle gH_1^+ : g \in \mathbf{J}_1 \rangle$), so clause 2.2(1)(b) holds. Furthermore, $\mathbf{I}_1^* \cap H_2^+ = L = \mathbf{I}_1'$ so clause 4.1(e) holds.

Next as $e_{G_0} \in \mathbf{J}_1$ and $e_{G_0} \in \mathbf{I}_1'$ clearly $e_{G_0} \in \mathbf{I}_1^*$, so $\mathbf{I}_1^*$ satisfies clause 2.2(1)(c). Also if $g \in \mathbf{I}_1^* \wedge b \in \mathbf{I}_1'$, then for some $g_1 \in \mathbf{J}_1, b_1 \in \mathbf{I}_1'$ we have $G_1 \models$ "$g = g_1 b_1$" hence $G_1 \models$ "$gb = (g_1 b_1)b = g_1(b_1 b)$" and so $g_1 \in \mathbf{J}_1$ and $b_1 b \in \mathbf{I}_1'$ as $\mathbf{I}_1' = L$ is closed under products. Thus together $gb \in \mathbf{I}_1^*$, hence clause 4.1(1)(f) is satisfied. So $\mathbf{I}_1^*, \mathbf{I}_2^*$ are as required in 2.2(1) and 4.1. Hence there is $\mathbf{x} \in \mathbf{X_K}$ such that $G_{\mathbf{x},\ell} = G_\ell$ for $\ell = 0, 1, 2$ and $\mathbf{I}_{\mathbf{x},\ell} = \mathbf{I}_\ell^*$ for $\ell = 1, 2$.

2) Consider clause (B) of 4.4, the "if $\mathbf{x} \in \ldots$" is not empty so G_3 is a well defined group. Easily $G_1 \subseteq G_3$ and $G_2 \subseteq G_3$ but is G_3 locally finite? This follows from the results in §2, in particular 2.10. That is, if G_ℓ' is finite, $G_0 \subseteq G_\ell' \subseteq G_\ell$ for $\ell = 1, 2$ then we have finitely many possible choices of $(\mathbf{I}_{\mathbf{x},1} \cap x_1 G_1', \mathbf{I}_{\mathbf{x},2} \cap x_2 G_2')$ for $x_1 \in G_1, x_2 \in G_2$ hence the group G_3 that we get is locally finite. Or note that G_3 is a homomorphic image of the G_3 of 2.10(1).

3) Should be clear. $\qquad\qquad\qquad\qquad\qquad\qquad\qquad\qquad\qquad\qquad\qquad\square_{4.6}$

4(B) Schemes and derived sets

Definition 4.7. 1) Let $\mathbf{X}_0$ be the set of $\mathbf{x}$ such that:

(a) $\mathbf{x}$ has the form $(K_1, K_2, \bar{a}_2, \bar{a}_1) = (K_1[\mathbf{x}], K_2[\mathbf{x}], \bar{a}_2[\mathbf{x}], \bar{a}_1[\mathbf{x}])$;

(b) $K_1 \subseteq K_2$ are finite groups;

(c) $\bar{a}_1$ is a finite sequence generating K_1;

(d) $\bar{a}_2$ is a finite sequence from K_2 such that $\bar{a}_2{}^\frown \bar{a}_1$ generates K_2 (if $\bar{a}_2 = \langle a_2 \rangle$ we may write just a_2);

(e) K_1 has trivial center.

2) Let $\mathbf{X}_1$ be the set of $\mathbf{x}$ such that:

(a) $\mathbf{x} = (K, \bar{a}) = (K[\mathbf{x}], \bar{a}[\mathbf{x}])$;

(b) $K \in \mathbf{K}$ is finite;

(c) $\bar{a}$ is a finite sequence from K generating $K, \ell g(\bar{a}) \geq 1$; let $a_* = a_*[\mathbf{x}] = a_0$, the first element of $\bar{a}$.

3) Let $\mathbf{X}_2$ be the set of $\mathbf{x} \in \mathbf{X}_1$ such that:

(∗) K has trivial center.

4) Let $\mathbf{X}_3$ be the set of $\mathbf{x} \in \mathbf{X}_1$ such that:

(∗) if f is a nontrivial automorphism of K <u>then</u> for some conjugate b of $a_* = a_*[\mathbf{x}] = a_0[\mathbf{x}]$ we have $f(b) \notin \langle a_* \rangle_K$; equivalently, for some conjugate b of a_*, $\langle b \rangle_K \neq \langle a \rangle_K$.

Observation 4.8. *If $m \in \{2, 3, \ldots\}$, then for some $\mathbf{x} \in \mathbf{X}_3$ the element $a_*[\mathbf{x}] \in K[\mathbf{x}]$ has order m.*

Claim 4.9. *If $\mathbf{x} \in \mathbf{X}_0$, <u>then</u> there is one and only one $\mathfrak{s}$, call it $\mathfrak{s}_{\mathrm{cm}} = \mathfrak{s}_{\mathrm{cm}}[\mathbf{x}]$ such that:*

(a) $\mathfrak{s} \in \Omega[\mathbf{K}_{\mathrm{lf}}]$;

(b) $k_{\mathfrak{s}} = \ell g(\bar{a}_1[\mathbf{x}])$ and $n_{\mathfrak{s}} = \ell g(\bar{a}_2[\mathbf{x}])$;

(c) $p_{\mathfrak{s}}(\bar{x}_{\mathfrak{s}}) = \mathrm{tp}_{\mathrm{bs}}(\bar{a}_1[\mathbf{x}], \emptyset, K[\mathbf{x}])$;

(d) if $G_1 \subseteq G_3 \in \mathbf{K}_{\mathrm{lf}}$ and $\mathrm{tp}_{\mathrm{bs}}(\bar{a}, \emptyset, G_1) = \mathrm{tp}_{\mathrm{bs}}(\bar{a}_1[\mathbf{x}], \emptyset, K[\mathbf{x}])$ and $\bar{c}$ realizes $q_{\mathfrak{s}}(\bar{a}, G_1)$ in G_3 <u>then</u> $\mathrm{NF}^3(\langle \bar{a} \rangle_{G_1}, G_1, \langle \bar{a} \hat{\ } \bar{c} \rangle_{G_3}, G_3)$.

Proof. As in §2 using §(4A). Let $K_\ell = K_\ell[x]$ and let $G_0 = K_1$ and $G_1 \in \mathbf{K}$ be existentially closed, extend K_1 and be such that $K_2 \cap G_1 = K_1$. Let $L = \mathbf{C}_{G_0}(G_1)$, so as $G_0 = K_1$ has trivial center (by 4.7(1)(e)), so we have $L \cap G_0 = \{e_{G_0}\}$ and let $H_1 = c\ell(G_0 \cup L, G_1)$, $H_0 = \{e_{K_1}\}$ and let $H_2 = G_2 := K_2$. Now we apply Claim 4.6(3), so there is $\mathbf{x}$ such that $(\mathbf{x}, H_1, H_2) \in \mathbf{X}_{\mathrm{lf}}^3$, see Definition 4.4. By it, the type $\mathrm{tp}_{\mathrm{bs}}(\bar{a}_2[\mathbf{x}], G_2, G_{\mathbf{x}})$ does not split over $G_0 = K_1$. From this it is easy to define $\mathfrak{s}$ and to prove it is as required. $\square_{4.9}$

4.10 Definition/Claim. For $\mathbf{x} \in \mathbf{X}_1$ let $\mathfrak{s} = \mathfrak{s}_{\mathrm{ab}}[\mathbf{x}]$ be such that:

(a) $\mathfrak{s} \in \Omega[\mathbf{K}_{\mathrm{lf}}]$;

(b) $k_{\mathfrak{s}} = 0$;

(c) if $\bar{c}$ realizes $q_2(<>, G_1)$ in G_2 so $G_1 \subseteq G_2$ <u>then</u> $\bar{c}$ realizes $\mathrm{tp}_{\mathrm{bs}}(\bar{a}[\mathbf{x}], \emptyset, K[\mathbf{x}])$ and commutes with G_1, and $\langle \bar{c} \rangle_{G_2} \cap G_1 = \{e\}$.

Proof. Easy. $\square_{4.11}$

4.11 Definition/Claim. For $\mathbf{x} \in \mathbf{X}_2$ we define $\mathfrak{s} = \mathfrak{s}_{\mathrm{gm}}[\mathbf{x}]$ such that:

(a) $\mathfrak{s} \in \Omega[\mathbf{K}_{\mathrm{lf}}]$;

(b) $k_\mathfrak{s} = 2\ell g(\bar{a}[\mathbf{x}])$ and $n_\mathfrak{s} = 1$;

(c) if $G_1 \subseteq G_2 \in \mathbf{K}_{\mathrm{lf}}$ and $\mathrm{tp}_{\mathrm{bs}}(\bar{a}_\ell, \emptyset, G_1) = \mathrm{tp}_{\mathrm{bs}}(\bar{a}[\mathbf{x}], \emptyset, K[\mathbf{x}])$ for $\ell = 1, 2$ and $\langle \bar{a}_1 \rangle_{G_1}, \langle \bar{a}_2 \rangle_{G_1}$ commute in G_1 and[18] have intersection $\{e_G\}$, then $p_\mathfrak{s}(\bar{x}_\mathfrak{s}) = \mathrm{tp}_{\mathrm{bs}}(\bar{a}_1 \hat{\ } \bar{a}_2, \emptyset, G_1)$;

(d) moreover, in clause (c), if $c \in G_2$ realizes $q_\mathfrak{s}(\bar{a}_1 \hat{\ } \bar{a}_2, G_1)$ in G_2 <u>then</u> conjugation by c interchanges $\bar{a}_1, \bar{a}_2$ and is the identity on $\mathbf{C}_{G_1}(\bar{a}_1 \hat{\ } \bar{a}_2)$.

Proof. Let $G_2 \in \mathbf{K}_{\mathrm{exlf}}$ be an extension of $K[\mathbf{x}]$ in which some c realizes $q_{\mathfrak{s}_{\mathrm{cg}}}(K_\mathbf{x})$; let $\bar{a}_1 = \bar{a}[\mathbf{x}], \bar{a}_2 = c^{-1}\bar{a}_1 c := \langle c^{-1}a_{1,\ell}c : \ell < \ell g(\bar{a}_1) \rangle$ in G_2.

Note that, by inspection, $G_0 = \langle \bar{a}_1 \hat{\ } \bar{a}_2 \rangle_{G_1}$ is finite with trivial center and let $G_0 \subseteq G_1 \in \mathbf{K}_{\mathrm{lf}}$. Now use 4.1 with

$$G_0, G_1, c\ell(\bar{a}_1 \hat{\ } \bar{a}_2 \hat{\ } \langle c \rangle, G_2), \mathbf{C}_{G_1}(G_0), \mathbf{C}_{G_1}(G_0), c\ell(\bar{a}_1 \hat{\ } \bar{a}_2 \hat{\ } \langle c \rangle, G_2)$$

here standing for $G_0, G_1, G_2, G_1, H_1, L, H_2$ there. $\square_{4.10}$

Definition 4.12. 1) For $\mathfrak{s} \in \Omega[\mathbf{K}]$ and $G_1 \subseteq G_2$ let $\mathrm{cp}_\mathfrak{s}(G_1, G_2) = \{c_0 : \bar{c} \in {}^{n(\mathfrak{s})}(G_2)$ realizes $q_t(G_1)$ where $t \in \mathrm{def}(G_1)$ satisfies $\mathfrak{s}_t = \mathfrak{s}\}$.
2) For $\mathbf{x} \in \mathbf{X}_1$ and $G_1 \subseteq G_2$ let $\mathrm{cp}_\mathbf{x}(G_1, G_2) = \mathrm{cp}_{\mathfrak{s}_{ab}[\mathbf{x}]}(G_1, G_2)$.
3) For $G_1 \subseteq G_2 \in \mathbf{K}_{\mathrm{lf}}$ and $\ell \in \{1, 2, 3\}$ let $\mathrm{cp}_\ell(G_1, G_2) = \cup\{\mathrm{cp}_{\mathfrak{s}_{ab}[\mathbf{x}]}(G_1, G_2) : \mathbf{x} \in \mathbf{X}_\ell\}$; if $\ell = 2$ we may omit it.

4(C)   Larger Definable Types

Definition 4.13. 1) For $G \in \mathbf{K}, \mathfrak{S} \subseteq \Omega[\mathbf{K}]$ and set I let $\mathrm{Def}_{I, <\kappa}(G, \mathfrak{S})$ be the set of t such that:

(a) $t = \langle t_u : u \subseteq I$ finite $\rangle$;

(b) $t_u \in \mathrm{def}_\mathfrak{S}(G)$ with $\bar{x}_{t_u} = \langle x_i : i \in u \rangle$ and $\bar{a}_{t_u} = \bar{a}_t$ or pedantically $\bar{a}_{t_u} = \bar{a}_t \upharpoonright w_u$ where $w_u \subseteq \ell g(\bar{a}_t)$ is finite;

(c) $\ell g(\bar{a}_t) := I$ has cardinality $< \kappa$ and $\mathrm{Rang}(\bar{a}_t) \subseteq G$;

---

[18]In fact this follows.

(d) if $G \subseteq H \subseteq L \in \mathbf{K}_{\mathrm{lf}}$ and $u \subseteq v \subseteq I$ are finite and $\bar{b} \in {}^{v}L$ realizes $q_{t_v}(H)$, then $\bar{b} \restriction u$ realizes $q_{t_u}(H)$.

2) We define $\Omega_{I,<\kappa}[\mathbf{K}, \mathfrak{S}]$ parallely and if $\mathfrak{S} = \Omega[\mathbf{K}]$ <u>then</u> we may omit it.
3) If $t \in \mathrm{Def}_{I,<\kappa}(G, \mathfrak{S})$, then $q_t(G) \in \mathbf{S}^I_{\mathrm{bs}}(G)$ is defined by $\cup \{q_{t_u}(\langle x_i : i \in u \rangle) : u \subseteq I \text{ finite}\}$.
4) Omitting κ means $\aleph_0$. We may replace "$< \kappa^+$" by κ and even a set I_1. We may replace I by "$< \mu$" meaning "some $\chi < \mu$". Similarly for "$\leq \mu$".
5) For $n < \omega$ and $\mathfrak{s}_0, \ldots, \mathfrak{s}_{n-1} \in \Omega_{<\mu,<\kappa}[\mathbf{K}]$ we define $\mathfrak{s}_0 \oplus \ldots \oplus \mathfrak{s}_{n-1}$ and $\mathfrak{s}_0 \otimes \ldots \otimes \mathfrak{s}_{n-1}$ naturally.

Claim 4.14. 1) If $G \in \mathbf{K}, \mathfrak{S} \subseteq \Omega[\mathbf{K}]$ and $t \in \mathrm{Def}_I(G, \mathfrak{S})$ <u>then</u> for some pair $(\bar{c}, H)$ we have $G \subseteq H \in \mathbf{K}_{\mathrm{lf}}, \bar{c} \in {}^I H, H = \langle G \cup \bar{c} \rangle_H$ and $\mathrm{tp}_{\mathrm{bs}}(\bar{c}, G, H) = q_t(G)$.
2) If $\mathfrak{S}$ is closed <u>then</u> above $G \leq_{\mathfrak{S}} H$.

Definition 4.15. Assume $\bar{H} = \langle H_i : i < \delta \rangle$ is $\subseteq$-increasing in $\mathbf{K}$ and $H_\delta = \cup \{H_i : i < \delta\}$. We say $\mathscr{A}$ is a one-step $(< \mu, < \kappa, \delta, \mathfrak{S}) - \otimes$-construction (if $\delta = \omega$ we may omit it) <u>when</u>: as in 3.16 except that

(c)' $t_{\alpha,i} \in \mathrm{Def}_{I_{\alpha,i},<\kappa}(H_i, \mathfrak{S})$ for some set $I_{\alpha,i}$ of cardinality $< \mu$.

The case we shall actually use in §5 is:

Claim 4.16. Assume $K \subseteq L \in \mathbf{K}_{\mathrm{lf}}, K$ is finite and f embeds K into $G_1 \in \mathbf{K}_{\mathrm{lf}}$ and $\langle c_i : i < \mu \rangle$ list the members of L and $\{c_\ell : \ell < n\}$ is the set of elements of K. <u>Then</u> there is $t \in \mathrm{Def}_{\leq \mu}(G_1, \Omega[\mathbf{K}])$ such that: if $\bar{c}^* = \langle c_i^* : i < \mu \rangle \in {}^\mu(G_2)$ realizes $q_t(G_1)$ in G_2, so $G_1 \subseteq G_2$, <u>then</u> $c_i \mapsto c_i^*$ (for $i < \mu$) is an embedding of L into G_2 extending f.

Proof. Straightforward by §2. $\square_{4.16}$

4.17 Discussion. Those definable types are still locally definable over finite sets.

5   Constructing complete existentially closed G

Theorem 5.1. *Assume if $G \in \mathbf{K}_{\mathrm{lf}}$ and $|G| \leq \mu = \mu^{\aleph_0}$.*
1) There is a complete $G' \in \mathbf{K}_{\mathrm{lf}}$ which extend G such that $|G'| = \mu^+$ and G' is existentially closed.
2) Moreover, $G \leq_{\Omega[\mathbf{K}_{\mathrm{lf}}]} G'$ and G' is full.
3) There is G' such that $G \leq_{\mathfrak{S}} G'$ and $G' \in \mathbf{K}_{\mu^+}^{\mathrm{exlf}}$ is complete and $\mathfrak{S}$-full provided that $\mathfrak{S}$ satisfies:

$(*)$ (α) $\mathfrak{S} \subseteq \Omega[\mathbf{K}_{\mathrm{lf}}]$

  (β) $\mathfrak{S}$ *is dense and $\otimes$-closed (for $\mathbf{K}_{\mathrm{lf}}$)*

  (γ) *some schemes introduced earlier belong to $\mathfrak{S}$, specifically:*

  - $\mathfrak{s}_{\mathrm{ab}(2)}$ *from Definition 2.20, used in the paragraph before $\boxplus_3$*
  - $\mathfrak{s}_{\mathrm{cm}}$ *from Definition 4.9, used in $(*)_{4.3}$*
  - $\mathfrak{s}_{\mathrm{cg}}$, *from Definition 2.17(1), 2.18(2) used after $\boxplus_7$ Stage E*
  - $\mathfrak{s}_{\mathrm{gl}}$ *from Definition 2.17(2), 2.18(3)*
  - $\mathfrak{s}_{\mathrm{gm}}$ *from Definition 4.10, see $(*)_{5.1}(f)$.*

Proof. Proof of 5.1
  We let $\mathfrak{S} = \Omega[\mathbf{K}_{\mathrm{lf}}]$ for parts (1),(2) and fix $\mathfrak{S}$ for part (3) as there.

Stage A: Without loss of generality the universe of G is an ordinal $\leq \mu$ and let $\lambda = \mu^+$.
  Let $S \subseteq S_{\aleph_0}^\lambda := \{\delta < \lambda : \mathrm{cf}(\delta) = \aleph_0\}$ be a stationary subset of λ such that also $S_{\aleph_0}^\lambda \backslash S$ is stationary in λ and $\alpha \in S \Rightarrow (\mu$ divides $\alpha)$. Let $\langle S_\zeta : \zeta < \lambda \rangle$ be a partition of S to stationary sets. Let $S_* \subseteq \lambda \backslash S$ be stationary and a set of limit ordinals.
  Let C_δ be an unbounded subset of δ of order type ω for $\delta \in S$ such that $\bar{C}_\zeta = \langle C_\delta : \delta \in S_\zeta \rangle$ guess clubs for each $\zeta < \lambda$, this means that for every club E of λ the set $\{\delta \in S_\zeta : C_\delta \subseteq E\}$ is a stationary subset of λ; such $\langle C_\delta : \delta \in S_\zeta \rangle$ exists by [Sh:g, Ch.III] = [Sh:365].
  Let $\alpha_\delta(n)$ be the n-th member of C_δ.
  Let $\bar{\tau}$ be such that:

- $\bar{\tau} = \langle \tau_\zeta : \zeta < \lambda \rangle$

- $\tau_\zeta \subseteq \mathscr{H}(\aleph_0)$ is a countable vocabulary

- if $\tau \subseteq \mathscr{H}(\aleph_0)$ is a countable vocabulary then $\{\zeta : \tau_\zeta = \tau\}$ has cardinality λ.

By [Sh:309, 3.26(3)=L6.11A, pg.31] there is $\mathbf{b}_\zeta$, a BB, black box for $(S_\zeta, \bar{C}_\zeta)$ say $\mathbf{b}_\zeta = \langle N_i^\delta : i \in \mathscr{T}_\delta, \delta \in S_\zeta \rangle$, that is:

$\boxplus_{0,\zeta}$ (a) N_i^δ is a model of cardinality $\aleph_0$ with universe $\subseteq \delta = \sup(N_i^\delta)$ and vocabulary $\tau_\zeta \subseteq \mathscr{H}(\aleph_0)$

(b) if N is a τ_ζ-model with universe λ <u>then</u> for stationarily many $\delta \in E_N \cap S_\zeta$ for some $i \in \mathscr{T}_\delta$ we have $C_\delta \subseteq E_N \backslash S$ where $E_N := \{\alpha : N\restriction\alpha \prec N\}$ and $N_i^\delta \prec N$; moreover

$(b)^+$ if $\tau = \tau_\zeta, \bar{N} = \langle N_\eta : \eta \in \mathscr{T} \rangle, \mathscr{T}$ a nonempty subtree of $^{\omega >}\lambda$ such that $\tau(N) = \tau_\zeta, \eta \vartriangleleft \nu \Rightarrow N_\eta \prec N_\nu$ and $|N_\eta| \in [\lambda]^{\aleph_0}$ and E a club of $\lambda, \eta \in \mathscr{T} \Rightarrow (\exists^\lambda \alpha)(\eta\hat{\ }\langle\alpha\rangle \in \mathscr{T})$ and $\eta \vartriangleleft \nu \in \mathscr{T} \Rightarrow \sup(N_\eta) < \sup(N_\nu)$ <u>then</u> for some $\delta \in S_\zeta \cap E$ we have $C_\delta \subseteq E, i \in \mathscr{T}_\delta$ and $\eta \in \lim_\omega(\mathscr{T})$ we have $N_i^\delta = \cup\{N_{\eta\restriction n} : n < \omega\}$;

(c) if $i \neq j \in \mathscr{T}_\delta$, then $N_i^\delta \cap N_j^\delta$ is bounded in δ (used just after $(*)_{5.5}$), moreover:

$(c)^+$ if $i \neq j \in \mathscr{T}_\delta$ <u>then</u> the set $\{\beta < \delta : \beta$ a limit ordinal such that $\sup(N_i^\delta \cap \beta) = \beta = \sup(N_j^\delta \cap \beta)\}$ is bounded in δ;

(d) $N_i^\delta \cap (\alpha_\delta(n), \alpha_\delta(n+1)) \neq \emptyset$ and $N_i^\delta\restriction\alpha_\delta(n) \prec N_i^\delta$ for $n < \omega, \delta \in S, i \in \mathscr{T}_\delta$;

(e) for notational simplicity we assume $\mathscr{T}_\delta \subseteq \mu$.

<u>Stage B</u>: By induction on $\gamma < \lambda$ we shall choose the following:

$\boxplus_1$ (a) $G_\gamma \in \mathbf{K}_{\mathrm{lf}}$ of cardinality μ and the universe of G_γ is an ordinal $< \lambda$;

(b) $G_0 = G$;

(c) $\langle G_\beta : \beta \leq \gamma \rangle$ is increasing continuous;

(d) if $\beta \in \gamma \backslash S$, then $G_\beta \leq_{\mathfrak{S}} G_\gamma$;

(e) if $\gamma = \beta + 1, \beta \notin S$, then:

(α) G_γ is generated by $\{\bar{c}_{\beta,i} : i \in \mathscr{T}_\beta\} \cup G_\beta$, where $\mathscr{T}_\beta$ is a set of cardinality $\leq \mu$ (to be chosen),

(β) $t_{\beta,i} \in \mathrm{Def}_{\leq\mu}(G_\beta, \mathfrak{S})$, nontrivial (see Definition 4.13(5)) for $i \in \mathscr{T}_\beta$,

(γ) $\mathrm{tp}_{\mathrm{bs}}(\bar{c}_{\beta,i}, G_\beta, G_\gamma) = q_{t_{\beta,i}}(G_\beta)$ for $i \in \mathscr{T}_\beta$,

(δ) if $n < \omega$ and $i(0), \ldots, i(n-1) \in \mathscr{T}_\beta$ are pairwise distinct, then $\mathrm{tp}_{\mathrm{bs}}(\bar{c}_{\beta,i(0)}\hat{\ }\ldots\hat{\ }\bar{c}_{\beta,i(n-1)}, G_\beta, G_\gamma) = q_t(G_\beta)$, where $t = t_{\beta,i(0)} \otimes \ldots \otimes t_{\beta,i(n-1)}$,

> (ε) if $t = (\mathfrak{s}, \bar{a}) \in \mathrm{def}_{\mathfrak{S}}(G_\beta)$ is nontrivial <u>then</u> for some $i \in \mathscr{T}_\beta$ we have $t_{\beta,i} = t$;

(f) if $\gamma = \delta + 1, \delta \in S$ <u>then</u>:

> (α) G_γ is generated by $\{\bar{c}_{\delta,i} : i \in \mathscr{T}_\delta\} \cup G_\delta$,
>
> (β) $\mathscr{A}_\gamma = (G_{\delta+1}, G_\delta, \langle \bar{c}_{\delta,i}, t_{\delta,i,n} : i \in \mathscr{T}_\delta \rangle)$ is a one-step ($< \aleph_0, <$ $\aleph_0, \mathfrak{S}) - \otimes$-construction over $\langle G_{\alpha_\delta(n)} : n < \omega \rangle$, see 3.16; used in $(*)_{5.2}$'s proof[19],

(g) $t_{\beta,i} = (\mathfrak{s}_{\beta,i}, \bar{a}_{\beta,i})$ for $\beta \in \gamma \backslash S$.

First we shall show:

$\boxplus_2$ we can carry the induction.

Why? For $\gamma = 0$ we have nothing to do by clause (b).

For γ limit we let $G_\gamma = \cup \{G_\beta : \beta < \gamma\}$.

For $\gamma = \beta + 1, \beta \notin S$ we have some freedom, as we have $t_{\beta,i} \in \mathrm{Def}_{\leq \mu}(G_\beta, \mathfrak{S})$ not just $\mathrm{def}(G_\beta, \mathfrak{S})$. So let $\mathscr{T}_\beta = \mu, \{t_{\beta,i} : i \in \mathscr{T}_\beta\} \subseteq \mathrm{Def}_{\leq \mu}(G_\beta, \mathfrak{S})$ be of cardinality μ and including $\mathrm{def}(G_\beta, \mathfrak{S})$ and so $\langle t_{\beta,i} = (\mathfrak{s}_{\beta,i}, \bar{a}_{\beta,i}) : i < \mu \rangle$, possibly with repetitions. Clearly $\boxplus_1(e)(\varepsilon)$ holds.

Now as in Claim 3.13 we can find $G_\gamma, \langle \bar{c}_{\beta,i} : i < \mu \rangle$ such that:

- $G_\beta \leq_{\mathfrak{S}} G_\gamma$;

- $G_\gamma = \langle \{\bar{c}_{\beta,i} : i < \mu\} \cup G_\beta \rangle_{G_\gamma}$;

- if $n < \omega$ and $i_k < \mu$ for $k < n$ and $\langle i_k : \ell < n \rangle$ is with no repetitions, then

$$\bar{c}_{\beta,i_0} {}^\frown \ldots {}^\frown \bar{c}_{\beta,i_{n+1}} \text{ realizes } q_t(G_\beta) \text{ where } t = t_{\beta,i_0} \otimes \ldots \otimes t_{\beta,i_{n-1}}.$$

If $\gamma = \delta + 1, \delta \in S$ we can let $\mathfrak{s}_{\delta,i} = \mathfrak{s}_{\mathrm{ab}(2)}$, clearly we satisfy clause (f); but we may act differently. Clearly, as in the previous case, there is some freedom left: what we do for $\gamma = \delta + 1, \delta \in S$ and this will depend on the $\langle N_i^\delta : i \in \mathscr{T}_\delta \rangle$ from $\boxplus_0$. During the rest of the proof we shall use (some of the freedom left) to guarantee that G_* (see below) is as required.

Of course, we let:

$\boxplus_3$ $G_* = G_\lambda = \cup \{G_\alpha : \alpha < \lambda\}$.

We now point out some useful properties of the construction:

---

[19] Actually can use a one-step ($\leq \mu, < \aleph_0, \mathfrak{S}) - \otimes$-construction.

$(*)_{3.1}$ there is a model N_* expanding G_*, so with universe λ, and a countable vocabulary such that for any $N \subseteq N_*$ we have:

(a) $G_* {\restriction} N$ is a subgroup of G_*;

(b) $\beta \in N$ iff $N \cap G_{\beta+1} \backslash G_\beta \neq \emptyset$ iff $\beta + 1 \in N$;

(c) if $\gamma = \beta + 1, \gamma \in N$, then $N \cap G_\gamma = \langle \cup \{\bar{c}_{\beta,i} : i \in N \cap \mathscr{T}_\beta\} \cup (N \cap G_\beta)\rangle_{G_\gamma}$;

(d) if $i \in N \cap \mathscr{T}_\beta$ and $\beta \in N$, <u>then</u> $|\ell g(\bar{c}_{\beta,i})| \leq \omega \Rightarrow \bar{c}_{\beta,i} \subseteq N \cap G_{\beta+1}$ and $|\ell g(\bar{a}_{t_{\beta,i}})| \leq \omega \Rightarrow \bar{a}_{t_{\beta,i}} \subseteq N \cap G_\beta$;

(e) $\tau(N_*) \subseteq \mathscr{H}(\aleph_0)$, but $\mathscr{H}(\aleph_0) \backslash \tau(G_*)$ is infinite;

(f) if $\delta \in N \cap S$, then $C_\delta \subseteq N$.

Now note

$(*)_{3.2}$ if $\alpha < \lambda$ is a limit ordinal, <u>then</u> $G_\alpha \in \mathbf{K}_{\mathrm{exlf}}$.

[Why? Recall clause $(e)(\varepsilon)$ of $\boxplus_1$ noting that S is a set of limit ordinals, hence $\alpha = \sup(\alpha \backslash S)$.]
We now assume:

$\boxplus_4$ **h** is an automorphism of G_*.

We shall eventually prove that (if we suitably use the freedom left in $\boxplus_1$, then) **h** is an inner automorphism, i.e., $b \in G_* \Rightarrow \mathbf{h}(b) = a^{-1}ba$ for some $a \in G_*$, this clearly suffices noting that G_* has no center as $\mathfrak{s}_{\mathrm{cg}} \in \mathfrak{S}$.
We shall often use

$(*)_{4.1}$ for limit $\beta \in \lambda \backslash S$ let $L_\beta^* = \mathrm{cp}(G_\beta, G_{\beta+\omega})$ (see Definition 4.12(3)), i.e., $c \in L_\beta^*$ iff for some finite $K \subseteq \mathbf{C}_{G_{\beta+\omega}}(G_\beta)$ with trivial center we have $c \in K$ and $K \cap G_\beta = \{e_{G_\beta}\}$.

Note that

$(*)_{4.2}$ the last demand in $(*)_{4.1}$, "$K \cap G_\beta = \{e_{G_*}\}$", is redundant.

[Why? Recalll β is a limit ordinal hence by $(*)_{3.2}$, G_β has trivial center.]
Note:

$(*)_{4.3}$ if $a \in L_\beta^*$ and K witnesses it, <u>then</u> $K \subseteq L_\beta^*, K \cap G_\beta = \{e\}$, and moreover, there is $L \in \mathbf{K}_{\mathrm{exlf}}$ included in L_β^* and including K.

[Why? We can choose $\bar{K} = \langle K_n : n < \omega \rangle$ such that $K_0 = K, K_n$ is a finite group with trivial center, $K_n \subseteq K_{n+1}$ and $\bigcup_n K_n \in \mathbf{K}_{\text{exlf}}$. We now choose by induction on n an embedding f_n of K_n into $G_{\beta+\omega}$ such that $f_0 = \text{id}_K, f_n \subseteq f_{n+1}$ and $\text{Rang}(f_n) \subseteq L_\beta^*$; the induction step is possible by 4.9. Now $\bigcup_n f_n(K_n)$ is as required.]

We shall use:

$(*)_{4.4}$ let $E_{\mathbf{h}} = \{\delta : \delta$ is a limit ordinal and $\mathbf{h}$ maps G_δ onto G_δ and $(N^* {\restriction} \delta, \mathbf{h}{\restriction}\delta) \prec (N^*, \mathbf{h})\}$.

Now

$(*)_{4.5}$ $E_{\mathbf{h}}$ is a club of λ.

[Why? Just look at $(*)_{4.4}$.]

Stage C: We shall prove

⊞₅ for some $\alpha(*) < \lambda$, for every $\beta \in S_* \cap E_{\mathbf{h}} \backslash \alpha(*)$ and $c \in L_\beta^*$ we have $\mathbf{h}(c) \in c\ell(G_{\alpha(*)} \cup \{c\}, G_*)$.

Why? If not, for every $\alpha < \lambda$ there are $\beta_\alpha \in S_* \cap E_{\mathbf{h}}\backslash\alpha, m(\alpha) = m_\alpha \in \{2, 3, \ldots\}$ and $c_\alpha \in L_{\beta_\alpha}^*$ of order m_α such that $\mathbf{h}(c_\alpha) \notin c\ell(G_\alpha \cup \{c_\alpha\}, G_*)$. Now let $\bar{c}_\alpha$ witness that $c_\alpha \in L_{\beta_\alpha}^*$ with $c_{\alpha,0} = c_\alpha$, i.e., $\bar{c}_\alpha$ list the members of a finite subgroup of $G_{\beta_\alpha+\omega}$ commuting with G_{β_α} with trivial center and so included in $L_{\beta_\alpha}^*$. Without loss of generality, $\mathbf{h}(c_\alpha) \notin c\ell(G_\alpha \cup \{c_\alpha\}, G_*)$. Let $\mathbf{x}_\alpha \in X_2$ be such that $\bar{c}_\alpha$ realizes $q_{\mathfrak{s}_{\text{ab}}[\mathbf{x}_\alpha]}(\langle\rangle, G_{\beta_\alpha})$, see 4.7(2) + 4.11. But if $\alpha_1 < \alpha_2$, then $(\beta_{\alpha_2}, c_{\alpha_2}, m_{\alpha_2})$ can serve as $(\beta_{\alpha_1}, c_{\alpha_1}, m_{\alpha_1})$, hence, without loss of generality, $\mathbf{x}_\alpha = \mathbf{x}, m_\alpha = m_*$ for every α.

$(*)_{5.1}$ (a) Let $\bar{b}_{\alpha,1} = \bar{c}_\alpha$; let $k_{\alpha,1} < \omega$ be such that $\bar{b}_{\alpha,1} \subseteq G_{\beta_\alpha+k_{\alpha,1}+1}, \bar{b}_{\alpha,1} \not\subseteq G_{\beta_\alpha+k_{\alpha,1}}$;

(b) let $k_{\alpha,*} \in (k_{\alpha,1} + 1, \omega)$ be such that: $\text{tp}_{\text{bs}}(\mathbf{h}(\bar{b}_{\alpha,1}), G_{\beta_\alpha+\omega}, G_*) = q_{\mathfrak{s}_\alpha}(\bar{a}_\alpha^\bullet, G_{\beta_\alpha+\omega})$ for some $\mathfrak{s}_\alpha \in \mathfrak{S}$ with $\bar{a}_\alpha^\bullet \subseteq G_{\beta_\alpha+k_{\alpha,*}}$;

(c) let $\bar{b}_{\alpha,2} \subseteq G_{\beta_\alpha+\omega}$ realize $q_{\mathfrak{s}_{\text{ab}}[\mathbf{x}]}(\langle\rangle, G_{\beta_\alpha+k_{\alpha,*}})$;

(d) let $k_{\alpha,2} < \omega$ be such that $\bar{b}_{\alpha,2} \subseteq G_{\beta_\alpha+k_{\alpha,2}+1}, \bar{b}_{\alpha,2} \not\subseteq G_{\beta_\alpha+k_{\alpha,2}}$, so actually without loss of generality $k_{\alpha,2} = k_{\alpha,*} + 1$;

(e) note that $\bar{b}_{\alpha,1}{}^\frown\bar{b}_{\alpha,2}$ realizes $p_{\mathfrak{s}_{\text{gm}}}(\bar{x})$, see 4.11;

(f) let $k_{\alpha,3} < \omega$ be $> k_{\alpha,1}, k_{\alpha,2}$ and let $b_{\alpha,3} \in G_{\beta_\alpha+k_{\alpha,3}+1}$ realizes $q_{\mathfrak{s}_{\text{gm}}[\mathbf{x}]}(\bar{b}_{\alpha,1}{}^\frown\bar{b}_{\alpha,2}, G_{\beta_\alpha+k_{\alpha,3}}, G_{\beta_\alpha})$, (see Definition 4.11); so it commutes with $\mathbf{C}_{G_{\beta_\alpha+k_{\alpha,2}+1}}(\bar{b}_{\alpha,1}{}^\frown\bar{b}_{\alpha,2})$, hence with G_{β_α} and conjugating by it interchange $\bar{b}_{\alpha,1}, \bar{b}_{\alpha,2}$;

(g) without loss of generality $\mathfrak{s}_\alpha = \mathfrak{s}_*$ and $(\ell g(\bar{b}_{\alpha,1}), k_{\alpha,2}, \ell g(\bar{b}_{\alpha,2}))$ does not depend on α.

Our intention (in this stage) is to find $\alpha_n < \lambda$ increasing with n satisfying $\beta_{\alpha_n} < \alpha_{n+1}$ and element d such that, on the one hand, conjugating with d maps $c_{\alpha_n} = b_{\alpha_n,1,0}$ to $b_{\alpha_n,2,0}$ for each n, and on the other hand, $\mathrm{tp}_{\mathrm{bs}}(d, G_{\beta_n+\omega}, G_\lambda)$ does not split over $G_{\beta_n} + b_{\alpha_n,3}$, a contradiction.

Let N be such that:

$(*)_{5.2}$ (a) N is a model with universe λ;

(b) N is with countable vocabulary;

(c) N expands N_* from $(*)_{3.1}$;

(d) • $F_0^N = \mathbf{h}$, so F_0 is a unary function symbol,

• $F_{1,\iota,\ell}^N(\alpha) = b_{\alpha,\iota,\ell}$ for $\iota = 1, 2$ and $\ell < \ell g(\bar{b}_{\alpha,\iota})$, (if $\ell = 0$ we may omit it),

• $F_{1,3}^N(\alpha) = b_{\alpha,3}$,

• $F_2^N(\alpha) = \beta_\alpha$,

• $F_{2,\iota}(\alpha) = \beta_\alpha + k_{\alpha,\iota}$ for $\iota = 1, 2, 3$,

• $F_3^N(\alpha) = \beta_\alpha + \omega$,

(e) $F_{4,n}^N$ is a $(n+1)$-place function such that: if $\alpha_0 < \ldots < \alpha_n, c_{\alpha_\ell} \in G_{\alpha_{\ell+1}}$, each α_ℓ is a limit ordinal, then $F_{4,n}^N(\alpha_0, \ldots, \alpha_n)$ is the product of $a_0 a_1 \ldots a_n$ where $a_k = F_{1,1,0}(\alpha_k)$;

(f) $P^N = \{(\alpha, c) : \alpha < \lambda \text{ and } c \in G_\alpha\}$.

Without loss of generality $\tau_N \subseteq \mathscr{H}(\aleph_0)$, choose $\zeta(1) < \lambda$ such that $\tau_{\zeta(1)} = \tau_N$ and for each $\delta \in S_{\zeta(1)}$ we use the amount of freedom we are left with (see before $\boxplus_3$), choosing $G_{\delta+1}$ such that:

$(*)_{5.3}$ if $\delta \in S_{\zeta(1)}, i \in \mathscr{T}_\delta$, letting $\alpha_{\delta,i,n} := \min(N_i^\delta \backslash \alpha_\delta(n))$ then $(a) \Rightarrow (b)$ where:

(a) • $\beta_{\delta,i,n} := F_2^{N_i^\delta}(\alpha_{\delta,i,n})$ is $\geq \alpha_{\delta,i,n}$ but $< \alpha_\delta(n+1)$,

• $F_3^{N_i^\delta}(\alpha_{\delta,i,n}) = \beta_{\delta,i,n} + \omega$,

• $b_{\delta,i,n,\iota} = F_{1,\iota,\ell}^N(\alpha_{\delta,i,n})$ for $\iota = 1, 2$ and $\ell = 0$,

• $k_{\delta,i,n,\iota} = F_{2,\iota}^{N_i^\delta}(\alpha_{\delta,i,n}) - \alpha_{\delta,i,n}$ for $\iota = 1, 2, 3$,

• $b_{\delta,i,n,3} = F_3^{N_i^\delta}(\beta_{\delta,i,n})$,

- $b_{\delta,i,n,\iota} \in G_{\beta_{\delta,i,n}+k_{\delta,i,n,\iota}+1}$ commute with $G_{\beta_{\delta,i,n}}$ and conjugating by $b_{\delta,i,n,3}$ interchange $b_{\delta,i,n,1,\ell}, b_{\delta,i,n,2,\ell}$,

- δ is the set of elements of G_δ, similarly $\alpha_{\delta,i,n}$ (as they $\in E_{\mathbf{h}}$),

- for every $\beta < \delta$ we have $(G_{\beta+1} \backslash G_\beta) \cap N_i^\delta \neq \emptyset \Leftrightarrow \beta \in N_{\delta,i}$,

- if $\beta \in N_i^\delta \backslash S$ and $\bar{c} \in {}^{\omega>}(N_i^\delta)$, so $\bar{c} \in {}^{\omega>}(G_\delta)$, then $\mathrm{tp}_{\mathrm{bs}}(\bar{c}, G_\beta, G_\delta) \in q_t(G_\beta)$ for some $t \in \mathrm{def}(G_\beta)$ satisfying $\bar{a}_t \in {}^{\omega>}(N_i^\delta \cap G_\beta)$,

(b) $\bar{c}_{\delta,i} = \langle c_{\delta,i} \rangle$ and $\mathrm{tp}_{\mathrm{bs}}(c_{\delta,i}, G_\delta, G_{\delta+1})$ is as in claim 2.19 with $G_{\alpha_\delta(n)}(n < \omega), G_\delta, b_{\beta_{\delta,i,n,3}} \in N_i^\delta(n < \omega)$ here standing for $G_n(n < \omega), G_\omega, a_n(n < \omega)$ there;

$(*)_{5.4}$ let $\mathscr{T}_\delta' = \{i \in \mathscr{T}_\delta \colon$ clause (a) of $(*)_{5.3}$ holds$\}$;

$(*)_{5.5}$ let $\beta_{\delta,i,n} = \alpha_{\delta,i,n}+\omega$ and $b_{\delta,i,n} = b_{\delta,i,n,\iota} \in G_{\beta_{\delta,i,n}+1}$ for $\iota = 1, 2, 3$ realizes $q_{\mathfrak{s}_{\mathrm{ab}(2)}}(\langle\rangle, G_{\beta_{\delta,i,n}})$ <u>when</u> the assumption of clause (a) fails.

Why can we fulfill $(*)_{5.3}$? Let $\langle i_\ell : \ell < \ell(*) \rangle$ be a finite sequence of members of $\mathscr{T}_\delta$. For $\ell < \ell(*)$ and $n < \omega$ let $d_{\ell,n} = b_{\delta,i_\ell,n,3}$.
    Now

$(*)_{5.6}$ $\langle d_{\ell,n} : n < \omega \rangle$ pairwise commute if $i(\ell) \in \mathscr{T}_\delta'$ for each $\ell < \ell(*)$.

[Why? As $b_{\delta,i(\ell),n,3} \in \mathbf{C}(G_{\beta_{\delta,i(\ell),n}}, G_{\beta_{\delta,i(\ell),n}+\omega})$ for $n < \omega$ and $\beta_{\delta,i(\ell),n} + \omega < \alpha_\delta(n+1) \le \alpha_{\delta,i(\ell),n+1} \le \beta_{\delta,i(\ell),n+1}$, recalling $N_\delta^i \restriction \alpha_\delta(n+1) \prec N_\delta^i$ and $N_\delta^i \cap (\alpha_\delta(n), \alpha_\delta(n+1)) \neq \emptyset$.]

$(*)_{5.7}$ $\langle d_{\ell,n} : n < \omega \rangle$ pairwise commute when $i(\ell) \notin \mathscr{T}_\delta'$.

[Why? Even easier.]

$(*)_{5.8}$ if $\ell(1) \neq \ell(2)$ <u>then</u> for every $n(1) < n(2)$ the elements $b_{\delta,i(\ell(1)),n(1),3}, b_{\delta,i(\ell(2)),n(2),3}$ commute.

[Why? Recall that $b_{\delta,i(\ell(1)),n(1),3} \in G_{\alpha_\delta(n(2))} \subseteq G_{\beta_{\delta,i(\ell(2)),n(2)}}$; note that $b_{\delta,i,n,\iota} \in G_{\beta_{\delta,i,n,\iota}+\omega}$ commute with $G_{\beta_{\delta,i,n}}$ rather than with $G_{\alpha_{\delta,i,n}}$ but not used.]

$(*)_{5.9}$ if $\ell(1), \ell(2) < \ell(*)$, <u>then</u> for n large enough, for every $n(1), n(2) \in (n, \omega)$ the elements $d_{\ell(1),n(1)}, d_{\ell(2),n(2)}$ of G_δ commute.

[Why? Similarly, as $N_{i_{\ell(1)}}^\delta \cap N_{i_{\ell(2)}}^\delta$ is bounded in δ, but not used.]

$(*)_{5.10}$ The conditions in 2.19 hold hence we can fulfill $(*)_{5.3}, (*)_{5.4}$, i.e., we can carry the induction in $\boxplus_1$.

[Why? Think.]
Next let

$(*)_{5.11}$ $E = \{\delta < \lambda : \delta$ a limit ordinal is the universe of G_δ and $N{\restriction}\delta \prec N$, hence $\mathbf{h}$ maps G_δ onto itself$\}$.

Clearly E is a club of λ, hence by $\boxplus_{0,\zeta(1)}$ from stage A, there is a pair $(\delta, i_*) = (\delta, i(*))$ such that

$(*)_{5.12}$ $\delta \in E \cap S_{\zeta(1)}$ and $i_* \in \mathscr{T}_\delta$ and $N_{i_*}^\delta \prec N$.

Let $d = \mathbf{h}(c_{\delta,i_*}) \in G_*$, so:

$(*)_{5.13}$ (a) the pair (δ, i_*) satisfies the demands in $(*)_{5.3}(a)$;

     (b) for some finite set $u_* \subseteq \mathscr{T}_\delta$ and $\bar{b}_* \in {}^{\omega>}(G_\delta)$, the type $\mathrm{tp}_{\mathrm{bs}}(d, G_{\delta+1}, G_*)$ does not split over $\{c_{\delta,i} : i \in u_*\} \cup \bar{b}_*$;

     (c) without loss of generality $i_* \in u_*$.

[Why? For clause (a), as $\delta \in E$ and $N_{i(*)}^\delta \prec N$, recalling the choice of N (including $\mathbf{h} = F_0^N$). For clause (b), apply properties of the construction in $\boxplus_1$, i.e., $G_{\delta+1} \leq_{\mathfrak{S}} G_*$.]

$(*)_{5.14}$ conjugating by d in G_* interchange $b_{\delta,i(*),n,1}$ with $b_{\delta,i(*),n,2}$ for $n < \omega$.

[Why? Should be clear as for $m \in \omega \setminus \{n\}$ and $\iota(1), \iota(2) \in \{1, 2, 3\}$, the element $b_{\delta,i(*),m,\iota(1)}$ commutes with $b_{\delta,i(*),m,\iota(2)}$.]
Recalling $\boxplus_0(c)$ there is $n(*) < \omega$ large enough such that:

$(*)_{5.15}$ $\bar{b}_* \subseteq G_{\beta_{\delta,i(*),n(*)}}$ and $j_1 \neq j_2 \in u_* \Rightarrow N_{j_1} \cap N_{j_2} \subseteq G_{\alpha_\delta(n(*))}$ and j_1, j_2 are like i, j as in $\boxplus_{0,\zeta(1)}(c)^+$.

Clearly for some $\beta(*) < \lambda$ we have $\mathbf{h}(b_{\delta,i(*),n(*),1}) \in G_{\beta(*)+1} \setminus G_{\beta(*)}$. As $\alpha_{\delta,i(*),n(*)} = \min(N_{i(*)}^\delta \setminus \alpha_\delta(n(*)) \in N_{i(*)}^\delta$, clause (d) of $\boxplus_0$ and $N_{i_*}^\delta \prec N$, clearly:

$(*)_{5.16}$ $\mathbf{h}$ maps $G_{\alpha_{\delta,i(*),n(*)}} \cap N_{i_*}^\delta$ onto itself and so $\beta(*) \in N_{i_*}^\delta \setminus \alpha_{\delta,i(*),n(*)}$.

Also,

$(*)_{5.17}$ if $\beta(*) < \beta_{\delta,i(*),n(*)} + \omega$, then $\beta(*) \leq \beta_{\delta,i(*),n(*)} + k_{\delta,i(*),n(*),2}$.

[Why? By $(*)_{5.1}$.]

Now,

$(*)_{5.18}$ there is $\beta \in N_{i_*}^\delta \cap (\beta(*) + 1) \backslash \alpha_\delta(n(*)) \backslash S$ such that $[\beta, \beta(*) + \omega) \cap N_{i_*}^\delta$ is disjoint from N_j^δ if $j \in u_*$ but $j \neq i_*$.

[Why? First assume $\beta(*) \notin S$, let $\beta = \beta(*)$, so clearly $\beta \in N_{i_*}^\delta$ by $(*)_{5.14}, \beta \in (\beta(*) + 1)$, also $\beta \notin \alpha_\delta(n(*))$ as by $(*)_{5.6}$ and the fact that $\beta \notin S$ by its choice. Also $[\beta, \beta(*) + \omega) = [\beta(*), \beta(*) + \omega) \subseteq N_{i_*}^\delta$ as N is closed under $\alpha \mapsto \alpha + 1$ by $(*)_{3.1}(b)$. If $j \in u_*$ but $j \neq i_*$, then $N_j^\delta \cap N_{i_*}^\delta \subseteq \alpha_\delta(n(*)) \leq \beta$, hence $[\beta, \beta(*) + \omega) \cap N_j^\delta = \emptyset$, so we are done.

Second, assume $\beta(*) \in S$, hence $\mathrm{cf}(\delta) = \aleph_0$, and by $(*)_{3.2}(f), \{\alpha_{\beta(*)}(n) : n < \omega\} \subseteq N_{i_*}^\delta$. But by $\boxplus_0(c)^+$ we have $j \in u_* \wedge j \neq i_* \Rightarrow \sup(N_j^\delta \cap \beta(*)) < \beta(*)$. As u_* is finite there is $\beta \in \{\alpha_{\beta(*)}(n) : n < \omega\}$ such that $(\beta, \beta(*)) \cap N_j^\delta = \emptyset$; hence as before also $(\beta, \beta(*) + \omega) \cap N_j^\delta = \emptyset$, whenever $j \in u_* \wedge j \neq i_*$. So $(*)_{5.16}$ holds indeed.]

We finish the proof of $\boxplus_5$ by getting a contradiction as follows.

<u>Case 1</u>: $\beta(*) \geq \beta_{\delta,i(*),n(*)} + \omega$.

So by the choice of β and the proof of $(*)_{5.3}$ the type $\mathrm{tp}_{\mathrm{bs}}(d, G_{\beta(*)+\omega}, G_*)$ does not split over G_β, and even over some finite subset of it.

Now by $\boxplus_1(e)$ in $G_{\beta(*)+\omega}$ there is $d' \neq \mathbf{h}(b_{\beta_{\delta,i(*),n(*),1}})$ realizing $\mathrm{tp}_{\mathrm{bs}}(\mathbf{h}(b_{\beta_{\delta,i(*),n(*),1}}), G_\beta, G_{\beta(*)+\omega})$ so $\mathbf{h}(b_{\beta_{\delta,i(*),n(*),3}}) \notin c\ell(G_\beta \cup \{d\})$. However, $G_* \models d^{-1} \mathbf{h}(c_{\beta_{\delta,i(*),n(*),1}})d = \mathbf{h}(c_{\beta_{\delta,i(*),n(*),2}})$, contradiction.

<u>Case 2</u>: $\beta(*) < \beta_{\delta,i(*),n(*)} + \omega$.

Hence $\beta(*) \leq \beta_{\delta,i(*),n(*)} + k_{\delta,i(*),n(*),2}$ and so $\mathrm{tp}_{\mathrm{bs}}(d, G_{\beta_{\delta,i(*),n(*)}+\omega}, G_*)$ does not split over $G_{\beta_{\delta,i(*),n(*)}} \cup \{b_{\delta,i(*),n(*),3}\})$ but $\mathrm{tp}(b_{\delta,i(*),n(*),3}, G_{\beta_{\delta,i(*),n(*)}+k_{\delta,i(*),n(*),3}}, G_*)$ does not split over $G_{\beta_{\delta,i(*),n(*)}} \cup \mathrm{Rang}(\bar{b}_{\delta,i(*),n(*),1}) \cup \mathrm{Rang}(\bar{b}_{\delta,i(*),n(*),2})$.

It follows that $\mathrm{tp}_{\mathrm{bs}}(d, G_{\beta_{\delta,i(*),n(*)}+k_{\delta,i(*),n(*),2}}, G_*)$ does not split over $G_{\beta_{\delta,i(*),n(*)}} \cup \bar{b}_{\delta,i(*),n(*),1}$, contradiction.

So we have finished proving $\boxplus_5$.

<u>Stage D</u>:

$\boxplus_6$　(a) for some stationary $S_1^* \subseteq S_* \subseteq \lambda \backslash S$ for every $\beta \in S_1^* \backslash \alpha(*)$ if $b \in L_\beta^*$ <u>then</u> $\mathbf{h}(b) = \sigma^{G_*}(b, \bar{a})$ for some $\bar{a} \in {}^{\omega>}(G_\beta)$ and group-term $\sigma(x, \bar{y})$,

　(b) moreover, $\mathbf{h}(b) = \sigma^{G_*}(b)$ if $b \in L_\beta^*$.

Why?

$(*)_{6.1}$　clause (a) of $\boxplus_6$ holds even for every $\beta \in S_2^* := S_* \cap E_{\mathbf{h}} \backslash \alpha(*)$.

[Why? By $\boxplus_5$.]

$(*)_{6.2}$ Without loss of generality if $\beta \in S_2^*$ and $b \in L_\beta^*$, <u>then</u> $\mathbf{h}(b) = \sigma_b(b)a_b$ for some $a_b \in G_\beta$.

[Why? This by $(*)_{6.1}$ because $\mathbf{h}$ maps G_β onto itself, b commutes with G_β whereas $\bar{a}_b \in {}^{\omega >}(G_\beta)$.]

$(*)_{6.3}$ (a) $b \mapsto \sigma_b(b)$ is a homomorphism from the set L_β^* into L_β^* (but we did not claim L_β^* is a subgroup);

   (b) $b \mapsto a_b$ induces a homomorphism from the set L_β^* into the group G_β, that is if $\sigma(x_0, \ldots, x_{n-1})$ is a group term and $b_0, \ldots, b_{n-1} \in L_\beta^*$ and $G_{\beta+\omega} \models \sigma(b_0, \ldots, b_{n-1}) = e$, then $G_\beta \models \sigma(a_{b_0}, \ldots, a_{b_{n-1}}) = e$.

[Why? As $\mathbf{h}$ is an automorphism of G_* and as $a_{b_1}, \sigma_{b_2}(b_2)$ commute for $b_1, b_2 \in L_\beta^*$.]

We try to get rid of the homomorphism from $(*)_{6.3}(b)$ in order to prove $\boxplus_6(b)$.

Toward contradiction assume (for the rest of this stage):

$(*)_{6.4}$ $\gamma \in S_2^* \subseteq \lambda \backslash S_1^*$ is a limit ordinal and $b_* \in L_\gamma^*$ and $a_{b_*} \neq e$.

Now as $\gamma \in S_* \subseteq \lambda \backslash S$ we can find a sequence $\bar{f}^\gamma = \langle f_\eta^\gamma : \eta \in {}^\omega \mu \rangle$ satisfying f_η^γ is a function from $\{\eta \restriction n : n < \omega\}$ into G_γ such that for every $f : {}^{\omega >}\mu \to G_\gamma$ for some $\eta \in {}^\omega \mu$ we have $f_\eta^\gamma \subseteq f$; i.e., a simple black box, see [Sh:309, §1], it exists as $\mu = \mu^{\aleph_0}$. Now generally for $\gamma \in \lambda \backslash S$ let $\mathcal{W}_\gamma = \{\eta \in {}^\omega \mu$: for some $c \in G_\gamma$ of order 2 we have $n < \omega \Rightarrow c^{-1} f_\eta^*(\eta \restriction (2n))c = f_\eta^*(\eta \restriction (2n+1))\}$.

Let K_* be the group of permutation of $I = {}^{\omega >}\mu \times \{0, 1\}$ with finite support, i.e., $\{f \in \mathrm{Sym}(I) : (\exists^{<\aleph_0} t \in I)(f(t) \neq t)\}$. For $\eta \in {}^{\omega >}\mu$ let $h_\eta \in K_*$ be such that $h_\eta((\eta, \iota)) \equiv (\eta, 1 - \iota)$, for $\iota = 0, 1$, and is the identity otherwise. Let K_γ be the group of permutations of $I = {}^{\omega >}\mu \times \{0, 1\}$ generated by $K_* \cup \{y_\eta : \eta \in {}^{\omega >}\mu\}$, where:

$(*)_{6.5}$ (a) if $\eta \in \mathcal{W}_\gamma$, then y_η interchanges $(\eta \restriction (2n+1), \iota), (\eta \restriction (2n+2), \iota)$ for $n < \omega, \iota = 0, 1$ and otherwise is the identity;

   (b) if $\eta \in {}^\omega \mu \backslash \mathcal{W}_\gamma$, then y_η interchanges $(\eta \restriction (2n), \iota)$ and $(\eta \restriction (2n+1), \iota)$ for $n < \omega, \iota = 0, 1$, and is the identity otherwise.

Let

- d be the permutation of I interchanging $(<>, 0), (<>, 1)$ and being the identity otherwise.

Now we shall use some of the amount of freedom left, clearly:

$(*)_{6.6}$ (a) there is $K \subseteq \mathbf{C}_{G_{\gamma+\omega}}(G_\beta)$ finite with trivial center such that $b_* \in K$;

(b) there is $\bar{b}$ which lists the member of K such that $d_0 = b_*$;

(c) there is $\bar{d}$, a finite sequence from K_γ realizing $\mathrm{tp}(\bar{b}, \emptyset, G_*)$;

(d) there is $n(*)$ such that $K \subseteq G_{\gamma+n(*)}$.

$(*)_{6.7}$ There is an embedding g_γ of K_γ into $\mathbf{C}_{G_{\gamma+n(*)+1}}(G_\gamma)$ mapping $\bar{d}$ to $\bar{b}$ hence d_0 to b_*;

$(*)_{6.8}$ $b \mapsto a_b$ (for $b \in g_\gamma(K_\gamma)$) is a homomorphism from $g_\gamma(K_\gamma)$ into G_γ;

$(*)_{6.9}$ let $f : {}^{\omega>}\lambda \to G_\beta$ be defined by $f(\eta) = a_{g_\gamma(h_\eta)}$.

By the choice of $\langle f_\eta : \eta \in {}^{\omega>}\mu \rangle$ for some $\eta \in {}^\omega\mu$ we have $n < \omega \Rightarrow f_\eta^\gamma(\eta{\restriction}n) = f(\eta{\restriction}n)$.

Now does $\eta \in \mathscr{W}_\gamma$? First, assume $\eta \notin \mathscr{W}_\gamma$, then (by the choice of K_γ) $(g_\gamma(y_\eta) \in G_\gamma$ and) conjugating by $g_\gamma(y_\eta)$ for each n, interchanges $g_\gamma(h_{\eta{\restriction}(2n)}), g_\gamma(h_{\eta{\restriction}(2n+1)})$ which means that in K_γ, conjugating by h_η interchanges $f_\eta^\gamma(\eta{\restriction}(2n)), f_\eta^\gamma(\eta{\restriction}(2n+1))$, but by the choice of $\mathscr{W}_\gamma$ this means $\eta \in \mathscr{W}_\gamma$. Second, assume $\eta \in \mathscr{W}_\gamma$, by the definition of $\mathscr{W}_\gamma$ there is $c \in G_\gamma$ of order 2 such that conjugating by c for each n interchanges $g_\gamma(h_{\eta{\restriction}(2n)}), g_\gamma(h_{\eta{\restriction}(2n+1)})$. But conjugating by $g_\gamma(y_\eta)$ for n interchange $g_\gamma(h_{\eta{\restriction}(2n+1)}), g_\gamma(h_{\eta{\restriction}(2n+2)})$ and we shall get a contradiction.

So in G_*, the subgroup generated by $\{c, g_\gamma(y_\eta), g_\gamma(h_{\eta{\restriction}1})\}$ includes $g_\gamma(h_{\eta{\restriction}n})$ for $\eta = 1, 2, \ldots$; why? just prove it by induction on n. But $\{g_\gamma(h_{\eta{\restriction}n}) : n = 1, 2, \ldots\} \subseteq G_*$ is infinite, contradiction.

<u>Stage E:</u>

$\boxplus_7$ there is a finite sequence $\bar{a}_*$ such that for every $b \in G_*$ we have $\mathbf{h}(b) \in c\ell(\bar{a}_* \cup \{(b, G_*)\})$.

[Why? For $\beta \in S_1^*$ let $d_\beta \in G_{\beta+1}$ realize $\mathfrak{s}_{\mathrm{cg}}(<>, G_\beta)$ in $G_{\beta+1}$. So for every $a \in G_\beta$ of order m as G_β is existentially closed there is a finite $K_a \subseteq G_\beta$ with trivial center to which a belongs. Hence, the element $d_\beta a d_\beta^{-1}$ commute with G_β and belongs to $G_{\beta+1}$ and moreover to L_β^*. Hence, by $\boxplus_6(b)$, for some $k(a) < m$ we have:

$(*)_{7.1}$ $\mathbf{h}(d_\beta^{-1} a d_\beta) = (d_\beta^{-1} a d_\beta)^{k(a)}$.

Hence,

$(*)_{7.2}$ $\mathbf{h}(a) = \mathbf{h}(d_\beta^{-1})\mathbf{h}(d_\beta^{-1} a d_\beta)\mathbf{h}(d_\beta^{-1}) = \mathbf{h}(d_\beta^{-1})(d_\beta^{-1} a d_\beta)^{k(a)}\mathbf{h}(d_\beta)$.

Also, as $\beta \notin S$, there is a finite $K_\beta \subseteq G_\beta$ such that $\mathrm{tp}_{\mathrm{bs}}(\langle \mathbf{h}(d_\beta), d_\beta \rangle, G_\beta, G_*; \mathbf{K}_{\mathrm{lf}})$ does not split over K_β. By $(*)_{7.2}$, $\mathrm{tp}_{\mathrm{bs}}(\mathbf{h}(a), G_\beta, G_*; \mathbf{K}_{\mathrm{lf}})$ does not split over $K_\beta \cup \{d\}$, but $\mathbf{h}(a) \in G_\beta$, hence $\mathbf{h}(a) \in \langle K \cup \{d\} \rangle_{G_*}$. By Fodor's lemma this is enough for $⊞_7$.

Clearly we are done by 2.23. $\square_{5.1}$

$$* \qquad * \qquad *$$

Question 5.2. 1) In 5.1 we can easily get 2^λ pairwise nonisomorphic groups G'. But can they be pairwise far? (i.e., no $G \in \mathbf{K}_\lambda$, can be embedded in two of them)?

2) Even more basically can we demand G_* has no uncountable Abelian subgroup (when G does not)? Or at least no Abelian group of cardinality λ?

3) Can we prove 5.1 for every $\lambda > \aleph_0$? or at least $\lambda \geq \beth_\omega$?

5.3 Discussion. 1) Concerning 5.2(1), the problem with our approach is using $p \in \mathbf{S}_\mathfrak{S}(G)$, so as λ is regular we will get subgroups generated by indiscernible sequences, but let us elaborate. Assume $G_* \in \mathbf{K}_\lambda, G_* = \cup\{G_\alpha : \alpha < \lambda\}, G_\alpha$ increases with α and $|G_\alpha| < \lambda$. Further, assume $\mathfrak{s} \in \Omega[\mathbf{K}]$ and $\bar{a} \in {}^{n(\mathfrak{s})}G_*$ and $S = \{\alpha < \lambda : \bar{a} \subseteq G_\alpha$ and the type $q_\mathfrak{s}(\bar{a}, G_\alpha)$ is realized in $G_*\}$ is unbounded in λ and thus it is an end segment. Let $\bar{c}_\alpha \in {}^{k(\mathfrak{s})}G_*$ realize $q_\mathfrak{s}(\bar{a}, G_\alpha)$ and so for some club E of $\lambda, \alpha \in S \cap E \Rightarrow \bar{c}_\alpha \in G_{\min(E\setminus(\alpha+1))}$. Now $\bar{c} = \langle \bar{c}_\alpha : \alpha \in S \cap E \rangle$ satisfies: if h is a partial increasing finite function from $S \cap E$ to $S \cap E$, then it induces a partial automorphism of $G_* : \bar{c}_\alpha \mapsto \bar{c}_{h(\alpha)}$. This is a case of indiscernible sequences. Hence, the isomorphism type of $cl(\cup\{\bar{c}_\alpha : \alpha \in S \cap E\}, G_*)$ depends only on $\mathfrak{s}$ (and $\mathrm{tp}_{\mathrm{bs}}(\bar{a}, \emptyset, G_*)$. Hence the number of pairwise far such G_*'s is $\leq |\mathfrak{S}| + \aleph_0$.

2) Concerning 5.2(2), the problem with our approach is that we use $\mathfrak{s} = \mathfrak{s}_{\mathrm{ab}(k)}$ and more generally $\mathfrak{s} \in \Omega[\mathbf{K}]$ such that if $q_\mathfrak{s}(\bar{a}, G) = \mathrm{tp}_{\mathrm{bs}}(\bar{c}, G, H)$ <u>then</u> some $c \in H\setminus G$ commute with every (or simply many) members of G. Hence in the construction above, G_* has Abelian subgroups of cardinality λ.

3) What about considering the class of $(G, F_h)_{h \in H}, F_h \in \mathrm{aut}(G), G \in \mathbf{K}_{\mathrm{lf}}, h \mapsto F_h$ a homomorphism? We intend to deal with it in [Sh:1098].

5.4 Discussion. 1) Naturally the construction in the proof of 5.1 is not unique, the class has many complicated models. In the construction in the proof of 5.1 we choose one where we realize many definable types.

2) We may like in $⊞_5$ of Stage C in the proof of 5.1 to consider $c \in G_\lambda$, not necessarily from $G_{\beta+\omega}$; (so later the role of $\mathfrak{s}_{\mathrm{cg}}$ in translating knowledge on $\mathbf{h}\restriction G_{\beta+\omega}$ to knowledge on G_β + use of Fodor is not necessary). Presently the way we combine $\langle b_{\delta,i(\ell),n,3} : n < \omega, \ell < \ell(*) \rangle$ to one n-type in $\mathbf{S}_{\mathrm{bs}}(G_\delta)$ works using 2.19.

Concerning the existence of complete groups in $\mathbf{K}_\lambda^{\mathrm{lf}}$ extending any $G \in \mathbf{K}_\lambda^{\mathrm{lf}}$ there are some restrictions.

Claim 5.5. Assume $\lambda > \mathrm{cf}(\lambda) = \aleph_0, \chi = \lambda^{\aleph_0}$.
1) If $G \in \mathbf{K}_\lambda^{\mathrm{lf}}$ is full, <u>then</u> its outer automorphism group has cardinality $\geq \chi$.
2) G has $\geq \chi$ outer automorphisms <u>when</u> $G \in \mathbf{K}_\lambda^{\mathrm{lf}}$ and for some sequence $\bar{a} = \langle a_\alpha : \alpha < \lambda \rangle$ listing the elements of G, letting $G_\alpha = c\ell(\{a_\beta : \beta < \alpha\}, G)$ we have:

   (a) for every $\alpha < \lambda$ for λ ordinals $\beta < \lambda, a_\beta$ commutes with G_α

   (b) for every $a \in G \backslash \{e_G\}$ some element $b \in G, a$ does not commute with b.

3) Like (2), but instead clause (b), G has center of cardinality $< \lambda$.
4) Instead of (a),(b) we can use:

   (a)$'$ for every $\alpha < \lambda$ we have $\lambda = |\{a/\mathrm{Cent}(G) : a \in G$ commute with $G_\alpha\}|$.

Proof. 1) We reduce it to part (2). Let $\bar{a} = \langle a_\alpha : \alpha < \lambda \rangle$ witness fullness (so $\lambda \geq 2^{\aleph_0}$). Now using the schemes $\mathfrak{s} = s_{\mathrm{ab}(2)}$, the pair $(G, \bar{a})$ satisfies clause (a) of part (2). Using, e.g., the scheme $\mathfrak{s} = \mathfrak{s}_{\mathrm{cg}}$ and the claim on noncommuting, 2.21, also clause (b) there holds.
2) Let $\lambda = \sum_n \lambda_n, \lambda_n < \lambda_{n+1}$. For each n, by clause (a) we have $|S_n^1| = \lambda$ where $S_n^1 := \{\alpha : a_\alpha$ commute with $c\ell(\{a_\beta : \beta < \lambda_n\}, G)\}$. Hence, for some $k_n > n$ we have $S_n^3 = \{\alpha < \lambda_{k_n} : \alpha \in S_n^1\}$ has cardinality $> \lambda_n$.

Replacing $\langle \lambda_n : n < \omega \rangle$ by a subsequence without loss of generality $\bigwedge_n k_{2n} = 2n + 1$. Let $\langle \alpha_{n,i} : i < \lambda_n \rangle$ be a sequence of pairwise distinct members of $S_{2n}^3 \backslash \lambda_{2n}$. Now for each $\eta \in \prod_{\ell < n} \lambda_{2\ell}$ let $b_\eta = a_{\eta(0)} a_{\eta(1)} \ldots a_{\eta(n-1)} \in G$ and $h_\eta := \Box_{b_\eta}$, conjugation by b_η, is an inner automorphism of H. Also $\nu \triangleleft \eta \in \prod_{\ell < n} \lambda_{2\ell} \Rightarrow \Box_{b_\eta}, \Box_{b_\nu}$ agree on $\{a_\beta : \beta < \lambda_{2\lg(\nu)}\}$.

Hence if $\eta \in \prod_n \lambda_{2n}$, then $\langle h_{\eta \restriction n} : n < \omega \rangle$ converge, i.e., for every $a \in G$, the sequence $\langle h_{\eta \restriction n}(a) : n < \omega \rangle$ is eventually constant and called the eventual value $h_\eta(a)$.

So h_η is an automorphism of G (for each $\eta \in \prod_n \lambda_{2n}$). Now if $\eta_1, \eta_2 \in \prod_n \lambda_{2n}, \eta_1(k) \neq \eta_2(k), \eta_1 \restriction k = \eta_2 \restriction k$ and for some $\alpha < \lambda_{2k}, a_\alpha$ does not commute with $a_{\eta_1(k)} a_{\eta_2(k)}^{-1}$, then $h_{\eta_1} \neq h_{\eta_2}$. Hence we can easily find $2^{\aleph_0}$ pairwise distinct h_η's. So if $\lambda < 2^{\aleph_0}$ we are done; otherwise, let $\mu = \min\{\mu : \mu^{\aleph_0} \geq \lambda$ equivalently $\mu^{\aleph_0} = \lambda^{\aleph_0}\}$, so $2^{\aleph_0} < \mu < \lambda$ and $\alpha < \mu \Rightarrow |\alpha|^{\aleph_0} < \mu$.

Choose $\bar{\mu} = \langle \mu_n : n < \omega \rangle$ such that $\sum_n \mu_n = \mu, \mu_n < \mu_{n+1}$; moreover, each μ_n regular and $\alpha < \mu_n \Rightarrow |\alpha|^{\aleph_0} < \mu_n$. Now for $n < k$ let $E_{n,k} = \{(i,j) : i, j < \mu_n$ and the conjugation $\Box_{a_{\alpha_{n,i}}}, \Box_{a_{\alpha_{n,j}}}$ agree on $\{a_\beta : \beta < \lambda_{2k}\}\}$, an equivalence relation. By clause (b) in the assumption, $\bigcap\limits_{k>n} E_{n,k}$ is the equality on μ_n, hence for some $k(n) > n, \mu_n/E_{n,k}$ has μ_n equivalence class. The rest should be clear.

3),4) Similarly. $\square_{5.5}$

6   Other classes

Note that

Theorem 6.1. *The results of §1 holds for any universal class* **K**; *see [Sh:300b].*

However, we cannot in general prove the existence of dense $\mathfrak{S} \subseteq \Omega[\mathbf{K}]$, in fact, possibly $\Omega[\mathbf{K}] = \emptyset$. We refer the reader to §0 before 0.13, and to 0.17, 2.1. We may expand an lf group by choosing representations for left cosets bK, for K a finite subgroup of $G, b \in G$. Then the density of $\Omega[\mathbf{K}]$ is easy.

Definition 6.2. 1) Let $\mathbf{K}_{\mathrm{clf}}$ be the class of structures M such that M is an expansion of an lf group $G = G_M$ by $F_n = F_n^M$ for $n \geq 1$ such that:

(a) F_n^M is a partial $(n+1)$-place function from G to G;

(b) if $(a_0, \ldots, a_n) \in \mathrm{Dom}(F_n^M)$, then $(a_0, \ldots, a_{n-1})$ list without repetitions the elements of a subgroup of G_M and $a_n \in G_M$, of course;

(c) if $F_n^M(a_0, \ldots, a_n) = b$, then $b \in \{a_n a_\ell : \ell < n\}$;

(d) if K is a finite subgroup of G_M with n elements and for some $(a_0, \ldots, a_{n-1})$ listing its elements with no repetitions and b we have $(a_0, \ldots, a_{n-1}, b) \in \mathrm{Dom}(F_n^M)$, then for every $(a'_0, \ldots, a'_{n-1})$ listing the members of K and $b' \in bK \subseteq G_M$ we have $(a'_0, \ldots, a'_{n-1}, b') \in \mathrm{Dom}(F_n^M)$ and $b'K = bK \Rightarrow F_n^M(a'_0, \ldots, a'_{n-1}, b') = F_n^M(a_0, \ldots, a_{n-1}, b)$;

(e) if K_1, K_2 are as in clause (d), then also $K_1 \cap K_2$ is;

(f) if $A \subseteq G_M$ is finite, then there is a minimal K as in clause (d) which contains A and if A is empty, then $K = \{e_{G_M}\}$.

Definition 6.3. Let $\mathbf{K}_{\mathrm{plf}}$ be the class of structures M such that: M expands a lf group G by P_n^M for $n < \omega$ and F_n^M for $n < \omega$ (actually definable from the rest) such that:

(a) P_n^M is an $(n+3)$-place relation;

(b) if $\bar{a} = (a_0, \ldots, a_{n+2}) \in P_n^M$, then $\{a_0, \ldots, a_{n-1}\}$ list with no repetitions the elements of a finite subgroup of G_M;

(c) if $\{a_0, \ldots, a_{n-1}\} = \{a'_0, \ldots, a'_{n-1}\}$ are as above and moreover, $b, b' \in M$ and $\{ba_0, \ldots, ba_{n-1}\} = \{b'a'_0, \ldots, b'a'_{n-1}\}$, then $M \models$ "$P_n(a_0, \ldots, a_{n-1}, b, c, d) = P_n(a'_0, \ldots, a'_{n-1}, b', c, d)$" for every $c, d \in M$;

(d) if $(a_0, \ldots, a_{n-1})$ list the members of a finite subgroup K of G with no repetitions and $b \in G$, then $\{(c, d) : (a_0, \ldots, a_{n-1}, b, c, d) \in P_n^M\}$ is a linear order on the right coset bK, which we denote by $<_{K,b}^M$;

(e) if the sequence $(a_0, \ldots, a_{n-1})$ is as above and $b \in G$, then $F_n^M(a_0, \ldots, a_{n-1}, b)$ is the first element by the order there in $\{ba_0, \ldots, ba_{n-1}\}$.

Definition 6.4. 1) For $M \in \mathbf{K}_{\mathrm{clf}}$ let fsb(M) be the set of finite subgroups K of G_M such that for some $a_0, \ldots, a_{n-1}$ listing with no repetitions the elements of K and for some $b \in G_M$ we have $(a_0, \ldots, a_{n-1}, b) \in \mathrm{Dom}(F_n^M)$, i.e., they are as in clause (d) of Definition 6.2.

2) In this case we may write $F_K^M(b) = F_n^M(a_0, \ldots, a_{n-1}, b)$.

3) For $M, N \in \mathbf{K}_{\mathrm{clf}}$ let $M \leq_{\mathrm{elf}} N$ or $M \subseteq N$ mean that $G_M \subseteq G_N$ and $F_n^M = F_n^N \restriction M$ hence $K \in \mathrm{sfb}(N) \wedge K \subseteq M \Rightarrow K \in \mathrm{fsb}(M)$. We define similarly $\leq_{\mathrm{plf}}, \leq_{\mathrm{olf}}$, see Definition 6.3, 0.15. We may write $M \leq_{\mathbf{K}} N$ for the appropriate $\mathbf{K}$, etc.

4) "$M \in \mathbf{K}_{\mathrm{clf}}$ is (existentially closed)" is defined as in 0.13(2).

5) Let clf-group mean a member of K_{clf} and similarly an olf-group.

6) Similarly for "olf-groups" and "plf-groups".

Convention 6.5. Let $\mathbf{K}$ denote one of the classes defined above, but let it be $\mathbf{K}_{\mathrm{clf}}$ if not said otherwise.

6.6 Definition/Claim. 1) For $M \in \mathbf{K}_{\mathrm{olf}}$ let $M^{[\mathrm{clf}]}$ be the unique $N \in \mathbf{K}_{\mathrm{clf}}$ such that: $G_N = G_M$ and fsb(N) $= \{K : K \subseteq G_M$ is finite$\}$ and $F_K^M(b)$ is the $<_M$-first member of bK $\subseteq G$ (well defined as bK is finite nonempty).

1A) For $M \in \mathbf{K}_{\mathrm{olf}}$ we define $M^{[\mathrm{plf}]}$ and for $M \in \mathbf{K}_{\mathrm{plf}}$ we define $M^{[\mathrm{clf}]}$ parallely.

2) For $M \in \mathbf{K}_{\mathrm{clf}}$ and $A \subseteq M$, there is $N \subseteq M$ from $\mathbf{K}_{\mathrm{clf}}$ with universe A iff for every finite $A \subseteq B$ there is $K \in \mathrm{fsb}(M)$ such that $A \subseteq K \subseteq B$.

2A) So if $M \in \mathbf{K}_{\mathrm{clf}}$ and $K \in \mathrm{fsb}(M)$, then $M \restriction K \in \mathbf{K}_{\mathrm{clf}}$ and is finite.

3) For $A \subseteq M \in \mathbf{K}$ let $c\ell(A, M)$ be the minimal $N \subseteq M$ such that $A \subseteq N$, equivalently $\cup \{K : K \in \mathrm{fsb}(M)$ and there is no $L \in \mathrm{fsb}(M)$ such that $A \cap K \subseteq L \subset K\}$.

4) For $A \subseteq M \in \mathbf{K}$ let $c\ell_{\mathrm{gr}}(A, M)$ be the closure of A under the group operations.

5) We call $M \in \mathbf{K}_{\mathrm{clf}}$ full when fsb(M) is the set of finite $K \subseteq G_M$.

Claim 6.7. 1) The objects in 6.6 are well defined (in the right class).

2) If $M \in \mathbf{K}_{\mathrm{olf}}$ or $M \in \mathbf{K}_{\mathrm{plf}}$ then $M^{[\mathrm{clf}]} \in \mathbf{K}_{\mathrm{clf}}$ is full.

3) $\mathfrak{S}(\mathbf{K}_{\mathrm{olf}})$ is dense.

4) $\mathfrak{S}(\mathbf{K}_{\mathrm{clf}})$ is dense.

Proof. 1) Straightforward, e.g., in part (3) for K_{clf} the closure is well defined because fsb(M) is closed under intersections.

2) Easy, too.

3),4) As in §2. $\square_{6.7}$

Remark 6.8. Call $M \in \mathbf{K}_{\mathrm{clf}}$ invariant <u>when</u> for every finite $K \subseteq G_M$ there is a function $F_K^M : G \to G$ such that $F_K^M(g) \in gK$ and is equal to $F_n^M(a_0, \ldots, a_{n-1})$ when $a_0, \ldots, a_{n-1}$ list the members of K with no repetitions. Restricting ourselves to such M seems to cause problems in amalgamations, whereas for $\mathbf{K}_{\mathrm{plf}}$ this is not so.

Definition 6.9. For $M \in \mathbf{K}$ and $n < \omega$ let $\mathbf{S}_{\mathrm{gd}}^n(M)$ be the set of good n-types $p(\bar{x}) \in \mathbf{S}_{\mathrm{bs}}^n(M)$ which means: $p = \mathrm{tp}(\bar{a}, M, N)$ where $M \subseteq N \in \mathbf{K}$ and $\bar{a} \in {}^n N$ and $c\ell_{\mathrm{gr}}(\bar{a} + M, N) = c\ell(\bar{a} + M, N)$.

Claim 6.10. The classes $\mathbf{K} = \mathbf{K}_{\mathrm{cfl}}, \mathbf{K}_{\mathrm{plf}}, \mathbf{K}_{\mathrm{olf}}$ have dense closed $\mathfrak{S} \subseteq \Omega[K]$.

Proof. Straightforward. $\qquad\qquad\qquad\qquad\qquad\qquad\qquad\qquad$ $\square_{6.10x}$

$$* \qquad * \qquad *$$

Definition 6.11. 1) Let $\mathbf{K}_{\mathrm{sl}}$ be the class of locally finite semi-groups, i.e., G, it has only one operation, binary which is associative.
2) Let $\mathbf{K}_{\mathrm{usl}}$ be defined similarly with an individual constant e such that $G \models ge_G = g = e_G g$ for every $g \in G \in \mathbf{K}_{\mathrm{usl}}$.

Bibliography

[BG03] Gábor Braun and Rüdiger Göbel, *Outer automorphisms of locally finite p-groups*, J. Algebra **264** (2003), no. 1, 55–65.

[DG93] Manfred Dugas and Rüdiger Göbel, *On locally finite p-groups and a problem of Philip Hall's*, J. Algebra **159** (1993), no. 1, 115–138.

[GgSh:83] Donato Giorgetta and Saharon Shelah, *Existentially closed structures in the power of the continuum*, Annals of Pure and Applied Logic **26** (1984), 123–148, Proceedings of the 1980/1 Jerusalem Model Theory year.

[GrSh:174] Rami Grossberg and Saharon Shelah, *On universal locally finite groups*, Israel Journal of Mathematics **44** (1983), 289–302.

[Hic78] Ken Hickin, *Complete universal locally finite groups*, Transactions of the American Mathematical Society **239** (1978), 213–227.

[KW73] Otto H. Kegel and Bertram A.F. Wehrfritz, *Locally finite groups*, xi+210.

[McSh:55] Angus Macintyre and Saharon Shelah, *Uncountable universal locally finite groups*, Journal of Algebra **43** (1976), 168–175.

[Sh:c] Saharon Shelah, *Classification theory and the number of nonisomorphic models*, Studies in Logic and the Foundations of Mathematics, vol. 92, North-Holland Publishing Co., Amsterdam, xxxiv+705 pp, 1990.

[Sh:f] _____, *Proper and improper forcing*, Perspectives in Mathematical Logic, Springer, 1998.

[Sh:g] _____, *Cardinal Arithmetic*, Oxford Logic Guides, vol. 29, Oxford University Press, 1994.

[Sh:h] _____, *Classification Theory for Abstract Elementary Classes I*, Studies in Logic: Mathematical logic and foundations, vol. 18, College Publications, 2009.

[Sh:i] _____, *Classification Theory for Abstract Elementary Classes II*, Studies in Logic: Mathematical logic and foundations, vol. 20, College Publications, 2009.

[Sh:3] ———, *Finite diagrams stable in power*, Annals of Mathematical Logic **2** (1970), 69–118.

[Sh:E62] Saharon Shelah, *Combinatorial background for Non-structure*, arxiv:math.LO/1512.04767.

[Sh:88r] Saharon Shelah, *Abstract elementary classes near* $\aleph_1$, Chapter I in [Sh:h]. arxiv:0705.4137.

[ShZi:96] Saharon Shelah and Martin Ziegler, *Algebraically closed groups of large cardinality*, The Journal of Symbolic Logic **44** (1979), 522–532, arxiv:arXiv.

[Sh:300] Saharon Shelah, *Universal classes*, Classification theory (Chicago, IL, 1985), Lecture Notes in Mathematics, vol. 1292, Springer, Berlin, 1987, Proceedings of the USA–Israel Conference on Classification Theory, Chicago, December 1985; ed. Baldwin, J.T., pp. 264–418.

[Sh:300a] ———, *Stability theory for a model*, Chapter V (A), in [Sh:i].

[Sh:300b] ———, *Universal Classes: Axiomatic Framework*, Chapter V (B) in [Sh:i].

[Sh:309] ———, *Black Boxes*, , 0812.0656. 0812.0656. arxiv:0812.0656.

[Sh:365] ———, *There are Jonsson algebras in many inaccessible cardinals*, Cardinal Arithmetic, Oxford Logic Guides, vol. 29, Oxford University Press, 1994.

[Sh:900] ———, *Dependent theories and the generic pair conjecture*, Communications in Contemporary Mathematics **17** (2015), 1550004 (64 pps.), arxiv:math.LO/0702292.

[Sh:950] ———, *Dependent dreams: recounting types*, arxiv:1202.5795.

[Sh:1098] ———, *Excl LF groups with few automorphisms*, preprint.

[Tho86] Simon Thomas, *Complete universal locally finite groups of large cardinality*, Logic colloquium '84 (Manchester, 1984), 277-301, Stud. Logic Found. Math., 120, North-Holland, Amsterdam, 1986.

[Woo72] Carol Wood, *Forcing for infinitary languages*, Z. Math. Logik Grundlagen Math. **18** (1972), 385–402.

Chapter 8

Analytic Zariski structures and non-elementary categoricity

Boris Zilber
University of Oxford
Oxford, UK

The notion of an analytic Zariski structure was introduced in [1] by the author and N. Peatfield in a form slightly different from the one presented here, and then in [4], Ch. 6, in the current form. Analytic Zariski generalises the previously known notion of a Zariski structure. The latter has been defined as a structure M with a Noetherian topology on all cartesian powers M^n of the universe, the closed sets of which are given by positive quantifier-free formulas. Any closed set is assigned a dimension which behaves in a certain way (modelled on algebraic geometry) with regards to projection maps $M^n \to M^m$, see the *addition formula* (AF) and the *fibre condition* (FC) in section 1 below.

In the definition of analytic Zariski structures, we drop the requirement of Noetherianity. This leads to a considerably more flexible and broader notion at the cost of a longer list of assumptions modelled on the properties of analytic subsets of complex manifolds.

In [1] we assumed that the Zariski structure is compact (or compactifiable); here we drop this assumption, which may be too restrictive in applications.

We remark that in the broad setting it is appropriate to consider the notion of a Zariski structure as belonging to *positive model theory* in the sense of Ben-Yaacov [5].

The class of analytic Zariski structures is much broader and geometrically richer than the class of Noetherian Zariski structures. The main examples come from two sources:

(i) structures which are constructed in terms of complex analytic functions and relations;

(ii) "new stable structures" introduced by Hrushovski's construction; in many cases, these objects exhibit properties similar to those of class (i).

However, although there are concrete examples for both (i) and (ii), in many cases we lack the technology to prove that the structure is analytic Zariski. In particular, despite some attempts the conjecture that $\mathbb{C}_{\exp}$ is analytic Zariski, assuming it satisfies axioms of pseudo-exponentiation (see [17]), is still open.

The aim of this chapter is to carry out a model-theoretic analysis of analytic Zariski structures in the appropriate language. Recall that if M is a Noetherian Zariski structure, the relevant key model-theoretic result states that its first-order theory allows elimination of quantifiers and is ω-stable of finite Morley rank. In particular, it is strongly minimal (and so uncountably categorical) if $\dim M = 1$ and M is irreducible.

For analytic Zariski 1-dimensional M, we carry out a model theoretic study in the spirit of the theory of *abstract elementary classes*. We start by introducing a suitable countable fragment of the family of basic Zariski relations and a correspondent substructure of constants over which all the further analysis is carried out. Then we proceed to the analysis of the notion of dimension of Zariski closed sets and define more delicate notions of the *predimension* and *dimension* of a tuple in M. In fact, by doing this, we reinterpret dimensions which are present in every analytic structure in terms familiar to many from Hrushovski's construction, thus establishing once again conceptual links between classes (i) and (ii).

Our main results are proved under assumption that M is one-dimensional (as an analytic Zariski structure) and irreducible. No assumption on presmoothness is needed. We prove for such an M, in the terminology of [16]:

(1) M *is a quasiminimal pregeometry structure with regards to a closure operator* cl *associated with the predimension;*

(2) M *has quantifier-elimination to the level of* ∃*-formulas in the following sense: every two tuples which are (first-order)* ∃*-equivalent over a countable submodel, are* $L_{\infty,\omega}$ *equivalent;*

(3) *The abstract elementary class associated with* M *is categorical in uncountable cardinals and is excellent.*

In fact, (3) is a corollary of (1) using the main result of [16], so the main work is in proving (1) which involves (2) as an intermediate step.

Note that the class of 1-dimensional Noetherian Zariski structures is essentially classifiable by the main result of [2], and in particular the class contains no instances of structures obtained by the proper Hrushovski construction. The class of analytic Zariski structures, in contrast, is consistent with Hrushovski's construction and at the same time, by the result above, has excellent model-theoretic properties. This gives a hope for a classification theory based on the relevant notions.

However, it must be mentioned that some natural questions in this context are widely open. In particular, we have no classification for presmooth analytic Zariski groups (with the graph of multiplication analytic). It is not known if a 1-dimensional irreducible presmooth analytic group has to be abelian. See related analysis of groups in [9].

Acknowledgment. I want to express my thanks to Assaf Hasson who saw a very early version of this work and made many useful comments, and also to Levon Haykazyan who through his own contributions to the theory of quasiminimality kept me informed in the recent developments in the field.

1   Analytic *L*-Zariski structures

Let M = $(M; L)$ be a structure with primitives (basic relations) L. We use also the extension $L(M_0)$ of the language L with names for points of a subset M_0 of M.

We introduce a topology on M^n, for all $n \geq 1$, by declaring a subset $P \subseteq M^n$ closed if there is an n-type p consisting of quantifier-free positive formulas with parameters in M such that

$$P = \{a \in M^n : \text{M} \vDash p(a)\}.$$

In other words, the sets defined by atomic $L(M)$-formulae form a basis for the topology.

We say P is L-closed ($L(M_0)$-closed) if p is over $\emptyset$ (over M_0).

1.1 Remark

Note that it follows that projections

$$\text{pr}_{i_1,\ldots,i_m} : M^n \to M^m, \quad \langle x_1, \ldots, x_n \rangle \mapsto \langle x_{i_1}, \ldots, x_{i_m} \rangle$$

are continuous in the sense that the inverse image of a closed set under a projection is closed. Indeed, $\text{pr}^{-1}_{i_1,\ldots,i_m} S = S \times M^{n-m}$.

We will drop the subscript in $\text{pr}_{i_1,\ldots,i_m}$ when it is clear from the context.

We write $X \subseteq_{op} V$ to say that X is open in V and $X \subseteq_{cl} V$ to say it is closed. The latter means that $X = V \cap S$, for some $S \subseteq M^n$ closed in M^n. The former, that $X = V \setminus S$.

We say that $P \subseteq M^n$ is **constructible** if P is a finite union of some sets S, such that $S \subseteq_{cl} U \subseteq_{op} M^n$.

A subset $P \subseteq M^n$ will be called **projective** if P is a union of finitely many sets of the form $\text{pr}\, S$, for some $S \subseteq_{cl} U \subseteq_{op} M^{n+k}$, $\text{pr} : M^{n+k} \to M^n$.

We say that P is L-constructible or L-projective if P is defined over L.

Note that any set S such that $S \subseteq_{cl} U \subseteq_{op} M^{n+k}$, is constructible, a projection of a constructible set is projective and that any constructible set is projective.

1.2 Dimension

To any nonempty projective S a non-negative integer $\dim S$, called **the dimension of** S, is attached.

We assume:

(SI) (**Strong Irreducibility**) For an irreducible set $S \subseteq_{cl} U \subseteq_{op} M^n$ (that is S is not a proper union of two closed in S subsets) and any closed subset $S' \subseteq_{cl} S$,

$$\dim S' = \dim S \Rightarrow S' = S;$$

(DP) (**Dimension of Points**) for a nonempty projective S, $\dim S = 0$ if and only if S is at most countable.

(CU) (**Countable Unions**) If $S = \bigcup_{i \in \mathbb{N}} S_i$, all projective, then $\dim S =$

$\max_{i \in \mathbb{N}} \dim S_i$;

(WP) (**Weak Properness**) Given an irreducible $S \subseteq_{cl} U \subseteq_{op} M^n$ and $F \subseteq_{cl} V \subseteq_{op} M^{n+k}$ with the projection $\mathrm{pr} : M^{n+k} \to M^n$ such that $\mathrm{pr}\, F \subseteq S$ and $\dim \mathrm{pr}\, F = \dim S$, there exists $D \subseteq_{op} S$ such that $D \subseteq \mathrm{pr}\, F$.

1.3 Remark

(CU) in the presence of the descending chain condition implies the *essential uncountability property* (EU) usually assumed for Noetherian Zariski structures.

We postulate further, for an irreducible $S \subseteq_{cl} U \subseteq_{op} M^{n+k}$, a projection $\mathrm{pr} : M^{n+k} \to M^n$ and its fibres $S_u := \mathrm{pr}^{-1}(u) \cap S$ on S over $u \in \mathrm{pr} S$:

(AF) $\dim \mathrm{pr}\, S = \dim S - \min_{u \in \mathrm{pr}\, S} \dim S_u$;

(FC) The set $\{a \in \mathrm{pr}\, S : \dim S_a \geq m\}$ is of the form $T \cap \mathrm{pr}\, S$ for some constructible T, and there exists an open set V such that $V \cap \mathrm{pr}\, S \neq \emptyset$ and
$$\min_{a \in \mathrm{pr}\, S} \dim S_a = \dim S_v, \text{ for any } v \in V \cap \mathrm{pr}\,(S).$$

The following helps to understand the dimension of projective sets.

1.4 Lemma

Let $P = \mathrm{pr}\, S \subseteq M^n$, for S irreducible constructible, and $U \subseteq_{op} M^n$ with $P \cap U \neq \emptyset$. Then
$$\dim P \cap U = \dim P.$$

Proof. We can write $P \cap U = \mathrm{pr}\, S' = P'$, where $S' = S \cap \mathrm{pr}^{-1} U$ constructible irreducible, $\dim S' = \dim S$ by (SI). By (FC), there is $V \subseteq_{op} M^n$ such that for all $c \in V \cap P$,
$$\dim S_c = \min_{a \in P} \dim S_a = \dim S - \dim P.$$

Note that $\mathrm{pr}^{-1} U \cap \mathrm{pr}^{-1} V \cap S \neq \emptyset$, since S is irreducible. Taking $s \in \mathrm{pr}^{-1} U \cap \mathrm{pr}^{-1} V \cap S$ and $c = \mathrm{pr}\, s$ we get, using (AF) for S',
$$\dim S'_c = \dim S_c = \min_{a \in P'} \dim S_a = \dim S - \dim P'.$$

So, $\dim P' = \dim P$. $\square$

1.5   Analytic subsets

A subset S, $S \subseteq_{cl} U \subseteq_{op} M^n$, is called **analytic in** U if for every $a \in S$ there is an open $V_a \subseteq_{op} U$ such that $a \in V_a$ and $S \cap V_a$ is the union of finitely many closed in V_a irreducible subsets. We write $S \subseteq_{an} U$ accordingly.

We postulate the following properties:

(INT) (**Intersections**) If $S_1, S_2 \subseteq_{an} U$ are irreducible, then $S_1 \cap S_2$ is analytic in U;

(CMP) (**Components**) If $S \subseteq_{an} U$ and $a \in S$, a closed point, then there is $S_a \subseteq_{an} U$, a finite union of irreducible analytic subsets of U, and some $S'_a \subseteq_{an} U$ such that $a \in S_a \setminus S'_a$ and $S = S_a \cup S'_a$;

Each of the irreducible subsets of S_a above is called an **irreducible component of S containing** a;

(CC) (**Countability of the number of components**) Any $S \subseteq_{an} U$ is a union of at most countably many irreducible components.

1.6   Remark

For S analytic and $a \in \text{pr} \, S$, the fibre S_a is analytic.

1.7   Lemma

If $S \subseteq_{an} U$ is irreducible, V open, then $S \cap V$ is an irreducible analytic subset of V and, if non-empty, $\dim S \cap V = \dim S$.

Proof. Immediate.□

1.8   Lemma

(i) $\emptyset$, any singleton and U are analytic in U;

(ii) If $S_1, S_2 \subseteq_{an} U$, then $S_1 \cup S_2$ is analytic in U;

(iii) If $S_1 \subseteq_{an} U_1$ and $S_2 \subseteq_{an} U_2$, then $S_1 \times S_2$ is analytic in $U_1 \times U_2$;

(iv) If $S \subseteq_{an} U$ and $V \subseteq U$ is open, then $S \cap V \subseteq_{an} V$;

(v) If $S_1, S_2 \subseteq_{an} U$, then $S_1 \cap S_2$ is analytic in U.

Proof. Immediate. □

1.9 Definition

Given a subset $S \subseteq_{cl} U \subseteq_{op} M^n$ we define the notion of the **analytic rank** of S in U, $\mathrm{ark}_U(S)$, which is a natural number satisfying

1. $\mathrm{ark}_U(S) = 0$ iff $S = \emptyset$;

2. $\mathrm{ark}_U(S) \leq k + 1$ iff there is a set $S' \subseteq_{cl} S$ such that $\mathrm{ark}_U(S') \leq k$ and with the set $S^0 = S \setminus S'$ being analytic in $U \setminus S'$.

Obviously, any nonempty analytic subset of U has analytic rank 1.

The next assumption guarantees that the class of analytic subsets explicitly determines the class of closed subsets in M.

(AS) [**Analytic stratification**] For any $S \subseteq_{cl} U \subseteq_{op} M^n$, $\mathrm{ark}_U S$ is defined and is finite.

We will justify this nonobvious property later in 3.10 and 3.11.

1.10 Lemma

For any $S \subseteq_{cl} U \subseteq_{op} M^n$,

$$\dim \mathrm{pr}\, S + \min_{a \in \mathrm{pr}\, S} \dim S_a \geq \dim S.$$

Proof. We use (AS) and prove the statement by induction on $\mathrm{ark}_U S \geq 1$.

For $\mathrm{ark}_U S = 1$, S is analytic in U and so by (CC) is the union of countably many irreducibles $S^{(i)}$. By (AF)

$$\dim \mathrm{pr}\, S^{(i)} + \min_{a \in \mathrm{pr}\, S^{(i)}} \dim S_a^{(i)} \geq \dim S^{(i)}$$

and so by (CU) lemma follows.□

1.11 Presmoothness

The following property (which we are not going to use in the context of the present chapter) is relevant.

(PS) [**Presmoothness**] If $S_1, S_2 \subseteq_{an} U \subseteq_{op} M^n$ and S_1, S_2 and U irreducible, then for any irreducible component S_0 of $S_1 \cap S_2$

$$\dim S_0 \geq \dim S_1 + \dim S_2 - \dim U.$$

1.12   Definition

An L-structure M is said to be **analytic L-Zariski** if

- M satisfies (SI), (WP), (CU), (INT), (CMP), (CC), (AS);

- the expansion $M^{\sharp}$ of M to the language $L(M)$ (names for points in M added) satisfies all the above with the dimension extending the one for M;

- $M^{\sharp}$ also satisfies (AF) and (FC) with V in (FC) being L-definable whenever S is.

An analytic Zariski structure will be called **presmooth** if it has the presmoothness property (PS).

---

2   Model theory of analytic Zariski structures

For the rest of the section, we assume that M be analytic L-Zariski and assume L is countable.

2.1   Lemma

There is a countable $M_0 \preccurlyeq M$ such that for any $L(M_0)$-closed set S any irreducible component P of S is $L(M_0)$-closed.

Proof. Use the standard Löwenheim-Skolem downward arguments. □

We call such M_0 a **core substructure (subset) of** M.

2.2   Assumption

By extending L to $L(M_0)$ we assume that the set of L-closed points is the core subset.

For finite subset X of M of size n we denote $\vec{X}$ an n-tuple with range X.

2.3   Definition

For finite $X \subseteq M$ we define the **predimension**

$$\delta(X) = \min\{\dim S : \ \vec{X} \in S, \ S \subseteq_{an} U \subseteq_{op} M^n, \ S \text{ is } L\text{-constructible}\}, \quad (8.1)$$

relative predimension for finite $X, Y \subseteq M$

$$\delta(X/Y) = \min\{\dim S : \vec{X} \in S, \ S \subseteq_{an} U \subseteq_{op} M^n, \ S \text{ is } L(Y)\text{-constructible}\}, \tag{8.2}$$

and **dimension of** X

$$\partial(X) = \min\{\delta(XY) : \text{ finite } Y \subset M\}.$$

(Here and below XY means $X \cup Y$ and $Xy = X \cup \{y\}$).

We call a minimal S as in (8.2) an **analytic locus of** X **over** Y.

For $X \subseteq M$ finite, we say that X is **self-sufficient** and write $X \leq M$, if $\partial(X) = \delta(X)$.

For infinite $A \subseteq M$ we say $A \leq M$ if for any finite $X \subseteq A$ there is a finite $X \subseteq X' \subseteq A$ such that $X' \leq M$.

2.4

For the rest of the chapter we assume that $\dim M = 1$ and M is irreducible. This is an analogue of an analytic curve.

Note that we then have

$$0 \leq \delta(Xy) \leq \delta(X) + 1, \text{ for any } y \in M,$$

since $\vec{X}y \in S \times M$.

2.5 Lemma

Given $F \subseteq_{an} U \subseteq_{op} M^k$, $\dim F > 0$, there is $i \leq k$ such that for $\mathrm{pr}_i :$ $(x_1, \ldots, x_k) \mapsto x_i$,

$$\dim \mathrm{pr}_i F > 0.$$

Proof. Use (AF) and induction on k. $\square$

2.6 Proposition

Let $P = \mathrm{pr}\, S$, for some L-constructible $S \subseteq_{an} U \subseteq_{op} M^{n+k}$, $\mathrm{pr} : M^{n+k} \to M^n$. Then

$$\dim P = \max\{\partial(x) : x \in P(M)\}. \tag{8.3}$$

Moreover, this formula is true when $S \subseteq_{cl} U \subseteq_{op} M^{n+k}$.

Proof. We use induction on $\dim S$.

We first note that by induction on $\mathrm{ark}_U S$, if (8.3) holds for all analytic S

of dimension less or equal to k then it holds for all closed S of dimension less or equal to k.

The statement is obvious for $\dim S = 0$ and so we assume that $\dim S > 0$ and for all analytic S' of lower dimension the statement is true.

By (CU) and (CMP) we may assume that S is irreducible. Then by (AF)

$$\dim P = \dim S - \dim S_c \tag{8.4}$$

for any $c \in P \cap V$ (such that S_c is of minimal dimension) for some open L-constructible V.

Claim 1. It suffices to prove the statement of the proposition for the projective set $P \cap V'$, for some L-open $V' \subseteq_{op} M^n$.

Indeed,

$$P \cap V' = \mathrm{pr}(S \cap \mathrm{pr}^{-1} V'), \quad S \cap \mathrm{pr}^{-1} V' \subseteq_{cl} \mathrm{pr}^{-1} V' \cap U \subseteq_{op} M^{n+k}.$$

And $P \setminus V' = \mathrm{pr}(S \cap T)$, $T = \mathrm{pr}^{-1}(M^n \setminus V') \in L$. So, $P \setminus V'$ is the projection of a proper analytic subset, of lower dimension. By induction, for $x \in P \setminus V'$, $\partial(x) \leq \dim P \setminus V' \leq \dim P$ and hence, using 1.4,

$$\dim P \cap V' = \max\{\partial(x) : \ x \in P \cap V'\} \Rightarrow \dim P = \max\{\partial(x) : \ x \in P\}.$$

Claim 2. The statement of the proposition holds if $\dim S_c = 0$ in (8.4).

Proof. Given $x \in P$ choose a tuple $y \in M^k$ such that $S(x^\frown y)$ holds. Then $\delta(x^\frown y) \leq \dim S$. So we have $\partial(x) \leq \delta(x^\frown y) \leq \dim S = \dim P$.

It remains to notice that there exists $x \in P$ such that $\partial(x) \geq \dim P$.

Consider the L-type

$$x \in P \ \& \{x \notin R : \ \dim R \cap P < \dim P \text{ and } R \text{ is projective}\}.$$

This is realised in M, since otherwise $P = \bigcup_R (P \cap R)$ which would contradict (CU).

For such an x let y be a tuple in M such that $\delta(x^\frown y) = \partial(x)$. By definition there exist $S' \subseteq_{an} U' \subseteq_{op} M^m$ such that $\dim S' = \delta(x^\frown y)$. Let $P' = \mathrm{pr}\, S'$, the projection into M^n. By our choice of x, $\dim P' \geq \dim P$. But $\dim S' \geq \dim P'$. Hence, $\partial(x) \geq \dim P$. Claim proved.

Claim 3. There is a L-constructible $R \subseteq_{an} S$ such that all the fibres R_c of the projection map $R \to \mathrm{pr}\, R$ are 0-dimensional and $\dim \mathrm{pr}\, R = \dim P$.

Proof. We have by construction $S_c \subseteq M^k$. Assuming $\dim S_c > 0$ on every open subset we show that there is a $b \in M_0$ such that (up to the order of coordinates) $\dim S_c \cap \{b\} \times M^{k-1} < \dim S_c$, for all $c \in P \cap V' \neq \emptyset$, for some open $V' \subseteq V$ and $\dim \mathrm{pr}\, S_c \cap \{b\} \times M^{k-1} = \dim P$. By induction on $\dim S$ this will prove the claim.

To find such a b choose $a \in P \cap V$ and note that by 2.5, up to the order of coordinates, $\dim \mathrm{pr}_1 S(a, M) > 0$, where $\mathrm{pr}_1 : M^k \to M$ is the projection on the first coordinate.

Consider the projection $\mathrm{pr}_{M^n,1} : M^{n+k} \to M^{n+1}$ and the set $\mathrm{pr}_{M^n,1}S$. By (AF) we have

$$\dim \mathrm{pr}_{M^n,1}S = \dim P + \dim \mathrm{pr}_1 S_a = \dim P + 1.$$

Using (AF) again for the projection $\mathrm{pr}_1 : M^{n+1} \to M$ with the fibres $M^n \times \{b\}$, we get, for all b in some open subset of M,

$$1 \geq \dim \mathrm{pr}_1 \mathrm{pr}_{M^n,1}S = \dim \mathrm{pr}_{M^n,1}S - \dim[\mathrm{pr}_{M^n,1}S] \cap [M^n \times \{b\}] =$$

$$= \dim P + 1 - \dim[\mathrm{pr}_{M^n,1}S] \cap [M^n \times \{b\}].$$

Hence $\dim[\mathrm{pr}_{M^n,1}S] \cap [M^n \times \{b\}] \geq \dim P$, for all such b, which means that the projection of the set $S_b = S \cap (M^n \times \{b\} \times M^{k-1})$ on M^n is of dimension $\dim P$, which finishes the proof if $b \in M_0$. But $\dim S_b = \dim S - 1$ for all $b \in M \cap V'$, some L-open V', so for any $b \in M_0 \cap V'$. The latter is not empty since (M_0, L) is a core substructure. This proves the claim.

Claim 4. Given R satisfying Claim 3,

$$P \setminus \mathrm{pr}\, R \subseteq \mathrm{pr}\, S', \text{ for some } S' \subseteq_{\mathrm{cl}} S, \dim S' < \dim S.$$

Proof. Consider the cartesian power

$$M^{n+2k} = \{x^\frown y^\frown z : x \in M^n, y \in M^k, z \in M^k\}$$

and its L-constructible subset

$$R\&S := \{x^\frown y^\frown z : x^\frown z \in R \,\&\, x^\frown y \in S\}.$$

Clearly $R\&S \subseteq_{an} W \subseteq_{op} M^{n+2k}$, for an appropriate L-constructible W.

Now notice that the fibres of the projection $\mathrm{pr}_{xy} : x^\frown y^\frown z \mapsto x^\frown y$ over $\mathrm{pr}_{xy}R\&S$ are 0-dimensional and so, for some irreducible component $(R\&S)^0$ of the analytic set $R\&S$, $\dim \mathrm{pr}_{xy}(R\&S)^0 = \dim S$. Since $\mathrm{pr}_{xy}R\&S \subseteq S$ and S irreducible, we get by (WP) $D \subseteq \mathrm{pr}_{xy}R\&S$ for some $D \subseteq_{op} S$. Clearly

$$\mathrm{pr}\, R = \mathrm{pr}\, \mathrm{pr}_{xy}R\&S \supseteq \mathrm{pr}\, D$$

and $S' = S \setminus D$ satisfies the requirement of the claim.

Now we complete the proof of the proposition: By Claims 2 and 3

$$\dim P = \max_{x \in \mathrm{pr}\, R} \partial(x).$$

By induction on $\dim S$, using Claim 4, for all $x \in P \setminus \mathrm{pr}\, R$,

$$\partial(x) \leq \dim \mathrm{pr}\, S' \leq \dim P.$$

The statement of the proposition follows. $\square$

In what follows a L-substructure of M is a L-structure on a subset $N \supseteq M_0$. Recall that L is purely relational.

Recall the following well-known fact, see [10].

2.7   Karp's characterisation of $\equiv_{\infty,\omega}$

Given $a, a' \in M^n$ the $L_{\infty,\omega}(L)$-types of the two n-tuples in M are equal if and only if they are *back and forth equivalent* that is there is a nonempty set I of isomorphisms of L-substructures of M such that $a \in \operatorname{Dom} f_0$ and $a' \in \operatorname{Range} f_0$, for some $f_0 \in I$, and

(forth) for every $f \in I$ and $b \in M$ there is a $g \in I$ such that $f \subseteq g$ and $b \in \operatorname{Dom} g$;

(back) For every $f \in I$ and $b' \in M$ there is a $g \in I$ such that $f \subseteq g$ and $b' \in \operatorname{Range} g$.

2.8   Definition

For $a \in M^n$, the **projective type of a over M** is

$$\{P(x): \ a \in P, \ P \text{ is a projective set over } L\}\cup$$

$$\cup\{\neg P(x): \ a \notin P, \ P \text{ is a projective set over } L\}.$$

2.9   Lemma

Suppose $X \leq M$, $X' \leq M$ and the (first-order) quantifier-free L-type of X is equal to that of X'. Then the $L_{\infty,\omega}(L)$-types of X and X' are equal.

Proof. We are going to construct a back-and-forth system for X and X'.

Let $S_X \subseteq_{an} V \subseteq_{op} M^n$, S_X irreducible, all L-constructible, and such that $X \in S_X(M)$ and $\dim S_X = \delta(X)$.

Claim 1. The quantifier-free L-type of X (and X') is determined by formulas equivalent to $S_X \cap V'$, for V' open such that $X \in V'(M)$.

Proof. Use the stratification of closed sets (AS) to choose L-constructible $S \subseteq_{cl} U \subseteq_{op} M^n$ such that $X \in S$ and $\operatorname{ark}_U S$ is minimal. Obviously then $\operatorname{ark}_U S = 0$, that is $S \subseteq_{an} U \subseteq_{op} M^n$. Now S can be decomposed into irreducible components, so we may choose S to be irreducible. Among all such S choose one which is of minimal possible dimension. Obviously $\dim S = \dim S_X$, that is we may assume that $S = S_X$. Now clearly any constructible set $S' \subseteq_{cl} U' \subseteq_{op} M^n$ containing X must satisfy $\dim S' \cap S_X \geq \dim S_X$, and this condition is also sufficient for $X \in S'$.

Let y be an element of M. We want to find a finite Y containing y and an Y' such that the quantifier-free type of XY is equal to that of $X'Y'$ and both are self-sufficient in M (recall that $XY := X \cup Y$). This, of course, extends the partial isomorphism $X \to X'$ to $XY \to X'Y'$ and will prove the lemma.

We choose Y to be a minimal set containing y and such that $\delta(XY)$ is also minimal, that is

$$1 + \delta(X) \geq \delta(Xy) \geq \delta(XY) = \partial(XY)$$

and $XY \leq M$.

We have two cases: $\delta(XY) = \partial(X)+1$ and $\delta(XY) = \partial(X)$. In the first case $Y = \{y\}$. By Claim 1 the quantifier-free L-type r_{Xy} of Xy is determined by the formulas of the form $(S_X \times M) \setminus T, T \subseteq_{cl} M^{n+1}, T \in L, \dim T < \dim(S_X \times M)$.

Consider

$$r_{Xy}(X', M) = \{z \in M : X'z \in (S_X \times M) \setminus T, \ \dim T < \dim S_X, \ \text{all } T\}.$$

We claim that $r_{Xy}(X', M) \neq \emptyset$. Indeed, otherwise M is the union of countably many sets of the form $T(X', M)$. But the fibres $T(X', M)$ of T are of dimension 0 (since otherwise $\dim T = \dim S_X + 1$, contradicting the definition of the T). This is impossible, by (CU).

Now we choose $y' \in r_{Xy}(X', M)$ and this is as required.

In the second case, by definition, there is an irreducible $R \subseteq_{an} U \subseteq_{op} M^{n+k}$, $n = |X|, k = |Y|$, such that $XY \in R(M)$ and $\dim R = \delta(XY) = \partial(X)$. We may assume $U \subseteq V \times M^k$.

Let $P = \operatorname{pr} R$, the projection into M^n. Then $\dim P \leq \dim R$. But also $\dim P \geq \partial(X)$, by 2.6. Hence, $\dim R = \dim P$. On the other hand, $P \subseteq S_X$ and $\dim S_X = \delta(X) = \dim P$. By axiom (WP) we have $S_X \cap V' \subseteq P$ for some L-constructible open V'.

Hence $X' \in S_X \cap V' \subseteq P(M)$, for P the projection of an irreducible analytic set R in the L-type of XY. By Claim 1 the quantifier-free L-type of XY is of the form

$$r_{XY} = \{R \setminus T : T \subseteq_{cl} R, \ \dim T < \dim R\}.$$

Consider

$$r_{XY}(X', M) = \{Z \in M^k : X'Z \in R \setminus T, \ T \subseteq_{cl} R, \ \dim T < \dim R\}.$$

We claim again that $r_{XY}(X', M) \neq \emptyset$. Otherwise the set $R(X', M) = \{X'Z : R(X'Z)\}$ is the union of countably many subsets of the form $T(X', M)$. But $\dim T(X', M) < \dim R(X', M)$ as above, by (AF).

Again, an $Y' \in r_{XY}(X', M)$ is as required. □

2.10 Corollary

There are at most countably many $L_{\infty, \omega}(L)$-types of tuples $X \leq M$.

Indeed, any such type is determined uniquely by the choice of a L-constructible $S_X \subseteq_{an} U \subseteq_{op} M^n$ such that $\dim S_X = \partial(X)$.

2.11 Lemma

Suppose, for finite $X, X' \subseteq M$, the projective L-types of X and X' coincide. Then the $L_{\infty, \omega}(L)$-types of the tuples are equal.

Proof. Choose finite Y such that $\partial(X) = \delta(XY)$. Then $XY \leq M$. Let $XY \in S \subseteq_{an} U \subseteq_{op} M^n$ be L-constructible and such that $\dim S$ is minimal possible, that is $\dim S = \delta(XY)$. We may assume that S is irreducible. Notice that for every proper closed L-constructible $T \subseteq_{cl} U$, $XY \notin T$ by dimension considerations.

By assumptions of the lemma $X'Y' \in S$, for some Y' in M. We also have $X'Y' \notin T$, for any T as above, since otherwise a projective formula would imply that $XY'' \in T$ for some Y'', contradicting that $\partial(X) > \dim T$.

We also have $\delta(X'Y') = \dim S$. But for no finite Z' it is possible that $\delta(X'Z') < \dim S$, for then again a projective formula will imply that $\delta(XZ) < \dim S$, for some Z.

It follows that $X'Y' \leq M$ and the quantifier-free types of XY and $X'Y'$ coincide, hence the $L_{\infty,\omega}(L)$-types are equal, by 2.9.$\square$

2.12   Definition

Set, for finite $X \subseteq M$,

$$\mathrm{cl}_L(X) = \{y \in M : \ \partial(Xy) = \partial(X)\}.$$

We fix L and omit the subscript below.

2.13   Lemma

The following two conditions are equivalent:

(a) $b \in \mathrm{cl}(A)$, for $\vec{A} \in M^n$;

(b) $b \in P(\vec{A}, M)$ for some projective first-order $P \subseteq M^{n+1}$ such that $P(\vec{A}, M)$ is at most countable.

In particular, $\mathrm{cl}(A)$ is countable for any finite A.

Proof. Let $d = \partial(A) = \delta(AV)$, and $\delta(AV)$ is minimal for all possible finite $V \subseteq M$. So by definition $d = \dim S_0$, some analytic irreducible S_0 such that $\vec{AV} \in S_0$ and S_0 of minimal dimension. This corresponds to a L-definable relation $S_0(x, v)$, where x, v strings of variables of length n, m

First assume (b), that is, that b belongs to a countable $P(\vec{A}, M)$. By definition

$$P(x, y) \equiv \exists w \, S(x, y, w),$$

for some analytic $S \subseteq M^{n+1+k}$, some tuples x, y, w of variables of length $n, 1$ and k respectively, and the fibre $S(\vec{A}, b, M^k)$ is nonempty. We also assume that P and S are of minimal dimension, answering this description. By (FC), (AS) and minimality we may choose S so that $\dim S(\vec{A}, b, M^k)$ is minimal among all the fibres $S(\vec{A}', b', M^k)$.

Consider the analytic set $S^\sharp \subseteq M^{n+m+1+k}$ given by $S_0(x, v) \,\&\, S(x, y, w)$. By (AF), considering the projection of the set on (x, v)-coordinates,

$$\dim S^\sharp \leq \dim S_0 + \dim S(\vec{A}, M, M^k),$$

since $S(\vec{A}, M, M^k)$ is a fibre of the projection. Now we note that by countability $\dim S(\vec{A}, M, M^k) = \dim S(\vec{A}, b, M^k)$, so

$$\dim S^\sharp \leq \dim S_0 + \dim S(\vec{A}, b, M^k).$$

Now the projection $\mathrm{pr}_w S^\sharp$ along w (corresponding to $\exists w\, S^\sharp$) has fibres of the form $S(\vec{X}, y, M^k)$, so by (AF)

$$\dim \mathrm{pr}_w S^\sharp \leq \dim S_0 = d.$$

Projecting further along v we get $\dim \mathrm{pr}_v \mathrm{pr}_w S^\sharp \leq d$, but $\vec{A}b \in \mathrm{pr}_v \mathrm{pr}_w S^\sharp$ so by Proposition 2.6 $\partial(\vec{A}b) \leq d$. The inverse inequality holds by definition, so the equality holds. This proves that $b \in \mathrm{cl}(A)$.

Now assume (a), that is $b \in \mathrm{cl}(A)$. So, $\partial(\vec{A}b) = \partial(\vec{A}) = d$. By definition there is a projective set P containing $\vec{A}b$, defined by the formula $\exists w\, S(x, y, w)$ for some analytic S, $\dim S = d$. Now $\vec{A}$ belongs to the projective set $\mathrm{pr}_y P$ (defined by the formula $\exists y \exists w\, S(x, y, w)$) so by Proposition 2.6 $d \leq \dim \mathrm{pr}_y P$, but $\dim \mathrm{pr}_y P \leq \dim P \leq \dim S = d$. Hence all the dimensions are equal and so, the dimension of the generic fibre is 0. We may assume, as above, without loss of generality that all fibres are of minimal dimension, so

$$\dim S(\vec{A}, M, M^k) = 0.$$

Hence, b belongs to a 0-dimensional set $\exists w\, S(\vec{A}, y, w)$, which is projective and countable. $\square$

2.14   Lemma

Suppose $b \in \mathrm{cl}(A)$ and the projective type of $\vec{A}b$ is equal to that of $\vec{A}'b'$. Then $b' \in \mathrm{cl}(A')$.

Proof. First note that, by (FC) and (AS), for analytic $R(u, v)$ and its fibre $R(a, v)$ of minimal dimension one has

$$\mathrm{tp}(a) = \mathrm{tp}(a') \Rightarrow \dim R(a, v) = \dim R(a', v).$$

By the second part of the proof of 2.13, the assumption of the lemma implies that for some analytic S we have $\models \exists w S(\vec{A}, b, w)$ and $\dim S(\vec{A}, M, M^k) = 0$. Hence $\models \exists w S(\vec{A}', b', w)$ and $\dim S(\vec{A}', M, M^k) = 0$. But this immediately implies $b' \in \mathrm{cl}(A')$. $\square$

2.15 Lemma

(i)
$$\mathrm{cl}(\emptyset) = \mathrm{cl}(M_0) = M_0.$$

(ii) Given finite $X \subseteq M$, $y, z \in M$,

$$z \in \mathrm{cl}(X, y) \setminus \mathrm{cl}(X) \Rightarrow y \in \mathrm{cl}(X, z).$$

(iii)
$$\mathrm{cl}(\mathrm{cl}(X)) = \mathrm{cl}(X).$$

Proof. (i) Clearly $M_0 \subseteq \mathrm{cl}(\emptyset)$, by definition.

We need to show the converse, that is if $\partial(y) = 0$, for $y \in M$, then $y \in M_0$. By definition $\partial(y) = \partial(\emptyset) = \min\{\delta(Y) : y \in Y \subset M\} = 0$. So, $y \in Y$, $\vec{Y} \in S \subseteq_{an} U \subseteq_{op} M^n$, $\dim S = 0$. The irreducible components of S are closed points (singletons) and $\{\vec{Y}\}$ is one of them, so must be in M_0, hence $y \in M_0$.

(ii) Assuming the left-hand side of (ii), $\partial(Xyz) = \partial(Xy) > \partial(X)$ and $\partial(Xz) > \partial(X)$. By the definition of ∂ then,

$$\partial(Xy) = \partial(X) + 1 = \partial(Xz),$$

so $\partial(Xzy) = \partial(Xz)$, $y \in \mathrm{cl}(Xz)$.

(iii) Immediate by 2.13. $\square$

Below, if not stated otherwise, we use the language $L^{\exists}$ the primitives of which correspond to relations $\exists$-definable in M. Also, we call a **submodel of** M any $L^{\exists}$-substructure closed under cl.

2.16 Theorem

(i) Every $L_{\infty,\omega}(L)$-type realised in M is equivalent to a projective type, that is a type consisting of existential (first-order) formulas and the negations of existential formulas.

(ii) There are only countably many $L_{\infty,\omega}(L)$-types realised in M.

(iii) $(M, L^{\exists})$ is quasiminimal ω-homogeneous over countable submodels, that is the following hold:

(a) for any countable (or empty) submodel G and any n-tuples X and X', both cl-independent over G, a bijection $\phi : X \to X'$ is a G-monomorphism;

(b) given any G-monomorphism $\phi : Y \to Y'$ for finite tuples Y, Y' in M and given a $z \in M$ we can extend ϕ so that $z \in \mathrm{Dom}\, \phi$.

Proof. (i) Immediate from 2.11.

(ii) By 2.10 there are only countably many types of finite tuples $Z \leq M$. Let $N \subseteq M_0$ be a countable subset of M such that any finite $Z \leq M$ is $L_{\infty,\omega}(L)$-equivalent to some tuple in N. Every finite tuple $X \subset M$ can be extended to $XY \leq M$, so there is a $L_{\infty,\omega}(L)$-monomorphism $XY \to N$. This monomorphism identifies the $L_{\infty,\omega}(L)$-type of X with one of a tuple in N, hence there are no more than countably many such types.

(iii) Lemma 2.15 proves that cl defines a pregeometry on M.

Consider first (a). Note that $GX \leq M$ and $GX' \leq M$ and so the types of X and X' over G are L-quantifier-free. But there is no proper L-closed subset $S \subseteq_{cl} M^n$ such that $\vec{X} \in S$ or $\vec{X}' \in S$. Hence the types are equal.

For (b) just use the fact that the G-monomorphism by our definition preserves $\exists$-formulas, so by 2.11 complete $L_{\infty,\omega}(L(G))$-types of X and X' coincide, so by 2.7 ϕ can be extended. $\square$

2.17   Theorem

M is a quasiminimal pregeometry structure (see [16]). In other words, the following properties of M hold:

(QM1) The pregeometry cl is determined by the language. That is, if $\text{tp}(x,Y) = \text{tp}(x',Y')$, then $x \in \text{cl}(Y)$ if and only if $x' \in \text{cl}(Y')$. (Here the types are first order).

(QM2) The structure M is infinite-dimensional with respect to cl.

(QM3) (Countable closure property). If $X \subset M$ is finite, then $\text{cl}(X)$ is countable.

(QM4) (Uniqueness of the generic type). Suppose that $H, H' \subset M$ are countable closed subsets, enumerated such that $\text{tp}(H) = \text{tp}(H')$. If $y \in M \setminus H$ and $y' \in M \setminus H'$, then $\text{tp}(H,y) = \text{tp}(H',y')$.

(QM5) (ω-homogeneity over closed sets and the empty set). Let $H, H' \subset M$ be countable closed subsets or empty, enumerated such that $\text{tp}(H) = \text{tp}(H')$, and let Y, Y' be finite tuples from M such that $\text{tp}(H,Y) = \text{tp}(H',Y')$, and let $z \in \text{cl}(H,Y)$. Then there is $z' \in M$ such that $\text{tp}(H,Y,z) = \text{tp}(H',Y',z')$.

Proof. (M, cl) is a pregeometry by 2.15. (QM1) is proved in 2.14. (QM3) is 2.13 and (QM2) follows from (QM3) and (CU). (QM4)&(QM5) is 2.16(iii). $\square$

Now we define an *abstract elementary class* $\mathcal{C}$ associated with M. We follow [4], Ch. 6 for this construction. Similar construction was used in [16].

Set

$$\mathcal{C}_0(M) = \{ \text{ countable } L^\exists\text{-structures N} : \text{ N} \cong \text{N}' \subseteq \text{M}, \text{ cl}(N') = N'\}$$

and define embedding $N_1 \preccurlyeq N_2$ in the class as an $L^\exists$-embedding $f : N_1 \to N_2$

such that there are isomorphisms $g_i : N_i \to N_i'$, $N_1' \subseteq N_2' \subseteq M$, all embeddings commuting and $cl(N_i') = N_i'$.

Now define $\mathcal{C}(M)$ to be the class of L-structures H with cl_L defined with respect to H and satisfying:

(i) $\mathcal{C}_0(H) \subseteq \mathcal{C}_0(M)$ as classes with embeddings

and

(ii) for every finite $X \subseteq H$ there is $N \in \mathcal{C}_0(H)$, such that $X \subseteq N$.

Given $H_1 \subseteq H_2$, $H_1, H_2 \in \mathcal{C}(M)$, we define $H_1 \preccurlyeq H_2$ to hold in the class, if for every finite $X \subseteq H_1$, $cl(X)$ is the same in H_1 and H_2. More generally, for $H_1, H_2 \in \mathcal{C}(M)$ we define $H_1 \preccurlyeq_f H_2$ to be an embedding f such that there are isomorphisms $H_1 \cong H_1'$, $H_2 \cong H_2'$ such that $H_1' \subseteq H_2'$, all embeddings commute, and $H_1 \preccurlyeq H_2$.

2.18 Lemma

$\mathcal{C}(M)$ is closed under the unions of ascending $\preccurlyeq$-chains.

Proof. Immediate from the fact that for infinite $Y \subseteq M$,

$$cl(Y) = \bigcup \{cl(X) : \ X \subseteq_{\text{finite}} Y\}.$$

□

2.19 Theorem

The class $\mathcal{C}(M)$ contains structures of any infinite cardinality and is categorical in uncountable cardinals.

Proof. This follows from 2.17 by the main result of [16].
□

2.20 Proposition

Any uncountable $H \in \mathcal{C}(M)$ is an analytic 1-dimensional irreducible Zariski structure in the language L. Also H is presmooth if M is.

Proof. We define $\mathcal{C}(H)$ to consist of the subsets of H^n of the form

$$P_a(H) := \{x \in H^n : H \vDash P(a^\frown x)\},$$

for $P \in L$ of arity $k + n$, $a \in H^k$. The assumption (L) is obviously satisfied.

Now note that the constructible and projective sets in $\mathcal{C}(H)$ are also of the form $P_a(H)$ for some L-constructible or L-projective P.

Define

$\dim P_a(H) := d$ if $\dim P_b(M) = d$, for some $b \in M^k$ such that the $L^{\exists}$-quantifier-free types of a and b are equal.

This is well defined by (FC) and the fact that any $L^{\exists}$-quantifier-free type realised in H is also realised in M. Moreover, we have the following.

Claim. The set of $L^{\exists}$-quantifier-free types realised in H is equal to that realised in M.

Indeed, this is immediate from the definition of the class $\mathcal{C}(M)$, stability of $\mathcal{C}(M)$ and the fact that the class is categorical in uncountable cardinalities.

The definition of dimension immediately implies (DP), (CU), (AF), and (FC) for H.

(SI): if $P'_{a_1}(H) \subseteq_{cl} P_{a_0}(H)$, $\dim P'_{a_1}(H) = \dim P_{a_0}(H)$ and the two sets arc not equal, then the same holds for $P'_{b_1}(M)$ and $P_{b_0}(M)$ for equivalent b_0, b_1 in M. Then, $P_{b_0}(M)$ is reducible, that is for some proper $P''_{b_2}(M) \subset_{cl} P_{b_0}(M)$ we have $P_{b_0}(M) = P'_{b_1}(M) \cup P''_{b_2}(M)$. Now, by homogeneity we can choose a_2 in H such that $P_{a_0}(H) = P'_{a_1}(H) \cup P''_{a_2}(H)$, a reducible representation.

This also shows that the notion of irreducibility is preserved by equivalent substitution of parameters. Then the same is true for the notion of analytic subset. Hence (INT), (CMP), (CC) and (PS) follow. For the same reason (AS) holds. Next, we notice that the axiom (WP) follows by the homogeneity argument. $\square$

---

3 Some examples

3.1 Universal covers of semiabelian varieties

Let $\mathbb{A}$ be a semiabelian variety of dimension d, e.g. $d = 1$ and $\mathbb{A}$ the algebraic torus $\mathbb{C}^{\times}$. Let V be the universal cover of $\mathbb{A}$, which classically can be identified as a complex manifold $\mathbb{C}^d$.

We define a structure with a (formal) topology on V and show that this is analytic Zariski.

By definition of universal cover there is a covering holomorphic map

$$\exp : V \to \mathbb{A}$$

(a generalisation of the usual exp on $\mathbb{C}$).

We will assume that $\mathcal{C}$ has no proper semiabelian subvarieties (is simple) and no complex multiplication.

We consider the two-sorted structure $(V, \mathbb{A})$ in the language that has all Zariski closed subsets of $\mathbb{A}^n$, all n, the addition $+$ on V and the map exp as the primitives.

This case was first looked at model-theoretically in [14] and the special case $\mathbb{A} = \mathbb{C}^\times$ in [12], [13] and in the DPhil thesis [15] of Lucy Smith.

Our aim here is to show that the structure on the sort V with a naturally given formal topology is analytic Zariski.

The positive quantifier-free definable subsets of V^n, $n = 1, 2, \ldots$ form a base of a topology which we call the PQF-topology. In other words,

3.2   Definition

A PQF-closed subset of V^n is defined as a finite union of sets of the form

$$L \cap m \cdot \ln W \qquad (8.1)$$

where $W \subseteq \mathbb{A}^n$, an algebraic subvariety, and L is a $\mathbb{Q}$-linear subspace of V^n, that is defined by a system of equations of the form $m_1 x_1 + \ldots + m_n x_n = a$, $m_i \in \mathbb{Z}$, $a \in V^n$.

The relations on V which correspond to PQF_ω-closed sets are the primitives of our language L.

PQF-closed subsets form a base for a topology on the cartesian powers of V which will underlie the analytic Zariski structure on V.

Remark. Among closed sets of the topology we have sets of the form

$$\cup_{a \in I}(S + a)$$

where S is of the form (8.1) and I a subset of $(\ker \exp)^n$.

Slightly rephrasing the quantifier-elimination statement proved in [14] Corollary 2 of section 3, we have the following result.

3.3   Proposition

(i) Projection of a PQF-closed set is PQF-constructible, that is a Boolean combination of PQF-closed sets.

(ii) The image of a constructible set under exponentiation is a Zariski-constructible (algebraic) subset of $\mathbb{A}^n$. The image of the set of the form (8.1) is Zariski closed.

We assign **dimension** to a closed set of the form (8.1)

$$\dim L \cap m \cdot \ln W := \dim \exp\left(L \cap m \cdot \ln W\right).$$

using the fact that the object on the right-hand side is an algebraic variety. We extend this to an arbitrary closed set assuming (CU), that is that the dimension of a countable union is the maximum dimension of its members. This immediately gives (DP). Using 3.3 we also get (WP).

The analysis of irreducibility below is more involved. Since $\exp L$ is definably and topologically isomorphic to $\mathbb{A}^k$, some $k \geq 1$, we can always reduce the analysis of a closed set of the form (8.1) to a one of the form $\ln W$ with $W \subseteq \mathbb{A}^k$ not contained in a coset of a proper algebraic subgroup.

For such a W consider its m-th "root"

$$W^{\frac{1}{m}} = \{\langle x_1, \ldots, x_n \rangle \in \mathbb{A}^k : \langle x_1^m, \ldots, x_k^m \rangle \in W\}.$$

Let $d = d_W(m)$ be the number of irreducible components of $W^{\frac{1}{m}}$.

It is easy to see that if $d > 1$, irreducible components $W_i^{\frac{1}{m}}$, $i = 1, \ldots, d$, of $W^{\frac{1}{m}}$ are shifts of each other by m-th roots of unity, and $m \cdot \ln W_i^{\frac{1}{m}}$ are proper closed subsets of $\ln W$ of the same dimension. It follows that $\ln W$ *is irreducible (in the sense of* (SI)*) if and only if $d_W(m) = 1$ for all $m \geq 1$.* In [18] W satisfying this condition is called **Kummer generic**. If $W \subset \exp L$ for some $\mathbb{Q}$-linear subspace $L \subset V^n$, then one uses the relative version of Kummer genericity.

We say that the sequence $W^{\frac{1}{m}}$, $m \in \mathbb{N}$, stops branching if the sequence $d_W(m)$ is eventually constant, that is if $W^{\frac{1}{m}}$ is Kummer generic for some $m \geq 1$.

The following is proved for $\mathbb{A} = \mathbb{C}^\times$ in [12], (Theorem 2, case $n = 1$ and its Corollary) and in general in [18].

3.4 Theorem

The sequence $W^{\frac{1}{m}}$ stops branching if and only if W is not contained in a coset of a proper algebraic subgroup of $\mathbb{A}^k$.

3.5 Corollary

Any irreducible closed subset of V^n is of the form $L \cap \ln W$, for W Kummer generic in $\exp L$.

Any closed subset of V^n is analytic in V^n.

It is easy now to check that the following.

3.6 Corollary

The structure $(V; L)$ is analytic Zariski and presmooth.

The reader may notice that the analysis above treats only *formal* notion of analyticity on the cover $\mathbb{C}$ of $\mathbb{C}^\times$, but does not address the classical one. In particular, the following question is in order: *is the formal analytic decomposition as described by 3.5 the same as the actual complex analytic one?* In a private communication F. Campana answered this question in positive, using a cohomological argument. M. Gavrilovich proved this and a much more general statement in his thesis (see [11], III.1.2) by a similar argument.

3.7 Covers in positive characteristic

Now we look into yet another version of a cover structure which is proven to be analytic Zariski, a **cover of the one-dimensional algebraic torus over an algebraically closed field of a positive characteristic.**

Let $(V, +)$ be a divisible torsion free abelian group and K an algebraically closed field of a positive characteristic p. We assume that V and K are both of the same uncountable cardinality. Under these assumptions it is easy to construct a surjective homomorphism

$$\mathrm{ex} : V \to K^\times.$$

The kernel of such a homomorphism must be a subgroup which is p-divisible but not q-divisible for each q coprime with p. One can easily construct ex so that

$$\ker \mathrm{ex} \cong \mathbb{Z}[\frac{1}{p}],$$

the additive group (which is also a ring) of rationals of the form $\frac{m}{p^n}$, $m, n \in \mathbb{Z}$, $n \geq 0$. In fact in this case it is convenient to view V and $\ker \mathrm{ex}$ as $\mathbb{Z}[\frac{1}{p}]$-modules.

In this new situation Lemma 3.3 is still true, with obvious alterations, and we can use the definition 3.2 to introduce a topology and the family L as above. The necessary version of Theorem 3.4 is proved in [18]. Hence the corresponding versions of 3.5 follows.

3.8 Remark

In all the above examples, the analytic rank of any nonempty closed subset is 1, that is any closed subset is analytic.

3.9 $\mathbb{C}_{\exp}$ and other pseudo-analytic structures

$\mathbb{C}_{\exp}$, the structure $(\mathbb{C}; +, \cdot, \exp)$, was a prototype of **the field with pseudo-exponentiation** studied by the current author in [17]. It was proved (with later corrections, see [19]) that this structure is quasiminimal and its (explicitly written) $L_{\omega_1, \omega}(Q)$-axioms are categorical in all uncountable cardinality. This result has been generalised to many other structures of analytic origin in [19], and in particular to the formal analogue of $\mathbb{C}_\mathfrak{P} = (\mathbb{C}; +, \cdot, \mathfrak{P})$, where $\mathfrak{P} = \mathfrak{P}(\tau, z)$ is the Weierstrass function of variable z with parameter τ. We call these structures **pseudo-analytic.**

It is a reasonable conjecture to assume that the pseudo-analytic structures of cardinality continuum are isomorphic to their complex prototypes. Nevertheless, even under this conjecture it is not known whether $\mathbb{C}_{\exp}$, $\mathbb{C}_\mathfrak{P}$ or any of the other pseudo-analytic structures (which do not satisfy 3.8) are analytic Zariski. One may start by defining the family of (formal) closed sets in the structure to coincide with the family of definable subsets which are closed in the metric topology of the complex manifold. The problem then is to conveniently classify such subsets. A suggestion for such a classification may come from the following notion.

3.10 Generalised analytic sets

In [6] we have discussed the following notion of generalised analytic subsets of $[\mathbf{P}^1(\mathbb{C})]^n$ and, more generally, of $[\mathbf{P}^1(K)]^n$ for K algebraically closed complete valued field.

Let $F \subseteq \mathbb{C}^2$ be a graph of an entire analytic function and $\bar{F}$ its closure in $[\mathbf{P}^1(\mathbb{C})]^2$. It follows from Picard's Theorem that $\bar{F} = F \cup \{\infty\} \times \mathbf{P}^1(\mathbb{C})$, in particular, $\bar{F}$ has analytic rank 2.

Generalised analytic sets are defined as the subsets of $[\mathbf{P}^1(\mathbb{C})]^n$ for all n, obtained from classical (algebraic) Zariski closed subsets of $[\mathbf{P}^1(\mathbb{C})]^n$ and some number of sets of the form $\bar{F}$ by applying the positive operations: Cartesian products, finite intersections, unions and projections. It is clear by definition that the complex generalised analytic sets are closed (but not obvious for the case of K, algebraically closed complete non-Archimedean valued field).

3.11 Theorem (see [6])

Any generalised analytic set is of finite analytic rank.

Bibliography

[1] N. Peatfield and B. Zilber, *Analytic Zariski structures and the Hrushovski construction,* Annals of Pure and Applied Logic, 132/2-3 (2004), 127–180.

[2] E. Hrushovski and B. Zilber, *Zariski Geometries,* Journal of AMS, 9(1996), 1–56.

[3] B. Zilber, *Model Theory and Algebraic Geometry.* In **Seminarberichte Humboldt Universitat zu Berlin**, Nr 93-1, Berlin 1993, 202–222.

[4] B. Zilber, **Zariski Geometries**, CUP, 2010.

[5] I. Ben-Yaacov, *Positive model theory and compact abstract theories,* Journal of Mathematical Logic 3(1) (2003), no. 1, 85–118.

[6] B. Zilber, *Generalized Analytic Sets* Algebra i Logika, Novosibirsk, v.36(4) (1997), 361–380. (Translation on the author's web page and published by Kluwer as 'Algebra and Logic'.)

[7] B. Zilber, *A categoricity theorem for quasi-minimal excellent classes.* In: **Logic and its Applications** eds. Blass and Zhang, Cont. Maths, 380(5) (2005), 297–306.

[8] J. Baldwin, **Categoricity**, University Lecture Series, v.50, AMS, Providence, 2009.

[9] T. Hyttinen, O. Lessmann, and S. Shelah, *Interpreting groups and fields in some nonelementary classes,* J. Math. Log., 5(1) (2005), 1–47.

[10] D. Kueker, *Back-and-forth arguments in infinitary languages,* In **Infinitary Logic: In Memoriam Carol Karp**, D. Kueker ed., Lecture Notes in Math 72, Springer-Verlag, 1975.

[11] M. Gavrilovich, **Model Theory of the Universal Covering Spaces of Complex Algebraic Varieties**, DPhil Thesis, Oxford 2006, http://misha.uploads.net.ru/misha-thesis.pdf.

[12] B. Zilber, *Covers of the multiplicative group of an algebraically closed field of characteristic zero,* J. London Math. Soc. (2), 74(1) (2006), 41–58.

[13] M. Bays and B. Zilber, *Covers of Multiplicative Groups of Algebraically Closed Fields of Arbitrary Characteristic*, Bull. London Math. Soc. 43(4) (2011), 689–702.

[14] B. Zilber, *Model theory, geometry and arithmetic of the universal cover of a semi-abelian variety* In: **Model Theory and Applications**, ed. L. Belair et al. (Proc. of Conf. in Ravello 2000) Quaderni di Matematica, v.11, Series edited by Seconda Univli, Caserta, 2002.

[15] L. Smith, **Toric Varieties as Analytic Zariski Structures**, DPhil Thesis, Oxford, 2008.

[16] M. Bays, B. Hart, T. Hyttinen, M. Kesala, J. Kirby, *Quasiminimal structures and excellence*, Bulletin of the London Mathematical Society, 46. (2014), 155–163.

[17] B. Zilber, *Pseudo-exponentiation on algebraically closed fields of characteristic zero*, Ann. Pure Appl. Logic, 132 (2005), 67–95.

[18] M. Bays, M. Gavrilovich and M. Hils, *Some Definability Results in Abstract Kummer Theory* Int. Math. Res. Not. IMRN, (14) (2014).

[19] M. Bays and J. Kirby, *Some pseudo-analytic functions on commutative algebraic groups*, arXiv:1512.04262.

Part III

Abstract Elementary Classes

Chapter 9

Hanf numbers and presentation theorems in AECs

John T. Baldwin

University of Illinois at Chicago
Chicago, Illinois, USA

Will Boney

Harvard University
Cambridge, Massachusetts, USA

Baldwin was partially supported by Simons travel grant G5402 and Boney was partially supported by the National Science Foundation under Grant No. DMS-1402191.

1   Introduction

This chapter addresses a number of fundamental problems in logic and the philosophy of mathematics by considering some more technical problems in model theory and set theory. The interplay between syntax and semantics is usually considered the hallmark of model theory. At first sight, Shelah's notion of abstract elementary class shatters that icon. As in the beginnings of the modern theory of structures ([Cor92]), Shelah studies certain classes of models and relations among them, providing an axiomatization in the Bourbaki ([Bou50]) as opposed to the Gödel or Tarski sense: mathematical requirements, not sentences in a formal language. This formalism-free approach ([Ken13]) was designed to circumvent confusion arising from the syntactical schemes of infinitary logic; if a logic is closed under infinite conjunctions, what is the sense of studying types? However, Shelah's Presentation Theorem and more strongly Boney's use [Bon] of AEC's as theories of $\mathbb{L}_{\kappa,\omega}$ (for κ strongly compact) reintroduce syntactical arguments. The issues addressed in this chapter trace to the failure of infinitary logics to satisfy the *upward* Löwenheim-Skolem theorem or more specifically the compactness theorem. The compactness theorem allows such basic algebraic notions as amalgamation and joint embedding to be easily encoded in first order logic. Thus, all complete first order theories have amalgamation and joint embedding in all cardinalities. In contrast these and other familiar concepts from algebra and model theory turn out to be heavily cardinal-dependent for infinitary logic and specifically for abstract elementary classes. This is especially striking as one of the most important contributions of modern model theory is the freeing of first order model theory from its entanglement with axiomatic set theory ([Bal15a], Chapter 7 of [Bal15b]).

Two main issues are addressed here. We consider not the interaction of syntax and semantics in the usual formal language/structure dichotomy but methodologically. What are reasons for adopting syntactic and/or semantic approaches to a particular topic? We compare methods from the very beginnings of model theory with semantic methods powered by large cardinal hypotheses. Secondly, what then are the connections of large cardinal axioms with the cardinal dependence of algebraic properties in model theory. Here we describe the opening of the gates for potentially large interactions between set theorists (and incidentally graph theorists) and model theorists. More precisely, can the combinatorial properties of small large cardinals be coded as structural properties of abstract elementary classes so as to produce Hanf numbers intermediate in cardinality between 'well below the first inaccessible' and 'strongly compact'?

Most theorems in mathematics are either true in a specific small cardinality (at most the continuum) or in all cardinals. For example all, *finite* division rings are commutative, thus all finite Desarguesian planes are Pappian. But *all* Pappian planes are Desarguesian and not conversely. Of course this stricture does not apply to set theory, but the distinctions arising in set theory are combinatorial. First order model theory, to some extent, and Abstract Elementary Classes (AEC) are beginning to provide a deeper exploration of Cantor's paradise: algebraic properties that are cardinality dependent. In this article, we explore whether certain key properties (amalgamation, joint embedding, and their relatives) follow this line. These algebraic properties are structural in the sense of [Cor04].

Much of this issue arises from an interesting decision of Shelah. Generalizing Fraïssé [Fra54] who considered only finite and countable stuctures, Jónsson laid the foundations for AEC by his study of universal and homogeneous relation systems [Jón56, Jón60]. Both of these authors assumed the amalgamation property (AP) and the joint embedding property (JEP), which in their context is cardinal independent. Variants such as disjoint or free amalgamation (DAP) are a well-studied notion in model theory and universal algebra. But Shelah omitted the requirement of amalgamation in defining AEC. Two reasons are evident for this: it is cardinal dependent in this context; Shelah's theorem (under weak diamond) that categoricity in κ and few models in κ^+ implies amalgamation in κ suggests that amalgamation might be a dividing line.

Grossberg [Gro02, Conjecture 9.3] first raised the question of the existence of Hanf numbers for joint embedding and amalgamation in Abstract Elementary Classes (AEC). We define four kinds of amalgamation properties (with various cardinal parameters) in Subsection 1.1 and a fifth at the end of Section 3.1. The first three notions are staples of the model theory and universal algebra since the fifties and treated for first order logic in a fairly uniform manner by the methods of Abraham Robinson. It is a rather striking feature of Shelah's Presentation Theorem that issues of disjointness require careful study for AEC, while disjoint amalgamation is trivial for complete first order theories.

Our main result is the following:

Theorem 1.1. *Let κ be strongly compact and **K** be an AEC with Löwenheim-Skolem number less than κ.*

*If **K** satisfies[1] AP/JEP/DAP/DJEP/ NDJEP for models of size $[\mu, < \kappa)$, then **K** satisfies AP/JEP/DAP/DJEP/ NDJEP for all models of size $\geq \mu$.*

We conclude with a survey of results showing the large gap for many properties between the largest cardinal where an 'exotic' structure exists and the smallest where eventual behavior is determined. Then we provide specific questions to investigate this distinction.

Our starting place for this investigation was the second author's work [Bon] that emphasized the role of large cardinals in the study of AEC. A key aspect of the definition of AEC is the absence of a formal syntax in it. However,

---

[1] This alphabet soup is decoded in Definition 1.3.

Shelah's Presentation Theorem says that AECs are expressible in infinitary languages, $\mathbb{L}_{\kappa,\omega}$, which allowed a proof via sufficiently complete ultraproducts that, assuming enough strongly compact cardinals, all AEC's were eventually tame in the sense of [GV06].

Thus we approached the problem of finding a Hanf number for amalgamation, etc. from two directions: using ultraproducts to give purely semantic arguments and using Shelah's Presentation Theorem to give purely syntactic arguments. However, there was a gap: although syntactic arguments gave characterizations similar to those found in first order, they required looking at the *disjoint* versions of properties, while the semantic arguments did not see this difference.

The requirement of disjointness in the syntactic arguments stems from a lack of canonicity in Shelah's Presentation Theorem: a single model has many expansions which means that the transfer of structural properties between an AEC **K** and its expansion can break down. To fix this problem, we developed a new presentation theorem, called the *relational presentation theorem* because the expansion consists of relations rather than the Skolem-like functions from Shelah's Presentation Theorem.

Theorem 1.2 (The relational presentation theorem, Theorem 3.9). *To each AEC* **K** *with* $LS(K) = \kappa$ *in vocabulary* τ, *there is an expansion of* τ *by predicates of arity* κ *and a theory* T^* *in* $\mathbb{L}_{(2^\kappa)^+,\kappa^+}$ *such that* **K** *is exactly the class of* τ *reducts of models of* T^*.

Note that this presentation theorem works in $\mathbb{L}_{(2^\kappa)^+,\kappa^+}$ and has symbols of arity κ, a far cry from the $\mathbb{L}_{(2^\kappa)^+,\omega}$ and finitary language of Shelah's Presentation Theorem. The benefit of this is that the expansion is canonical or functorial (see Definition 3.1). This functoriality makes the transfer of properties between **K** and $(\text{Mod}\,T^*, \subset_{\tau^*})$ trivial (see Proposition 3.2). This allows us to formulate natural syntactic conditions for our structural properties.

Comparing the relational presentation theorem to Shelah's, another well-known advantage of Shelah's is that it allows for the computation of Hanf numbers for existence (see Section 4) because these exist in $\mathbb{L}_{\kappa,\omega}$. However, there is an advantage of the relational presentation theorem: Shelah's Presentation Theorem works with a sentence in the logic $\mathbb{L}_{(2^{LS(K)})^+,\omega}$ and there is little hope of bringing that cardinal down[2]. On the other hand, the logic and size of theory in the relational presentation theorem can be brought down by putting structure assumptions on the class **K**, primarily on the number of non-isomorphic extensions of size $LS(\mathbf{K})$, $|\{(M, N)/ \cong: M \prec_{\mathbf{K}} N \text{ from } \mathbf{K}_{LS(\mathbf{K})}\}|$.

We would like to thank Spencer Unger and Sebastien Vasey for helpful discussions regarding these results.

---

[2]Indeed an AEC **K** where the sentence is in a smaller logic would likely have to satisfy the very strong property that there are $< 2^{LS(\mathbf{K})}$ many $\tau(\mathbf{K})$ structures that **are not** in **K**.

1.1 Preliminaries

We discuss the relevant background of AECs, especially for the case of disjoint amalgamation.

Definition 1.3. We consider several variations on the joint embedding property, written as JEP or JEP$[\mu, \kappa)$.

1. Given a class of cardinals $\mathcal{F}$ and an AEC $\mathbf{K}$, $\mathbf{K}_{\mathcal{F}}$ denotes the collection of $M \in \mathbf{K}$ such that $|M| \in \mathcal{F}$. When $\mathcal{F}$ is a singleton, we write $\mathbf{K}_{\kappa}$ instead of $\mathbf{K}_{\{\kappa\}}$. Similarly, when $\mathcal{F}$ is an interval, we write $< \kappa$ in place of $[\mathrm{LS}(\mathbf{K}), \kappa)$; $\leq \kappa$ in place of $[\mathrm{LS}(\mathbf{K}), \kappa]$; $> \kappa$ in place of $\{\lambda \mid \lambda > \kappa\}$; and $\geq \kappa$ in place of $\{\lambda \mid \lambda \geq \kappa\}$.

2. An AEC $(\mathbf{K}, \prec_{\mathbf{K}})$ has the joint embedding property, JEP, (on the interval $[\mu, \kappa)$) if any two models (from $\mathbf{K}_{[\mu, \kappa)}$) can be $\mathbf{K}$-embedded into a larger model.

3. If the embeddings witnessing the joint embedding property can be chosen to have disjoint ranges, then we call this the *disjoint embedding property* and write *DJEP*.

4. An AEC $(\mathbf{K}, \prec_{\mathbf{K}})$ has the amalgamation property, AP, (on the interval $[\mu, \kappa)$) if, given any triple of models $M_0 \prec M_1, M_2$ (from $\mathbf{K}_{[\mu, \kappa)}$), M_1 and M_2 can be $\mathbf{K}$-embedded into a larger model by embeddings that agree on M_0.

5. If the embeddings witnessing the amalgamation property can be chosen to have disjoint ranges except for M_0, then we call this the *disjoint amalgamation property* and write *DAP*.

Definition 1.4. 1. A *finite diagram* or *$EC(T, \Gamma)$-class* is the class of models of a first order theory T which omit all types from a specified collection Γ of complete types in finitely many variables over the empty set.

2. Let Γ be a collection of first order types in finitely many variables over the empty set for a first order theory T in a vocabulary τ_1. A *$PC(T, \Gamma, \tau)$ class* is the class of reducts to $\tau \subset \tau_1$ of models of a first order τ_1-theory T which omit all members of the specified collection Γ of partial types.

---

2 Semantic arguments

It turns out that the Hanf number computation for the amalgamation properties is immediate from Boney's "Łoś' Theorem for AECs" [Bon, Theorem 4.3]. We will sketch the argument for completeness. For convenience here, we take the following of the many equivalent definitions of strongly compact; it is the most useful for ultraproduct constructions.

Definition 2.1 ([Jec06].20). The cardinal κ is strongly compact iff for every S and every κ-complete filter on S can be extended to a κ-complete ultrafilter. Equivalently, for every $\lambda \geq \kappa$, there is a fine[3], κ-complete ultrafilter on $P_\kappa \lambda = \{\sigma \subset \lambda : |\sigma| < \kappa\}$.

For this chapter, "essentially below κ" means "$LS(K) < \kappa$."

Fact 2.2 (Łoś' Theorem for AECs). Suppose K is an AEC essentially below κ and U is a κ-complete ultrafilter on I. Then K and the class of K-embeddings are closed under κ-complete ultraproducts and the ultrapower embedding is a K-embedding.

The argument for Theorem 2.3 has two main steps. First, use Shelah's presentation theorem to interpret the AEC into $\mathbb{L}_{\kappa,\omega}$ and then use the fact that $\mathbb{L}_{\kappa,\omega}$ classes are closed under ultraproduct by κ-*complete* ultraproducts.

Theorem 2.3. *Let κ be strongly compact and* **K** *be an AEC with Löwenheim-Skolem number less than κ.*

- *If* **K** *satisfies $AP(< \kappa)$, then* **K** *satisfies AP.*
- *If* **K** *satisfies $JEP(< \kappa)$, then* **K** *satisfies JEP.*
- *If* **K** *satisfies $DAP(< \kappa)$, then* **K** *satisfies DAP.*

Proof: We first sketch the proof for the first item, AP, and then note the modifications for the other two.

Suppose that **K** satisfies $AP(< \kappa)$ and consider a triple of models (M, M_1, M_2) with $M \prec_{\mathbf{K}} M_1, M_2$ and $|M| \leq |M_1| \leq |M_2| = \lambda \geq \kappa$. Now we will use our strongly compact cardinal. An *approximation* of (M, M_1, M_2) is a triple $\mathbf{N} = (N^{\mathbf{N}}, N_1^{\mathbf{N}}, N_2^{\mathbf{N}}) \in (K_{<\kappa})^3$ such that $N^{\mathbf{N}} \prec M, N_\ell^{\mathbf{N}} \prec M_\ell, N^{\mathbf{N}} \prec N_\ell^{\mathbf{N}}$ for $\ell = 1, 2$. We will take an ultraproduct indexed by the set X below of approximations to the triple (M, M_1, M_2). Set

$$X := \{\mathbf{N} \in (\mathbf{K}_{<\kappa})^3 : \mathbf{N} \text{ is an approximation of } (M, M_1, M_2)\}$$

For each $\mathbf{N} \in X$, $AP(< \kappa)$ implies there is an amalgam of this triple. Fix $f_\ell^{\mathbf{N}} : N_\ell^{\mathbf{N}} \to N_*^{\mathbf{N}}$ to witness this fact. For each $(A, B, C) \in [M]^{<\kappa} \times [M_1]^{<\kappa} \times [M_2]^{<\kappa}$, define

$$G(A, B, C) := \{\mathbf{N} \in X : A \subset N^{\mathbf{N}}, B \subset N_1^{\mathbf{N}}, C \subset N_2^{\mathbf{N}}\}$$

These sets generate a κ-complete filter on X, so it can be extended to a κ-complete ultrafilter U on X; note that this ultrafilter will satisfy the appropriate generalization of fineness, namely that $G(A, B, C)$ is always a U-large set.

We will now take the ultraproduct of the approximations and their amalgam. In the end, we will end up with the following commuting diagram, which provides the amalgam of the original triple.

---

[3]U is fine iff $G(\alpha) := \{z \in P_\kappa(\lambda) | \alpha \in z\}$ is an element of U for each $\alpha < \lambda$.

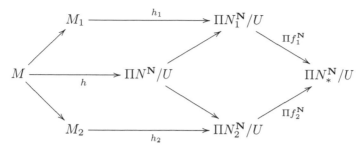

First, we use Łoś' Theorem for AECs to get the following maps:

$$h : M \to \Pi N^{\mathbf{N}}/U$$
$$h_\ell : M_\ell \to \Pi N_\ell^{\mathbf{N}}/U \qquad \text{for } \ell = 1, 2$$

h is defined by taking $m \in M$ to the equivalence class of constant function $\mathbf{N} \mapsto x$; this constant function is not always defined, but the fineness-like condition guarantees that it is defined on a U-large set (and h_1, h_2 are defined similarly). The uniform definition of these maps implies that $h_1 \restriction M = h \restriction M = h_2 \restriction M$.

Second, we can average the $f_\ell^{\mathbf{N}}$ maps to get ultraproduct maps

$$\Pi f_\ell^{\mathbf{N}} : \Pi N_\ell^{\mathbf{N}}/U \to \Pi N_*^{\mathbf{N}}/U$$

These maps agree on $\Pi N^{\mathbf{N}}/U$ since each of the individual functions does. As each M_ℓ embeds in $\Pi N_\ell^{\mathbf{N}}/U$ the composition of the f and h maps gives the amalgam.

There is no difficulty if one of M_0 or M_1 has cardinality $< \kappa$; many of the approximating triples will have the same first or second coordinates, but this causes no harm. Similarly, we get the JEP transfer if $M_0 = \emptyset$. And we can transfer disjoint amalgamation since in that case each $N_1^{\mathbf{N}} \cap N_2^{\mathbf{N}} = N^{\mathbf{N}}$ and this is preserved by the ultraproduct.

---

3 Syntactic approaches

The two methods discussed in this section both depend on expanding the models of **K** to models in a larger vocabulary. We begin with a concept introduced in Vasey [Vasa, Definition 3.1].

Definition 3.1. A *functorial expansion* of an AEC **K** in a vocabulary τ is an AEC $\hat{\mathbf{K}}$ in a vocabulary $\hat{\tau}$ extending τ such that

1. each $M \in \mathbf{K}$ has a *unique* expansion to a $\hat{M} \in \hat{\mathbf{K}}$,

2. if $f : M \cong M'$, then $f : \hat{M} \cong \hat{M}'$, and

3. if M is a strong substructure of M' for **K**, then $\hat{M}$ is strong substructure of $\hat{M}'$ for $\hat{\mathbf{K}}$.

This concept unifies a number of previous expansions: Morley's adding a predicate for each first order definable set, Chang's adding a predicate for each $\mathbb{L}_{\omega_1,\omega}$ definable set, T^{eq}, Cherlin, Harrington, and Lachlan's [CHL85] adding predicates $R_n(\mathbf{x}, y)$ for closure (in an ambient geometry) of $\mathbf{x}$, and the expansion by naming the orbits in a Fraïssé model[4].

An important point in both [Vasa] and our relational presentation is that the process does not just reduce the complexity of already definable sets (as in the case of Morley and Chang), but adds new definable sets. But the crucial distinction here is that the expansion in Shelah's presentation theorem is not 'functorial' in the sense here: each model has several expansions, rather than a single expansion. That is why there is an extended proof for amalgamation transfer in Section 3.1, while the transfer in Section 3.2 follows from the following result which is easily proved by chasing arrows.

Proposition 3.2. *Let* **K** *to* $\hat{\mathbf{K}}$ *be a functorial expansion.* $(K, \prec)$ *has* λ-*amalgamation [joint embedding, etc.] iff* $\hat{\mathbf{K}}$ *has* λ-*amalgamation [joint embedding, etc.].*

3.1   Shelah's Presentation Theorem

In this section, we provide syntactic characterizations of the various amalgamation properties in a finitary language. Our first approach to these results stemmed from the realization that the amalgamation property has the same syntactic characterization for $\mathbb{L}_{\kappa,\kappa}$ as for first order logic if κ is strongly compact, i.e., the compactness theorem holds for $\mathbb{L}_{\kappa,\kappa}$. Combined with Boney's recognition that one could code each AEC with Löwenheim-Skolem number less than κ in $\mathbb{L}_{\kappa,\kappa}$ this seemed a path to showing amalgamation. Unfortunately, this path leads through the trichotomy in Fact 3.3. The results depend directly (or with minor variations) on Shelah's Presentation Theorem and illustrate its advantage (finitary language) and disadvantage (lack of canonicity).

Fact 3.3 (Shelah's presentation theorem). If **K** is an AEC (in a vocabulary τ with $|\tau| \leq \mathrm{LS}(\mathbf{K})$) with Löwenheim-Skolem number $\mathrm{LS}(\mathbf{K})$, there is a vocabulary $\tau_1 \supseteq \tau$ with cardinality $|\mathrm{LS}(\mathbf{K})|$, a first order τ_1-theory T_1 and a set Γ of at most $2^{\mathrm{LS}(\mathbf{K})}$ partial types such that

1. $\mathbf{K} = \{M' \upharpoonright \tau : M' \models T_1 \text{ and } M' \text{ omits } \Gamma\}$;

2. if M' is a τ_1-substructure of N' where M', N' satisfy T_1 and omit Γ, then $M' \upharpoonright \tau \prec_{\mathbf{K}} N' \upharpoonright \tau$; and

---

[4]This has been done for years but there is a slight wrinkle in, e.g., [BKL15] where the orbits are not first order definable.

3. if $M \prec N \in \mathbf{K}$ and $M' \in EC(T_1, \Gamma)$ such that $M' \upharpoonright \tau = M$, then there is $N' \in EC(T_1, \Gamma)$ such that $M' \subset N'$ and $N' \upharpoonright \tau = N$.

The exact assertion for part 3 is new in this chapter; we do not include the slight modification in the standard proofs (e.g., [Bal09, Theorem 4.15]). Note that we have a weakening of Definition 3.1 caused by the possibility of multiple 'good' expansion of a model M.

Here are the syntactic conditions equivalent to DAP and DJEP.

Definition 3.4. • Ψ has $< \lambda$-*DAP satisfiability* iff for any expansion by constants $\mathbf{c}$ and all sets of atomic and negated atomic formulas (in $\tau(\Psi) \cup \{\mathbf{c}\}$) $\delta_1(\mathbf{x}, \mathbf{c})$ and $\delta_2(\mathbf{y}, \mathbf{c})$ of size $< \lambda$, if $\Psi \wedge \exists \mathbf{x} (\bigwedge \delta_1(\mathbf{x}, \mathbf{c}) \wedge \bigwedge x_i \neq c_j)$ and $\Psi \wedge \exists \mathbf{y} (\bigwedge \delta_2(\mathbf{y}, \mathbf{c}) \wedge \bigwedge y_i \neq c_j)$ are separately satisfiable, then so is

$$\Psi \wedge \exists \mathbf{x}, \mathbf{y} \left(\bigwedge \delta_1(\mathbf{x}, \mathbf{c}) \wedge \bigwedge \delta_2(\mathbf{y}, \mathbf{c}) \wedge \bigwedge_{i,j} x_i \neq y_j \right)$$

• Ψ has $< \lambda$-*DJEP satisfiability* iff for all sets of atomic and negated atomic formulas (in $\tau(\Psi)$) $\delta_1(\mathbf{x})$ and $\delta_2(\mathbf{y})$ of size $< \lambda$, if $\Psi \wedge \exists \mathbf{x} \bigwedge \delta_1(\mathbf{x})$ and $\Psi \wedge \exists \mathbf{y} \bigwedge \delta_2(\mathbf{y})$ are separately satisfiable, then so is

$$\Psi \wedge \exists \mathbf{x}, \mathbf{y} \left(\bigwedge \delta_1(\mathbf{x}) \wedge \bigwedge \delta_2(\mathbf{y}) \wedge \bigwedge_{i,j} x_i \neq y_j \right)$$

We now outline the argument for $DJEP$; the others are similar. Note that $(2) \to (1)$ for the analogous result with DAP replacing DJEP has been shown by Hyttinen and Kesälä [HK06, 2.16].

Lemma 3.5. *Suppose that $\mathbf{K}$ is an AEC, $\lambda > LS(\mathbf{K})$, and T_1 and Γ are from Shelah's Presentation Theorem. Let Φ be the $\mathbb{L}_{LS(\mathbf{K})^+, \omega}$ theory that asserts the satisfaction of T_1 and omission of each type in Γ. Then the following are equivalent:*

1. $\mathbf{K}_{<\lambda}$ *has DJEP.*

2. $(EC(T_1, \Gamma), \subset)_{<\lambda}$ *has DJEP.*

3. Φ *has $< \lambda$-DJEP-satisfiability.*

Proof:

$(1) \leftrightarrow (2)$: First, suppose that $\mathbf{K}_{<\lambda}$ has DJEP. Let $M_0^*, M_1^* \in EC(T_1, \Gamma)_{<\lambda}$ and set $M_\ell := M_\ell^* \upharpoonright \tau$. By disjoint embedding for $\ell = 0, 1$, there is $N \in \mathbf{K}$ such that each $M_\ell \prec N$. Our goal is to expand N to be a member of $EC(T_1, \Gamma)$ in a way that respects the already existing expansions.

Recall from the proof of Fact 3.3 that expansions of $M \in \mathbf{K}$ to models $M^* \in EC(T_1, \Gamma)$ exactly come from writing M as a directed union of $LS(\mathbf{K})$-sized models indexed by $P_\omega|M|$, and then enumerating the models in the union. Thus, the expansion of M_ℓ to M_ℓ^* come from $\{M_{\ell,\mathbf{a}} \in \mathbf{K}_{LS(\mathbf{K})} \mid \mathbf{a} \in M_\ell\}$, where $|M_{\ell,\mathbf{a}}| = \{\left(F_{|\mathbf{a}|}^i\right)^{M_\ell^*}(\mathbf{a}) \mid i < LS(\mathbf{K})\}$ and the functions F_n^i are from the expansion. Because M_1 and M_2 are disjoint strong submodels of N, we can write N as a directed union of $\{N_\mathbf{a} \in \mathbf{K}_{LS(\mathbf{K})} \mid \mathbf{a} \in N\}$ such that $\mathbf{a} \in M_\ell$ implies that $M_{\ell,\mathbf{a}} = N_\mathbf{a}$. Now, any enumeration of the universes of these models of order type $LS(\mathbf{K})$ will give rise to an expansion of N to $N^* \in EC(T_1, \Gamma)$ by setting $\left(F_{|\mathbf{a}|}^i\right)^{N^*}(\mathbf{a})$ to be the ith element of $|N_\mathbf{a}|$.

Thus, choose an enumeration of them that agrees with the original enumerations from M_ℓ^*; that is, if $\mathbf{a} \in M_\ell$, then the ith element of $|N_\mathbf{a}| = |M_{\ell,\mathbf{a}}|$ is $\left(F_{|\mathbf{a}|}^i\right)^{M_\ell^*}(\mathbf{a})$ (note that, as used before, the disjointness guarantees that there is at most one ℓ satisfying this). In other words, our expansion N^* will have

$$\mathbf{a} \in M_\ell \implies \left(F_{|\mathbf{a}|}^i\right)^{M_\ell^*}(\mathbf{a}) = \left(F_{|\mathbf{a}|}^i\right)^{N^*}(\mathbf{a}) \text{ for all } i < LS(\mathbf{K})$$

This precisely means that $M_\ell^* \subset N^*$, as desired. Furthermore, we have constructed the expansion so $N^* \in EC(T_1, \Gamma)$. Thus, $(EC(T_1, \Gamma), \subset)_{<\lambda}$ has DJEP.

Second, suppose that $EC(T_1, \Gamma)$ has λ-DJEP. Let $M_0, M_1 \in \mathbf{K}$; WLOG, $M_0 \cap M_1 = \emptyset$. Using Shelah's Presentation Theorem, we can expand to $M_0^*, M_1^* \in EC(T_1, \Gamma)$. Then we can use disjoint embedding to find $N^* \in EC(T_1, \Gamma)$ such that $M_1^*, M_2^* \subset N^*$. By Shelah's Presentation Theorem 3.3.(1), $N := N^* \upharpoonright \tau$ is the desired model.

$(2) \leftrightarrow (3)$: First, suppose that Φ has $< \lambda$-DJEP satisfiability. Let $M_0^*, M_1^* \in EC(T_1, \Gamma)$ be of size $< \lambda$. Let $\delta_0(\mathbf{x})$ be the quantifier-free diagram of M_0^*, and let $\delta_1(\mathbf{y})$ be the quantifier-free diagram of M_1^*. Then $M_0^* \models \Phi \wedge \exists \mathbf{x} \bigwedge \delta_0(\mathbf{x})$; similarly, $\Phi \wedge \exists \mathbf{y} \bigwedge \delta_1(\mathbf{y})$ is satisfiable. By the satisfiability property, there is N^* such that

$$N^* \models \Psi \wedge \exists \mathbf{x}, \mathbf{y} \left(\bigwedge \delta_0(\mathbf{x}) \wedge \bigwedge \delta_1(\mathbf{y}) \wedge \bigwedge_{i,j} x_i \neq y_j \right)$$

Then $N^* \in EC(T_1, \Gamma)$ and contains disjoint copies of M_0^* and M_1^*, represented by the witnesses of $\mathbf{x}$ and $\mathbf{y}$, respectively.

Second, suppose that $(EC(T_1, \Gamma), \subset)_{<\lambda}$ has DJEP. Let $\Phi \wedge \exists \mathbf{x} \bigwedge \delta_1(\mathbf{x})$ and $\Phi \wedge \exists \mathbf{y} \bigwedge \delta_2(\mathbf{y})$ be as in the hypothesis of $< \lambda$-DJEP satisfiability. Let M_0^* witness the satisfiability of the first and M_1^* witness the satisfiability of the second; note both of these are in $EC(T_1, \Gamma)$. By DJEP, there is

$N \in EC(T_1, \Gamma)$ that contains both as substructures. This witnesses

$$\Psi \wedge \exists \mathbf{x}, \mathbf{y} \left(\bigwedge \delta_1(\mathbf{x}) \wedge \bigwedge \delta_2(\mathbf{y}) \wedge \bigwedge_{i,j} x_i \neq y_j \right)$$

Note that the formulas in δ_1 and δ_2 transfer up because they are atomic or negated atomic.

The following is a simple use of the syntactic characterization of strongly compact cardinals.

Lemma 3.6. *Assume κ is strongly compact and let $\Psi \in \mathbb{L}_{\kappa,\omega}(\tau_1)$ and $\lambda > \kappa$. If Ψ has $< \kappa$-DJEP-satisfiability, then Ψ has $< \lambda$-DJEP-satisfiability.*

Proof: $< \lambda$-DJEP satisfiability hinges on the consistency of a particular $\mathbb{L}_{\kappa,\omega}$ theory. If Ψ has $< \kappa$-DJEP-satisfiability, then every $< \kappa$ sized subtheory is consistent, which implies the entire theory is by the syntactic version of strong compactness we introduced at the beginning of this section.

Obviously, the converse (for $\Psi \in \mathbb{L}_{\infty,\omega}$) holds without any large cardinals.

Proof of Theorem 1.1 for DAP and $DJEP$: We first complete the proof for DJEP. By Lemma 3.5, $< \kappa$-DJEP implies that Φ has $< \kappa$-DJEP satisfiability. By Lemma 3.6, Φ has $< \lambda$-DJEP satisfiability for every $\lambda \geq \kappa$. Thus, by Lemma 3.5 again, $\mathbf{K}$ has DJEP. The proof for DAP is exactly analogous.

3.2   The relational presentation theorem

We modify Shelah's Presentation Theorem by eliminating the two instances where an arbitrary choice must be made: the choice of models in the cover and the choice of an enumeration of each covering model. Thus the new expansion is functorial (Definition 3.1). However, there is a price to pay for this canonicity. In order to remove the choices, we must add predicates of arity $LS(K)$ and the relevant theory must allow $LS(K)$-ary quantification, potentially putting it in $\mathbb{L}_{(2^\kappa)^+,\kappa^+}$, where $\kappa = LS(K)$; contrast this with a theory of size $\leq 2^\kappa$ in $\mathbb{L}_{\kappa^+,\omega}$ for Shelah's version. As a possible silver lining, these arities can actually be brought down to $\mathbb{L}_{(I(\mathbf{K},\kappa)+\kappa)^+,\kappa^+}$. Thus, properties of the AEC, such as the number of models in the Löwenheim-Skolem cardinal are reflected in the presentation, while this has no effect on the Shelah version.

We fix some notation. Let $\mathbf{K}$ be an AEC in a vocabulary τ and let $\kappa = LS(\mathbf{K})$. We assume that $\mathbf{K}$ contains no models of size $< \kappa$. The same arguments could be done with $\kappa > LS(\mathbf{K})$, but this case reduces to applying our result to $\mathbf{K}_{\geq \kappa}$.

We fix a collection of compatible enumerations for models $M \in K_\kappa$. The term *compatible enumerations* means that each M has an enumeration of its universe, denoted $\mathbf{m}^M = \langle m_i^M : i < \kappa \rangle$, and, if $M \cong M'$, there is some fixed isomorphism $f_{M,M'} : M \cong M'$ such that $f_{M,M'}(m_i^M) = m_i^{M'}$ and if $M \cong M' \cong M''$, then $f_{M,M''} = f_{M',M''} \circ f_{M,M'}$.

For each isomorphism type $[M]_\cong$ and $[M \prec N]_\cong$ with $M, N \in K_\kappa$, we add to τ

$$R_{[M]}(\mathbf{x}) \text{ and } R_{[M \prec N]}(\mathbf{x}; \mathbf{y})$$

as κ-ary and $\kappa 2$-ary predicates to form τ^*.

A skeptical reader might protest that we have made many arbitrary choices so soon after singing the praises of our choiceless method. The difference is that all choices are made prior to defining the presentation theory, T^*.

Once T^* is defined, no other choices are made.

The goal of the theory T^* is to recognize every strong submodel of size κ and every strong submodel relation between them via our predicates. This is done by expressing in the axioms below concerning sequences $\mathbf{x}$ of length at most κ the following properties connecting the canonical enumerations with structures in $\mathbf{K}$.

$$R_{[M]}(\mathbf{x}) \text{ holds iff } x_i \mapsto m_i^M \text{ is an isomorphism}$$
$$R_{[M \prec N]}(\mathbf{x}, \mathbf{y}) \text{ holds iff } x_i \mapsto m_i^M \text{ and } y_i \mapsto m_i^N \text{ are isomorphisms and}$$
$$x_i = y_j \text{ iff } m_i^M = m_j^N$$

Note that, by the coherence of the isomorphisms, the choice of representative from $[M]_\cong$ does not matter. Also, we might have $M \cong M'$; $N \cong N'$; $M \prec N$ and $M' \prec N'$; but not $(M, N) \cong (M', N')$. In this case $R_{[M \prec N]}$ and $R_{[M' \prec N']}$ are different predicates.

We now write the axioms for T^*. *A priori* they are in the logic $\mathbb{L}_{(2^\kappa)^+, \kappa^+}(\tau^*)$, but the theorem states a slightly finer result. To aid in understanding, we include a description prior to the formal statement of each property.

Definition 3.7. The theory T^* in $\mathbb{L}_{(I(\mathbf{K}, \kappa) + \kappa)^+, \kappa^+}(\tau^*)$ is the collection of the following schema:

1. *If $R_{[M]}(\mathbf{x})$ holds, then $x_i \mapsto m_i^M$ should be an isomorphism.*
   If $\phi(z_1, \ldots, z_n)$ is an atomic or negated atomic τ-formula that holds of $m_{i_1}^M, \ldots, m_{i_n}^M$, then include

   $$\forall \mathbf{x} \left(R_{[M]}(\mathbf{x}) \to \phi(x_{i_1}, \ldots, x_{i_n}) \right)$$

2. *If $R_{[M \prec N]}(\mathbf{x}, \mathbf{y})$ holds, then $x_i \mapsto m_i^M$ and $y_i \mapsto m_i^N$ should be isomorphisms and the correct overlap should occur.*
   If $M \prec N$ and $i \mapsto j_i$ is the function such that $m_i^M = m_{j_i}^N$, then include

   $$\forall \mathbf{x}, \mathbf{y} \left(R_{[M \prec N]}(\mathbf{x}, \mathbf{y}) \to \left(R_{[M]}(\mathbf{x}) \wedge R_{[N]}(\mathbf{y}) \wedge \bigwedge_{i < \kappa} x_i = y_{j_i} \right) \right)$$

3. *Every κ-tuple is covered by a model.*
   Include the following where $\lg(\mathbf{x}) = \lg(\mathbf{y}) = \kappa$

   $$\forall \mathbf{x} \exists \mathbf{y} \left(\bigvee_{[M]_\cong \in K_\kappa/\cong} R_{[M]}(\mathbf{y}) \wedge \bigwedge_{i < \kappa} \bigvee_{j < \kappa} x_i = y_{j_i} \right)$$

4. If $R_{[N]}(\mathbf{x})$ holds and $M \prec N$, then $R_{[M \prec N]}(\mathbf{x}^\circ, \mathbf{x})$ should hold for the appropriate subtuple $\mathbf{x}^\circ$ of $\mathbf{x}$.
   If $M \prec N$ and $\pi : \kappa \to \kappa$ is the unique map so $m_i^M = m_{\pi(i)}^N$, then denote $\mathbf{x}^\pi$ to be the subtuple of $\mathbf{x}$ such that $x_i^\pi = x_{\pi(i)}$ and include

$$\forall \mathbf{x} \left(R_{[N]}(\mathbf{x}) \to R_{[M \prec N]}(\mathbf{x}^\pi, \mathbf{x}) \right)$$

5. *Coherence: If $M \subset N$ are both strong substructures of the whole model, then $M \prec N$.*
   If $M \prec N$ and $m_i^M = m_{j_i}^N$, then include

$$\forall \mathbf{x}, \mathbf{y} \left(R_{[M]}(\mathbf{x}) \wedge R_{[N]}(\mathbf{y}) \wedge \bigwedge_{i < \kappa} x_i = y_{j_i} \to R_{[M \prec N]}(\mathbf{x}, \mathbf{y}) \right)$$

Remark 3.8. We have intentionally omitted the converse to Definition 3.7.(1), namely

$$\forall \mathbf{x} \left(\bigwedge_{\phi(z_{i_1}, \dots, z_{i_n}) \in tp_{qf}(M/\emptyset)} \phi(x_{i_1}, \dots, x_{i_n}) \to R_{[M]}(\mathbf{x}) \right)$$

because it is not true. The "toy example" of a nonfinitary AEC–the $L(Q)$-theory of an equivalence relation where each equivalence class is countable–gives a counter-example.

For any $M^* \vDash T^*$, denote $M^* \restriction \tau$ by M.

Theorem 3.9 (Relational Presentation Theorem). *1. If $M^* \vDash T^*$ then $M^* \restriction \tau \in K$. Further, for all $M_0 \in K_\kappa$, we have $M^* \vDash R_{[M_0]}(\mathbf{m})$ implies that $\mathbf{m}$ enumerates a strong substructure of M.*

2. *Every $M \in K$ has a unique expansion M^* that models T^*.*

3. *If $M \prec N$, then $M^* \subset N^*$.*

4. *If $M^* \subset N^*$ both model T^*, then $M \prec N$.*

5. *If $M \prec N$ and $M^* \vDash T$ such that $M^* \restriction \tau = M$, then there is $N^* \vDash T$ such that $M^* \subset N^*$ and $N^* \restriction \tau = N$.*

Moreover, this is a functorial expansion in the sense of Vasey [Vasa, Definition 3.1] and $(\operatorname{Mod} T^*, \subset)$ is an AEC except that it allows κ-ary relations.

Note that although the vocabulary τ^* is κ-ary, the structure of objects and embeddings from $(\operatorname{Mod} T^*, \subset)$ still satisfies all of the category theoretic conditions on AECs, as developed by Lieberman and Rosický [LR]. This is because $(\operatorname{Mod} T^*, \subset)$ is equivalent to an AEC, namely $\mathbf{K}$, via the forgetful functor.

Proof: (1): We will build a $\prec$-directed system $\{M_\mathbf{a} \subset M : \mathbf{a} \in {}^{<\omega} M\}$ that are members of K_κ. We do not (and cannot) require in advance that $M_\mathbf{a} \prec M$, but this will follow from our argument.

For singletons $a \in M$, taking $\mathbf{x}$ to be $\langle a : i < \kappa \rangle$ in (3.7.3), implies that there is $M'_a \in K_\kappa$ and $\mathbf{m}^a \in {}^\kappa M$ with $a \in \mathbf{m}^a$ such that $M \vDash R_{[M'_a]}(\mathbf{m}^a)$. By (1), this means that $m_i^a \mapsto m_i^{M'_a}$ is an isomorphism. Set $M_a := \mathbf{m}^a$.[5]

Suppose $\mathbf{a}$ is a finite sequence in M and $M_{\mathbf{a}'}$ is defined for every $\mathbf{a}' \subsetneq \mathbf{a}$. Using the union of the universes as the $\mathbf{x}$ in (3.7.3), there is some $N \in K_\kappa$ and $\mathbf{m}^\mathbf{a} \in {}^\kappa M$ such that

- $|M_{\mathbf{a}'}| \subset \mathbf{m}^\mathbf{a}$ for each $\mathbf{a}' \subsetneq \mathbf{a}$.
- $M \vDash R_{[N]}(\mathbf{m}^\mathbf{a})$.

By (3.7.4), this means that $M \vDash R_{\bar{M}_{\mathbf{a}'} \prec \bar{N}}(\mathbf{m}^{\mathbf{a}'}, \mathbf{m}^\mathbf{a})$, after some permutation of the parameters. By (2) and (1), this means that $M_{\mathbf{a}'} \prec N$; set $M_\mathbf{a} := \mathbf{m}^\mathbf{a}$.

Now that we have finished the construction, we are done. AECs are closed under directed unions, so $\cup_{\mathbf{a} \in M} M_\mathbf{a} \in K$. But this model has the same universe as M and is a substructure of M; thus $M = \cup_{\mathbf{a} \in M} M_\mathbf{a} \in K$.

For the further claim, suppose $M^* \vDash R_{[M_0]}(\mathbf{m})$. We can redo the same proof as above with the following change: whenever $\mathbf{a} \in M$ is a finite sequence such that $\mathbf{a} \subset \mathbf{m}$, then set $\mathbf{m}^\mathbf{a} = \mathbf{m}$ directly, rather than appealing to (3.7.3) abstractly. Note that $\mathbf{m}$ witnesses the existential in that axiom, so the rest of the construction can proceed without change. At the end, we have

$$\mathbf{m} = M_\mathbf{a} \prec \bigcup_{\mathbf{a}' \in {}^{<\omega} M} M_{\mathbf{a}'} = M$$

(2): First, it is clear that $M \in K$ has an expansion; for each $M_0 \prec M$ of size κ, make $R_{[M_0]}(\langle m_i^{M_0} : i < \kappa \rangle)$ hold and, for each $M_0 \prec N_0 \prec M$ of size κ, make $R_{[M_0 \prec N_0]}(\langle m_i^{M_0} : i < \kappa \rangle, \langle m_i^{N_0} : i < \kappa \rangle)$ hold. Now we want to show this expansion is the unique one.
Suppose $M^+ \vDash T^*$ is an expansion of M. We want to show this is in fact the expansion described in the above paragraph. Let $M_0 \prec M$. By (3.7.3) and (1) of this theorem, there is $N_0 \prec M$ and $\mathbf{n} \in {}^\kappa M$ such that

- $M^+ \vDash R_{[N_0]}(\mathbf{n})$
- $|M_0| \subset \mathbf{n}$

By coherence, $M_0 \prec \mathbf{n}$. Since $n_i \mapsto m_i^{N_0}$ is an isomorphism, there is $M_0^* \cong M_0$ such that $M_0^* \prec N_0$. Note that $T^* \vDash \forall \mathbf{x} R_{[M_0^*]}(\mathbf{x}) \leftrightarrow R_{[M_0]}(\mathbf{x})$. By (3.7.4),

$$M^+ \vDash R_{[M_0^* \prec N_0]}(\langle m_i^{M_0} : i < \kappa \rangle, \mathbf{n})$$

---

[5] We mean that we set M_a to be τ-structure with universe the range of $\mathbf{m}^a$ and functions and relations inherited from M'_a via the map above.

By (3.7.2), $M^+ \vDash R_{[M_0^*]}(\langle m_i^{M_0} : i < \kappa \rangle)$, which gives us the conclusion by the further part of (1) of this theorem.

Similarly, if $M_0 \prec N_0 \prec M$, it follows that

$$M^+ \vDash R_{[M_0 \prec N_0]}(\langle m_i^{M_0} : i < \kappa \rangle, \langle m_i^{N_0} : i < \kappa \rangle)$$

Thus, this arbitrary expansion is actually the intended one.

(3): Apply the uniqueness of the expansion and the transitivity of $\prec$.

(4): As in the proof of (1), we can build $\prec$-directed systems $\{M_\mathbf{a} : \mathbf{a} \in {}^{<\omega}M\}$ and $\{N_\mathbf{b} : \mathbf{b} \in {}^{<\omega}N\}$ of submodels of M and N, so that $M_\mathbf{a} = N_\mathbf{a}$ when $\mathbf{a} \in {}^{<\omega}M$. From the union axioms of AECs, we see that $M \prec N$.

(5): This follows from (3), (4) of this theorem and the uniqueness of the expansion.

Recall that the map $M^* \in \operatorname{Mod} T^*$ to $M^* \restriction \tau \in \mathbf{K}$ is a an abstract Morleyization if it is a bijection such that every isomorphism $f : M \cong N$ in $\mathbf{K}$ lifts to $f : M^* \cong N^*$ and $M \prec N$ implies $M^* \subset N^*$. We have shown that this is true of our expansion.

Remark 3.10. *The use of infinitary quantification might remind the reader of the work on the interaction between AECs and $\mathbb{L}_{\infty,\kappa^+}$ by Shelah [She09, Chapter IV] and Kueker [Kue08] (see also Boney and Vasey [BV] for more in this area). The main difference is that, in working with $\mathbb{L}_{\infty,\kappa^+}$, those authors make use of the semantic properties of equivalence (back and forth systems and games). In contrast, particularly in the following transfer result we look at the syntax of $\mathbb{L}_{(2^\kappa)^+,\kappa^+}$.*

The functoriality of this presentation theorem allows us to give a syntactic proof of the amalgamation, etc. transfer results without assuming disjointness (although the results about disjointness follow similarly). We focus on amalgamation and give the details only in this case, but indicate how things are changed for other properties.

Proposition 3.2 applied to this context yields the following result.

Proposition 3.11. *$(K, \prec)$ has λ-amalgamation [joint embedding, etc.] iff $(\operatorname{Mod} T^*, \subset)$ has λ-amalgamation [joint embedding, etc.].*

Now we show the transfer of amalgamation between different cardinalities using the technology of this section.

Notation 3.12. Fix an AEC $\mathbf{K}$ and the language τ^* from Theorem 3.9.

1. Given τ^*-structures $M_0^* \subset M_1^*, M_2^*$, we define the *amalgamation diagram* $AD(M_1^*, M_2^*/M_0^*)$ to be

$$\{\phi(\mathbf{c}_{\mathbf{m}_0}, \mathbf{c}_{\mathbf{m}_1})) : \phi \text{ is quantifier-free from } \tau^* \text{ and for } \ell = 0 \text{ or } 1,$$
$$M_\ell^* \vDash \phi(\mathbf{c}_{\mathbf{m}_0}, \mathbf{c}_{\mathbf{m}_1}), \text{ with } \mathbf{m}_0 \in M_0^* \text{ and } \mathbf{m}_1 \in M_\ell^* \}$$

in the vocabulary $\tau^* \cup \{c_m : m \in M_1^* \cup M_2^*\}$ where each constant is distinct except for the common submodel M_0 and $\mathbf{c_m}$ denotes the finite sequence of constants $c_{m_1}, \ldots, c_{m_n}$.

The *disjoint amalgamation diagram* $DAD(M_1^*, M_2^*/M_0^*)$ is

$$AD(M_1^*, M_2^*/M_0^*) \cup \{c_{m_1} \neq c_{m_2} : m_\ell \in M_\ell^* - M_0^*\}$$

2. Given τ^*-structures M_0^*, M_1^*, we define the *joint embedding diagram* $JD(M_0^*, M_1^*)$ to be

$$\{\phi(\mathbf{c_m}) \quad : \quad \phi \text{ is quantifier-free from } \tau^* \text{ and}$$
$$\text{for } \ell = 0 \text{ or } 1, M_\ell^* \vDash \phi(\mathbf{c_m}) \text{ with } \mathbf{m} \in M_\ell^*\}$$

in the vocabulary $\tau^* \cup \{c_m : m \in M_1^* \cup M_2^*\}$ where each constant is distinct.

The *disjoint amalgamation diagram* $DJD(M_0^*, M_1^*)$ is

$$AD(M_1^*, M_2^*/M_0^*) \cup \{c_{m_1} \neq c_{m_2} : m_\ell \in M_\ell^* - M_0^*\}$$

The use of this notation is obvious.

Claim 3.13. Any amalgam of M_1 and M_2 over M_0 is a reduct of a model of

$$T^* \cup AD(M_1^*, M_2^*/M_0^*)$$

Proof: An amalgam of $M_0 \prec M_1, M_2$ is canonically expandable to an amalgam of $M_0^* \subset M_1^*, M_2^*$, which is precisely a model of $T^* \cup AD(M_1^*, M_2^*/M_0^*)$. Conversely, a model of that theory will reduct to a member of $\mathbf{K}$ with embeddings of M_1 and M_2 that fix M_0.

There are similar claims for other properties. Thus, we have connected amalgamation in $\mathbf{K}$ to amalgamation in $(\operatorname{Mod} T^*, \subset)$ to a syntactic condition, similar to Lemma 3.5. Now we can use the compactness of logics in various large cardinals to transfer amalgamation between cardinals. To do this, recall the notion of an amalgamation base.

Definition 3.14. For a class of cardinals $\mathcal{F}$, we say $M \in K_{\mathcal{F}}$ is a $\mathcal{F}$-*amalgamation base* ($\mathcal{F}$-a.b.) if any pair of models from $K_{\mathcal{F}}$ extending M can be amalgamated over M. We use the same rewriting conventions as in Definition 1.3.(1), e.g., writing $\leq \lambda$-a.b. for $[LS(K), \lambda]$-amalgamation base.

We need to specify two more large cardinal properties.

Definition 3.15. 1. A cardinal κ is weakly compact if it is strongly inaccessible and every set of κ sentence in $\mathbb{L}_{\kappa,\kappa}$ that is $< \kappa$-satisfiable is satisfiable[6].

---

[6]At one time strong inaccessiblity was not required, but this is the current definition.

2. A cardinal κ is measurable if there exists a κ-additive, nontrivial, $\{0,1\}$-valued measure on the power set of κ.

3. κ is (δ, λ)-strongly compact for $\delta \le \kappa \le \lambda$ if there is a δ-complete, fine ultrafilter on $P_\kappa(\lambda)$.

   κ is λ-strongly compact if it is (κ, λ)-strongly compact.

This gives us the following results syntactically.

Proposition 3.16. *Suppose* $LS(K) < \kappa$.

- *Let κ be weakly compact and $M \in K_\kappa$. If M can be written as an increasing union $\cup_{i<\kappa} M_i$ with each $M_i \in K_{<\kappa}$ being a $< \kappa$-a.b., then M is a κ-a.b.*

- *Let κ be measurable and $M \in K$. If M can be written as an increasing union $\cup_{i<\kappa} M_i$ with each M_i being a λ_i-a.b., then M is a $(\sup_{i<\kappa} \lambda_i)$-a.b.*

- *Let κ be λ-strongly compact and $M \in K$. If M can be written as a directed union $\cup_{x \in P_\kappa \lambda} M_x$ with each M_x being a $< \kappa$-a.b., then M is a $\le \lambda$-a.b.*

Proof: The proof of the different parts are essentially the same: take a valid amalgamation problem over M and formulate it syntactically via Claim 3.13 in $\mathbb{L}_{\kappa,\kappa}(\tau^*)$. Then use the appropriate syntactic compactness for the large cardinal to conclude the satisfiability of the appropriate theory.

First, suppose κ is weakly compact and $M = \cup_{i<\kappa} M_i \in K_\kappa$ where $M_i \in K_{<\kappa}$ is a $< \kappa$-a.b. Let $M \prec M^1, M^2$ is an amalgamation problem from K_κ. Find resolutions $\langle M_i^\ell \in K_{<\kappa} : i < \kappa \rangle$ with $M_i \prec M_i^\ell$ for $\ell = 1, 2$. Then

$$T^* \cup AD(M^{1*}, M^{2*}/M^*) = \bigcup_{i<\kappa} \left(T^* \cup AD(M_i^{1*}, M_i^{2*}/M_i^*) \right)$$

and is of size κ. Each member of the union is satisfiable (by Claim 3.13 because M_i is a $< \kappa$-a.b.) and of size $< \kappa$, so $T^* \cup AD(M^{1*}, M^{2*}/M^*)$ is satisfiable. Since $M^1, M^2 \in K_\kappa$ were arbitrary, M is a κ-a.b.

Second, suppose that κ is measurable and $M = \cup_{i<\kappa} M_i$ where M_i is a λ_i-a.b. Set $\lambda = \sup_{i<\kappa} \lambda_i$ and let $M \prec M^1, M^2$ is an amalgamation problem from K_λ. Find resolutions $\langle M_i^\ell \in K : i < \kappa \rangle$ with $M_i \prec M_i^\ell$ for $\ell = 1, 2$ and $\|M_i^\ell\| = \lambda_i$. Then

$$T^* \cup AD(M^{1*}, M^{2*}/M^*) = \bigcup_{i<\kappa} \left(T^* \cup AD(M_i^{1*}, M_i^{2*}/M_i^*) \right)$$

Each member of the union is satisfiable because M_i is a λ_i-a.b. By the syntactic characterization of measurable cardinals (see [CK73, Exercise 4.2.6]), the union is satisfiable. Thus, M is λ-a.b.

Third, suppose that κ is λ-strongly compact and $M = \cup_{x \in P_\kappa \lambda} M_x$ with each M_x being a $< \kappa$-a.b. Let $M \prec M^1, M^2$ be an amalgamation problem from K_λ. Find directed systems $\langle M_x^\ell \in K_{<\kappa} \mid x \in P_\kappa \lambda \rangle$ with $M_x \prec M_x^\ell$ for $\ell = 1, 2$. Then

$$T^* \cup AD(M^{1*}, M^{2*}/M^*) = \bigcup_{x \in P_\kappa \lambda} \left(T^* \cup AD(M_x^{1*}, M_x^{2*}/M_x^*) \right)$$

Every subset of the left side of size $< \kappa$ is contained in a member of the right side because $P_\kappa \lambda$ is $< \kappa$-directed, and each member of the union is consistent because each M_x is an amalgamation base. Because κ is λ-strongly compact, this means that the entire theory is consistent. Thus, M is a λ-a.b.

From this, we get the following corollaries computing upper bounds on the Hanf number for the $\leq \lambda$-AP.

Corollary 3.17. *Suppose $LS(K) < \kappa$.*

- *If κ is weakly compact and K has $< \kappa$-AP, then K has $\leq \kappa$-AP.*
- *If κ is measurable, cf $\lambda = \kappa$, and K has $< \lambda$-AP, then K has $\leq \lambda AP$.*
- *If κ is λ-strongly compact and K has $< \kappa$-AP, then K has $\leq \lambda$-AP.*

Moreover, when κ is strongly compact, we can imitate the proof of [MS90, Corollary 1.6] to show that being an amalgamation base follows from being a $< \kappa$-existentially closed model of T^*. This notion turns out to be the same as the notion of $< \kappa$-universally closed from [Bon], and so this is an alternate proof of [Bon, Lemma 7.2].

---

4   The Big Gap

This section concerns examples of 'exotic' behavior in small cardinalities as opposed to behavior that happens unboundedly often or even eventually. We discuss known work on the spectra of existence, amalgamation of various sorts, tameness, and categoricity.

Intuitively, Hanf's principle is that if a certain property can hold for only set-many objects, then it is eventually false. He refines this twice. First, if $\mathcal{K}$ a *set* of collections of structures $\mathbf{K}$ and $\phi_P(X, y)$ is a formula of set theory such $\phi(\mathbf{K}, \lambda)$ means some member of $\mathbf{K}$ with cardinality λ satisfies P, then there is a cardinal κ_P such that for any $\mathbf{K} \in \mathcal{K}$, if $\phi(\mathbf{K}, \kappa')$ holds for some $\kappa' \geq \kappa_P$, then $\phi(\mathbf{K}, \lambda)$ holds for arbitrarily large λ. Secondly, he observed that if the property P is closed down for sufficiently large members of each $\mathbf{K}$, then 'arbitrarily large' can be replaced by 'on a tail' (i.e., eventually).

Existence: Morley (plus the Shelah Presentation Theorem) gives a decisive concrete example of this principle to AEC's. Any AEC in a countable vocabulary with countable Löwenheim-Skolem number with models up to $\beth_{\omega_1}$ has arbitrarily large models. And Morley [Mor65] gave easy examples showing this bound was tight for arbitrary sentences of $\mathbb{L}_{\omega_1,\omega}$. But it was almost 40 years later that Hjorth [Hjo02, Hjo07] showed this bound is also tight for *complete*-sentences of $\mathbb{L}_{\omega_1,\omega}$. And a fine point in his result is interesting.

We say a ϕ *characterizes* κ, if there is a model of ϕ with cardinality κ but no larger. Further, ϕ *homogeneously* [Bau74] characterizes κ if ϕ is a complete sentence of $\mathbb{L}_{\omega_1,\omega}$ that characterizes κ, contains a unary predicate U such that if M is the countable model of ϕ, every permutation of $U(M)$ extends to an automorphism of M (i.e., $U(M)$ is a set of absolute indiscernibles) and there is a model N of ϕ with $|U(N)| = \kappa$.

In [Hjo02], Hjorth found, by an inductive procedure, for each $\alpha < \omega_1$, a countable (finite for finite α) set S_α of complete $\mathbb{L}_{\omega_1,\omega}$-sentences such that some $\phi_\alpha \in S_\alpha$ characterizes $\aleph_\alpha$[7]. This procedure was nondeterministic in the sense that he showed one of (countably many if α is infinite) sentences worked at each $\aleph_\alpha$; it is conjectured [Sou13] that it may be impossible to decide in ZFC which sentence works. In [BKL15], we show a modification of the Laskowski-Shelah example (see [LS93, BFKL16]) gives a family of $\mathbb{L}_{\omega_1,\omega}$-sentences ϕ_r, such that ϕ_r homogeneously characterizes $\aleph_r$ for $r < \omega$. Thus for the first time [BKL15] establishes in ZFC, the existence of specific sentences ϕ_r characterizing $\aleph_r$.

Amalgamation: In this chapter, we have established a similar upper bound for a number of amalgamation-like properties. Moreover, although it is not known beforehand that the classes are eventually downward closed, that fact falls out of the proof. In all these cases, the known lower bounds (i.e., examples where AP holds initially and eventually fails) are far smaller. We state the results for countable Löwenheim-Skolem numbers, although the [BKS09, KLH14] results generalize to larger cardinalities.

The best lower bounds for the disjoint amalgamation property is $\beth_{\omega_1}$ as shown in [KLH14] and [BKS09]. In [BKS09], Baldwin, Kolesnikov, and Shelah gave examples of $\mathbb{L}_{\omega_1,\omega}$-definable classes that had disjoint embedding up to $\aleph_\alpha$ for every countable α (but did not have arbitrarily large models). Kolesnikov and Lambie-Hanson [KLH14] show that for the collection of all coloring classes (again $\mathbb{L}_{\omega_1,\omega}$-definable when α is countable) in a vocabulary of a fixed size κ, the Hanf number for amalgamation (equivalently in this example disjoint amalgamation) is precisely $\beth_{\kappa^+}$ (and many of the classes have arbitrarily large models). In [BKL15], Baldwin, Koerwein, and Laskowski construct, for each $r < \omega$, a *complete* $\mathbb{L}_{\omega_1,\omega}$-sentence ϕ^r that has disjoint 2-amalgamation up to and including $\aleph_{r-2}$; disjoint amalgamation and even amalgamation fail in $\aleph_{r-1}$ but hold (trivially) in $\aleph_r$; there is no model in $\aleph_{r+1}$.

The joint embedding property and the existence of maximal models are

---

[7]Malitz [Mal68] (under GCH) and Baumgartner [Bau74] had earlier characterized the $\beth_\alpha$ for countable α.

closely connected[8]. The main theorem of [BKS16] asserts: If $\langle \lambda_i : i \leq \alpha < \aleph_1 \rangle$ is a strictly increasing sequence of characterizable cardinals whose models satisfy JEP($< \lambda_0$), there is an $\mathbb{L}_{\omega_1,\omega}$-sentence ψ such that

1. The models of ψ satisfy JEP($< \lambda_0$), while JEP fails for all larger cardinals and AP fails in all infinite cardinals.

2. There exist $2^{\lambda_i^+}$ nonisomorphic maximal models of ψ in λ_i^+, for all $i \leq \alpha$, but no maximal models in any other cardinality; and

3. ψ has arbitrarily large models.

Thus, a lower bound on the Hanf number for either maximal models of the joint embedding property is again $\beth_{\omega_1}$. Again, the result is considerably more complicated for complete sentences. But Baldwin and Souldatos [BS15b] show that there is a sentence ϕ in a vocabulary with a predicate X such that if $M \models \phi$, $|M| \leq |X(M)|^+$ and for every κ there is a model with $|M| = \kappa^+$ and $|X(M)| = \kappa$. Further they note that if there is a sentence ϕ that homogenously characterizes κ, then there is a sentence ϕ' with a new predicate B such that ϕ' also characterizes κ, B defines a set of absolute indiscernibles in the countable model, and there are models M_λ for $\lambda \leq \kappa$ such that $(|M|, |B(M_\lambda)|) = (\kappa, \lambda)$. Combining these two with earlier results of Souldatos [Sou13], one obtains several different ways to show the lower bound on the Hanf number for a complete $\mathbb{L}_{\omega_1,\omega}$-sentence having maximal models is $\beth_{\omega_1}$. In contrast to [BKS16], all of these examples have no models beyond $\beth_{\omega_1}$.

No maximal models: Baldwin and Shelah [BS15a] have announced that the exact Hanf number for the nonexistence of maximal models is the first measurable cardinal. Souldatos observed that this implies the lower bound on the Hanf number for **K** has joint embedding of models at least μ is the first measurable.

Tameness: Note that the definition of a Hanf number for tameness is more complicated, as tameness is fundamentally a property of two variables: **K** is $(< \chi, \mu)$-tame if for any $N \in \mathbf{K}_\mu$, if the Galois types p and q over N are distinct, there is an $M \prec N$ with $|M| < \chi$ and $p \upharpoonright M \neq q \upharpoonright M$.

Thus, we define the *Hanf number for $< \kappa$-tameness* to be the minimal λ such that the following holds:

if **K** is an AEC with $LS(\mathbf{K}) < \kappa$ that is $(< \kappa, \mu)$-tame for *some* $\mu \geq \lambda$, then it is $(< \kappa, \mu)$-tame for arbitrarily large μ.

The results of [Bon] show that Hanf number for $< \kappa$-tameness is κ when κ is strongly compact[9]. However, this is done by showing a much stronger "global tameness" result that ignores the hypothesis: *every* AEC **K** with $LS(\mathbf{K}) < \kappa$ is $(< \kappa, \mu)$-tame for all $\mu \geq \kappa$. Boney and Unger [BU], building on earlier work of Shelah [Shec], have shown that this global tameness result is actually an

---

[8]Note that, under joint embedding, the existence of a maximal model is equivalent to the nonexistence of arbitrarily large models.

[9]This can be weakened to almost strongly compact; see Brooke-Taylor and Rosický [BTR15] or Boney and Unger [BU].

equivalence (in the almost strongly compact form). Also, due to monotonicity results for tameness, the Boney results show that the Hanf number for $< \lambda$-tameness is at most the first almost strongly compact above λ (if such a thing exists). The results [BU, Theorem 4.9] put a large restriction on the structure of the tameness spectrum for any ZFC Hanf number. In particular, the following

Fact 4.1. Let $\sigma = \sigma^\omega < \kappa \leq \lambda$. Every AEC **K** with $LS(\mathbf{K}) = \sigma$ is $\left(< \kappa, \sigma^{(\lambda^{<\kappa})} \right)$-tame iff κ is (σ^+, λ)-strongly compact.

This means that a ZFC (i.e., not a large cardinal) Hanf number for $< \kappa$-tameness would consistently have to avoid cardinals of the form $\sigma^{(\lambda^{<\kappa})}$ (under GCH, all cardinals are of this form except for singular cardinals and successors of singulars of cofinality less than κ).

One could also consider a variation of a Hanf number for $< \kappa$ that requires $(< \kappa, \mu)$-tameness *on a tail* of μ, rather than for arbitrarily large μ. The argument above shows that that is exactly the first strongly compact above κ.

Categoricity: Another significant instance of Hanf's observation is She-lah's proof in [She99a] that if $\mathcal{K}$ is taken as all AEC's **K** with $LS_\mathbf{K}$ bounded by a cardinal κ, then there is such an eventual Hanf number for categoricity in a successor. Boney [Bon] places an upper bound on this Hanf number as the first strongly compact above κ. This depended on the results on tameness discussed in the previous paragraphs.

Building on work of Shelah [She09, She10], Vasey [Vasb] proves that if a universal class (see [She87]) is categorical in a λ at least the Hanf number for existence, then it has amalgamation in all $\mu \geq \kappa$. Then he shows that for universal class in a countable vocabulary, *that satisfies amalgamation*, the Hanf number for categoricity is at most $\beth_{\beth_{(2^\omega)^+}}$. Note that the lower bound for the Hanf number for categoricity is $\aleph_\omega$, ([HS90, BK09]).

Question 4.2. 1. Can one calculate in ZFC an upper bound on these Hanf numbers for 'amalgamation'? Can[10] the gaps in the upper and lower bounds of the Hanf numbers reported here be closed in ZFC? Will smaller large cardinal axioms suffice for some of the upper bounds? Does categoricity help?

2. (Vasey) Are there any techniques for downward transfer of amalgamation[11]?

3. Does every AEC have a functional expansion to a $PC\Gamma$ class? Is there a natural class of AEC with this property — e.g., solvable groups?

---

[10] Grossberg initiated this general line of research.

[11] Note that there is an easy example in [BKS09] of a sentence in $\mathbb{L}_{\omega_1,\omega}$ that is categorical and has amalgamation in every uncountable cardinal but it fails both in $\aleph_0$.

4. Can[12] one define in ZFC a sequence of sentences ϕ_α for $\alpha < \omega_1$, such that ϕ_α characterizes $\aleph_\alpha$?

5. (Shelah) If $\aleph_{\omega_1} < 2^{\aleph_0}$ $\mathbb{L}_{\omega_1,\omega}$-sentence has models up to $\aleph_{\omega_1}$, must it have a model in $2^{\aleph_0}$? (He proves this statement is consistent in [She99b]).

6. (Souldatos) Is any cardinal except $\aleph_0$ characterized by a complete sentence of $\mathbb{L}_{\omega_1,\omega}$ but not homogeneously?

---

[12]This question seems to have originated from discussions of Baldwin, Souldatos, Laskowski, and Koerwien.

Bibliography

[Bal09]   John T. Baldwin. *Categoricity*. Number 51 in University Lecture Notes. American Mathematical Society, Providence, USA, 2009. www.math.uic.edu/~jbaldwin.

[Bal15a]   J.T. Baldwin. The entanglement of model theory and set theory. lecture slides and audio `http://homepages.math.uic.edu/`, 2015.

[Bal15b]   J.T. Baldwin. Formalization without foundationalism; model theory and the philosophy of mathematics practice. Book manuscript available on request, 2015.

[Bau74]   J. Baumgartner. The Hanf number for complete $L_{\omega_1,\omega}$-sentences (without GCH). *Journal of Symbolic Logic*, 39:575–578, 1974.

[BFKL16]   J.T. Baldwin, Sy Friedman, M. Koerwien, and C. Laskowski. Three red herrings around Vaught's conjecture. *Transactions of the American Math Society*, 368:22, 2016. Published electronically: November 6, 2015.

[BK09]   John T. Baldwin and Alexei Kolesnikov. Categoricity, amalgamation, and tameness. *Israel Journal of Mathematics*, 170:411–443, 2009. Also at `www.math.uic.edu/\~\jbaldwin`.

[BKL15]   J.T. Baldwin, M. Koerwien, and C. Laskowski. Amalgamation, characterizing cardinals, and locally finite aec. *Journal of Symbolic Logic*, 81:1142–1162. Preprint: `http://homepages.math.uic.edu/~jbaldwin/`.

[BKS09]   J.T. Baldwin, A. Kolesnikov, and S. Shelah. The amalgamation spectrum. *Journal of Symbolic Logic*, 74:914–928, 2009.

[BKS16]   J.T. Baldwin, M. Koerwien, and I. Souldatos. The joint embedding property and maximal models. *Archive for Mathematical Logic*, 55:545–565, 2016.

[Bon]   W. Boney. Tameness from large cardinal axioms. Tameness from large cardinal axioms. *Journal of Symbolic Logic*, 79:1092–1119.

[Boo76]   Boos. Infinitary compactness without strong inaccessibility. *The Journal of Symbolic Logic*, 41:33–38, 1976.

[Bou50]   Nicholas Bourbaki. The architecture of mathematics. *The American Mathematical Monthly*, 57:221–232, 1950.

[BS15a]   J.T. Baldwin and S. Shelah.   Hanf numbers for maximality.
           Preprint. Shelah F1533, 2015.

[BS15b]   J.T. Baldwin and I. Souldatos.   Complete $L_{\omega_1,\omega}$-sentences with
           maximal models in multiple cardinalities. Preprint, 2015.

[BTR15]   Andrew Brooke-Taylor and Jiří Rosický. Accessible images revis-
           ited. Preprint, arXiv:1506.01986, 2015.

[BU]   W. Boney and S. Unger. Large cardinal axioms from tameness in
           AECs. In preparation: arXiv:1509.01191.

[BV]   W. Boney and S. Vasey.   Categoricity and infinitary logics.
           Preprint.

[CCHJ]   B. Cody, S. Cox, J. Hamkins, and T. Johnston. The weakly com-
           pact embedding property. In preparation.

[CHL85]   G.L. Cherlin, L. Harrington, and A.H. Lachlan. $\aleph_0$-categorical, $\aleph_0$-
           stable structures. *Annals of Pure and Applied Logic*, 28:103–135,
           1985.

[CK73]   C. C. Chang and H. J Keisler. *Model theory*. North-Holland, 1973.
           3rd edition 1990.

[Cor92]   Leo Corry. Nicolas bourbaki and the concept of mathematical
           structure. *Synthese*, 92:315–348, 1992.

[Cor04]   Leo Corry. *Modern Algebra and the Rise of Mathematical Struc-
           tures*. Birkhäuser Verlag, 2004.

[Fra54]   R. Fraïssé. Sur quelques classifications des systèmes de relations.
           Publ. Sci. Univ. Algeria Sèr. A, 1:35–182, 1954.

[Gro02]   R. Grossberg. Classification theory for non-elementary classes. In
           Yi Zhang, editor, *Logic and Algebra*, pages 165–204. AMS, 2002.
           Contemporary Mathematics 302.

[GV06]   R. Grossberg and M. VanDieren. Galois stability for tame abstract
           elementary classes. *Journal of Mathematical Logic*, 6:1–24, 2006.

[Hjo02]   Greg Hjorth. Knight's model, its automorphism group, and charac-
           terizing the uncountable cardinals. *Journal of Mathematical Logic*,
           113–144, 2002.

[Hjo07]   Greg Hjorth. A note on counterexamples to Vaught's conjecture.
           Notre Dame Journal of Formal Logic, 2007.

[HK06]   T. Hyttinen and M. Kesälä. Independence in finitary abstract
           elementary classes. *Ann. Pure Appl. Logic*, 143(1-3):103–138, 2006.

[HS90]   B. Hart and S. Shelah. Categoricity over P for first order T or cate-
           goricity for $\phi \in l_{\omega_1\omega}$ can stop at $\aleph_k$ while holding for $\aleph_0, \cdots, \aleph_{k-1}$.
           Israel Journal of Mathematics, 70: 219–235, 1990.

[Jec06]   T. Jech. *Set Theory: 3rd Edition*. Springer Monographs in Math-
           ematics. Springer, 2006.

[Jón56]   B. Jónsson. Universal relational systems. *Mathematica Scandinavica*, 4:193–208, 1956.

[Jón60]   B. Jónsson. Homogeneous universal relational systems. *Mathematica Scandinavica*, 8:137–142 ,1960.

[Ken13]   Juliette Kennedy. On formalism freeness: Implementing Gödel's 1946 Princeton Bicentennial Lecture. *Bulletin of Symbolic Logic*, 19:351–393, 2013.

[KLH14]   Alexei Kolesnikov and Christopher Lambie-Hanson. Hanf numbers for amalgamation of coloring classes. Preprint, 2014.

[Kue08]   D. W. Kueker. Abstract elementary classes and infinitary logics. *Annals of Pure and Applied Logic*, 156:274–286, 2008.

[LR]   M. Lieberman and Jirí Rosický. Classification theory of accessible categories. *Journal of Symbolic Logic*, 81:151–165.

[LS93]   Michael C. Laskowski and Saharon Shelah. On the existence of atomic models. *J. Symbolic Logic*, 58:1189–1194, 993.

[Mal68]   J. Malitz. The Hanf number for complete $L_{\omega_1,\omega}$ sentences. In J. Barwise, editor, *The syntax and semantics of infinitary languages*, Lecture Notes in Mathematics 72,166–181. Springer-Verlag, 1968.

[Mor65]   M. Morley. Omitting classes of elements. In Addison, Henkin, and Tarski, editors, *The Theory of Models*, pages 265–273. North-Holland, Amsterdam, 1965.

[MS90]   Michael Makkai and Saharon Shelah. Categoricity of theories in $l_{\kappa,\omega}$, with κ a compact cardinal. *Annals of Pure and Applied Logic*, 47:41–97, 1990.

[She]   S. Shelah. Maximal failures of sequence locality in a.e.c. Preprint on Shelah archive.

[She87]   Saharon Shelah. Universal classes, part I. In J.T. Baldwin, editor, *Classification theory (Chicago, IL, 1985)*, pages 264–419. Springer, Berlin, 1987. Paper 300: Proceedings of the USA–Israel Conference on Classification Theory, Chicago, December 1985; volume 1292 of *Lecture Notes in Mathematics*.

[She99a]   S. Shelah. Categoricity for abstract classes with amalgamation. *Annals of Pure and Applied Logic*, 98:261–294, 1999. Paper 394. Consult Shelah for post-publication revisions.

[She99b]   Saharon Shelah. Borel sets with large squares. *Fundamenta Mathematica*, 159:1–50, 1999.

[She09]   S. Shelah. *Classification Theory for Abstract Elementary Classes*. Studies in Logic. College Publications www.collegepublications.co.uk, 2009. Binds together papers 88r, 600, 705, 734 with introduction E53.

[She10]   S. Shelah.   *Classification Theory for Abstract Elementary Classes: II.*   Studies in Logic. College Publications <www. collegepublications.co.uk>, 2010. Binds together papers 300 A-G, E46, 838.

[Sou13]   Ioannis Souldatos.   Characterizing the powerset by a complete (Scott) sentence. *Fundamenta Mathematica*, 222:131–154, 2013.

[Vasa]   Sebastien Vasey. Infinitary stability theory. Preprint: http:// arxiv.org/abs/1412.3313.

[Vasb]   Sebastien Vasey. Shelah's eventual categoricity conjecture in universal classes. To appear in *Annals of Pure and Applied Logic*.

Chapter 10

A survey on tame abstract elementary classes

Will Boney

Harvard University
Cambridge, Massachussets, USA

Sebastien Vasey

Carnegie Mellon University
Pittsburgh, Pennsylvania, USA

Boney was partially supported by National Science Foundation under Grant No. DMS-1402191 and Vasey was supported by the Swiss National Science Foundation under Grant No. 155136.

1   Introduction

Abstract elementary classes (AECs) are a general framework for nonelementary model theory. They encompass many examples of interest while still
allowing some classification theory, as exemplified by Shelah's recent two-
volume book [She09b, She09c] titled *Classification Theory for Abstract Elementary Classes.*

So why study the classification theory of *tame* AECs in particular? Before
going into a technical discussion, let us simply say that several key results
can be obtained assuming tameness that provably cannot be obtained (or are
very hard to prove) without assuming it. Among these results are the construction, assuming a stability or superstability hypothesis, of certain global
independence notions akin to first-order forking. Similarly to the first-order
case, the existence of such notions allows us to prove several more structural
properties of the class (such as a bound on the number of types or the fact that
the union of a chain of saturated models is saturated). After enough of the
theory is developed, categoricity transfers (in the spirit of Morley's celebrated
categoricity theorem [Mor65]) can be established.

A survey of such results (with an emphasis on forking-like independence)
is in Section 5. However, we did not want to overwhelm the reader with a long
list of sometimes technical theorems, so we thought we would first present an
application: the categoricity transfer for universal classes from the abstract
(Section 4). We chose this result for several reasons. First, its statement is
simple and does not mention tameness or even abstract elementary classes.
Second, the proof revolves around several notions (such as good frames) that
might seem overly technical and ill-motivated if one does not see them in action
first. Third, the result improves on earlier categoricity transfers in several
ways, for example not requiring that the categoricity cardinal be a successor
and not assuming the existence of large cardinals. Finally, the method of
proof leads to Theorem 5.47, the currently best known ZFC approximation to
Shelah's eventual categoricity conjecture (which is the main test question for
AECs, see below).

Let us go back to what tameness is. Briefly, tameness is a property of AECs
saying that Galois (or orbital) types are determined locally: two distinct Galois
types must already be distinct when restricted to some small piece of their
domain. This holds in elementary classes: types as sets of formulas can be
characterized in terms of automorphisms of the monster model and distinct
types can be distinguished by a finite set of parameters. However, Galois types
in general AECs are not syntactic and their behavior can be wild, with "new"
types springing into being at various cardinalities and increasing chains of
Galois types having no unique upper bound (or even no upper bound at all).
This wild behavior makes it very hard to transfer results between cardinalities.

For a concrete instance, consider the problem of developing a forking-like independence notion for AECs. In particular, we want to be able to extend each (Galois) type p over M to a unique nonforking[1] extension over every larger set N. If the AEC, **K**, is nice enough, one might be able to develop an independence notion that allows one to obtain a nonforking extension q of p over N. But suppose that **K** is not tame and that this nontameness is witnessed by q. Then there is another type q' over N that has all the same small restrictions as q. In particular it extends p and (assuming our independence notion has a reasonable continuity property) is a nonforking extension. In this case the quest to have a unique nonforking extension is (provably, see Example 3.2.1.(4)) doomed.

This failure has led, in part, to Shelah's work on a local approach where the goal is to build a structure theory cardinal by cardinal without any "traces of compactness" (see [She01a, p. 5]). The central concept there is that of a good λ-frame (the idea is, roughly, that if an AEC **K** has a good λ-frame, then **K** is "well behaved in λ"). Multiple instances of categoricity together with non-ZFC hypotheses (such as the weak generalized continuum hypothesis: $2^\mu < 2^{\mu^+}$ for all μ) are used to build a good λ-frame [She01a], to push it up to models of size λ^+ (changing the class in the process) [She09b, Chapter II], and finally to push it to models of sizes $\lambda^{+\omega}$ and beyond in [She09b, Chapter III] (see Section 2.5).

In contrast, the amount of compactness offered by tameness and other locality properties has been used to prove similar results in simpler ways and with fewer assumptions (after tameness is accounted for). In particular, the work can often be done in ZFC.

In tame AECs, Galois types are determined by their small restrictions and the behavior of these small restrictions can now influence the behavior of the full type. An example can be seen in uniqueness results for nonsplitting extensions: in general AECs, uniqueness results hold for non-μ-splitting extensions to models of size μ, but no further (Theorem 2.24). However, in μ-tame AECs, uniqueness results hold for non-μ-splitting extensions to models *of all sizes* (Theorem 5.15). Indeed, the parameter μ in non-μ-splitting becomes irrelevant. Thus, tameness can replace several extra assumptions. Compared to the good frame results above, categoricity in a single cardinal, tameness, and amalgamation are enough to show the existence of a good frame (Theorem 5.44) and tameness and amalgamation are enough to transfer the frame upwards without any change of the underlying class (Theorem 5.26).

Although tameness seems to only be a weak approximation to the nice properties enjoyed by first-order logic, it is still *strong enough* to import more of the model-theoretic intuition and technology from first-order. When dealing with tame AECs, a type can be identified with the set of all of its restrictions to small domains, and these small types play a role similar to formulas. This can be made precise: one can even give a sense in which types are sets of (in-

---

[1] In the sense of the independence notion mentioned above. This will often be different from the first-order definition.

finitary) formulas (see Theorem 3.5). This allows several standard arguments
to be painlessly repeated in this context. For instance, the proof of the properties of $< \kappa$-satisfiability and the equivalence between Galois-stability and
no order property in this context are similar to their first-order counterparts
(see Section 5.2). On the other hand, several arguments from the theory of
tame AECs have no first-order analog (see for example the discussion around
amalgamation in Section 4).

On the other side, while tameness is in a sense a form of the first-order
compactness theorem, it is *sufficiently weak* that several examples of nonelementary classes are tame. Section 3.2.1 goes into greater depth, but diverse
classes like locally finite groups, rank one valued fields, and Zilber's pseudoex-
ponentiation all turn out to be tame. Tameness can also be obtained for free
from a large cardinal axiom, and a weak form of it follows from model-theoretic
hypotheses such as the combination of amalgamation and categoricity in a
high-enough cardinal.

Indeed, examples of *non*-tame AECs are in short supply (Section 3.2.2). All
known examples are set-theoretic in nature, and it is open whether there are
nontame "natural" mathematical classes (see (5) in Section 6). The focus on
ZFC results for tame AECs allows us to avoid situations where, for example,
conclusions about rank one valued fields depend on whether $2^{\aleph_0} < 2^{\aleph_1}$. This
replacing of set-theoretic hypotheses with model-theoretic ones suggests that
developing a classification theory for tame AECs is possible *within ZFC*.

Thus, tame AECs seem to strike an important balance: they are *general* enough to encompass several nonelementary classes and yet *well be-
haved/specific* enough to admit a classification theory. Even if one does not
believe that tameness is a justified assumption, it can be used as a first approximation and then one can attempt to remove (or weaken) it from the
statement of existing theorems. Indeed, there are several results in the literature (see the end of Section 2.4) which do not directly assume tameness, but
whose proof starts by deducing some weak amount of tameness from the other
assumptions, and then use this tameness in crucial ways.

We now highlight some results about tame AECs that will be discussed
further in the rest of this survey. We first state two motivating test questions.
The first is the well known categoricity conjecture which can be traced back
to an open problem in [She78]. The following version appears as [She09b,
Conjecture N.4.2]:

Conjecture 1.1 (Shelah's eventual categoricity conjecture). There exists
a function $\mu \mapsto \lambda(\mu)$ such that if **K** is an AEC categorical in *some* $\lambda \geq
\lambda(\mathrm{LS}(\mathbf{K}))$, then **K** is categorical in *all* $\lambda' \geq \lambda(\mathrm{LS}(\mathbf{K}))$.

Shelah's categoricity conjecture is the main test question for AECs and
remains the yardstick by which progress is measured. Using this yardstick,
tame AECs are well developed. Grossberg and VanDieren [GV06b] isolated
tameness from Shelah's proof of a downward categoricity transfer in AECs
with amalgamation [She99]. Tameness was one of the key (implicit) prop-

erties there (in the proof of [She99, Main claim 2.3], where Shelah proves that categoricity in a high-enough successor implies that types over Galois-saturated models are determined by their small restrictions[2], this property has later been called weak tameness). Grossberg and VanDieren defined tameness without the assumption of saturation and developed the theory in a series of papers [GV06b, GV06c, GV06a], culminating in the proof of Shelah's eventual categoricity conjecture from a successor in tame AECs with amalgamation[3].

Progress towards other categoricity transfers often proceed by first proving tameness and then using it to transfer categoricity. One of the achievements of developing the classification theory of tame AECs is the following result, due to the second author [Vasf]:

Theorem 1.2. *Shelah's eventual categoricity conjecture is true when* **K** *is a universal class with amalgamation. In this case, one can take* $\lambda(\mu) := \beth_{(2^\mu)^+}$. *Moreover, amalgamation can be derived from categoricity in cardinals of arbitrarily high cofinality.*

The proof starts by observing that every universal class is tame (a result of the first author [Bonc], see Theorem 3.11).

The second test question is more vague and grew out of the need to generalize some of the tools of first-order stability theory to AECs.

Question 1.3. Let **K** be an AEC categorical in a high-enough cardinal. Does there exist a cardinal χ such that $\mathbf{K}_{\geq \chi}$ admits a notion of independence akin to first-order forking?

The answer is positive for universal classes with amalgamation (see Theorem 5.41), and more generally for classes with amalgamation satisfying a certain strengthening of tameness:

Theorem 5.53. *Let* **K** *be a fully* $< \aleph_0$-*tame and -type short AEC with amalgamation. If* **K** *is categorical in a* $\lambda > \mathrm{LS}(\mathbf{K})$, *then* $\mathbf{K}_{\geq \beth_{\left(2^{\mathrm{LS}(\mathbf{K})}\right)^+}}$ *has (in a precise sense) a superstable forking-like independence notion.*

Varying the locality assumption, one can obtain weaker, but still powerful, conclusions that are key in the proof of Theorem 1.2.

One of the big questions in developing classification theory outside of first-order is which of the characterizations of dividing lines to take as the definition (see Section 2.1). This is especially true when dealing with superstability. In the first-order context, this is characterizable by forking having certain properties or the union of saturated models being saturated or one of several other properties. In tame AECs, these characterizations have recently been proven to also be equivalent! See Theorem 5.23.

This survey is organized as follows: Section 2 reviews concepts from the

---

[2]In [She01a, Definition 0.24], Shelah defines a type to be *local* if it is defined by all its $\mathrm{LS}(\mathbf{K})$-sized restrictions.

[3]The work on [GV06b] was done in 2000-2001 and preprints of [GV06c, GV06a] were circulated in 2004.

study of general AEC. This begins with definitions and basic notions (Galois type, etc.) that are necessary for work in AECs. Subsection 2.1 is a review of classification theory without tameness. The goal here is to review the known results that do not involve tameness in order to emphasize the strides that assuming tameness makes. Of course, we also set up notation and terminology. Previous familiarity with the basics of AECs as laid out in, e.g., [Bal09, Chapter 4] would be helpful. We also assume that the reader knows the basics of first-order model theory.

Section 3 formally introduces tameness and related principles. Subsection 3.2.1 reviews the known examples of tameness and nontameness.

Section 4 outlines the proof of Shelah's Categoricity Conjecture for universal classes. The goal of this outline is to highlight several of the tools that exist in the classification theory of tame AECs and tie them together in a single result. After whetting the reader's appetite, Section 5 goes into greater detail about the classification-theoretic tools available in tame AECs.

This introduction has been short on history and attribution and the historical remarks in Section 7 fill this gap. We have written this survey in a somewhat informal style where theorems are not attributed when they are stated: the reader should also look at Section 7, where proper credits are given. *It should not be assumed that an unattributed result is the work of the authors.*

Let us say a word about what is *not* discussed: We have chosen to focus on recent material which is not already covered in Baldwin's book [Bal09], so while its interest cannot be denied, we do not for example discuss the proof of the Grossberg-VanDieren upward categoricity transfer [GV06c, GV06a]. Also, we focus on tame AECs, so tameness-free results (such as Shelah's study of Ehrenfeucht-Mostowski models in [She09b, Chapter IV], [Shea], or the work of VanDieren and the second author on the symmetry property [Vanb, Van16, VV]) are not emphasized. Related frameworks which we do not discuss much are homogeneous model theory (see Example 3.2.1.(7)), tame finitary AECs (Example 3.2.1.(6)), and tame metric AECs (see Example 3.2.1.(9)).

Finally, let us note that the field is not a finished body of work but is still very much active. Some results may become obsolete soon after, or even before, this survey is published[4] . Still, we felt there was a need to write this chapter, as the body of work on tame AECs has grown significantly in recent years and there is, in our opinion, a need to summarize the essential points.

1.1   Acknowledgments

This chapter was written while the second author was working on a Ph.D. thesis under the direction of Rami Grossberg at Carnegie Mellon University

---

[4]Indeed, since this chapter was first circulated (in December 2015) the amalgamation assumption has been removed from Theorem 1.2 [Vasg] and Question 2.37 has been answered positively [Vasd].

and he would like to thank Professor Grossberg for his guidance and assistance in his research in general and in this work specifically.

We also thank Monica VanDieren and the referee for useful feedback that helped us improve the presentation of this chapter.

---

2 A primer in abstract elementary classes without tameness

In this section, we give an overview of some of the main concepts of the study of abstract elementary classes. This is meant both as a presentation of the basics and as a review of the "pre-tameness" literature, with an emphasis on the difficulties that were faced. By the end of this section, we give several state-of-the-art results on Shelah's categoricity conjecture. While tameness is not assumed, deriving a weak version from categoricity is key in the proofs.

We only sketch the basics here and omit most of the proofs. The reader who wants a more thorough introduction should consult [Gro02], [Bal09], or the upcoming [Gro]. We are light on history and motivation for this part; interested readers should consult one of the references or Section 7.

Abstract elementary classes (AECs) were introduced by Shelah in the mid-seventies. The original motivation was axiomatizing classes of models of certain infinitary logics ($\mathbb{L}_{\omega_1,\omega}$ and $\mathbb{L}(Q)$), but the definition can also be seen as extracting the category-theoretic essence of first-order model theory (see [Lie11a]).

Definition 2.1. An abstract elementary class (AEC) is a pair $(\mathbf{K}, \leq)$ satisfying the following conditions:

1. $\mathbf{K}$ is a class of L-structures for a fixed language $L := L(\mathbf{K})$.

2. $\leq$ is a reflexive and transitive relation on $\mathbf{K}$.

3. Both $\mathbf{K}$ and $\leq$ are closed under isomorphisms: If $M, N \in \mathbf{K}$, $M \leq N$, and $f : N \cong N'$, then $f[M], N' \in \mathbf{K}$ and $f[M] \leq N'$.

4. If $M \leq N$, then M is an L-substructure of N (written[5] $M \subseteq N$).

5. Coherence axiom: If $M_0, M_1, M_2 \in \mathbf{K}$, $M_0 \subseteq M_1 \leq M_2$, and $M_0 \leq M_2$, then $M_0 \leq M_1$.

6. Tarski-Vaught chain axioms: If δ is a limit ordinal and $\langle M_i : i < \delta \rangle$ is an increasing chain (that is, for all $i < j < \delta$, $M_i \in \mathbf{K}$ and $M_i \leq M_j$), then:

   (a) $M_\delta := \bigcup_{i<\delta} M_i \in \mathbf{K}$.

---

[5] We write $|M|$ for the universe of an L-structure M and $\|M\|$ for the cardinality of the universe. Thus $M \subseteq N$ means M is a substructure of N while $|M| \subseteq |N|$ means that the universe of M is a subset of the universe of N.

(b) $M_i \leq M_\delta$ for all $i < \delta$.

(c) If $N \in \mathbf{K}$ and $M_i \leq N$ for all $i < \delta$, then $M_\delta \leq N$.

7. Löwenheim-Skolem-Tarski axiom[6]: There exists a cardinal $\mu \geq |L(\mathbf{K})| + \aleph_0$ such that for every $M \in \mathbf{K}$ and every $A \subseteq |M|$, there exists $M_0 \leq M$ so that $A \subseteq |M_0|$ and $\|M_0\| \leq \mu + |A|$. We define the Löwenheim-Skolem-Tarski number of $\mathbf{K}$ (written $\mathrm{LS}(\mathbf{K})$) to be the least such cardinal.

We often will not distinguish between $\mathbf{K}$ and the pair $(\mathbf{K}, \leq)$. We write $M < N$ when $M \leq N$ and $M \neq N$.

Example 2.2. $(\mathrm{Mod}(T), \preceq)$ for T a first-order theory, and more generally $(\mathrm{Mod}(\psi), \preceq_\Phi)$ for ψ an $\mathbb{L}_{\lambda,\omega}$ sentence and Φ a fragment containing ψ are among the motivating examples. The Löwenheim-Skolem-Tarski numbers in those cases are, respectively, $|L(T)| + \aleph_0$ and $|\Phi| + |L(\Phi)| + \aleph_0$. In the former case, we say that the class is *elementary*. See the aforementioned references for more examples.

Notation 2.3. For $\mathbf{K}$ an AEC, we write $\mathbf{K}_\lambda$ for the class of $M \in \mathbf{K}$ with $\|M\| = \lambda$, and similarly for variations such as $\mathbf{K}_{\geq\lambda}$, $\mathbf{K}_{<\lambda}$, $\mathbf{K}_{[\lambda,\theta)}$, etc.

Remark 2.4 (Existence of resolutions). Let $\mathbf{K}$ be an AEC and let $\lambda > \mathrm{LS}(\mathbf{K})$. If $M \in \mathbf{K}_\lambda$, it follows directly from the axioms that there exists an increasing chain $\langle M_i : i \leq \lambda \rangle$ which is continuous[7] and so that $M_\lambda = M$ and $M_i \in \mathbf{K}_{<\lambda}$ for all $i < \lambda$; such a chain is called a *resolution* of M. We also use this name to refer to the initial segment $\langle M_i : i < \lambda \rangle$ with $M_\lambda = M = \bigcup_{i<\lambda} M_i$ left implicit.

Remark 2.5. Let $\mathbf{K}$ be an AEC. A few quirks are not ruled out by the definition:

- $\mathbf{K}$ could be empty.

- It could be that $\mathbf{K}_{<\mathrm{LS}(\mathbf{K})}$ is nonempty. This can be remedied by replacing $\mathbf{K}$ with $\mathbf{K}_{\geq\mathrm{LS}(\mathbf{K})}$ (also an AEC with the same Löwenheim-Skolem-Tarski number as $\mathbf{K}$). Note however that in some examples, the models below $\mathrm{LS}(\mathbf{K})$ give a lot of information on the models of size $\mathrm{LS}(\mathbf{K})$, see Baldwin, Koerwein, and Laskowski [BKL].

Most authors implicitly assume that $\mathbf{K}_{<\mathrm{LS}(\mathbf{K})} = \emptyset$ and $\mathbf{K}_{\mathrm{LS}(\mathbf{K})} \neq \emptyset$, and the reader can safely make these assumptions throughout. However, we will try to be careful about these details when stating results.

An AEC $\mathbf{K}$ may not have certain structural properties that always hold in the elementary case:

---

[6]This axiom was initially called the Löwenheim-Skolem axiom, which explains why it is written $\mathrm{LS}(\mathbf{K})$. However, later works have referred to it this way (and sometimes written $\mathrm{LST}(\mathbf{K})$) as an acknowledgment of Tarski's role in the corresponding first-order result.

[7]That is, for every limit i, $M_i = \bigcup_{j<i} M_j$.

Definition 2.6. Let **K** be an AEC.

1. **K** has *amalgamation* if for any $M_0, M_1, M_2 \in \mathbf{K}$ with $M_0 \leq M_\ell$, $\ell = 1, 2$, there exists $N \in \mathbf{K}$ and $f_\ell : M_\ell \xrightarrow[M_0]{} N$, $\ell = 1, 2$.

$$
\begin{array}{ccc}
M_1 & \xdashrightarrow{\ f_1\ } & N \\
\uparrow & & \wedge \\
& & \vdots\ f_2 \\
M_0 & \longrightarrow & M_2
\end{array}
$$

2. **K** has *joint embedding* if for any $M_1, M_2 \in \mathbf{K}$, there exists $N \in \mathbf{K}$ and $f_\ell : M_\ell \to N$, $\ell = 1, 2$.

3. **K** has *no maximal models* if for any $M \in \mathbf{K}$ there exists $N \in \mathbf{K}$ with $M < N$.

4. **K** has *arbitrarily large models* if for any cardinal λ, $\mathbf{K}_{\geq \lambda} \neq \emptyset$.

We define localizations of these properties in the expected way. For example, we say that $\mathbf{K}_\lambda$ *has amalgamation* or **K** has *amalgamation in* λ (or λ-*amalgamation*) if the definition of amalgamation holds when all the models are required to be of size λ.

There are several easy relationships between these properties. We list here a few:

Proposition 2.7. *Let* **K** *be an AEC,* $\lambda \geq \mathrm{LS}(\mathbf{K})$.

1. *If* **K** *has joint embedding and arbitrarily large models, then* **K** *has no maximal models.*

2. *If* **K** *has joint embedding in* λ, $\mathbf{K}_{<\lambda}$ *has no maximal models, and* $\mathbf{K}_{\geq \lambda}$ *has amalgamation, then* **K** *has joint embedding.*

3. *If* **K** *has amalgamation in every* $\mu \geq \mathrm{LS}(\mathbf{K})$, *then* $\mathbf{K}_{\geq \mathrm{LS}(\mathbf{K})}$ *has amalgamation.*

In a sense, joint embedding says that the AEC is "complete". Assuming amalgamation, it is possible to partition the AEC into disjoint classes each of which has amalgamation and joint embedding.

Proposition 2.8. *Let* **K** *be an AEC with amalgamation. For* $M_1, M_2 \in \mathbf{K}$, *say* $M_1 \sim M_2$ *if and only if* M_1 *and* M_2 *embed inside a common model (i.e., there exists* $N \in \mathbf{K}$ *and* $f_\ell : M_\ell \to N$). *Then* $\sim$ *is an equivalence relation, and its equivalence classes partition* **K** *into at most* $2^{\mathrm{LS}(\mathbf{K})}$-*many AECs with joint embedding and amalgamation.*

Thus if **K** is an AEC with amalgamation and arbitrarily large models, we can find a sub-AEC of it which has amalgamation, joint embedding, and no maximal models. In that sense, global amalgamation implies all the other properties (see also Corollary 2.12).

Using the existence of resolutions, it is not difficult to see that an AEC $(\mathbf{K}, \leq)$ is determined by its restriction to size λ ($\mathbf{K}_\lambda, \leq \cap(\mathbf{K}_\lambda \times \mathbf{K}_\lambda)$). Thus, there is only a set of AECs with a fixed Löwenheim-Skolem-Tarski number and hence there is a Hanf number for the property that the AEC has arbitrarily large models.

While this analysis only gives an existence proof for the Hanf number, Shelah's Presentation Theorem actually allows a computation of the Hanf number by establishing a connection between $\mathbf{K}$ and $\mathbb{L}_{\infty,\omega}$.

Theorem 2.9 (Shelah's presentation theorem). *If $\mathbf{K}$ is an AEC with $L(\mathbf{K}) = L$, there exists a language $L' \supseteq L$ with $|L'| + \mathrm{LS}(\mathbf{K})$, a first-order L'-theory T', and a set of T'-types Γ such that*

$$\mathbf{K} = PC(T', \Gamma, L) := \{M' \restriction L \mid M' \models T' \text{ and } M' \text{ omits all the types in } \Gamma\}$$

The proof proceeds by adding $\mathrm{LS}(\mathbf{K})$-many functions of each arity. For each M, we can write it as the union of a directed system $\{N_{\bar{a}} \in \mathbf{K}_{\mathrm{LS}(\mathbf{K})} : \bar{a} \in {}^{<\omega}|M|\}$ with $\bar{a} \in N_{\bar{a}}$. Then, the intended expansion M' of M is where the universe of $N_{\bar{a}}$ is enumerated by new functions of arity $\ell(\bar{a})$ applied to $\bar{a}$. The types of Γ are chosen such that M' omits them if and only if the reducts of the substructures derived in this way actually form a directed system[8].

In particular, $\mathbf{K}$ is the reduct of a class of models of an $\mathbb{L}_{\mathrm{LS}(\mathbf{K})^+,\omega}$-theory. An important caveat is that if $\mathbf{K}$ was given by the models of some $\mathbb{L}_{\mathrm{LS}(\mathbf{K})^+,\omega}$-theory, the axiomatization given by Shelah's Presentation Theorem is different and uninformative. However, it is enough to allow the computation of the Hanf number for existence.

Corollary 2.10. *If $\mathbf{K}$ is an AEC such that $\mathbf{K}_{\geq \chi} \neq \emptyset$ for all $\chi < \beth_{(2^{\mathrm{LS}(\mathbf{K})})^+}$, then $\mathbf{K}$ has arbitrarily large models.*

The cardinal $\beth_{(2^{\mathrm{LS}(\mathbf{K})})^+}$ appears frequently in studying AECs, so has been given a name:

Notation 2.11. For λ an infinite cardinal, write $h(\lambda) := \beth_{(2^\lambda)^+}$. When $\mathbf{K}$ is a fixed AEC, we write $H_1 := h(\mathrm{LS}(\mathbf{K}))$ and $H_2 := h(h(\mathrm{LS}(\mathbf{K})))$.

We obtain for example that any AEC with amalgamation and joint embedding in a single cardinal eventually has all the structural properties of Definition 2.6.

Corollary 2.12. *Let $\mathbf{K}$ be an AEC with amalgamation. If $\mathbf{K}$ has joint embedding in some $\lambda \geq \mathrm{LS}(\mathbf{K})$, then there exists $\chi < H_1$ so that $\mathbf{K}_{\geq \chi}$ has amalgamation, joint embedding, and no maximal models. More precisely, there exists an AEC $\mathbf{K}^*$ such that:*

1. $\mathbf{K}^* \subseteq \mathbf{K}$.

---

[8]Note that there are almost always the maximal number of types in Γ.

2. $\mathrm{LS}(\mathbf{K}^*) = \mathrm{LS}(\mathbf{K})$.

3. $\mathbf{K}^*$ *has amalgamation, joint embedding, and no maximal models.*

4. $\mathbf{K}_{\geq\chi} = (\mathbf{K}^*)_{\geq\chi}$.

Proof sketch. First use Proposition 2.8 to decompose the AEC into at most $2^{\mathrm{LS}(\mathbf{K})}$ many subclasses, each of which has amalgamation and joint embedding. Now if one of these partitions does not have arbitrarily large models, then there must exist a $\chi_0 < H_1$ in which it has no models. Take the sup of all such χ_0s and observe that $\mathrm{cf}(H_1) = (2^{\mathrm{LS}(\mathbf{K})})^+ > 2^{\mathrm{LS}(\mathbf{K})}$. $\qquad\square$

If $\mathbf{K}$ is an AEC with joint embedding, amalgamation, and no maximal models, we may build a proper-class[9] sized model-homogeneous universal model $\mathfrak{C}$, where:

Definition 2.13. Let $\mathbf{K}$ be an AEC, let $M \in \mathbf{K}$, and let λ be a cardinal.

1. M is λ-*model-homogeneous* if for every $M_0 \leq M$, $M_0' \geq M_0$ with $\|M\| < \lambda$, there exists $f : M_0' \xrightarrow[M_0]{} M$. When $\lambda = \|M\|$, we omit it.

2. M is *universal* if for every $M' \in \mathbf{K}$ with $\|M'\| \leq \|M\|$, there exists $f : M' \to M$.

Definition 2.14. We say that an AEC $\mathbf{K}$ *has a monster model* if it has a model $\mathfrak{C}$ as above. Equivalently, it has amalgamation, joint embedding, and arbitrarily large models.

Remark 2.15. Even if $\mathbf{K}$ only has amalgamation and joint embedding, we can construct a monster model, but it may not be proper-class sized. If in addition joint embedding fails, for any $M \in \mathbf{K}$ we can construct a big model-homogeneous model $\mathfrak{C} \geq M$.

Note that if $\mathbf{K}$ were in fact an elementary class, then the monster model constructed here is the same as the classical concept.

When $\mathbf{K}$ has a monster model $\mathfrak{C}$, we can define a semantic notion of type[10] by working inside $\mathfrak{C}$ and specifying that $\bar{b}$ and $\bar{c}$ have the same type over A if and only if there exists an automorphism of $\mathfrak{C}$ taking $\bar{b}$ to $\bar{c}$ and fixing A. In fact, this can be generalized to arbitrary AECs:

Definition 2.16 (Galois types). Let $\mathbf{K}$ be an AEC.

1. For an index set I, an I-*indexed Galois triple* is a triple $(\bar{b}, A, N)$, where $N \in \mathbf{K}$, $A \subseteq |N|$, and $\bar{b} \in {}^I|N|$.

---

[9]To make sense of this, we have to work in Gödel-Von Neumann-Bernays set theory. Alternatively, we can simply ask for the monster model to be bigger than any sizes involved in our proofs. In any case, the way to make this precise is the same as in the elementary theory, so we do not elaborate.

[10]A semantic (as opposed to syntactic) notion of type is the only one that makes sense in a general AEC as there is no natural logic to work in. Even in AECs axiomatized in a logic such as $\mathbb{L}_{\omega_1,\omega}$, syntactic types do not behave as they do in the elementary framework; see the discussion of the Hart-Shelah example in Section 3.2.1.

2. We say that the I-indexed Galois triples $(\bar{b}_1, A_1, N_1)$, $(\bar{b}_2, A_2, N_2)$ are *atomically equivalent* and write $(\bar{b}_1, A_1, N_1) E^I_{\mathrm{at}} (\bar{b}_2, A_2, N_2)$ if $A_1 = A_2$, and there exists $N \in \mathbf{K}$ and $f_\ell : N_\ell \xrightarrow[A_\ell]{} N$ so that $f_1(\bar{b}_1) = f_2(\bar{b}_2)$. When I is clear from context, we omit it.

3. Note that E_{at} is a symmetric and reflexive relation. We let E be its transitive closure.

4. For an I-indexed Galois triple $(\bar{b}, A, N)$, we let $\mathrm{gtp}(\bar{b}/A; N)$ (the *Galois type* of $\bar{b}$ over A in N) be the E-equivalence class of $(\bar{b}, A, N)$.

5. For $N \in \mathbf{K}$ and $A \subseteq |N|$, we let $\mathrm{gS}^I(A; N) := \{ \mathrm{gtp}(\bar{b}/A; N) \mid \bar{b} \in {}^I|N| \}$. We also let $\mathrm{gS}^I(N) := \bigcup_{N' \geq N} \mathrm{gS}^I(N; N')$. When I is omitted, this means that $|I| = 1$, e.g., $\mathrm{gS}(N)$ is $\mathrm{gS}^1(N)$.

6. We can define restrictions of Galois types in the natural way: for $p \in \mathrm{gS}^I(A; N)$, $I_0 \subseteq I$ and $A_0 \subseteq A$, write $p \restriction A_0$ for the restriction of p to A_0 and p^{I_0} for the restriction of p to I_0. For example, if $p = \mathrm{gtp}(\bar{b}/A; N)$ and $A_0 \subseteq A$, $p \restriction A_0 := \mathrm{gtp}(\bar{b}/A_0; N)$ (this does not depend on the choice of representative for p).

7. Given $p \in \mathrm{gS}^I(M)$ and $f : M \cong M'$, we can also define $f(p)$ in the natural way.

Remark 2.17.

1. If $M \leq N$, then $\mathrm{gtp}(\bar{b}/A; M) = \mathrm{gtp}(\bar{b}/A; N)$. Similarly, if $f : M \cong_A N$, then $\mathrm{gtp}(\bar{b}/A; M) = \mathrm{gtp}(f(\bar{b})/A; N)$. Equivalence of Galois types is the coarsest equivalence relation with these properties.

2. If $\mathbf{K}$ has amalgamation, then $E = E_{\mathrm{at}}$.

3. If $\mathfrak{C}$ is a monster model for $\mathbf{K}$, $\bar{b}_1, \bar{b}_2 \in {}^{<\infty}|\mathfrak{C}|$, $A \subseteq |\mathfrak{C}|$, then $\mathrm{gtp}(\bar{b}_1/A; \mathfrak{C}) = \mathrm{gtp}(\bar{b}_2/A; \mathfrak{C})$ if and only if there exists $f \in \mathrm{Aut}_A(\mathfrak{C})$ so that $f(\bar{b}_1) = \bar{b}_2$. When working inside $\mathfrak{C}$, we just write $\mathrm{gtp}(\bar{b}/A)$ for $\mathrm{gtp}(\bar{b}/A; \mathfrak{C})$, but in general, the model in which the Galois type is computed is important.

4. The cardinality of the index set is all that is important. However, when discussing type shortness later, it is convenient to allow the index set to be arbitrary.

When dealing with Galois types, one has to be careful about distinguishing between types over *models* and types over *sets*. Most of the basic definitions work the same for types over sets and models, and both require just amalgamation *over models* to make the transitivity of atomic equivalence work. Allowing types over sets gives slightly more flexibility in the definitions. For example, we can say what is meant to be $< \aleph_0$-tame or to be $(< \mathrm{LS}(\mathbf{K}))$-tame in $\mathbf{K}_{\geq \mathrm{LS}(\mathbf{K})}$. See the discussion around Definition 3.1.

On the other hand, several basic results–such as the construction of κ-saturated models–require amalgamation over the sort of object (set or model) desired in the conclusion. For instance, the following is true.

Proposition 2.18. *Suppose that* **K** *is an AEC with amalgamation*[11].

1. *The following are equivalent.*

   - *A is an amalgamation base*[12].
   - *For every* $p \in \mathrm{gS}^1(A; N)$ *and* $M \supseteq A$, *there is an extension of* p *to* M.

2. *The following are equivalent.*

   - **K** *has amalgamation over sets.*
   - *For every* M *and* κ, *there is an extension* $N \geq M$ *with the following property:*

     For every $A \subseteq |N|$ *and* $|M^*| \supseteq A$ *with* $|A| < \kappa$, *any*
     $$p \in \mathrm{gS}^{<\kappa}(A; M^*) \text{ is realized in } N.$$

A more substantial result is [She99, Claim 3.3], which derives a local character for splitting in stable AECs (see Lemma 2.23 below), but only in the context of Galois types over models.

One can give a natural definition of saturation in terms of Galois types.

Definition 2.19. A model $M \in \mathbf{K}$ is λ-*Galois-saturated* if for any $A \subseteq |M|$ with $|A| < \lambda$, any $N \geq M$, any $p \in \mathrm{gS}(A; N)$ is realized in M. When $\lambda = \|M\|$, we omit it.

Note the difference between this definition and Proposition 2.18.(2) above. When **K** does not have amalgamation or when $\lambda \leq \mathrm{LS}(\mathbf{K})$, it is not clear that this definition is useful. But if **K** has amalgamation and $\lambda > \mathrm{LS}(\mathbf{K})$, the following result of Shelah is fundamental:

Theorem 2.20. *Assume that* **K** *is an AEC with amalgamation and let* $\lambda > \mathrm{LS}(\mathbf{K})$. *Then* $M \in \mathbf{K}$ *is* λ-*Galois-saturated if and only if it is* λ-*model-homogeneous.*

2.1 Classification Theory

One theme of the classification theory of AECs is what Shelah has dubbed the "schizophrenia" of superstability (and other dividing lines) [She09b, p. 19]. Schizophrenia here refers to the fact that, in the elementary framework, dividing lines are given by several equivalent characterizations (e.g., stability is no order property or few types), typically with the existence of a definable, combinatorial object on the "high" or bad side and some good behavior of forking on the "low" or good side. However, this equivalence relies heavily on compactness or other ideas central to first-order and breaks down when dealing with general AECs. Thus, the search for stability, superstability, etc.

---

[11] Recall that this is defined to mean over models.
[12] This should be made precise, for example by considering the embedding of A inside a fixed monster model.

is in part a search for the "right" characterization of the dividing line and in part a search for equivalences between the different faces of the dividing line.

One can roughly divide approaches towards the classification of AECs into two categories: local approaches and global approaches[13]. Global approaches typically assume one or more structural properties (such as amalgamation or no maximal models) as well as a classification property (such as categoricity in a high-enough cardinal or Galois stability in a particular cardinality), and attempt to derive good behavior on a tail of cardinals. The local approach is a more ambitious strategy pioneered by Shelah in his book [She09b]. The idea is to first show (assuming, e.g., categoricity in a proper class of cardinals) that the AEC has good behavior in some suitable cardinal λ. Shelah precisely defines "good behavior in λ" as having a good λ-frame (see Section 2.5). In particular, this implies that the class is superstable in λ. The second step in the local approach is to argue that good behavior in some λ transfers upward to λ^+ and, if the behavior is good enough, to all cardinals above λ. Having established global good behavior, one can rely on the tools of the global approach to prove the categoricity conjecture.

The local approach seems more general but comes with a price: increased complexity, and often the use of non-ZFC axioms (like the weak GCH: $2^\lambda < 2^{\lambda^+}$ for all λ), as well as stronger categoricity hypotheses. The two approaches are not exclusive. In fact, in recent years, tools from local approach have been used and studied in a more global framework. We now briefly survey results in both approaches that do not use tameness.

2.2 Stability

Once Galois types have been defined, one can define *Galois-stability*:

Definition 2.21. An AEC **K** is *Galois-stable in λ* if for any $M \in \mathbf{K}$ with $\|M\| \leq \lambda$, we have $|\operatorname{gS}(M)| \leq \lambda$.

One can ask whether there is a notion like forking in stable AECs. The next sections discuss this problem in detail. A first approximation is μ-*splitting*:

Definition 2.22. Let **K** be an AEC, $\mu \geq \operatorname{LS}(\mathbf{K})$. Assume that **K** has amalgamation in μ. Let $M \leq N$ both be in $\mathbf{K}_{\geq\mu}$. A type $p \in \operatorname{gS}^{<\infty}(N)$ μ-*splits* over M if there exists $N_1, N_2 \in \mathbf{K}_\mu$ with $M \leq N_\ell \leq N$, $\ell = 1, 2$, and $f : N_1 \cong_M N_2$ so that $f(p \upharpoonright N_1) \neq p \upharpoonright N_2$.

One of the early results was that μ-splitting has a local character property in stable AECs:

Lemma 2.23. *Let **K** be an AEC, $\mu \geq \operatorname{LS}(\mathbf{K})$. Assume that **K** has amalgamation in μ and is Galois-stable in μ. For any $N \in \mathbf{K}_{\geq\mu}$ and $p \in \operatorname{gS}(N)$, there exists $N_0 \in \mathbf{K}_\mu$ with $N_0 \leq N$ so that p does not μ-split over N_0.*

---

[13]Monica VanDieren suggested set-theoretic scaffolding and model-theoretic scaffolding as alternate names for the local and global approaches.

With stability and amalgamation, we also get that there are unique non-μ-splitting extensions to universal models *of the same size.*

Theorem 2.24. *Let $\mathbf{K}$ be an AEC, $\mu \geq \mathrm{LS}(\mathbf{K})$. Assume that $\mathbf{K}$ has amalgamation in μ and is Galois-stable in μ. If $M_0, M_1, M_2 \in \mathbf{K}_\mu$ with M_1 universal over M_0[14], then each $p \in \mathrm{gS}(M_1)$ that does not μ-split over M_0 has a unique extension $q \in \mathrm{gS}(M_2)$ that does not split over M_0. Moreover, p is algebraic if and only if q is.*

Note in passing that stability gives existence of universal extensions:

Lemma 2.25. *Let $\mathbf{K}$ be an AEC and let $\lambda \geq \mathrm{LS}(\mathbf{K})$ be such that $\mathbf{K}$ has amalgamation in λ and is Galois-stable in λ. For any $M \in \mathbf{K}_\lambda$, there exists $N \in \mathbf{K}_\lambda$ which is universal over M.*

Similar to first-order model theory, there is a notion of an order property in AECs. The order property is more parametrized due to the lack of compactness. In the elementary framework, the order property is defined as the existence of a definable order of order type ω. However, the essence of it is that any order type can be defined. Thus the lack of compactness forces us to make the order property in AECs longer in order to be able to build complicated orders:

Definition 2.26.

1. $\mathbf{K}$ has the *κ-order property of length α* if there exists $N \in \mathbf{K}$, $p \in \mathrm{gS}^{<\kappa}(\emptyset; N)$, and $\langle \bar{a}_i \subset {}^{<\kappa}|M| : i < \alpha \rangle$ such that:

$$i < j \Leftrightarrow \mathrm{gtp}(\bar{a}_i \bar{a}_j / \emptyset; N) = p$$

2. $\mathbf{K}$ has the *κ-order property* if it has the κ-order property of all lengths.

3. $\mathbf{K}$ has the *order property* if it has the κ-order property for some κ.

From the presentation theorem, having the κ-order property of all lengths less than $h(\kappa)$ is enough to imply the full κ-order property. In this case, one can show that α above can be replaced by any linear ordering.

2.3 Superstability

A first-order theory T is superstable if it is stable on a tail of cardinals. One might want to adapt this definition to AECs, but it is not clear that it is enough to derive the property that we really want here: an analog of $\kappa(T) = \aleph_0$. A possible candidate is to say that a class is superstable if every type does not μ-split over a finite set. However, splitting is only defined for models, and, as remarked above, types over arbitrary sets are not too well behaved. Instead, as with Galois types, we take an implication of the desired property as the new definition: no long splitting chains.

---

[14]That is, for every $M' \in \mathbf{K}_\mu$ with $M_0 \leq M'$, there exists $f : M' \xrightarrow[M_0]{} M_1$.

Definition 2.27. An AEC $\mathbf{K}$ is μ-*superstable* (or *superstable in* μ) if:

1. $\mu \geq \mathrm{LS}(\mathbf{K})$.

2. $\mathbf{K}_\mu$ is nonempty, has amalgamation[15], joint embedding, and no maximal models.

3. $\mathbf{K}$ is Galois-stable in μ.

4. for all limit ordinal $\delta < \mu^+$ and every increasing continuous sequence $\langle M_i : i \leq \delta \rangle$ in $\mathbf{K}_\mu$ with M_{i+1} universal over M_i for all $i < \delta$, if $p \in \mathrm{gS}(M_\delta)$, then there exists $i < \delta$ so that p does not μ-split over M_i.

If $\mathbf{K}$ is the class of models of a first-order theory T, then $\mathbf{K}$ is μ-superstable if and only if T is stable in every $\lambda \geq \mu$.

Remark 2.28. In (4), note that we ask for M_{i+1} to be universal over M_i rather than only for $M_{i+1} \geq M_i$. Let us denote the stronger variation of (4) where M_{i+1} is not required to be universal over M_i by (4+). For reasons that we do not completely understand, it is unknown whether (4+) follows from categoricity (but (4) does, see Theorem 2.36). On the other hand, (4) seems to be sufficient for many purposes. In the tame case, the good frames derived from superstability (see Theorem 5.22) will actually satisfy (4+).

Another possible definition of superstability in AECs is the uniqueness of limit models:

Definition 2.29. Let $\mathbf{K}$ be an AEC and let $\mu \geq \mathrm{LS}(\mathbf{K})$.

1. A model $M \in \mathbf{K}_\mu$ is (μ, δ)-*limit* for limit $\delta < \mu^+$ if there exists a strictly increasing continuous chain $\langle M_i \in \mathbf{K}_\mu : i \leq \delta \rangle$ such that $M_\delta = M$ and for all $i < \delta$, M_{i+1} is universal over M_i. If we do not specify the δ, it means that there is one. We say that M is *limit over* N when such a chain exists with $M_0 = N$.

2. $\mathbf{K}$ has *uniqueness of limit models in* μ if whenever $M_0, M_1, M_2 \in \mathbf{K}_\mu$ and both M_1 and M_2 are limit over M_0, then $M_1 \cong_{M_0} M_2$.

3. $\mathbf{K}$ has *weak uniqueness of limit models in* μ if whenever $M_1, M_2 \in \mathbf{K}_\mu$ are limit models, then $M_1 \cong M_2$ (the difference is that the isomorphism is not required to fix M_0).

Limit models and their uniqueness have come to occupy a central place in the study of superstability of AECs. (μ^+, μ^+)-limit models are Galois-saturated, so even weak uniqueness of limit models in μ^+ implies that (μ^+, ω)-limit models are Galois-saturated. This tells us that Galois-saturated models can be built in fewer steps than expected, which is reminiscent of first-order characterizations of superstability. As an added benefit, the analysis of limit models can be carried out in a single cardinal (as opposed to Galois-saturated

---

[15]This requirement is not made in several other variations of the definition but simplifies notation. See the historical remarks for more.

models, which typically need smaller models) and, thus, lends itself well to the local analysis[16].

The following question is still open (the answer is positive for elementary classes):

Question 2.30. Let $\mathbf{K}$ be an AEC and let $\mu \geq \mathrm{LS}(\mathbf{K})$. If $\mathbf{K}_\mu$ is nonempty, has amalgamation, joint embedding, no maximal models, and is Galois-stable in μ, do we have that $\mathbf{K}$ has uniqueness of limit models in μ if and only if $\mathbf{K}$ is superstable in μ?

This phenomenon of having two potentially nonequivalent definitions of superstability that are equivalent in the first-order case is an example of the "schizophrenia" of superstability mentioned above.

Shelah and Villaveces [SV99] started the investigation of whether superstability implies the uniqueness of limit models. Eventually, VanDieren introduced a symmetry property for μ-splitting to show the following.

Theorem 2.31. *If* $\mathbf{K}$ *is a* μ*-superstable*[17] *AEC such that* μ*-splitting has symmetry, then* $\mathbf{K}$ *has uniqueness of limit models in* μ.

2.4 Categoricity

For an AEC $\mathbf{K}$, let us denote by $I(\mathbf{K}, \lambda)$ the number of nonisomorphic models in $\mathbf{K}_\lambda$. We say that $\mathbf{K}$ is *categorical in* λ if $I(\mathbf{K}, \lambda) = 1$. One of Shelah's motivation for introducing AECs was to make progress on the following test question:

Conjecture 2.32 (Shelah's categoricity conjecture for $\mathbb{L}_{\omega_1, \omega}$). If a sentence $\psi \in \mathbb{L}_{\omega_1, \omega}$ is categorical in *some* $\lambda \geq \beth_{\omega_1}$, then it is categorical in *all* $\lambda' \geq \beth_{\omega_1}$.

Note that the lower bound is the Hanf number of this class. One of the best results toward the conjecture is:

Theorem 2.33. *Let* $\psi \in \mathbb{L}_{\omega_1, \omega}$ *be a sentence in a countable language. Assume*[18] $\mathbf{V} = \mathbf{L}$. *If* ψ *is categorical in all* $\aleph_n$, $1 \leq n < \omega$, *then* ψ *is categorical in all uncoutable cardinals.*

Shelah's categoricity conjecture for $\mathbb{L}_{\omega_1, \omega}$ can be generalized to AECs, either by requiring only "eventual" categoricity (Conjecture 1.1) or by asking for a specific Hanf number.

This makes a difference: using the axiom of replacement and the fact that every AEC $\mathbf{K}$ is determined by its restrictions to models of size at most $\mathrm{LS}(\mathbf{K})$, it is easy to see that Shelah's eventual categoricity conjecture is equivalent to

---

[16]For example, it gives a way to define what it means for a model of size $\mathrm{LS}(\mathbf{K})$ to be saturated.

[17]Recall that the definition includes amalgamation and no maximal models in μ.

[18]Much weaker set-theoretic hypotheses suffice.

the following statement: If an AEC is categorical in a proper class of cardinals, then it is categorical on a tail of cardinals. Thus requiring that the Hanf number can in some sense be explicitly computed makes sure that one cannot "cheat" and automatically obtain a free upward transfer.

When the Hanf number is H_1 (recall Notation 2.11), we call the resulting statement *Shelah's categoricity conjecture for AECs*. This is widely recognized as the main test question[19] in the study of AECs.

Conjecture 2.34. If an AEC **K** is categorical in *some* $\lambda > H_1$, then it is categorical in *all* $\lambda' \geq H_1$.

One of the milestone results in the global approach to this conjecture is Shelah's downward transfer from a successor in AECs with amalgamation.

Theorem 2.35. *Let* **K** *be an AEC with amalgamation. If* **K** *is categorical in a successor* $\lambda \geq H_2$, *then* **K** *is categorical in all* $\mu \in [H_2, \lambda]$.

The structure of the proof involves first deriving a weak version of tameness from the categoricity assumption (see Example 3.2.1.(2)). A striking feature of this result (and several other categoricity transfers) is the successor requirement, which is, of course, missing from similar results in the first-order case. Removing it is a major open question, even in the tame framework (see Shelah [She00, Problem 6.14]). We see at least three difficulties when working with an AEC categorical in a limit cardinal $\lambda > LS(\mathbf{K})$:

1. It is not clear that the model of size λ should be Galois-saturated, see Question 2.37.

2. It is not clear how to transfer "internal characterizations" of categoricity such as no Vaughtian pairs or unidimensionality. In the first-order framework, compactness is a key tool to achieve this.

3. It is not clear how to even get that categoricity implies such an internal characterization (assuming $\lambda = \lambda_0^+$ is a successor, there is a relatively straightforward argument for the nonexistence of Vaughtian pairs in λ_0). In the first-order framework, all the arguments we are aware of use in some way *primary models* but here we do not know if they exist or are well behaved. For example, we cannot imitate the classical argument that primary models are primes (this relies on the compactness theorem).

Assuming tameness, the first two issues can be solved (see Theorem 5.44 and the proof of Theorem 5.50). It is currently not known how to solve the third in general, but adding the assumption that **K** has prime models over sets of the form $M \cup \{a\}$ is enough. See Theorem 5.47.

A key tool in the proof of Theorem 2.35 is the existence of Ehrenfeucht-Mostowski models which follow from the presentation theorem. In AECs with

---

[19]It is not expected that solving it will produce a useful lemma in solving other problems. Rather, like Morley's Theorem, it is expected that the solution will necessitate the development of ideas that will be useful in solving other problems.

amalgamation and no maximal models, several structural properties can be derived below the categoricity cardinal. For example:

Theorem 2.36 (The Shelah-Villaveces theorem, [SV99]). *Let* **K** *be an AEC with amalgamation and no maximal models. Let* $\mu \geq \mathrm{LS}(\mathbf{K})$. *If* **K** *is categorical in a* $\lambda > \mu$, *then* **K** *is* μ-*superstable.*

Note that Theorem 2.36 fails to generalize to $\lambda \geq \mu$. In general, **K** may not even be Galois-stable in λ, see the Hart-Shelah example (Section 3.2.2 below). In the presence of tameness, the difficulty disappears: superstability can be transferred all the way up (see Theorem 5.22). This seems to be a re-curring feature of the study of AECs without tameness: some structure can be established below the categoricity cardinal (using tools such as Ehrenfeucht-Mostowski models), but transferring this structure upward is hard due to the lack of locality. For example, in the absence of tameness, the following question is open:

Question 2.37. Let **K** be an AEC with amalgamation and no maximal models. If **K** is categorical in a $\lambda > \mathrm{LS}(\mathbf{K})$, is the model of size λ Galois-saturated?

It is easy to see that (if $\mathrm{cf}(\lambda) > \mathrm{LS}(\mathbf{K})$), the model of size λ is $\mathrm{cf}(\lambda)$-Galois-saturated. Recently, it has been shown that categoricity in a high-enough cardinal implies some degree of saturation:

Theorem 2.38. *Let* **K** *be an AEC with amalgamation and no maximal models. Let* $\lambda \geq \chi > \mathrm{LS}(\mathbf{K})$. *If* **K** *is categorical in* λ *and* $\lambda \geq h(\chi)$, *then the model of size* λ *is* χ-*Galois-saturated.*

What about the uniqueness of limit models? In the course of establishing Theorem 2.35, Shelah proves that categoricity in a successor λ implies weak uniqueness of limit models in all $\mu < \lambda$. Recently, VanDieren and the second author have shown:

Theorem 2.39. *Let* **K** *be an AEC with amalgamation and no maximal models. Let* $\mu \geq \mathrm{LS}(\mathbf{K})$. *If* **K** *is categorical in a* $\lambda \geq h(\mu)$, *then* **K** *has uniqueness of limit models in* μ.

2.5 Good frames

Roughly speaking, an AEC **K** *has a good* λ-*frame* if it is well behaved in λ (i.e., it is nonempty, has amalgamation, joint embedding, no maximal model, and is Galois-stable, all in λ) and there is a forking-like notion for types of length one over models in $\mathbf{K}_\lambda$ that behaves like forking in superstable first-order theories[20]. In particular, it is λ-superstable. One motivation for good frames was the following (still open) question:

Question 2.40. If an AEC is categorical in λ and λ^+, does it have a model of size λ^{++}?

Now it can be shown that if **K** has a good λ-frame (or even just λ-superstable), then it has a model of size λ^{++}. Thus, it would be enough to obtain a good frame to solve the question. Shelah has shown the following:

Theorem 2.41. *Assume* $2^\lambda < 2^{\lambda^+} < 2^{\lambda^{++}}$.
  Let **K** *be an AEC and let* $\lambda \geq \mathrm{LS}(\mathbf{K})$. *If:*

1. **K** *is categorical in* λ *and* λ^+.
2. $0 < I(\mathbf{K}, \lambda^{++}) < \mu_{unif}(\lambda^{++}, 2^{\lambda^+})$[21].

  Then **K** *has a good* λ^+-*frame.*

Corollary 2.42. *Assume* $2^\lambda < 2^{\lambda^+} < 2^{\lambda^{++}}$. *If* **K** *is categorical in* λ, λ^+, *and* λ^{++}, *then* **K** *has a model of size* λ^{+++}.

Note the non-ZFC assumptions as well as the strong categoricity hypothesis[22]. We will see that this can be removed in the tame framework, or even already by making some weaker (but global) assumptions than tameness.

Recall from the beginning of Section 2.1 that Shelah's local approach aims to transfer good behavior in λ upward. The successor step is to turn a good λ-frame into a good λ^+ frame . Shelah says a good λ-frame is *successful* if it satisfies a certain (strong) technical condition that allows it to extend it to a good λ^+-frame.

Theorem 2.43. *Assume* $2^\lambda < 2^{\lambda^+} < 2^{\lambda^{++}}$. *If an AEC* **K** *has a good* λ-*frame* $\mathfrak{s}$ *and* $0 < I(\mathbf{K}, \lambda^{++}) < \mu_{unif}(\lambda^{++}, 2^{\lambda^+})$, *then there exists a good* λ^+-*frame* $\mathfrak{s}^+$ *with underlying class the Galois-saturated models of size* λ^+ *(the ordering will also be different).*

The proof goes by showing that the weak GCH and few models assumptions imply that any good frame is successful.

So assuming weak GCH and few models in every λ^{+n}, one obtains an increasing sequence $\bar{\mathfrak{s}} = \mathfrak{s}, \mathfrak{s}^+, \mathfrak{s}^{++}, \ldots$ of good frames. One of the main results of Shelah's book is that the natural limit of $\bar{\mathfrak{s}}$ is also a good frame (the strategy is to show that a good frame in the sequence is *excellent*). Let us say that

---

[20]There is an additional parameter, the set of *basic* types. These are a dense set of types over models of size λ such that forking is only required to behave well with respect to them. However, basic types play little role in the discussion of tameness (and eventually are eliminated in most cases even in general AECs), so we do not discuss them here, see the historical remarks.

[21]The cardinal $\mu_{unif}(\lambda^{++}, 2^{\lambda^+})$ should be interpreted as $2^{\lambda^{++}}$; this is true when $\lambda \geq \beth_\omega$ and there is no example of inequality when $2^{\lambda^+} < 2^{\lambda^{++}}$.

[22]It goes without saying that the proof is also long and complex, see the historical remarks.

a good frame is ω-*successful* if $\mathfrak{s}^{+n}$ is successful for all $n < \omega$. At the end of Chapter III of his book, Shelah claims the following result and promises a proof in [Sheb]:

Claim 2.44. Assume $2^{\lambda^{+n}} < 2^{\lambda^{+(n+1)}}$ for all $n < \omega$. If an AEC **K** has an ω-successful good λ-frame, is categorical in λ, and $\mathbf{K}^{\lambda^{+\omega}\text{-sat}}$ (the class of $\lambda^{+\omega}$-Galois-saturated models in **K**) is categorical in a $\lambda' > \lambda^{+\omega}$, then $\mathbf{K}^{\lambda^{+\omega}\text{-sat}}$ is categorical in all $\lambda'' > \lambda^{+\omega}$.

Can one build a good frame in ZFC? In Chapter IV of his book, Shelah proves:

Theorem 2.45. *Let* **K** *be an AEC categorical in cardinals of arbitrarily high cofinality. Then there exists a cardinal* λ *such that* **K** *is categorical in* λ *and* **K** *has a good* λ-*frame.*

Theorem 2.46. *Let* **K** *be an AEC with amalgamation and no maximal models. If* **K** *is categorical in a* $\lambda \geq h(\aleph_{\mathrm{LS}(\mathbf{K})^+})$, *then there exists* $\mu < \aleph_{\mathrm{LS}(\mathbf{K})^+}$ *such that* $\mathbf{K}^{\mu\text{-}sat}$ *has a good* μ-*frame.*

The proofs of both theorems first get some tameness (and amalgamation in the first case), and then use it to define a good frame in λ by making use of the lower cardinals (as in Theorem 5.17).

Assuming amalgamation and weak GCH, Shelah shows that the good frame can be taken to be ω-successful. Combining this with Claim 2.44, Shelah deduces the eventual categoricity conjecture in AECs with amalgamation:

Theorem 2.47. *Assume Claim 2.44 and* $2^\theta < 2^{\theta^+}$ *for all cardinals* θ. *Let* **K** *be an AEC with amalgamation. If* **K** *is categorical in some* $\lambda \geq h(\aleph_{\mathrm{LS}(\mathbf{K})^+})$, *then* **K** *is categorical in all* $\lambda' \geq h(\aleph_{\mathrm{LS}(\mathbf{K})^+})$.

Note that the first steps in the proof are again proving enough tameness to make the construction of an ω-successful good frame.

---

3 Tameness: what and where

3.1 What – Definitions and basic results

Syntactic types have nice locality properties: different types must differ on a formula and this difference can be seen by restricting the type to the finite[23] set of parameters in such a formula. Galois types do not necessarily have this property. Indeed, assuming the existence of a monster model $\mathfrak{C}$, this would

---

[23] Or larger if the logic allows infinitely many free variables.

imply a strong closure property on $\text{Aut}(\mathfrak{C})$. Nonetheless, a generalization of this idea, called *tameness*, has become a key tool in the study of AECs.

For a set A, we write $P_\kappa A$ for the collection of subsets of A of size less than κ. We also define an analog notation for models: for $M \in \mathbf{K}_{\geq \kappa}$:

$$P_\kappa^* M := \{M_0 \in \mathbf{K}_{<\kappa} : M_0 \leq M\}$$

Definition 3.1. $\mathbf{K}$ is $< \kappa$-tame if, for all $M \in \mathbf{K}$ and $p \neq q \in \text{gS}^1(M)$, there is $A \in P_\kappa |M|$ such that $p \restriction A \neq q \restriction A$.

For $\kappa > \text{LS}(\mathbf{K})$, it is equivalent if we quantify over $P_\kappa^* M$ (models) rather than $P_\kappa^* |M|$ (sets). Quantifying over sets is useful to isolate notions such as $< \aleph_0$-tameness. Several parametrizations (e.g., of the length of type) and variations exist. Below we list a few that we use; note that, in all cases, writing "κ" in place of "$< \kappa$" should be interpreted as "$< \kappa^+$".

Definition 3.2. Suppose $\mathbf{K}$ is an AEC with $\kappa \leq \lambda$.

1. $\mathbf{K}$ is $(< \kappa, \lambda)$-*tame* if for any $M \in \mathbf{K}_\lambda$ and $p \neq q \in \text{gS}^1(M)$, there is some $A \in P_\kappa |M|$ such that $p \restriction A \neq q \restriction A$.

2. $\mathbf{K}$ is $< \kappa$-*type short* if for any $M \in \mathbf{K}$, index set I, and $p \neq q \in \text{gS}^I(M)$, there is some $I_0 \in P_\kappa I$ such that $p^{I_0} \neq q^{I_0}$.

3. $\mathbf{K}$ is κ-*local* if for any increasing, continuous $\langle M_i \in \mathbf{K} : i \leq \kappa \rangle$ and any $p \neq q \in \text{gS}(M_\kappa)$, there is $i_0 < \kappa$ such that $p \restriction M_{i_0} \neq q \restriction M_{i_0}$.

4. $\mathbf{K}$ is κ-*compact* if for any increasing, continuous $\langle M_i : i \leq \kappa \rangle$ and increasing $\langle p_i \in \text{gS}(M_i) : i < \kappa \rangle$, there is $p \in \text{gS}(M)$ such that $p_i \leq p$ for all $i < \kappa$.

5. $\mathbf{K}$ is *fully* $< \kappa$-*tame and -type short* if for any $M \in \mathbf{K}$, index set I, and $p \neq q \in \text{gS}^I(M)$, there are $A \in P_\kappa |M|$ and $I_0 \in P_\kappa I$ such that $p^{I_0} \restriction A \neq q^{I_0} \restriction A$.

When κ is omitted, we mean that there exists κ such that the property holds at κ. For example, "$\mathbf{K}$ is tame" means that there exists κ such that $\mathbf{K}$ is $< \kappa$-tame. Note that definitions of locality and compactness implicitly assume κ is regular.

These types of properties are often called locality properties for AECs because they assert, in different ways, that Galois types are locally defined.

Each of these notions also has a *weak* version: weak $< \kappa$-tameness, etc. This variation means that the property holds when the domain is Galois-saturated.

A brief summary of the ideas is below. In each (and throughout this chapter), "small" is used to mean "of size less than κ".

- $< \kappa$-tameness says that different types differ over some small subset of the domain.

- $< \kappa$-type shortness says that different types differ over some small subset of their length.

- κ-locality says that each increasing chain of Galois types of length κ has *at most* one upper bound.

- κ-compactness says that each increasing chain of Galois types of length κ has *at least* one upper bound.[24]

A combination of tameness and type shortness allows us to conceptualize Galois types as sets of smaller types.

There are several relations between the properties:

Proposition 3.3.

1. *For $\kappa > \mathrm{LS}(\mathbf{K})$, $< \kappa$-type shortness implies $< \kappa$-tameness.*

2. *$< \mathrm{cf}(\kappa)$-tameness implies κ-locality.*

3. *μ-locality for all $\mu < \lambda$ implies $(\mathrm{LS}(\mathbf{K}), \lambda)$-tameness.*

4. *μ-locality for all $\mu < \lambda$ implies λ-compactness.*

As discussed, one of the draws of working in a short and tame AEC is that Galois types behave much more like first-order syntactic types in the sense that a Galois type $p \in \mathrm{gS}(M)$ can be identified with the collection $\{p^{I_0} \upharpoonright M_0 : I_0 \in P_\kappa \ell(p) \text{ and } M_0 \in P_\kappa^* M\}$ of its small restrictions:

Proposition 3.4. $\mathbf{K}$ *is fully $< \kappa$-tame and -type short if and only if the map:*

$$p \in \mathrm{gS}(M) \mapsto \{p^{I_0} \upharpoonright M_0 : I_0 \in P_\kappa I, M_0 \in P_\kappa^* M\}$$

is injective.

In fact, one can see these small restrictions as formulas (this will be used later to generalize heir and coheir to AECs). This productive intuition can be made exact using *Galois Morleyization*. Start with an AEC $\mathbf{K}$ and add to the language an α-ary predicate R_p for each $N \in \mathbf{K}$, each $p \in \mathrm{gS}^\alpha(\emptyset; N)$, and each $\alpha < \kappa$. This gives us an infinitary language $\widehat{L}$. Then expand each $M \in \mathbf{K}$ to a $\widehat{L}$-structure $\widehat{M}$ by setting $R_p(\bar{a})$ to be true in M if and only if $\mathrm{gtp}(\bar{a}/\emptyset; M) = p$. We obtain a class $\widehat{K}^{<\kappa} := \{\widehat{M} \mid M \in \mathbf{K}\}$. $\widehat{K}$ has relations of infinite arity but it still behaves like an AEC. We call $\widehat{K}^{<\kappa}$ the *$< \kappa$-Galois Morleyization* of $\mathbf{K}$. The connection between tameness and $\widehat{K}$ is given by the following theorem:

Theorem 3.5. *Let $\mathbf{K}$ be an AEC. The following are equivalent:*

1. $\mathbf{K}$ *is fully $< \kappa$-tame and -type short.*

2. *The map $\mathrm{gtp}(\bar{b}/M; N) \mapsto tp_{qf\text{-}\mathbb{L}_{\kappa,\kappa}(\widehat{L})}(\bar{b}/\widehat{M}; \widehat{N})$ is an injection.*

---

[24] All AECs are ω-compact and global compactness statements have large cardinal strength; see [Shec, Section 2].

Here, the Galois type is computed in **K**, and the type on the right is the (syntactic) quantifier-free $\mathbb{L}_{\kappa,\kappa}$-type in the language $\widehat{L}$. Note that the locality hypothesis in (1) can be weakened to $< \kappa$-tameness if in (2) we ask that $\ell(\bar{b}) = 1$. Several other variations are possible.

The Galois Morleyization gives a way to directly use syntactic tools (such as the results of stability theory inside a model, see for example [She09c, Chapter V.A]) in the study of tame AECs. See for example Theorem 5.14.

Another way to see tameness is as a topological separation principle: consider the set X_M of Galois types over M. For a fixed κ, we can give a topology on X_M with basis given by sets of the form $U_{p,A} := \{q \in \mathrm{gS}(M) \mid A \subseteq |M| \wedge q \upharpoonright A = p\}$, for p a Galois type over A and $|A| < \kappa$. This is the same topology as that generated by quantifier-free $\mathbb{L}_{\kappa,\kappa}$-formulas in the $< \kappa$-Galois Morleyization. Thus, one can show:

Theorem 3.6. *Let* **K** *be an AEC and let* $\lambda \geq \mathrm{LS}(\mathbf{K})$. **K** *is* $(< \kappa, \lambda)$-*tame if and only if for any* $M \in \mathbf{K}_\lambda$, *the topology on* X_M *defined above is Hausdorff.*

3.2 Where – Examples and counterexamples

3.2.1 Examples

Several "mathematically interesting" classes turn out to be tame. Moreover, there are several general ways to *derive* tameness from structural assumptions. We list some here, roughly in decreasing order of generality.

1. **Locality from large cardinals**

   Large cardinals κ allow the generalization of compactness results from first-order logic to $\mathbb{L}_{\kappa,\omega}$ in various ways (see, for instance, [Jec03, Lemma 20.2]). Since tameness is a weak form of compactness, these generalizations correspond to compactness results in AECs that can be "captured" by $\mathbb{L}_{\kappa,\omega}$. We state a simple version of these results here:

   Theorem 3.7. *Suppose* **K** *is an AEC with* $\mathrm{LS}(\mathbf{K}) < \kappa$.

   - *If* κ *is weakly compact, then* **K** *is* $(< \kappa, \kappa)$-*tame.*
   - *If* κ *is measurable, then* **K** *is* κ-*local.*
   - *If* κ *is strongly compact, then* **K** *is fully* $< \kappa$-*tame and -type short.*

   These results can be strengthened in various ways. First, they apply also to AECs that are explicitly axiomatized in $\mathbb{L}_{\kappa,\omega}$. The key fact is that ultraproducts by κ-complete ultrafilters preserve the AEC (the proof uses the presentation theorem, Theorem 2.9). Second, each large cardinal can be replaced by its "almost" version: for example, almost strongly compact means that, *for each* $\delta < \kappa$, $\mathbb{L}_{\delta,\delta}$ *is* κ-compact; equivalently, given a κ-complete filter, *for each* $\delta < \kappa$, it can be extended to a δ-*complete* ultrafilter. See [BU, Definition 2.1] for a full list of the "almost" versions. Note that other structural properties (such as amalgamation) follow from the combination of large cardinals with categoricity. Thus, these large cardinals make the development of a structure theory (culminating for example in the existence of a well-behaved independence notion, see Corollary 5.54) much easier.

2. **Weak tameness from categoricity under amalgamation**
Recall from Section 3.1 that an AEC **K** is $(\chi_0, < \chi)$-weakly tame if for every *Galois-saturated* $M \in \mathbf{K}_{<\chi}$, every $p \neq q \in \mathrm{gS}(M)$, there exists $M_0 \leq M$ with $M_0 \in \mathbf{K}_{\leq \chi_0}$ such that $p \upharpoonright M_0 \neq q \upharpoonright M_0$. It is known that, in AECs with amalgamation categorical in a sufficiently high cardinal, weak tameness holds below the categoricity cardinal. More precisely:

Theorem 3.8. *Assume that* **K** *is an AEC with amalgamation and no maximal models which is categorical in a* $\lambda > \mathrm{LS}(\mathbf{K})$.

(a) *Let* χ *be a limit cardinal such that* $\mathrm{cf}(\chi) > \mathrm{LS}(\mathbf{K})$. *If the model of size* λ *is* χ-*Galois-saturated, then there exists* $\chi_0 < \chi$ *such that* **K** *is* $(\chi_0, < \chi)$-*weakly tame.*

(b) *If the model of size* λ *is* H_1-*Galois-saturated, then there exists* $\chi_0 < H_1$ *such that whenever* $\chi \geq H_1$ *is so that the model of size* λ *is* χ-*Galois-saturated, we have that* **K** *is* $(\chi_0, < \chi)$-*weakly tame*[25].

Remark 3.9. The model in the categoricity cardinal λ is χ-Galois-saturated whenever $\mathrm{cf}(\lambda) \geq \chi$ (e.g., if λ is a successor) or (by Theorem 2.38) if[26] $\lambda \geq h(\chi)$.

The proof of Theorem 3.8 heavily uses Ehrenfeucht-Mostowski models to transfer the behavior below H_1 to a larger model that is generated by a nice enough linear order. Then the categoricity assumption is used to embed every model of size χ into such a model of size λ.

Theorem 3.8 is key to prove several of the categoricity transfers listed in Section 2.4.

3. **Tameness from categoricity and large cardinals**
The hypotheses in the last two examples can be combined advantageously.

Theorem 3.10. *Let* **K** *be an AEC and let* $\kappa > \mathrm{LS}(\mathbf{K})$ *be a measurable cardinal. If* **K** *is categorical in a* $\lambda \geq \kappa$, *then* $\mathbf{K}_{[\kappa, \lambda)}$ *has amalgamation and is* $(\kappa, < \lambda)$-*tame.*

In particular, if there exists a proper class of measurable cardinals and **K** is categorical in a proper class of cardinals, then **K** is tame. It is conjectured that the large cardinal hypothesis is not necessary. Note that the tameness here is "full", i.e., not the weak tameness in Theorem 3.8.

4. **Tameness from stable forking**

---

[25]Note that χ_0 does not depend on χ.

[26]A more clever application of Theorem 2.38 shows that it is enough to have $\lambda \geq \sup_{\theta < \chi} h(\theta^+)$.

Suppose that the AEC **K** has amalgamation and a stable "forking-like" relation $\underset{\smile}{\perp}$ (see Definition 5.7). That is, we ask that there is a notion "$p \in \mathrm{gS}(N)$ does not fork over M" for $M \leq N$ satisfying the usual monotonicity properties, uniqueness, and local character[27]: there exists a cardinal $\bar{\kappa} = \bar{\kappa}(\underset{\smile}{\perp})$ such that for every $p \in \mathrm{gS}(N)$, there is $M \leq N$ of size less than $\bar{\kappa}$ such that p does not fork over M (see more on such relations in Section 5).

Then, given any two types $p, q \in \mathrm{gS}(N)$ we can find $M \leq N$ over which both types do not fork over and so that $\|M\| < \bar{\kappa}$. If $p \upharpoonright M = q \upharpoonright M$, then uniqueness implies $p = q$. Thus, **K** is $(< \bar{\kappa})$-tame.

5. **Universal classes**

A *universal class* is a class **K** of structures in a fixed language $L(\mathbf{K})$ that is closed under isomorphism, substructure, and unions of increasing chains. In particular, $(\mathbf{K}, \subseteq)$ is an AEC with Löwenheim-Skolem-Tarski number $|L(\mathbf{K})| + \aleph_0$.

In a universal class, any partial isomorphism extends uniquely to an isomorphism (just take the closure under the functions). This fact is key in the proof of:

Theorem 3.11. *Any universal class is fully $(< \aleph_0)$-tame and short.*

Thus, for instance, the class of locally finite groups (ordered with subgroup) is tame. Theorem 3.11 generalizes to any AEC **K** equipped with a notion of "generated by" which is (in a sense) canonical (for universal classes, this notion is just the closure under the functions). Note that this does not need to assume that **K** has amalgamation.

6. **Tame finitary AECs**

A finitary AEC **K** is defined by several properties (including amalgamation and $\mathrm{LS}(\mathbf{K}) = \aleph_0$), but the key notion is that the strong substructure relation $\leq$ has *finite character*. This means that, for $M, N \in \mathbf{K}$, we have $M \leq N$ if and only if $M \subseteq N$ and:

> For every $\bar{a} \in {}^{<\omega}M$, we have that $\mathrm{gtp}(\bar{a}/\emptyset; M) = \mathrm{gtp}(\bar{a}/\emptyset; N)$.

This means that there is a finitary test for when $\leq$ holds between two models that are already known to be members of **K**. This definition is motivated by the observation that this condition holds for any AEC axiomatized in a countable fragment of $\mathbb{L}_{\omega_1,\omega}$ by the Tarski-Vaught test[28]. Homogeneous model theory can be seen as a special case of the study of finitary AECs. Hyttinen and Kësala [HK06, 3.12] have shown that every $\aleph_0$-stable $\aleph_0$-tame finitary AEC is $(< \aleph_0)$-tame. These classes seem very amenable to some classification theory. For example, an $\aleph_0$-tame finitary AEC categorical in some uncountable λ is categorical in *all* $\lambda' \geq \min(\lambda, H_1)$. Recent work has even developed some geometric

---

[27]The extension property is not needed here.

[28]Kueker [Kue08] has asked whether any finitary AEC must be $\mathbb{L}_{\infty,\omega}$-axiomatizable.

stability theory in a larger class (Finite U-Rank classes, which included quasiminimal classes below) [HK16].

7. **Homogeneous model theory**
   Homogeneous model theory takes place in the context of a large "monster model" (for a first-order theory T) that omits a set of types D, but is still as saturated as possible with respect to this omission. The notion of "as saturated as possible" is captured by requiring it to be sequentially homogeneous rather than model homogeneous. Note that the particular case when $D = \emptyset$ is the elementary case. In this context, amalgamation, joint embedding, and no maximal models hold for free and Galois types are first-order syntactic types. This identification means that the AEC of models of T omitting D (ordered with elementary substructure) is fully $(< \aleph_0)$-tame and short. Homogeneous model theory has a rich classification theory in its own right, with connections to continuous first-order logic (see the historical remarks).

8. **Averageable classes**
   Averageable classes are type omitting classes $\mathrm{EC}(T, \Gamma)$ (ordered with a relation $\leq$) that are nice enough to have a relativized ultraproduct that preserves the omission of types in Γ and satisfies enough of Łoś' Theorem to interact well with $\leq$. This relativized ultraproduct gives enough compactness to show that types are syntactic (and much more), which implies that an averageable class is fully $(< \aleph_0)$-tame and short. Examples of averageable classes include torsion modules over PIDs and densely ordered abelian groups with a cofinal and coinitial $\mathbb{Z}$-chain.

9. **Continuous first-order logic**
   Continuous first-order logic can be studied in a fragment of $\mathbb{L}_{\omega_1, \omega}$ by using the infinitary logic to have a standard copy of $\mathbb{Q}$ and then studying dense subsets of complete metric spaces. Although the logic $\mathbb{L}_{\omega_1, \omega}$ is incompact, the fragment necessary to code this information is compact (as evidenced by the metric ultrapower and compactness results in continuous first-order logic), so the classes are fully $(< \aleph_0)$-tame and short. Beyond first-order, continuous model theory can be done in the so-called metric AECs, where a notion of tameness (d-tameness) can also be defined.

10. **Quasiminimal classes**
    A quasiminimal class is an AEC satisfying certain additional axioms; most importantly, the structures carry a pregeometry with certain nice properties. The axioms directly imply that Galois types over *countable* models are quantifier-free first-order types, and the excellence axiom can be used to transfer this to uncountable models. Therefore, quasiminimal classes are $< \aleph_0$-tame. Examples of quasiminimal classes include covers of $\mathbb{C}^\times$ and Zilber fields with pseudoexponentiation. Note that it can be shown (from the countable closure axiom) that these classes are strictly

$\mathbb{L}_{\omega_1,\omega}(Q)$-definable. This gives important examples of categorical AECs that are not finitary.

11. **λ-saturated models of a superstable first-order theory**
Let T be a first-order superstable theory. We know that unions of increasing chains of λ-saturated models are λ-saturated and that models of size λ have saturated extensions of size at most $\lambda + 2^{|T|}$. Thus, the class of λ-saturated models of T (ordered with elementary substructure) forms an AEC $\mathbf{K}_\lambda^T$ with $\mathrm{LS}(\mathbf{K}_\lambda^T) \le \lambda + 2^{|T|}$. Furthermore, this class has a monster model and is fully $(< \aleph_0)$-tame and short.

12. **Superior AECs**
Superior AECs are a generalization of what some call excellent classes. An AEC is *superior* if it carries an axiomatic notion of forking for which one can state multi-dimensional uniqueness and extension properties. A combination of these gives some tameness:

Theorem 3.12. *Let $\mathbf{K}$ be a superior AEC with weak $(\lambda, 2)$-uniqueness and λ-extension for some $\lambda \ge \mathrm{LS}(\mathbf{K}) + \bar{\kappa}(\mathbf{K})$. Then $\mathbf{K}$ is λ^+-local. In particular, it is (λ, λ^+)-tame.*

13. **Hrushovski fusions**
Villaveces and Zambrano [Zam12] have studied Hrushovski's method of fusing pregeometries over disjoint languages as an AEC with strong substructure being given by self-sufficient embedding. They show that these classes satisfy a weakening of independent 3-amalgamation. This weakening is still enough to show, as with superior AECs, that the classes are $\mathrm{LS}(\mathbf{K})$-tame.

14. **$^\perp N$ when N is an abelian group**
Given a module N, $^\perp N$ is the class of modules $\{M : \mathrm{Ext}^n(M, N) = 0$ for all $1 \le n < \omega\}$. We make this into an AEC by setting $M \le_\perp M'$ if and only if $M'/M \in {}^\perp N$. If N is an abelian group, then $^\perp N$ is set of all abelian groups that are p-torsion free for all p in some collection of primes P.

Theorem 3.13. *If N is an abelian group, then $^\perp N$ is $< \aleph_0$-tame.*

Moreover, such a $^\perp N$ is Galois-stable in exactly the cardinals $\lambda = \lambda^\omega$.

15. **Algebraically closed, rank one valued fields**
Let $\mathrm{ACVF}_\mathbb{R}$ be the $\mathbb{L}_{\omega_1,\omega}$-theory of an algebraically closed valued field such that the value group is Archimedean; equivalently, the value group can be embedded into $\mathbb{R}$. After fixing the characteristic, this AEC has a monster model and Galois types are determined by syntactic types. Thus, the class is fully $< \aleph_0$-tame and -type short. This determination of Galois types can be seen either through algebraic arguments or the construction of an appropriate ultraproduct.

Such a class cannot have an uncountable ordered sequence, so it has the $\aleph_0$-order property of length α for every $\alpha < \omega_1$, but it does not have the $\aleph_0$-order property of length ω_1.

3.2.2 Counterexamples

Life would be too easy if all AECs were tame. Above we have seen that several natural mathematical classes are tame; in contrast, all the known counterexamples to tameness are pathological[29], with the most natural being the Baldwin-Shelah example of short exact sequences. We list the known ones below in increasing "strength".

1. **The Hart-Shelah example**

   The Hart-Shelah examples are a family of examples axiomatized by complete sentences in $\mathbb{L}_{\omega_1,\omega}$.

   Theorem 3.14. *For each $n < \omega$, there is an AEC $\mathbf{K}_n$ that is axiomatized by a complete sentence in $\mathbb{L}_{\omega_1,\omega}$ with $\mathrm{LS}(\mathbf{K}_n) = \aleph_0$ and disjoint amalgamation such that:*

   (a) $\mathbf{K}_n$ is $(\aleph_0, \aleph_{n-1})$-tame (in fact, the types are first-order syntactic);

   (b) $\mathbf{K}_n$ is categorical in $[\aleph_0, \aleph_n]$;

   (c) $\mathbf{K}_n$ is Galois-stable in μ for $\mu \in [\aleph_0, \aleph_{n-1}]$; and

   (d) Each of these properties is sharp. That is:

       i. $\mathbf{K}_n$ is not $(\aleph_0, \aleph_n)$-tame,

       ii. $\mathbf{K}_n$ is not categorical in $\aleph_{n+1}$[30].

       iii. $\mathbf{K}_n$ is not Galois-stable in $\aleph_n$.

   Each model $M \in \mathbf{K}_n$ begins with an index set I (called the spine); the direct sum $G := \oplus_{[I]^{n+2}} \mathbb{Z}_2$; $G^* \subseteq [I]^k \times G$ with a projection $\pi : G^* \to [I]^{n+2}$ such that each stalk $G_u^* = \pi^{-1}\{u\}$ has a regular, transitive action of G on it; and, similarly, $H^* = [I]^{n+2} \times H$ with a projection $\pi' : H^* \to [I]^k$ such that each stalk has an action of $\mathbb{Z}_2$ on it. So far, the structure described (along with the extra information required to code it) is well behaved and totally categorical. Added to this is a $n + 3$-ary relation $Q \subseteq (G^*)^{n+2} \times H^*$ such that $Q(u_1, \ldots, u_{n+2}, v)$ is intended to code

   - there are exactly $n + 3$ elements of I that make up the projections of $u_1, \ldots, u_{n+2}, v$ (so each $(n + 2)$-element subset shows up exactly once in the projections); and

   - the sum of the second coordinates evaluated at $\pi'(v)$ is equal to some fixed function of the $n + 3$ elements of the projections.

   This coding allows one to "hide the zeros" and find nontameness at $\aleph_n$. The example shows that Theorem 2.33 is sharp.

   It should be noted that the ideas used in constructing the Hart-Shelah examples come from constructions that characterize various cardinals. Thus, although the construction takes place in ZFC, it still involves

---

[29]In the dictionary sense that they were constructed as counterexamples.

[30]Note that this follows from (1(d)i) by (the proof of) the upward categoricity transfer of Grossberg and VanDieren [GV06a].

set-theoretic ideas. Work in preparation by Shelah and Villaveces [SV] contains an extension of the Hart-Shelah example to larger cardinals, proving:

Theorem 3.15. *Assume the generalized continuum hypothesis. For each λ and $k < \omega$, there is $\psi_k^\lambda \in \mathbb{L}_{(2^\lambda)^+,\omega}$ that is categorical in $\lambda^{+2}, \ldots, \lambda^{+(k-1)}$ but not in $\beth_{k+1}(\lambda)^+$.*

As for the countable case, this example is likely not to be tame.

2. The Baldwin-Shelah example

The Baldwin-Shelah example **K** consists of several short exact sequences, each beginning with $\mathbb{Z}$.

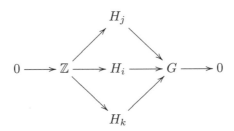

Formally this consists of sorts Z, G, I, and H with a projection $\pi :$ $H \to I$ and group operations and embeddings such that each fiber $H_i :=$ $\pi^{-1}(\{i\})$ is a group that is in the middle of a short exact sequence.

The locality properties of Galois types over a model depend heavily on the group G used. The key observation is that, given $i, j \in I$, their Galois types are equal precisely when there is an isomorphism of the fibers $\pi^{-1}\{i\}$ and $\pi^{-1}\{j\}$ that commute with the rest of the short exact sequence. Thus, Baldwin and Shelah consider an $\aleph_1$-free, not free, not Whitehead group[31] G^* of size $\aleph_1$. With G^* in hand, we can construct a counterexample to $(\aleph_0, \aleph_1)$-tameness: set i_0 and i_1 to be in a short exact sequence that ends in G^* such that $\pi^{-1}\{i_0\} = G^* \oplus \mathbb{Z}$ and $\pi^{-1}\{i_1\} = H$ is not isomorphic to $G^* \oplus \mathbb{Z}$; such a group exists exactly because H is not Whitehead. Then, by the observation above, i_0 and i_1 have different types over the entire uncountable set G^*. However, any countable approximation G_0 of G^* will see that i_0 and i_1 have the same Galois type over it: the countable approximation will have that the fibers over i_0 and i_1 are both the middle of a short exact ending in G_0. By the choice of G^*, G_0 is free, thus Whitehead, so these fibers are both isomorphic to $G_0 \oplus \mathbb{Z}$. This isomorphism witnesses the equality of the Galois types of i_0 and i_1 over the countable approximation.

---

[31]It is a ZFC theorem that such a group exists at $\aleph_1$. Having such a group at κ (κ-free, not free, not Whitehead of size κ) is, in the words of Baldwin and Shelah, "sensitive to set theory". The known sensitivities are summarized in [Bon14c, Section 8], primarily drawing on work in [MS94] and [EM02].

Given a κ version G_κ^* of this group allows one to construct a counterexample to $(< \kappa, \kappa)$-tameness. Indeed, all that is necessary is that G_κ^* is 'almost Whitehead:' it is not Whitehead, but every strictly smaller subgroup of it is.

3. The Shelah-Boney-Unger example

While the Baldwin-Shelah example reveals a connection between tameness and set theory, the Shelah-Boney-Unger example shows an outright equivalence between certain tameness statements and large cardinals. For each cardinal $\sigma^\omega = \sigma$, there is an AEC $\mathbf{K}_\sigma$ that consists of an index predicate J with a projection $Q : H \to J$ such that each fiber $Q^{-1}\{j\}$ has a specified structure and a projection $\pi : H \to I^{32}$. Given some partial order $(\mathcal{D}, \lhd)$ and set of functions $\mathcal{F}$ with domain $\mathcal{D}$, filtrations $\{M_{\ell,d} : d \in \mathcal{D}\}$ of a larger model $M_{\ell,\mathcal{D}}$, all from $\mathbf{K}_\sigma$, are built, for $\ell = 1, 2$. Similar to the Baldwin-Shelah example, types p_d and q_d are defined such that the types are equal if and only if there is a nice isomorphism between $M_{1,d}$ and $M_{2,d}$; the same is true of $p_\mathcal{D}$ and $q_\mathcal{D}$. Thus, various properties of type locality ($p_\mathcal{D} = q_\mathcal{D}$ following from $p_d = q_d$ for all $d \in \mathcal{D}$) is once more coded by "isomorphism locality".

In turn, the structure was built so that a nice isomorphism between $M_{1,\mathcal{D}}$ and $M_{2,\mathcal{D}}$ is equivalent to a combinatorial property $\#(\mathcal{D}, \mathcal{F})$.

Definition 3.16.

- Given functions f and g with the same domain, we define $f \leq^* g$ to hold if and only if there is some $e : \operatorname{ran} g \to \operatorname{ran} f$ such that $f = e \circ g$.
- Given a function f with a domain D that is partially ordered by $\leq_D$, we define $\operatorname{ran}^* f = \bigcap_{d \in D} \operatorname{ran} (f \restriction \{d' \in D : d \leq d'\})$ to be the eventual range of f.
- $\#(\mathcal{D}, \mathcal{F})$ holds if and only if there are $f^* \in \mathcal{F}$ and a collection of nonempty finite sets $\{u_f \subseteq \operatorname{ran}^* f : f^* \leq^* f\}$ such that, given any e witnessing $f^* \leq^* f$, $e \restriction u_f$ is a bijection from u_f to u_{f^*}.

So $\#(\mathcal{D}, \mathcal{F})$ eventually puts some kind of structure on functions in $\mathcal{F}$. Shockingly, this principle can, under the right assumption on $\mathcal{D}$ and $\mathcal{F}$, define a very complete (ultra)filter on $\mathcal{D}$: for each f with $f \leq^* f^*$, set $i_f \in u_f$ to be the unique image of $\min u_f$ by some e witnessing $f^* \leq^* f$. Then, for $A \subseteq \mathcal{D}$,

$$A \in U \iff \exists d \in \mathcal{D}, f \in \mathcal{F} \left(f^{-1}\{i_f\} \cap \{d' \in \mathcal{D} : d \lhd d'\} \subseteq A \right)$$

Thus, we get that type locality in $\mathbf{K}_\sigma$ implies the existence of filters and ultrafilters used in the definitions of large cardinals; the converse is mentioned above.

This argument can be used to give the following theorems.

---

[32] Although there are two projections, they are used differently: the projection Q is a technical device to code isomorphisms of structures via equality of Galois types, while the interaction of (a fiber of) H and I is more interesting.

Theorem 3.17. *Let κ such that $\mu^\omega < \kappa$ for all $\mu < \kappa$.*

(a) If [33] $\kappa^{<\kappa} = \kappa$ and every AEC $\mathbf{K}$ with $\mathrm{LS}(\mathbf{K}) < \kappa$ is $(< \kappa, \kappa)$-tame, then κ is almost weakly compact.

(b) If every AEC $\mathbf{K}$ with $\mathrm{LS}(\mathbf{K}) < \kappa$ is κ-local, then κ is almost measurable.

(c) If every AEC $\mathbf{K}$ with $\mathrm{LS}(\mathbf{K}) < \kappa$ is $< \kappa$-tame, then κ is almost strongly compact.

We obtain a characterization of the statement "all AECs are tame" in terms of large cardinals.

Corollary 3.18. *All AECs are tame if and only if there is a proper class of almost strongly compact cardinals.*

Note that Corollary 3.18 says nothing about "well behaved" classes of AECs, such as AECs categorical in a proper class of cardinals. In fact, Theorem 3.10 shows that the consistency strength of the statement "all AECs are tame" is much higher than that of the statement "all unboundedly categorical AECs are tame".

4　Categoricity transfer in universal classes: an overview

In this section, we sketch a proof of Theorem 1.2, emphasizing the role of tameness in the argument:

Theorem 4.1. *Let $\mathbf{K}$ be a universal class. If $\mathbf{K}$ is categorical in cardinals of arbitrarily high cofinality[34], then $\mathbf{K}$ is categorical on a tail of cardinals.*

The arguments in this section are primarily from Vasey [Vasf].

Note that (as pointed out in Section 2.4), we can replace the categoricity hypotheses of Theorem 4.1 by categoricity in a *single* "high-enough" cardinal of "high-enough" cofinality.

We avoid technical definitions in this section, instead referring the reader to Section 2 or Section 5.

So let $\mathbf{K}$ be a universal class categorical in cardinals of arbitrarily high cofinality. To prove the categoricity transfer, we first show that $\mathbf{K}$ has several structural properties that hold in elementary classes. As we have seen, amalgamation is one such property.

---

[33]The additional cardinal arithmetic here can be dropped at the cost of only concluding (κ is weakly compact)^L.

[34]This cofinality restriction is only used to obtain amalgamation. See the historical remarks for more.

4.1 Step 1: Getting amalgamation

It is not clear how to directly prove amalgamation in all cardinals, but Theorem 2.45 is a deep result of Shelah which says (since good frames must have amalgamation) that it holds for models of *some* suitable size[35].

This leads to a new fundamental question:

Question 4.2. If an AEC **K** has amalgamation in a cardinal λ, under what condition does it have amalgamation above λ?

One such condition is *excellence* (briefly, excellence asserts strong uniqueness of n-dimensional amalgamation results). However, it is open whether it follows from categoricity, even for classes of models of an uncountable first-order theory. Excellence also gives much more, and (for now) we are only interested in amalgamation. Another condition would be the existence of *large cardinals*. For example, a strongly compact κ with $\mathrm{LS}(\mathbf{K}) < \kappa \leq \lambda$ would be enough.

At that point, we recall a key theme in the study of tameness: when large cardinals appear in a model-theoretic result, tameness[36] can often replace them. For the purpose of an amalgamation transfer, it is not clear that this suffices. For one thing, one can ask what tameness really means without amalgamation (of course, its definition makes sense, but how do we get a handle on the transitive closure of atomic equality, Definition 2.16.(2)). In the case of universal classes, this question has a nice answer: even without amalgamation, equality of Galois types is witnessed by an isomorphism and, in fact, tameness holds for free! This is Theorem 3.11. From its proof, we isolate a technical weakening of amalgamation:

Definition 4.3. An AEC **K** has *weak amalgamation* if whenever $\mathrm{gtp}(a_1/M; N_1) = \mathrm{gtp}(a_2/M; N_2)$, there exists $M_1 \in \mathbf{K}$ with $M \leq M_1 \leq N_1$ and $a_1 \in |M_1|$ such that (a_1, M, M_1) is *atomically equivalent* to (a_2, M, N_2).

It turns out that universal classes have weak amalgamation: we can take M_1 to be the closure of $|M_1| \cup \{a\}$ under the functions of N_1 and expand the definition of equality of Galois types.

We now rephrase Question 4.2 as follows:

Question 4.4. Let **K** be an AEC which has amalgamation in a cardinal λ. Assume that **K** is λ-tame and has weak amalgamation. Under what condition does it have amalgamation above λ?

To make progress, a characterization of amalgamation will come in handy (this lemma is reminiscent of Proposition 2.18.(1)):

---

[35]This is the only place where we use the cofinality assumptions on the categoricity cardinals.

[36]Or really, a "tameness-like" property like full tameness and shortness.

Lemma 4.5. *Let* **K** *be an AEC with weak amalgamation. Then* **K** *has amalgamation if and only if for any* $M \in \mathbf{K}$, *any Galois type* $p \in \mathrm{gS}(M)$, *and any* $N \geq M$, *there exists* $q \in \mathrm{gS}(N)$ *extending* p.

The proof is easy if one assumes that atomic equivalence of Galois types is transitive. Weak amalgamation is a weakening of this property, but allows us to iterate the argument (when atomic equivalence is transitive) and obtain full amalgamation.

Now, it would be nice if we could not only extend Galois types, but also extend them canonically. This is reminiscent of first-order *forking*, a basic property of which is that every type has a (unique under reasonable conditions) nonforking extension. Thus, out of the apparently very set-theoretic problem of obtaining amalgamation, forking, a model-theoretic notion, appears in the discussion. What is an appropriate generalization of forking to AECs? Shelah's answer is that the *bare-bone* generalization is the notion of *good* λ-*frames*; see Section 2.5. There are several nonelementary setups where a good frame exists (see the next section). For example, Theorem 2.45 tells us that a good frame exists in our setup.

Still with the question of transferring amalgamation up in mind, one can ask whether it is possible to transfer an *entire good frame* up. In particular, given a notion of forking for models of size λ, is there one for models of size above λ? This is where tameness starts playing a very important role:

Theorem 4.6. *Let* **K** *be an AEC with amalgamation. Let* $\mathfrak{s}$ *be a good* λ-*frame with underlying AEC* **K**. *Then* $\mathfrak{s}$ *extends to a good* $(\geq \lambda)$-*frame (i.e., all the properties hold for models in* $\mathbf{K}_{\geq \lambda}$*) if and only if* **K** *is* λ-*tame.*

We give a sketch of the proof in Theorem 5.26. Let us also note that not only do the properties of forking transfer, but also the structural properties of **K**. Thus, $\mathbf{K}_{\geq \lambda}$ has no maximal models (roughly, this is obtained using the extension property and the fact that nonforking extensions are nonalgebraic).

Even better, it turns out that not too much amalgamation is needed for the proof of the frame transfer to go through: weak amalgamation is enough! Moreover, types can be extended by simply taking their nonforking extension. Thus, we obtain:

Theorem 4.7. *Let* **K** *be an AEC with weak amalgamation. If there is a good* λ-*frame* $\mathfrak{s}$ *with underlying AEC* **K** *and* **K** *is* λ-*tame, then* $\mathfrak{s}$ *extends to a good* $(\geq \lambda)$-*frame. In particular* $\mathbf{K}_{\geq \lambda}$ *has amalgamation.*

Corollary 4.8. *Let* **K** *be a universal class categorical in cardinals of arbitrarily high cofinality. Then there exists* λ *such that* $\mathbf{K}_{\geq \lambda}$ *has amalgamation.*

Proof. By Theorem 2.45, there is a cardinal λ such that **K** has a good λ-frame with underlying class $\mathbf{K}_\lambda$. By Theorem 3.11, **K** is λ-tame and (it is easy to see), **K** has weak amalgamation. Now apply Theorem 4.7. $\square$

Remark 4.9. Theorem 4.7 will be used even in the next steps, see the proof of Theorem 4.18.

4.2 Step 2: Global independence and orthogonality calculus

From the results so far, we see that we can replace $\mathbf{K}$ by $\mathbf{K}_{\geq \lambda}$ if necessary to assume without loss of generality that $\mathbf{K}$ is a universal class[37] categorical in a proper class of cardinals that has amalgamation. Other structural properties such as joint embedding and no maximal models follow readily. In fact, we have just pointed out that we can assume there is a good ($\geq \mathrm{LS}(\mathbf{K})$)-frame with underlying class $\mathbf{K}$. In particular, $\mathbf{K}$ is Galois-stable in all cardinals and has a superstable-like forking notion for types of length one.

What is the next step to get a categoricity transfer? The classical idea is to show that all big-enough models are Galois-saturated (note that by the above we have stability everywhere, so the model in the categoricity cardinal is Galois-saturated). Take M a model in a categoricity cardinal λ and p a nonalgebraic type over M. Assume that there exists $N > M$ of size λ such that p is omitted in N. If we can iterate this property λ^+-many times, we obtain a non-λ^+-Galois-saturated model. If $\mathbf{K}$ was categorical in λ^+, this gives a contradiction. More generally, if we can iterate longer to find $N > M$ of size $\mu > \lambda$ such that N omits p and $\mathbf{K}$ is categorical in μ, we also get a contradiction. This is reminiscent of a Vaughtian pair argument and more generally of Shelah's theory of unidimensionality. Roughly speaking, a class is *unidimensional* if it has essentially only one Galois type. Then a model cannot have arbitrarily large extensions omitting the type. Conversely, if the class is not unidimensional, then it has two "orthogonal" types and a model would be able to grow by adding more realizations of one type without realizing the other.

So we want to give a sense in which our class $\mathbf{K}$ is unidimensional. If $\mathbf{K}$ is categorical in a successor, this can be done much more easily than for the limit case using Vaughtian pairs. In fact a classical result of Grossberg and VanDieren for tame AECs says:

Theorem 4.10. *Suppose $\mathbf{K}$ has amalgamation and no maximal models. If $\mathbf{K}$ is a λ-tame AEC categorical in λ and λ^+, then $\mathbf{K}$ is categorical in all $\mu \geq \lambda$.*

To study general unidimensionality, we will use a notion of orthogonality. As for forking, we focus on developing a theory of orthogonality for types of length one over models of a single size.

We already have a good ($\geq \mathrm{LS}(\mathbf{K})$)-frame available, but for our purpose this is not enough. We will also use a notion of primeness:

Definition 4.11. We say an AEC $\mathbf{K}$ *has primes* if whenever $M \leq N$ are in $\mathbf{K}$ and $a \in |N| \backslash |M|$, there is a prime model $M' \leq N$ over $|M| \cup \{a\}$. This

---

[37]There is a small wrinkle here: if $\mathbf{K}$ is a universal class, $\mathbf{K}_{\geq \lambda}$ is not necessarily a universal class. We ignore this detail here since $\mathbf{K}_{\geq \lambda}$ will have enough of the properties of a universal class to carry the argument through.

means that if $\mathrm{gtp}(b/M; N') = \mathrm{gtp}(a/M; N)$, then[38] there exists $f : M' \xrightarrow[M]{} N'$ so that $f(b) = a$. We call (a, M, M') a *prime triple*.

Note that this makes sense even if the AEC does not have amalgamation. Some computations give us that:

Proposition 4.12. *If* $\mathbf{K}$ *is a universal class, then* $\mathbf{K}$ *has primes.*

Definition 4.13. Let $\mathbf{K}$ be an AEC with a good λ-frame. Assume that $\mathbf{K}$ has primes (at least for models of size λ). Let $M \in \mathbf{K}_\lambda$ and let $p, q \in \mathrm{gS}(M)$. We say that p and q are *weakly orthogonal* if there exists a prime triple (a, M, M') such that $\mathrm{gtp}(a/M; M') = q$ and p has a unique extension to $\mathrm{gS}(M')$. We say that p and q are *orthogonal* if for any $N \geq M$, the nonforking extensions p', q' to N of p and q, respectively, are weakly orthogonal.

Orthogonality and weak orthogonality coincide assuming categoricity:

Theorem 4.14. *Let* $\mathbf{K}$ *be an AEC which has primes and a good* λ-*frame. Assume that* $\mathbf{K}$ *is categorical in* λ*. Then weak orthogonality and orthogonality coincide.*

We have arrived to a definition of unidimensionality (we say that a good λ-frame is categorical when the underlying class is categorical in λ):

Definition 4.15. Let $\mathbf{K}$ be an AEC which has primes and a categorical good λ-frame. $\mathbf{K}_\lambda$ is *unidimensional* if there does *not* exist $M \in \mathbf{K}$, and types $p, q \in \mathrm{gS}(M)$ such that p and q are orthogonal.

Theorem 4.16. *Let* $\mathbf{K}$ *be an AEC which has primes and a categorical good* λ-*frame. If* $\mathbf{K}$ *is unidimensional, then* $\mathbf{K}$ *is categorical in* λ^+*.*

Using the result of Grossberg and VanDieren, if in addition $\mathbf{K}$ is λ-tame, $\mathbf{K}$ will be categorical in every cardinal above λ. Therefore, it is enough to prove unidimensionality. While step 2 was only happening locally in λ and did not use tameness, tameness will again have a crucial use in the next step.

4.3   Step 3: Proving unidimensionality

Let us make a slight diversion from unidimensionality. Recall that we work in a universal class $\mathbf{K}$ categorical in a proper class of cardinals with a lot of structural properties (amalgamation and existence of good frames, even global). We want to show that all big-enough models are Galois-saturated. Let M be a big model and assume it is not Galois-saturated, say it omits $p \in \mathrm{gS}(M_0)$, $M_0 \leq M$. Consider the class $\mathbf{K}_{\neg p}$ of all models $N \geq M_0$ that omit p. After adding constant symbols for M_0 and closing under isomorphisms,

---

[38]Why the formulation using Galois types? We have to make sure that the types of Ma in N and N' are the same.

this is an AEC. We would like to show that it has arbitrarily large models, for then this means that there exists a categoricity cardinal $\lambda > \|M_0\|$ and $N \in \mathbf{K}$ of size λ omitting p. This is a contradiction since we know that the model in the categoricity cardinal is Galois-saturated.

What are methods to show that a class has arbitrarily large models? A powerful one is again based on good frames: by definition, a good λ-frame has no maximal models in λ. If we can expand it to a good $(\geq \lambda)$-frame, then its underlying class $\mathbf{K}_{\geq\lambda}$ has no maximal models and hence models of arbitrarily large size. Recall that Theorem 4.7 gave mild conditions (tameness and weak amalgamation) under which a good frame can be transferred. Tameness was the key property there.

Moreover, we know already that $\mathbf{K}$ itself has a good $LS(\mathbf{K})$-frame, amalgamation, and is $LS(\mathbf{K})$-tame. But it is not so clear that these properties transfer to $\mathbf{K}_{\neg p}$. Consider for example amalgamation: let $M_0 \leq M_\ell$, $\ell = 1, 2$, and assume that $p \in gS(M_0)$ is omitted in both M_1 and M_2. Even if in $\mathbf{K}$ there exists amalgams of M_1 and M_2 over M_0, it is not clear that any such amalgams will omit p. Similarly, even if $q_1, q_2 \in gS(M)$ are Galois types in $\mathbf{K}_{\neg p}$ and they are equal in $\mathbf{K}$, there is no guarantee that they will be equal in $\mathbf{K}_{\neg p}$ (the amalgam witnessing it may not be a member of $\mathbf{K}_{\neg p}$). So it is not clear that $\mathbf{K}_{\neg p}$ is tame.

However, we are interested in universal classes, so consider the last property if $\mathbf{K}$ is a universal class. Say $q_1 = \text{gtp}(a_1/M; N_1)$, $q_2 = \text{gtp}(a_2/M; N_2)$. If q_1 is equal to q_2 in $\mathbf{K}$, then since $\mathbf{K}$ has primes there exists $M_1 \leq N_1$ containing a_1 and M and $f : M_1 \xrightarrow{M} N_2$ so that $f(a_1) = a_2$. If N_1 omits p, then M_1 also omits p and so q_1 and q_2 are equal also in $\mathbf{K}_{\neg p}$. By the same argument, $\mathbf{K}_{\neg p}$ also has primes (in fact, it is itself universal). Thus, it has weak amalgamation. Similarly, since $\mathbf{K}$ is tame, $\mathbf{K}_{\neg p}$ is also tame[39].

The last problem to solve is therefore whether a good λ-frame in $\mathbf{K}$ is also a good λ-frame in $\mathbf{K}_{\neg p}$. This is where we use orthogonality calculus and unidimensionality. However, the class we consider is slightly different than $\mathbf{K}_{\neg p}$: for $p \in gS(M_0)$ nonalgebraic, we let $\mathbf{K}_{\neg^* p}$ be the class of M such that $M_0 \leq M$ and p has a *unique* extension to M (we add constant symbols for M_0 to make $\mathbf{K}_{\neg^* p}$ closed under isomorphisms). Note that $\mathbf{K}_{\neg^* p} \subseteq \mathbf{K}_{\neg p}$, as the unique extension must be the nonforking extension, which is nonalgebraic if p is. Using orthogonality calculus, we can show:

Theorem 4.17. *Let $\mathbf{K}$ be an AEC which has primes and a categorical good λ-frame $\mathfrak{s}$ for types at most λ. If $\mathbf{K}$ is not unidimensional, then there exists $M \in \mathbf{K}_\lambda$ and a nonalgebraic $p \in gS(M)$ such that $\mathfrak{s}$ restricted to $\mathbf{K}_{\neg^* p}$ is still a good λ-frame.*

---

[39] A technical remark: if we only knew that $\mathbf{K}$ was *weakly* tame (i.e., tame for types over Galois-saturated models), we would not be able to conclude that $\mathbf{K}_{\neg p}$ was weakly tame: models that omit p of size larger than $|\text{dom}(p)|$ are not Galois-saturated. Thus, while many arguments in the study of tame AECs can be adapted to the weakly tame context, this one cannot.

We are ready to conclude:

Theorem 4.18. *Suppose that* **K** *is an AEC which has primes and a categorical good λ-frame for types at most λ. If* **K** *is categorical in some $\mu > \lambda$ and is λ-tame, then* $\mathbf{K}_\lambda$ *is unidimensional, and therefore* **K** *is categorical in all $\mu' > \lambda$.*

Proof. The last "therefore" follows from combining Theorem 4.16 and the Grossberg-VanDieren transfer (using tameness heavily) Theorem 4.10. To show that $\mathbf{K}_\lambda$ is unidimensional, suppose not. By Theorem 4.17, there exists $M \in \mathbf{K}_\lambda$ and a nonalgebraic $p \in \mathrm{gS}(M)$ such that there is a good λ-frame on $\mathbf{K}_{\neg^* p}$. By the argument above, $\mathbf{K}_{\neg^* p}$ is λ-tame and has weak amalgamation. But is $\mathbf{K}_{\neg^* p}$ an AEC? Yes! The only problematic part is if $\langle M_i : i < \delta \rangle$ is increasing in $\mathbf{K}_{\neg^* p}$, and we want to show that $M_\delta := \bigcup_{i < \delta} M_i$ is in $\mathbf{K}_{\neg^* p}$. Let $q_1, q_2 \in \mathrm{gS}(M_\delta)$ be extensions of p, we want to see that $q_1 = q_2$. By tameness, the good λ-frame of **K** transfers to a good $(\geq \lambda)$-frame. So we can fix $i < \delta$ such that q_1, q_2 do not fork over M_i. By definition of $\mathbf{K}_{\neg^* p}$, $q_1 \upharpoonright M_i = q_2 \upharpoonright M_i$. By uniqueness of nonforking extension, $q_1 = q_2$.

By Theorem 4.7 (using that $\mathbf{K}_{\neg^* p}$ is tame), $\mathbf{K}_{\neg^* p}$ has a good $(\geq \lambda)$-frame. In particular, it has arbitrarily large models. Thus, **K** has non-Galois-saturated models in every $\mu > \lambda$, hence cannot be categorical in any $\mu > \lambda$. $\square$

We wrap up:

Proof of Theorem 4.1. Let **K** be a universal class categorical in cardinals of arbitrarily high cofinality.

1. Just because it is a universal class, **K** has primes and is $\mathrm{LS}(\mathbf{K})$-tame (recall Example 3.2.1.(5)).

2. By Theorem 2.45, there exists a good λ-frame on **K**.

3. By the upward frame transfer (Theorem 4.7), $\mathbf{K}_{\geq \lambda}$ has amalgamation and in fact a good $(\geq \lambda)$-frame. This step uses λ-tameness.

4. By orthogonality calculus, if $\mathbf{K}_\lambda$ is not unidimensional then there exists a type p such that $\mathbf{K}_{\neg^* p}$ has a good λ-frame.

5. Since **K** has primes, $\mathbf{K}_{\neg^* p}$ is also λ-tame and has weak amalgamation, so by the upward frame transfer again (using tameness) it must have arbitrarily large models. So arbitrarily large models omit p, hence **K** has no Galois-saturated models of size above λ, so cannot be categorical above λ (by stability, **K** has a Galois-saturated model in every categoricity cardinal). This is a contradiction, therefore $\mathbf{K}_\lambda$ is unidimensional.

6. By Theorem 4.16, **K** is categorical in λ^+.

7. By the upward transfer of Grossberg and VanDieren (Theorem 4.10), **K** is categorical in all $\mu \geq \lambda$. This again uses tameness in a key way.

$\square$

Remark 4.19. The proof can be generalized to abstract elementary classes which are tame and have primes. See Theorem 5.47.

5 Independence, stability, and categoricity in tame AECs

We have seen that good frames are a crucial tool in the proof of Shelah's eventual categoricity conjecture in universal classes. In this section, we give the precise definition of good frames in a more general axiomatic independence framework. We survey when good frames and more global independence notions are known to exist (i.e., the best known answers to Question 1.3).

We look at what can be said in both strictly stable and superstable AECs. Along the way we look at stability transfers, and the equivalence of various definitions of superstability in tame AECs.

Finally, we survey the theory of categorical tame AECs and give the best known approximations to Shelah's categoricity conjecture in this framework.

5.1 Abstract independence relations

To allow us to state precise results, we first fix some terminology. The terms used should be familiar to readers with experience in working with forking, either in the elementary or nonelementary context. One potentially unfamiliar notation: we sometimes refer to the pair $i = (\mathbf{K}, \underset{\smile}{\smile})$ as an independence relation. This is particularly useful to deal with multiple classes as we can differentiate between the behavior of a possible forking relation on the class $\mathbf{K}$ compared to its behavior on the class $\mathbf{K}^{\lambda\text{-sat}}$ of λ-Galois-saturated models of $\mathbf{K}$.

Definition 5.1. An *independence relation* is a pair $i = (\mathbf{K}, \underset{\smile}{\smile})$, where:

1. $\mathbf{K}$ is an AEC[40] with amalgamation (we say that i is *on* $\mathbf{K}$ and write $\mathbf{K}_i = \mathbf{K}$).

2. $\underset{\smile}{\smile}$ is a relation on quadruples of the form (M, A, B, N), where $M \leq N$ are all in $\mathbf{K}$ and $A, B \subseteq |N|$. We write $\underset{\smile}{\smile}(M, A, B, N)$ or $A \underset{M}{\overset{N}{\smile}} B$ instead of $(M, A, B, N) \in \underset{\smile}{\smile}$.

3. The following properties hold:

   (a) <u>Invariance</u>: If $f : N \cong N'$ and $A \underset{M}{\overset{N}{\smile}} B$, then $f[A] \underset{f[M]}{\overset{N'}{\smile}} f[B]$.

   (b) <u>Monotonicity</u>: Assume $A \underset{M}{\overset{N}{\smile}} B$. Then:

---

[40] We may look at independence relations where $\mathbf{K}$ is not an AEC (e.g., it could be a class of Galois-saturated models in a strictly stable AEC).

i. Ambient monotonicity: If $N' \geq N$, then $A \underset{M}{\overset{N'}{\downarrow}} B$. If $M \leq N_0 \leq N$

and $A \cup B \subseteq |N_0|$, then $A \underset{M}{\overset{N_0}{\downarrow}} B$.

ii. Left and right monotonicity: If $A_0 \subseteq A$, $B_0 \subseteq B$, then $A_0 \underset{M}{\overset{N}{\downarrow}} B_0$.

iii. Base monotonicity: If $A \underset{M}{\overset{N}{\downarrow}} B$ and $M \leq M' \leq N$, $|M'| \subseteq B \cup |M|$,

then $A \underset{M'}{\overset{N}{\downarrow}} B$.

(c) <u>Left and right normality</u>: If $A \underset{M}{\overset{N}{\downarrow}} B$, then $AM \underset{M}{\overset{N}{\downarrow}} BM$.

When there is only one relation to consider, we sometimes write "$\downarrow$ is an independence relation on **K**" to mean "$(\mathbf{K}, \downarrow)$ is an independence relation".

Definition 5.2. Let $i = (\mathbf{K}, \downarrow)$ be an independence relation. Let $M \leq N$, $B \subseteq |N|$, and $p \in gS^{<\infty}(B; N)$ be given. We say that p *does not i-fork over* M if whenever $p = \text{gtp}(\bar{a}/B; N)$, we have that $\text{ran}(\bar{a}) \underset{M}{\overset{N}{\downarrow}} B$. When i is clear from context, we omit it.

Remark 5.3. By the ambient monotonicity and invariance properties, this is well defined (i.e., the choice of $\bar{a}$ and N does not matter).

An independence relation can satisfy several natural properties:

Definition 5.4 (Properties of independence relations). Let $i = (\mathbf{K}, \downarrow)$ be an independence relation.

1. i has *disjointness* if $A \underset{M}{\overset{N}{\downarrow}} B$ implies $A \cap B \subseteq |M|$.

2. i has *symmetry* if $A \underset{M}{\overset{N}{\downarrow}} B$ implies $B \underset{M}{\overset{N}{\downarrow}} A$.

3. i has *existence* if $A \underset{M}{\overset{N}{\downarrow}} M$ for any $A \subseteq |N|$.

4. i has *uniqueness* if whenever $M_0 \leq M \leq N_\ell$, $\ell = 1, 2$, $|M_0| \subseteq B \subseteq |M|$, $q_\ell \in gS^{<\infty}(B; N_\ell)$, $q_1 \upharpoonright M_0 = q_2 \upharpoonright M_0$, and q_ℓ does not fork over M_0, then $q_1 = q_2$.

5. i has *extension* if whenever $p \in gS^{<\infty}(MB; N)$ does not fork over M and $B \subseteq C \subseteq |N|$, there exists $N' \geq N$ and $q \in gS^{<\infty}(MC; N')$ extending p such that q does not fork over M.

6. i has *transitivity* if whenever $M_0 \leq M_1 \leq N$, $A \underset{M_0}{\overset{N}{\downarrow}} M_1$ and $A \underset{M_1}{\overset{N}{\downarrow}} B$ imply

$A \underset{M_0}{\overset{N}{\downarrow}} B$.

7. i has the $< \kappa$-*witness property* if whenever $M \leq N$, $A, B \subseteq |N|$, and $A_0 \underset{M}{\overset{N}{\downarrow}} B_0$ for all $A_0 \subseteq A$, $B_0 \subseteq B$ of size strictly less than κ, then $A \underset{M}{\overset{N}{\downarrow}} B$. The λ-*witness property* is the $(< \lambda^+)$-witness property.

The following cardinals are also important objects of study:

Definition 5.5 (Locality cardinals). Let $i = (\mathbf{K}, \downarrow)$ be an independence relation and let α be a cardinal.

1. Let $\bar{\kappa}_\alpha(i)$ be the minimal cardinal $\mu \geq \alpha^+ + \mathrm{LS}(\mathbf{K})^+$ such that for any $M \leq N$ in $\mathbf{K}$, any $A \subseteq |N|$ with $|A| \leq \alpha$, there exists $M_0 \leq M$ in $\mathbf{K}_{<\mu}$ with $A \underset{M_0}{\overset{N}{\downarrow}} M$. When μ does not exist, we set $\bar{\kappa}_\alpha(i) = \infty$.

2. Let $\kappa_\alpha(i)$ be the minimal cardinal $\mu \geq \alpha^+ + \aleph_0$ such that for any regular $\delta \geq \mu$, any increasing continuous chain $\langle M_i : i \leq \delta \rangle$ in $\mathbf{K}$, any $N \geq M_\delta$, and any $A \subseteq |N|$ of size at most α, there exists $i < \delta$ such that $A \underset{M_i}{\overset{N}{\downarrow}} M_\delta$. When μ does not exist, we set $\kappa_{\alpha_0}(i) = \infty$.

We also let $\bar{\kappa}_{<\alpha}(i) := \sup_{\alpha_0 < \alpha} \bar{\kappa}_{\alpha_0}(i)$. Similarly define $\kappa_{<\alpha}(i)$. When clear, we may write $\kappa_\alpha(\downarrow)$, etc., instead of $\kappa_\alpha(i)$.

Definition 5.6. Let us say that an independence relation i has *local character* if $\bar{\kappa}_\alpha(i) < \infty$ for all cardinals α.

Compared to the elementary framework, we differentiate between two local character cardinals, κ and $\bar{\kappa}$. The reason is that we do not in general (but see Theorem 5.41) know how to make sense of when a type does not fork over an arbitrary set (as opposed to a model). Thus, we cannot (for example) define superstability by requiring that every type does not fork over a finite set: looking at unions of chains is a replacement.

We make precise when an independence relation is "like forking in a first-order stable theory":

Definition 5.7. We say that i is a *stable independence relation* if it is an independence relation satisfying uniqueness, extension, and local character.

We could also define the meaning of a *superstable independence relation*, but here several nuances arise so to be consistent with previous terminology we will call it a *good* independence relation, see Definition 5.28.

As defined above, independence relations are *global objects*: they define an independence notion "p does not fork over M" for M of any size and p of any length. This is a strong requirement. In fact, the following refinement of Question 1.3 is still open:

Question 5.8. Let $\mathbf{K}$ be a fully tame and short AEC with amalgamation. Assume that $\mathbf{K}$ is categorical in a proper class of cardinals. Does there exist a λ and a stable independence relation i on $\mathbf{K}_{\geq \lambda}$?

It is known that one can construct such an i with local character and uniqueness, but proving that it satisfies extension seems hard in the absence of compactness. Note in passing that i as above must be unique:

Theorem 5.9 (Canonicity of stable independence). *If i and i' are stable independence relations on $\mathbf{K}$, then $i = i'$.*

As seen in Example 3.2.1.(4), we know that uniqueness and local character are enough to conclude some tameness and there are several relationships between the properties. We give one example:

Proposition 5.10. *Assume that $\underset{}{\downarrow}$ is a stable independence relation on $\mathbf{K}$.*

1. $\underset{}{\downarrow}$ *has symmetry, existence, and transitivity.*

2. *If $\mathbf{K}$ is fully $< \kappa$-tame and -type short, then $\underset{}{\downarrow}$ has the $< \kappa$-witness property.*

3. *For every α, $\kappa_\alpha(\underset{}{\downarrow}) \leq \bar{\kappa}_\alpha(\underset{}{\downarrow})$.*

4. *$\underset{}{\downarrow}$ has disjointness over sufficiently Galois-saturated models: if M is $LS(\mathbf{K})^+$-Galois-saturated and $A \underset{M}{\overset{N}{\downarrow}} B$, then $A \cap B \subseteq |M|$.*

Proof sketch for (2). By symmetry and extension, it is enough to show that for a given A, $A_0 \underset{M_0}{\overset{N}{\downarrow}} M$ for all $A_0 \subseteq A$ of size less than κ implies $A \underset{M_0}{\overset{N}{\downarrow}} M$. By extension, pick $N' \geq N$ and $A' \subseteq |N'|$ so that $A' \underset{M_0}{\overset{N'}{\downarrow}} M$ and $\mathrm{gtp}(\bar{a}'/M_0; N') = \mathrm{gtp}(\bar{a}/M_0; N)$ (where $\bar{a}$, $\bar{a}'$ are enumerations of A and A', respectively). By the uniqueness property, $\mathrm{gtp}(\bar{a}' \restriction I/M; N') = \mathrm{gtp}(\bar{a} \restriction I/M_0; N)$ for all $I \subseteq \mathrm{dom}(\bar{a})$ of size less than κ. Now use by shortness this implies $\mathrm{gtp}(\bar{a}/M; N) = \mathrm{gtp}(\bar{a}'/M; N')$, hence by invariance $A \underset{M_0}{\overset{N}{\downarrow}} M$. $\square$

In what follows, we consider several approximations to Question 5.8 in the stable and superstable contexts. We also examine consequences on categorical AECs. It will be convenient to *localize* Definition 5.1 so that:

1. The relation $\underset{}{\downarrow}$ is only defined on types of certain lengths (that is, the size of the left-hand side is restricted).

2. The relation $\underset{}{\downarrow}$ is only defined on types over domains of certain sizes (that is, the size of the right-hand side and base is restricted).

More precisely:

Notation 5.11. Let $\mathcal{F} = [\lambda, \theta)$ be an interval of cardinals. We say that $i = (\mathbf{K}, \underset{}{\downarrow})$ is a $(< \alpha, \mathcal{F})$-*independence relation* if it satisfies Definition 5.1 localized to types of length less than α and models in $\mathbf{K}_{\mathcal{F}}$ (so only amalgamation in $\mathcal{F}$ is required). We always require that $\theta \geq \alpha$ and $\lambda \geq LS(\mathbf{K})$.

($\leq \alpha, \mathcal{F}$) means ($< \alpha^+, \mathcal{F}$), and if $\mathcal{F} = [\lambda, \lambda^+)$, then we say that i is a ($\leq \alpha, \lambda$)-independence relation. Similar variations are defined as expected, e.g., ($\leq \alpha, \geq \lambda$) means ($\leq \alpha, [\lambda, \infty)$).

We often say that i is a ($< \alpha$)-ary independence relation on $\mathbf{K}_{\mathcal{F}}$ rather than a ($< \alpha, \mathcal{F}$)-independence relation. We write α-ary rather than ($\leq \alpha$)-ary.

The properties in Definition 5.4 can be adapted to such localized independence relations. For example, we say that i has *symmetry* if $A \underset{M}{\overset{N}{\downarrow}} B$ implies that for every $B_0 \subseteq B$ of size less than α, $B_0 \underset{M}{\overset{N}{\downarrow}} A$.

Using this terminology, we can give the definition of a *good λ-frame* (see Section 2.5), and more generally of a good $\mathcal{F}$-frame for $\mathcal{F}$ an interval of cardinals[41]:

Definition 5.12. Let $\mathcal{F} = [\lambda, \theta)$ be an interval of cardinals. A *good $\mathcal{F}$-frame* is a 1-ary independence relation i on $\mathbf{K}_{\mathcal{F}}$ such that:

1. i satisfies disjointness, symmetry, existence, uniqueness, extension, transitivity, and $\kappa_1(\mathfrak{i}) = \aleph_0$.

2. $\mathbf{K}_{\mathcal{F}}$ has amalgamation in $\mathcal{F}$, joint embedding in $\mathcal{F}$, no maximal models in $\mathcal{F}$, and is Galois-stable in every $\mu \in \mathcal{F}$. Also of course $\mathbf{K}_{\mathcal{F}} \neq \emptyset$.

When $\mathcal{F} = [\lambda, \lambda^+)$, we talk of a good λ-frame, and when $\mathcal{F} = [\lambda, \infty)$, we talk of a good ($\geq \lambda$)-frame. As is customary, we may use the letter $\mathfrak{s}$ rather than i to denote a good frame.

5.2 Stability

We compare results for stability in tame classes with those in general classes, summarized in Section 2.2. At a basic level, tameness strongly connects types over domains of different cardinalities. While a general AEC might be Galois-stable in λ but not in λ^+ (see the Hart-Shelah example in Section 3.2.2), this cannot happen in tame classes:

Theorem 5.13. *Suppose that* $\mathbf{K}$ *is an AEC with amalgamation which is λ-tame[42] and Galois-stable in λ. Then:*

1. $\mathbf{K}$ *is Galois-stable in λ^+.*

2. $\mathbf{K}$ *is Galois-stable in every $\mu > \lambda$ such that $\mu = \mu^\lambda$.*

There is also a partial stability spectrum theorem for tame AECs:

Theorem 5.14. *Let* $\mathbf{K}$ *be an AEC with amalgamation that is* $\mathrm{LS}(\mathbf{K})$*-tame. The following are equivalent:*

---

[41]Note that the definition here is different (but equivalent to) Shelah's notion of a *type-full* good λ-frame, see the historical remarks for more.

[42]For the first part, weak tameness suffices.

1. $\mathbf{K}$ *is Galois-stable in some cardinal* $\lambda \geq \mathrm{LS}(\mathbf{K})$.
2. $\mathbf{K}$ *does not have the order property (see Definition 2.26).*
3. *There are* $\mu \leq \lambda_0 < \beth_1$ *such that* $\mathbf{K}$ *is Galois-stable in any* $\lambda = \lambda^{<\mu} + \lambda_0$.

The proof makes heavy use of the Galois Morleyization (Theorem 3.5) to connect "stability theory inside a model" (results about formal, syntactic types within a particular model) to Galois types in an AEC. This allows the translation of classical proofs connecting the order property and stability.

This achieves two important generalizations from the elementary framework. First, it unites the characterizations of stability in terms of counting types and no order property from first-order, a connection still lacking in general AECs. Second, it gives one direction of the stability spectrum theorem by showing that, given stability in any one place, there are many stability cardinals, and some of the stability cardinals are given by satisfying some cardinal arithmetic above the first stability cardinal. Still lacking from this is a converse saying that the stability cardinals are exactly characterized by some cardinal arithmetic.

Another important application of the Galois Morleyization in stable tame AECs is that *averages* of suitable sequences can be analyzed. Roughly speaking, we can work inside the Galois Morleyization of a monster model and define the χ-*average over* A *of a sequence* $\mathbf{I}$ to be the set of formulas ϕ over A so that strictly less than χ-many elements of $\mathbf{I}$ satisfy ϕ. If χ is big enough and under reasonable conditions on $\mathbf{I}$ (i.e., it is a Morley sequence with respect to nonsplitting), we can show that the average is complete and (if $\mathbf{I}$ is long enough), realized by an element of $\mathbf{I}$. Unfortunately, a detailed study is beyond the scope of this chapter; see the historical remarks for references.

Turning to independence relations in stable AECs, there are two main candidates. The first is the familiar notion of splitting (see Definition 2.22). Tameness simplifies the discussion of splitting by getting rid of the cardinal parameter: it is impossible for a type to λ^+-split over M and also not λ-split over M in a (λ, λ^+)-tame AEC, as the witness to λ^+-splitting could be brought down to size λ. This observation allows for a stronger uniqueness result in nonsplitting. Rather than just having unique extensions in the same cardinality as in Theorem 2.24, we get a cardinal-free uniqueness result.

Theorem 5.15. *Suppose* $\mathbf{K}$ *is a* $\mathrm{LS}(\mathbf{K})$-*tame AEC with amalgamation and that* $M_0 \leq M_1 \leq M_2$ *are in* $\mathbf{K}_{\geq \mathrm{LS}(\mathbf{K})}$ *with* M_1 *universal over* M_0. *If* $p, q \in \mathrm{gS}(M_2)$ *do not split over* M_0 *and* $p \upharpoonright M_1 = q \upharpoonright M_1$, *then* $p = q$.

Proof sketch. If $p \neq q$, then there is a small $M^- \leq M_2$ with $p \upharpoonright M^- \neq q \upharpoonright M^-$; Without loss of generality pick M^- to contain M_0. By universality, we can find $f : M^- \xrightarrow[M_0]{} M_1$. By the nonsplitting,

$$p \upharpoonright f(M^-) = f(p \upharpoonright M^-) \neq f(q \upharpoonright M^-) = q \upharpoonright f(M^-)$$

Since $f(M^-) \leq M_1$, this contradicts the assumption they have equal restrictions. $\square$

Attempting to use splitting as an independence relation for **K** runs into the issue that several theorems require that the extension be *universal* (such as the above theorem). This can be mitigated by moving to the class of saturated enough models and looking at a localized version of splitting.

Definition 5.16.

1. Let **K** be an AEC with amalgamation. For $\mu > \mathrm{LS}(\mathbf{K})$, let $\mathbb{K}^{\mu\text{-sat}}$ denote the class of μ-Galois-saturated models in $\mathbf{K}_{\geq \mu}$.

2. Let **K** be an AEC with amalgamation and let $\mu \geq \mathrm{LS}(\mathbf{K})$ be such that **K** is Galois-stable in μ. For $M_0 \leq M$ both in $\mathbf{K}^{\mu^+\text{-sat}}$ and $p \in \mathrm{gS}(M)$, we say that *p does not μ-fork over M_0* if there exists $M_0' \leq M_0$ with $M_0' \in \mathbf{K}_\mu$ such that p does not μ-split over M_0' (see Definition 2.22).

Note that, by the μ^+-saturation of M_0, we have guaranteed that M_0 is a universal extension of M_0'. This gives us the following result.

Theorem 5.17. *Let* **K** *be an AEC with amalgamation and let* $\mu \geq \mathrm{LS}(\mathbf{K})$ *be such that* **K** *is Galois-stable in* μ *and* **K** *is μ-tame. Let* $\overset{\mu}{\underset{}{\downharpoonleft}}$ *be the μ-nonforking relation restricted to the class* $\mathbf{K}^{\mu^+\text{-sat}}$. *Then*

1. $\overset{\mu}{\underset{}{\downharpoonleft}}$ *is a 1-ary independence relation that further satisfies disjointness, existence, uniqueness, and transitivity when all models are restricted to* $\mathbf{K}^{\mu^+\text{-sat}}$ *(in the precise language of Section 5.1, this says that* $(\mathbf{K}^{\mu^+\text{-sat}}, \overset{\mu}{\underset{}{\downharpoonleft}})$ *is an independence relation with these properties).*

2. $\overset{\mu}{\underset{}{\downharpoonleft}}$ *has set local character in* $\mathbf{K}^{\mu^+\text{-sat}}$: *Given* $p \in \mathrm{gS}(M)$, *there is* $M_0 \in \mathbf{K}^{\mu^+\text{-sat}}$ *such that* $M_0 \leq M$ *and* p *does not μ-fork over* M_0.

3. $\overset{\mu}{\underset{}{\downharpoonleft}}$ *has a local extension property: If* $M_0 \leq M$ *are both Galois-saturated and* $\|M_0\| = \|M\| \geq \mu^+$ *and* $p \in \mathrm{gS}(M_0)$, *then there exists* $q \in \mathrm{gS}(M)$ *extending* p *and not μ-forking over* M_0.

Proof sketch. Tameness ensures that μ-splitting and λ-splitting coincide when $\lambda \geq \mu$. The local extension property uses the extension property of splitting (see Theorem 2.24). Local character and uniqueness are also translations of the corresponding properties of splitting. Disjointness is a consequence of the moreover part in the extension property of splitting. Finally, transitivity is obtained by combining the extension and uniqueness properties of splitting. □

The second candidate for an independence relation, drawing from stable first-order theories, is a notion of coheir, which we call $< \kappa$-satisfiability.

Definition 5.18. Let $M \leq N$ and $p \in \mathrm{gS}^{<\infty}(N)$.

1. We say that p is a $< \kappa$-*satisfiable over* M if for every $I \subseteq \ell(p)$ and $A \subseteq |N|$ both of size strictly less than κ, we have that $p^I \restriction A$ is realized in M.

2. We say that p is a $< \kappa$-*heir over* M if for every $I \subseteq \ell(p)$ and every $A_0 \subseteq |M|$, $N_0 \leq N$, with $A_0 \subseteq |N_0|$ and I, A_0, N_0 all of size less than κ, there is some $f : N_0 \xrightarrow[A_0]{} M$ such that

$$f(p^I \restriction N_0) = p^I \restriction f[N_0]$$

$< \kappa$-satisfiable is also called κ-*coheir*. As expected from first-order, these notions are dual[43] and they are equivalent under the κ-order property of length κ.

$< \kappa$-satisfiability turns out to be an independence relation in the stable context.

Theorem 5.19. *Let* **K** *be an AEC and* $\kappa > \mathrm{LS}(\mathbf{K})$. *Assume:*

1. **K** *has a monster model and is fully* $< \kappa$-*tame and -type short.*

2. **K** *does not have the* κ-*order property of length* κ.

Let $\underset{\smile}{\perp}$ *be the independence relation induced by* $< \kappa$-*satisfiability on the* κ-*Galois-saturated models of* **K**. *Then* $\underset{\smile}{\perp}$ *has disjointness, symmetry, local character, transitivity, and the* κ-*witness property. Thus, if* $\underset{\smile}{\perp}$ *also has extension, then it is a stable independence relation on the* κ-*Galois-saturated models of* **K**.

If $\kappa = \beth_\kappa$, then it turns out that not having the κ-order property of length κ is equivalent to not having the order property, which by Theorem 5.14 is equivalent to stability.

Note that the conclusion gives already that the AEC is stable. Similarly, the $< \kappa$-satisfiability relation analyzes a type by breaking it up into its κ-sized components, so the tameness and type shortness assumptions seem natural[44].

Theorem 5.19 does not tell us if $< \kappa$-satisfiability has the extension property. At first glance, it seems to be a compactness result about Galois types. In fact:

Theorem 5.20. *Under the hypotheses of Theorem 5.19, if* κ *is a strongly compact cardinal, then* $< \kappa$-*satisfiability has the extension property.*

Extension also holds in some nonelementary classes (such as averageable classes) and we will see that it "almost" follows from superstability (see Section 5.4).

The existence of a reasonable independence notion for stable classes can be combined with averages to obtain a result on chains of Galois-saturated models:

Theorem 5.21. *Let* **K** *be a* $\mathrm{LS}(\mathbf{K})$-*tame AEC with amalgamation. If* **K** *is Galois-stable in some* $\mu \geq \mathrm{LS}(\mathbf{K})$, *then there exists* $\chi < \beth_1$ *satisfying the following property:*

---

[43] A must be a model for the question to make sense.

[44] Although it is open if they are necessary.

If $\lambda \geq \chi$ is such that $\mathbf{K}$ is Galois-stable in μ for unboundedly many $\mu <$ λ, then whenever $\langle M_i : i < \delta \rangle$ is a chain of λ-Galois-saturated models and $\operatorname{cf}(\delta) \geq \chi$, we have that $\bigcup_{i<\delta} M_i$ is λ-Galois-saturated.

Proof sketch. First note that Theorem 5.13.(2) and tameness imply that $\mathbf{K}$ is Galois-stable in stationary many cardinals. Then, develop enough of the theory of averages (and also investigate their relationship with forking) to be able to imitate Harnik's first-order proof [Har75]. □

We will see that this can be vastly improved in the superstable case: the hypothesis that $\mathbf{K}$ be Galois-stable in μ for unboundedly many $\mu < \lambda$ can be removed and the Hanf number improved. Moreover, there is a proof of a version of the above theorem using only independence calculus and not relying on averages. Nevertheless, the use of averages has several other applications (for example getting solvability from superstability, see Theorem 5.23).

5.3 Superstability

As noted at the beginning of Section 2.1, Shelah has famously stated that superstability in AECs suffers from "schizophrenia". However, superstability is much better behaved in tame AECs than in general. Recall Definition 2.27 which gave a definition of superstability in a single cardinal using local character of splitting. Recall also that there are several other local candidates such as the uniqueness of limit models (Definition 2.29) and the existence of a good frame (Section 2.5 and Definition 5.12). Theorem 5.21 suggests another definition saying that the union of a chain of μ-Galois-saturated models is μ-Galois-saturated. As noted before, it is unclear whether these definitions are equivalent cardinal by cardinal, that is, μ-superstability and λ-superstability for $\mu \neq \lambda$ are potentially different notions and it is not easy to combine them. With tameness, this difficulty disappears:

Theorem 5.22. *Assume that $\mathbf{K}$ is μ-superstable, μ-tame, and has amalgamation. Then for every $\lambda > \mu$:*

1. *$\mathbf{K}$ is λ-superstable.*

2. *If $\langle M_i : i < \delta \rangle$ is an increasing chain of λ-Galois-saturated models, then $\bigcup_{i<\delta} M_i$ is λ-Galois-saturated.*

3. *There is a good λ-frame with underlying class $\mathbf{K}^{\lambda\text{-sat}}$.*

4. *$\mathbf{K}$ has uniqueness of limit models in λ. In fact, $\mathbf{K}$ also has uniqueness of limit models in μ.*

Proof sketch. Fix $\lambda > \mu$. We can first prove an approximation to (3) by defining forking as in Definition 5.16 and following the proof of Theorem 5.17. We obtain an independence relation i on 1-types whose underlying class is $\mathbf{K}^{\lambda\text{-sat}}$ (at that point, we do not yet know yet if it is an AEC), and which satisfies all the properties from the definition of a good frame[45] (including the structural properties on **K**) except perhaps symmetry.

Still we can use this to prove that **K** satisfies (4) in Definition 2.27. Using just this together with uniqueness, we can show that **K** is Galois-stable in λ. Joint embedding follows from amalgamation and no maximal models holds by a variation on a part of the proof of Theorem 4.7. Therefore (1) holds: **K** is λ-superstable. We can prove the symmetry property of the good λ-frame by proving that a failure of it implies the order property. This also give the symmetry property for splitting, and hence by Theorem 2.31 the condition (4), uniqueness of limit models in λ, holds. Uniqueness of limit models can in turn be used to obtain (2), hence the underlying class of i is really an AEC so (3) holds.   □

Strikingly, a converse to Theorem 5.22 holds. That is, several definitions of superstability are eventually equivalent in the tame framework:

Theorem 5.23. *Let* **K** *be a tame AEC with a monster model and assume that* **K** *is Galois-stable in unboundedly many cardinals. The following are equivalent:*

1. *For all high enough* λ, *the union of a chain of* λ-*Galois-saturated models is* λ-*Galois-saturated.*

2. *For all high enough* λ, **K** *has uniqueness of limit models in* λ.

3. *For all high enough* λ, **K** *has a superlimit model of size* λ.

4. *There is* θ *such that, for all high enough* λ, **K** *is* (λ, θ)-*solvable.*

5. *For all high enough* λ, **K** *is* λ-*superstable.*

6. *For all high enough* λ, *there is* $\kappa = \kappa_\lambda \leq \lambda$ *such that there is a good* λ-*frame on* $\mathbf{K}_\lambda^{\kappa\text{-}sat}$.

Any of these equivalent statements also implies that **K** *is Galois-stable in all high enough* λ.

Note that the "high enough" threshold can potentially vary from item to item. Also, note that the stability assumption in the hypothesis is not too important: in several cases, it follows from the assumption and, in others (such as the uniqueness of limit models), it is included to ensure that the condition is not vacuous. Finally, if **K** is LS(**K**)-tame, we can add in each of that $\lambda < H_1$ in each of the conditions (except in (4) where we can say that $\theta < H_1$).

Superlimit models and solvability both capture the notion of the AEC **K** having a "categorical core", a sub-AEC $\mathbf{K}_0$ that is categorical in some κ. In the case of superlimits, $M \in \mathbf{K}_\kappa$ is superlimit if and only if M is universal[46] and the class of models isomorphic to M generates a nontrivial AEC. That is, the class:

$$\{N \in \mathbf{K}_{\geq \kappa} \mid \forall N_0 \in P_{\kappa^+}^*(N) \exists N_1 \in P_{\kappa^+}^* N : N_0 \leq N_1 \wedge N_1 \cong M\}$$

---

[45] Note that tameness was crucial to obtain the uniqueness property.
[46] That is, every model of size κ embeds into M.

is an AEC with a model of size κ^{+}[47]. (λ, κ)-solvability further assumes that this superlimit is isomorphic to $\mathrm{EM}_{L(\mathbf{K})}(I, \Phi)$ for some proper Φ of size κ and *any* linear order I of size λ.

Note that although we did not mention them in Section 2, superlimits and especially solvable AECs play a large role in the study of superstability in general AECs (see the historical remarks).

The proof that superstability implies solvability relies on a characterization of Galois-saturated models using averages (essentially, a model M is Galois-saturated if and only if for every type $p \in \mathrm{gS}(M)$, there is a long enough Morley sequence $\mathbf{I}$ inside M whose average is p). We give the idea of the proof that a union of Galois-saturated models being Galois-saturated implies superstability. This can also be used to derive superstability from categoricity in the tame framework (without using the much harder proof of the Shelah-Villaveces Theorem 2.36).

Lemma 5.24. *Let* $\mathbf{K}$ *be an AEC with a monster model. Assume that* $\mathbf{K}$ *is* $\mathrm{LS}(\mathbf{K})$-*tame and let* $\kappa = \beth_{\kappa} > \mathrm{LS}(\mathbf{K})$ *be such that* $\mathbf{K}$ *is Galois-stable in* κ. *Assume that for all* $\lambda \geq \kappa$ *and all limit* δ, *if* $\langle M_i : i < \delta \rangle$ *is an increasing chain of* λ-*Galois-saturated models, then* $\bigcup_{i < \delta} M_i$ *is* λ-*Galois-saturated.*
Then $\mathbf{K}$ *is* κ^{+}-*superstable.*

Proof sketch. By tameness and Theorem 5.13.(1), we have that $\mathbf{K}$ is Galois-stable in κ^{+}. Thus, we only have to show Definition 2.27.(4), that there are no long splitting chains. There is a Galois-saturated model in κ^{+} and, by a back and forth argument, it is enough to show Definition 2.27.(4) when all the models are κ^{+}-Galois-saturated.

Let $\delta < \kappa^{++}$ be a limit ordinal and let $\langle M_i : i \leq \delta \rangle$ be an increasing continuous chain of Galois-saturated models in $\mathbf{K}_{\kappa^{+}}$; that we can make the models at limit stages Galois-saturated crucially uses the assumption. Let $p \in \mathrm{gS}(M_\delta)$. We need to show that there is $i < \delta$ such that p does not κ^{+}-split over M_i. By standard means, one can show that there is an $i < \delta$ such that p is $< \kappa^{+}$ satisfiable in M_i. Tameness gives the uniqueness of $< \kappa$-satisfiability, which allows us to conclude that p is $< \kappa$-satisfiable in M_i, which in turn implies that p does not κ^{+}-split over M_i, as desired. $\square$

Remark 5.25. From the argument, we obtain the following intriguing consequence in first-order model theory[48]: if T is a stable first-order theory, $\langle M_i : i \leq \delta \rangle$ is an increasing continuous chain of $\aleph_1$-saturated models (so M_i is $\aleph_1$-saturated also for limit i), then for any $p \in S(M_\delta)$, there exists $i < \delta$ so that p does not fork over M_i. This begs the question of whether any such chain exists in strictly stable theories.

---

[47] An equivalent definition: $M \in \mathbf{K}_\kappa$ is superlimit if and only if it is universal, has a proper extension isomorphic to it, and for any limit $\delta < \kappa^{+}$, and any increasing continuous chain $\langle M_i : i \leq \delta \rangle$, if $M_i \cong M$ for all $i < \delta$, then $M_\delta \cong M$.

[48] Hence showing that perhaps the study of AEC can also lead to new theorems in first-order model theory.

We now go back to the study of good frames. One can ask when instead of a good λ-frame, we obtain a good $(\geq \lambda)$-frame (i.e., forking is defined for types over models of all sizes). It turns out that the proof of Theorem 5.22 gives a good $(\geq \mu^+)$-frame on $\mathbf{K}^{\mu^+\text{-sat}}$. This still has the disadvantage of looking at Galois-saturated models. The next result starts from a good μ-frame and shows that μ-tameness can transfer it up (note that this was already stated as Theorem 4.6):

Theorem 5.26. *Assume* $\mathbf{K}$ *is an AEC with* $\mathrm{LS}(\mathbf{K}) \leq \lambda$ *and* $\mathfrak{s}$ *is a good* λ-*frame on* $\mathbf{K}$. *If* $\mathbf{K}$ *has amalgamation, then* $\mathbf{K}$ *is* λ-*tame if and only if there is a good* $(\geq \lambda)$-*frame* $\geq \mathfrak{s}$ *on* $\mathbf{K}$ *that extends* $\mathfrak{s}$.

Proof sketch. That tameness is necessary is discussed in Example 3.2.1.(4).

For the other direction, it is easy to check that if there is any way to extend forking to models of size at least λ, the definition must be the following:

$p \in \mathrm{gS}(M)$ does not fork over M_0 if and only if there exists $M_0' \leq M_0$ with $M_0' \in \mathbf{K}_\lambda$ and $p \upharpoonright M'$ does not fork over M_0' for all $M' \leq M$ with $M' \in \mathbf{K}_\lambda$.

Several frame properties transfer without tameness; however, the key properties of uniqueness, extension, stability, and symmetry can fail. λ-tameness can be easily seen to be equivalent to the transfer of uniqueness from $\mathfrak{s}$ to $\geq \mathfrak{s}$. Using uniqueness, extension and stability can easily be shown to follow. Symmetry is harder and the proof goes through independent sequences (see below and the historical remarks).

As an example, we show how to prove the extension property. Note that one of the key difficulties in proving extension in general is that upper bounds of types need not exist; while this is trivial in first-order, such AECs are called compact (see Definition 3.2). To solve this problem, we use the forking machinery of the frame to build a chain of types with a canonical extension at each step. This canonicity provides the existence of types.

Let $M \in \mathbf{K}_{\geq\lambda}$ and let $p \in \mathrm{gS}(M)$. Let $N \geq M$. We want to find a nonforking extension of p to N. By local character and transitivity, without loss of generality $M \in \mathbf{K}_\lambda$. We now work by induction on $\mu := \|N\|$. If $\mu = \lambda$, we know that p can be extended to N by definition of a good frame, so assume $\mu > \lambda$. Write $N = \bigcup_{i<\mu} N_i$, where $N_i \in \mathbf{K}_{\lambda+|i|}$. By induction, let $p_i \in \mathrm{gS}(N_i)$ be the nonforking extension of p to N_i. Note that by uniqueness $p_j \upharpoonright N_i = p_i$ for $i \leq j < \mu$. We want to take the "direct limit" of the p_i's: build $\langle f_i : i < \mu \rangle$, $\langle N_i' : i < \mu \rangle$, $\langle a_i : i < \mu \rangle$ such that $p_i = \mathrm{gtp}(a_i/N_i; N_i')$, $f_i : N_i' \xrightarrow{N_i} N_{i+1}'$ such that $f_i(a_i) = a_{i+1}$. If this can be done, then taking the direct limit of the system induced by $\langle f_i, a_i N_i' : i < \mu \rangle$, we obtain a_μ, N_μ' such that $\mathrm{gtp}(a_\mu/N_\mu; N_\mu')$ is a nonforking extension of p. How can we build such a system? The base and successor cases are no problem, but at limits, we want to take the direct limit and prove that everything is still preserved. This will be the case because of the local character and uniqueness property. $\qquad\square$

This should be compared to Theorem 2.43 which achieves the more modest goal of transferring $\mathfrak{s}$ to λ^+ (over Galois-saturated models and with a different ordering) with assumptions on the number of models and some non-ZFC hypotheses.

An interesting argument in the proof of Theorem 5.26 is the transfer of the symmetry property. One could ignore that issue and use that failure of the order property implies symmetry; however, this would make the argument nonlocal in the sense that we require knowledge about the AEC near the Hanf of λ to conclude good property at λ. A more local (but harder) approach is to study *independent sequences*.

Given a good $(\geq \lambda)$-frame and $M_0 \leq M \leq N$, we want to say that a sequence $\langle a_i \in N : i < \alpha \rangle$ is independent in (M_0, M, N) if and only if $\mathrm{gtp}(a_i/|M| \cup \{a_j : j < i\}; N)$ does not fork over M_0. However, forking behaves better for types over models so instead, we require that there is a sequence of models $M \leq N_i \leq N$ growing with the sequence $\langle a_i : i < \alpha \rangle$ such that $a_i \in |N_{i+1}| \backslash |N_i|$ and require $\mathrm{gtp}(a_i/N_i; N)$ does not fork over M_0.

The study of independent sequences shows that under tameness they themselves form (in a certain technical sense) a good frame. That is, from an independence relation for types of length one, we obtain an independence relation for types *of independent sequences* of all lengths. One other ramification of the study of independence sequence is the isolation of a good notion of dimension: inside a fixed model, any two infinite maximal independent sets must have the same size.

Theorem 5.27. *Let $\mathbf{K}$ be an AEC, $\lambda \geq \mathrm{LS}(\mathbf{K})$. Assume that $\mathbf{K}$ is λ-tame and has amalgamation. Let $\mathfrak{s}$ be a good $(\geq \lambda)$-frame on $\mathbf{K}$. Let $M_0 \leq M \leq N$ all be in $\mathbf{K}_{\geq \lambda}$.*

1. *Symmetry of independence: For a fixed set I, I is independent in (M_0, M, N) if and only if all enumerations are independent in (M_0, M, N).*

2. *Let $p \in \mathrm{gS}(M)$. Assume that I_1 and I_2 are independent in (M_0, M, N) and every $a \in I_1 \cup I_2$ realizes p. If both I_1 and I_2 are $\subseteq$-maximal with respect to that property and I_1 is infinite, then $|I_1| = |I_2|$.*

5.4 Global independence and superstability

Combined with Theorem 5.22, Theorem 5.26 shows that every tame superstable AEC has a good $(\geq \lambda)$-frame. It is natural to ask whether this frame can also be extended in the other direction: to types of length larger than one. More precisely, we want to build a superstability-like global independence relation (i.e., the global version of a good frame):

Definition 5.28. We say an independence relation $\underset{\smile}{\perp}$ on $\mathbf{K}$ is *good* if:

1. $\mathbf{K}$ is an AEC with amalgamation, joint embedding, and arbitrarily large models.

2. **K** is Galois-stable in all $\mu \geq \mathrm{LS}(\mathbf{K})$.

3. $\underset{\smile}{\perp}$ has disjointness, symmetry, existence, uniqueness, extension, transitivity, and the $\mathrm{LS}(\mathbf{K})$-witness property.

4. For all cardinals $\alpha > 0$:

    (a) $\bar{\kappa}_\alpha(\underset{\smile}{\perp}) = (\alpha + \mathrm{LS}(\mathbf{K}))^+$.

    (b) $\kappa_\alpha(\underset{\smile}{\perp}) = \alpha^+ + \aleph_0$.

We say that an AEC **K** is *good* if there exists a good independence relation on **K**.

We would like to say that if **K** is a $\mathrm{LS}(\mathbf{K})$-superstable AEC with amalgamation that is fully tame and short, then there exists λ such that $\mathbf{K}_{\geq \lambda}$ is good. At present, we do not know if this is true (see Question 5.8). All we can conclude is a weakening of good:

Definition 5.29. We say an independence relation $\underset{\smile}{\perp}$ is *almost good* if it satisfies all the conditions of Definition 5.28 except it only has the following weakening of extension: If $p \in \mathrm{gS}^\alpha(M)$ and $N \geq M$, we can find $q \in \mathrm{gS}^\alpha(N)$ extending p and not forking over M provided that at least one of the following conditions hold:

1. M is Galois-saturated.

2. $M \in \mathbf{K}_{\mathrm{LS}(\mathbf{K})}$.

3. $\alpha < \mathrm{LS}(\mathbf{K})^+$.

An AEC **K** is *almost good* if there is an almost good independence relation on **K**.

Remark 5.30. Assume that i is an independence relation on **K** which satisfies all the conditions in the definition of good except extension, and it has extension for types over Galois-saturated models. Then we can restrict i to $\mathbf{K}^{\mathrm{LS}(\mathbf{K})^+-\mathrm{sat}}$ and obtain an almost good independence relation. Thus extension over Galois-saturated models is the important condition in Definition 5.29.

We can now state a result on existence of global independence relation:

Theorem 5.31. *Let* **K** *be a fully* $\mathrm{LS}(\mathbf{K})$-*tame and short AEC with amalgamation. Let* $\lambda := \left(2^{\mathrm{LS}(\mathbf{K})}\right)^{+4}$. *If* **K** *is* $\mathrm{LS}(\mathbf{K})$-*superstable, then* $\mathbf{K}^{\lambda-sat}$ *is almost good.*

We try to describe the proof. For simplicity, we will work with $< \kappa$-satisfiability, so will obtain a Hanf number approximately equal to a fixed point of the beth function. The better bound is obtained by looking at splitting but this makes the proof somewhat more complicated. So let $\kappa = \beth_\kappa > \mathrm{LS}(\mathbf{K})$. We know that the $< \kappa$-satisfiability independence relation is an independence relation on $\mathbf{K}^{\kappa-\mathrm{sat}}$ with uniqueness, local character, and symmetry (but not extension). Let i denote this relation independence relation. Furthermore, we can show that $\kappa_1(\mathfrak{i}) = \aleph_0$. In fact, $\mathfrak{i}$ restricted to types of length one induces a good κ-frame $\mathfrak{s}$ on $\mathbf{K}_\kappa^{\kappa\text{-sat}}$. We would like to extend $\mathfrak{s}$ to types of length at most κ.

To do this, we need to make use of the notion of domination and successful frames[49]:

Definition 5.32. Suppose $\underset{\smile}{\perp}$ is an independence relation on $\mathbf{K}$. Work inside a monster model[50].

1. For $M \leq N$ κ-Galois-saturated and $a \in |N|$, a *dominates* N *over* M if for any B, $a \underset{M}{\perp} B$ implies $N \underset{M}{\perp} B$.

2. $\mathfrak{s}$ is *successful* if for every Galois-saturated $M \in \mathbf{K}_\kappa$, every nonalgebraic type $p \in \mathrm{gS}(M)$, there exists $N \geq M$ and $a \in |N|$ with $N \in \mathbf{K}_\kappa$ Galois-saturated such that a dominates N over M.

3. $\mathfrak{s}$ is *ω-successful* if $\mathfrak{s}^{+n}$ is successful for all $n < \omega$. Here, $\mathfrak{s}^{+n}$ is the good κ^{+n} induced on the Galois-saturated models of size κ^{+n} by $< \kappa$-satisfiability.

An argument of Makkai and Shelah [MS90, Proposition 4.22] shows that $\mathfrak{s}$ is successful (in fact, ω-successful), and a deep result of Shelah shows that if $\mathfrak{s}$ is successful, then we can extend $\mathfrak{s}$ to a κ-ary independence relation $\mathfrak{i}'$ which has extension, uniqueness, symmetry, and for all $\alpha \leq \kappa$, $\kappa_\alpha(\mathfrak{i}') = \alpha^+ + \aleph_0$. This completes the first step of the proof. Note that we have taken $\mathfrak{i}$ (which was built on $< \kappa$-satisfiability), restricted it to 1-types and then "lengthened" it to κ-ary types. However, we do not necessarily get $< \kappa$-satisfiability back! We do get, however, an independence relation with a better local character property.

From ω-successfulness, we could extend the frame $\mathfrak{s}$ to models of size κ^{+n}. Now we would like to extend $\mathfrak{i}'$ to models of all sizes above κ. However, the continuity of $\mathfrak{i}'$ is not strong enough. The missing property is:

Definition 5.33. An independence relation $\mathfrak{i} = (\mathbf{K}, \underset{\smile}{\perp})$ has *full model continuity* if for any limit ordinal δ, for any increasing continuous chain $\langle M_i^\ell : i \leq \delta \rangle$ with $\ell < 4$, and $M_i^0 \leq M_i^k \leq M_i^3$ for $k = 1, 2$ and $i \leq \delta$, if $M_i^1 \underset{M_i^0}{\overset{M_i^3}{\perp}} M_i^2$ for all $i < \delta$, then $M_\delta^1 \underset{M_\delta^0}{\overset{M_\delta^3}{\perp}} M_\delta^2$.

Let us say that $\mathfrak{i}$ is *fully good* [*almost fully good*] if it is good [almost good] and has full model continuity. As before, $\mathbf{K}$ is *[almost] fully good* if there is an [almost] fully good independence relation on $\mathbf{K}$.

---

[49]Note that the definitions here do not coincide with Shelah's, although they are equivalent in our context. The equivalence uses tameness again, including a result of Adi Jarden. See the historical remarks for more.

[50]So if $\mathfrak{C}$ is the monster model, $a \underset{M}{\perp} B$ means $a \underset{M}{\overset{\mathfrak{C}}{\perp}} B$.

Another powerful result of Shelah [She09b, III.8.19] connects ω-successful good frames with full model continuity. Suppose that $\mathfrak{s}$ is an ω-successful good κ-frame (as we have). We do not know that $\mathfrak{i}'$ defined above has full model continuity, but if we move to the (still ω-successful) good κ^{+3}-frame $\mathfrak{s}^{+3}$ and "lengthen" this to an independence relation $\mathfrak{i}'_{+3}$ on κ^{+3}-ary types, then $\mathfrak{i}'_{+3}$ has full model continuity!

This allows us to transfer all of the nice properties of $\mathfrak{i}'_{+3}$ to a κ^{+3}-ary independence relation $\mathfrak{i}''$ on models of all sizes above κ^{+3}. To get a truly global independence relation, we can define an independence relation $\mathfrak{i}'''$ on types of *all* lengths by specifying that $p \in \mathrm{gS}^\alpha(M)$ do not $\mathfrak{i}'''$-fork over $M_0 \leq M$ if and only if $p \restriction I$ does not $\mathfrak{i}''$-fork over M_0 for every $I \subseteq \alpha$ with $|I| \leq \kappa^{+3}$. With some work, we can show that $\mathfrak{i}'''$ is almost fully good (thus "fully" can be added to the conclusion of Theorem 5.31).

What about getting the extension over property over all models (not just the Galois-saturated models). It is known how to do it by making one more locality hypothesis:

Definition 5.34 (Type locality).

1. Let δ be a limit ordinal, and let $\bar{p} := \langle p_i : i < \delta \rangle$ be an increasing chain of Galois types, where for $i < \delta$, $p_i \in \mathrm{gS}^{\alpha_i}(M)$ and $\langle \alpha_i : i \leq \delta \rangle$ are increasing. We say $\bar{p}$ is κ-*type-local* if $\mathrm{cf}(\delta) \geq \kappa$ and whenever $p, q \in \mathrm{gS}^{\alpha_\delta}(M)$ are such that $p^{\alpha_i} = q^{\alpha_i} = p_i$ for all $i < \delta$, then $p = q$.

2. We say $\mathbf{K}$ is κ-*type-local* if every $\bar{p}$ as above is κ-type-local.

We think of κ-type-locality as the dual to κ-locality (Definition 3.2.(3)) in the same sense that shortness is the dual to tameness.

Remark 5.35. If κ is a regular cardinal and $\mathbf{K}$ is $< \kappa$-type short, then $\mathbf{K}$ is κ-type-local. In particular, if $\mathbf{K}$ is fully $< \aleph_0$-tame and -type short, then $\mathbf{K}$ is $\aleph_0$-type-local.

Remark 5.36. If there is a good λ-frame on $\mathbf{K}$, then $\mathbf{K}_\lambda$ is $\aleph_0$-local (use local character and uniqueness), and thus assuming λ-tameness $\mathbf{K}$ is $\aleph_0$-local. This is used in the transfer of a good λ-frame to a good $(\geq \lambda)$-frame. Unfortunately, an analog for this fact is missing when looking at $\aleph_0$-type-locality, i.e., it is not clear that even a fully good AEC is $\aleph_0$-type-local.

Using type-locality, we can start from a fully good $\mathrm{LS}(\mathbf{K})$-ary independence relation on $\mathbf{K}$ and prove extension for types of all lengths. Thus we obtain the following variation of Theorem 5.31:

Theorem 5.37. *Let $\mathbf{K}$ be a fully $\mathrm{LS}(\mathbf{K})$-tame and short AEC with amalgamation. Assume that $\mathbf{K}$ is $\aleph_0$-type-local. Let $\lambda := \left(2^{\mathrm{LS}(\mathbf{K})}\right)^{+4}$. If $\mathbf{K}$ is $\mathrm{LS}(\mathbf{K})$-superstable, then $\mathbf{K}^{\lambda\text{-}sat}$ is fully good.*

Remark 5.38. It is enough to assume that $\aleph_0$-type-locality holds "densely" in a certain technical sense. See the historical remarks.

Finally, we know of at least two other ways to obtain extension: using total categoricity and large cardinals. We collect all the results of this section in a corollary:

Corollary 5.39. *Let* **K** *be an AEC. Assume that* **K** *is* LS(**K**)-*superstable and fully* LS(**K**)-*tame and short.*

1. *If* $\kappa > $ LS(**K**) *is a strongly compact cardinal, then* $\mathbf{K}^{\kappa\text{-}sat}$ *is fully good.*

2. *If either* **K** *is* $\aleph_0$-*type-local (e.g., it is fully* $(< \aleph_0)$-*tame and short) or* **K** *is totally categorical, then* $\mathbf{K}^{\lambda\text{-}sat}$ *is fully good, where* $\lambda := \left(2^{\text{LS}(\mathbf{K})}\right)^{+4}$.

Proof sketch. By Theorem 5.31, $\mathbf{K}^{\lambda\text{-}sat}$ is almost good, and in fact (as we have discussed), almost fully good. If **K** is totally categorical, all the models are Galois-saturated and hence by definition of almost fully good, **K** is fully good. If **K** is $\aleph_0$-type-local, then apply Theorem 5.37. Finally, if $\kappa > $ LS(**K**) is strongly compact, then the extension property for $< \kappa$-satisfiability holds (see Theorem 5.20) and using a canonicity result similar to Theorem 5.9 one can conclude that $\mathbf{K}^{\kappa\text{-}sat}$ is fully good. □

Since the existence of a strongly compact cardinal implies full tameness and shortness (see Theorem 3.7), we can state a version of the first part of Corollary 5.39 as follows:

Theorem 5.40. *If* **K** *is an AEC which is superstable in every* $\mu \geq $ LS(**K**) *and* $\kappa > $ LS(**K**) *is a strongly compact cardinal, then* $\mathbf{K}^{\lambda\text{-}sat}$ *is fully good, where* $\lambda := \left(2^{\text{LS}(\mathbf{K})}\right)^{+4}$.

Note that in all of the results above, we are restricting ourselves to classes of sufficiently saturated models. This is related to the fact that the uniqueness property is required in the definition of a good independence relation, i.e., all types must be stationary. But what if we relax this requirement? Can we obtain an independence relation that specifies what it means to fork over an arbitrary set? A counterexample of Shelah [HL02, Section 4] shows that this cannot be done in general. However, this is possible for universal classes:

Theorem 5.41. *If* **K** *is an almost fully good universal class, then:*

1. **K** *is fully good.*

2. *We can define* $A \underset{A_0}{\overset{N}{\downarrow}} B$ *(for* A_0 *an arbitrary set) to hold if and only if* $\text{cl}^N(A_0 A) \underset{\text{cl}^N(A_0)}{\overset{N}{\downarrow}} \text{cl}^N(A_0 B)$. *Here* cl^N *is the closure under the functions of* N. *This has the expected properties (extension, existence, local character).*

3. *This also has the finite witness property:* $A \underset{A_0}{\overset{N}{\downarrow}} B$ *if and only if* $A' \underset{A_0}{\overset{N}{\downarrow}} B'$ *for all* $A' \subseteq A$, $B' \subseteq B$ *finite.*

Remark 5.42. It is enough to assume that $\mathbf{K}$ admits intersections, i.e., for any $N \in \mathbf{K}$ and any $A \subseteq |M|$, $\bigcap \{M \leq N \mid A \subseteq |M|\} \leq N$.

5.5 Categoricity

One of the first marks made by tame AEC was the theorem by Grossberg and VanDieren [GV06a] that tame AECs (with amalgamation) satisfy an *upward* categoricity transfer from a successor (see Theorem 4.10). Combining it with Theorem 2.35, we obtain that tame AECs satisfy Shelah's eventual categoricity conjecture from a successor:

Theorem 5.43. *Let $\mathbf{K}$ be an H_2-tame AEC with amalgamation. If $\mathbf{K}$ is categorical in some successor $\lambda \geq H_2$, then $\mathbf{K}$ is categorical in all $\lambda' \geq H_2$.*

Recall that categoricity implies superstability below the categoricity cardinal (Theorem 2.36). A powerful result is that assuming tameness, superstability also holds above, while this need not be true without tameness; recall the discussion after Theorem 2.36. In particular, Question 2.38 has a positive answer: the model in the categoricity cardinal is Galois-saturated.

Theorem 5.44. *Let $\mathbf{K}$ be a $\mathrm{LS}(\mathbf{K})$-tame AEC with amalgamation and no maximal models. If $\mathbf{K}$ is categorical in some $\lambda > \mathrm{LS}(\mathbf{K})$, then:*

1. $\mathbf{K}$ *is superstable in every $\mu \geq \mathrm{LS}(\mathbf{K})$.*

2. *For every $\mu > \mathrm{LS}(\mathbf{K})$, there is a good μ-frame with underlying class $\mathbf{K}^{\mu\text{-}sat}$.*

3. *The model of size λ is Galois-saturated.*

Proof.

1. By Theorem 2.36, $\mathbf{K}$ is superstable in $\mathrm{LS}(\mathbf{K})$. Now apply Theorem 5.22.

2. As above, using Theorem 5.22.

3. $\mathbf{K}$ is λ-superstable, so in particular Galois-stable in λ. It is not hard to build a μ^+-Galois-saturated model in λ for every $\mu < \lambda$ so the result follows from categoricity.

$\square$

Theorem 5.44 allows one to show that a tame AEC categorical in some cardinal is categorical in a closed unbounded set of cardinals of a certain form. This already plays a key role in Shelah's proof of Theorem 2.35. The key is what we call Shelah's omitting type theorem, a refinement of Morley's omitting type theorem. Note that a version of this theorem is also true without tameness, but removing the tameness assumption changes the condition on p being omitted to requiring that the small approximations to p be omitted[51].

---

[51] In the sense that each element omits *some* small approximation of p.

Theorem 5.45 (Shelah's omitting type theorem). *Let* **K** *be a* LS(**K**)-*tame AEC with amalgamation. Let* $M_0 \leq M$ *and let* $p \in \mathrm{gS}(M_0)$. *Assume that* p *is omitted in* M. *If* $\|M_0\| \geq \mathrm{LS}(\mathbf{K})$ *and* $\|M\| \geq \beth_{\left(2^{\mathrm{LS}(\mathbf{K})}\right)^+}(\|M_0\|)$, *then there is a* non-LS(**K**)$^+$-*Galois-saturated model in every cardinal.*

Corollary 5.46. *Let* **K** *be a* LS(**K**)-*tame AEC with amalgamation and no maximal models. If* **K** *is categorical in some* $\lambda > \mathrm{LS}(\mathbf{K})$, *then* **K** *is categorical in all cardinals of the form* $\beth_\delta$, *where* $\left(2^{\mathrm{LS}(\mathbf{K})}\right)^+$ *divides* δ.

Proof. Let δ be divisible by $\left(2^{\mathrm{LS}(\mathbf{K})}\right)^+$. If there is a model $M \in \mathbf{K}_{\beth_\delta}$ which is not Galois-saturated, then by Shelah's omitting type theorem we can build a non-LS(**K**)$^+$-Galois-saturated model in λ. This contradicts Theorem 5.44. □

In Section 4, a categoricity transfer was proven *without* assuming that the categoricity cardinal is a successor. As hinted at there, this generalizes to tame AECs that *have primes* (recall from Definition 4.11 that an AEC has primes if there is a prime model over every set of the form $M \cup \{a\}$):

Theorem 5.47. *Let* **K** *be an AEC with amalgamation and no maximal models. Assume that* **K** *is* LS(**K**)-*tame and has primes. If* **K** *is categorical in some* $\lambda > \mathrm{LS}(\mathbf{K})$, *then* **K** *is categorical in all* $\lambda' \geq \min(\lambda, H_1)$.

Remark 5.48. A partial converse is true: if a fully tame and short AEC with amalgamation and no maximal models is categorical on a tail, then it has primes on a tail.

We deduce Shelah's categoricity conjecture in homogeneous model theory (see Section 3.2.1.(7)):

Corollary 5.49. *Let* D *be a homogeneous diagram in a first-order theory* T. *If* D *is categorical in a* $\lambda > |T|$, *then* D *is categorical in all* $\lambda' \geq \min(\lambda, h(|T|))$.

Using a similar argument, we can also get rid of the hypothesis that **K** has primes if the categoricity cardinal is a successor. This allows us to obtain a downward transfer for tame AECs which improves on Theorem 5.43 (there H_1 was H_2). The price to pay is to assume more tameness.

Theorem 5.50. *Let* **K** *be a* LS(**K**)-*tame AEC with amalgamation and no maximal models. If* **K** *is categorical in a successor* $\lambda > \mathrm{LS}(\mathbf{K})^+$, *then* **K** *is categorical in all* $\lambda' \geq \min(\lambda, H_1)$.

Proof sketch. Let us work in a good $(\geq \mathrm{LS}(\mathbf{K})^+)$-frame $\mathfrak{s}$ on $\mathbf{K}^{\mathrm{LS}(\mathbf{K})^+\text{-sat}}$ (this exists by Theorems 2.36, 5.22, and 5.26). As in Section 4.2, we can define what it means for two types p and q to be orthogonal (written $p \perp q$) and say that $\mathfrak{s}$ is *μ-unidimensional*[52] if no two types are orthogonal. We can show

---

[52]In this framework, this definition need not be equivalent to categoricity in the next successor but we use it for illustrative purpose.

that $\mathfrak{s}$ is μ-unidimensional if and only if $\mathbf{K}^{\mathrm{LS}(\mathbf{K})^+\text{-sat}}$ is categorical in μ^+, and argue by studying the relationship between forking and orthogonality that $\mathfrak{s}$ is unidimensional in some μ if and only if it is unidimensional in all μ' (this uses tameness, since we are moving across cardinals). Thus $\mathbf{K}^{\mathrm{LS}(\mathbf{K})^+\text{-sat}}$ is categorical in every successor cardinal, hence also (by a straightforward argument using Galois-saturated models) in every limit. We conclude by using a version of Morley's omitting type theorem to transfer categoricity in $\mathbf{K}^{\mathrm{LS}(\mathbf{K})^+\text{-sat}}$ to categoricity in $\mathbf{K}$ (this is where the H_1 comes from). $\square$

What if we do not want to assume that the AEC has primes or that it is categorical in a successor? Then the best known results are essentially Shelah's results from Section 2.5. We show how to obtain a particular case using the results presented in this section:

Theorem 5.51. *Assume Claim 2.44 and $2^\theta < 2^{\theta^+}$ for every cardinal θ. Let $\mathbf{K}$ be a fully $\mathrm{LS}(\mathbf{K})$-tame and short AEC with amalgamation and no maximal models. If $\mathbf{K}$ is categorical in some $\lambda \geq H_1$, then $\mathbf{K}$ is categorical in all $\lambda' \geq H_1$.*

Proof sketch. By Theorem 2.36, $\mathbf{K}$ is $\mathrm{LS}(\mathbf{K})$-superstable. By the proof of Theorem 5.31, we can find an ω-successful good μ-frame (see Definition 5.32) on $\mathbf{K}_\mu^{\mu\text{-sat}}$ for some $\lambda < H_1$. By Claim 2.44, $\mathbf{K}^{\mu^{+\omega}\text{-sat}}$ is categorical in every $\mu' > \mu^{+\omega}$. Using a version of Morley's omitting type theorem, we get that $\mathbf{K}$ must be categorical on a tail of cardinals, hence in a successor above H_1. By Theorem 5.50, $\mathbf{K}$ is categorical in all $\lambda' \geq H_1$. $\square$

Remark 5.52. Slightly different arguments show that the locality assumption can be replaced by only $\mathrm{LS}(\mathbf{K})$-tameness. Moreover, it can be shown that categoricity in some $\lambda > \mathrm{LS}(\mathbf{K})$ implies categoricity in all $\lambda' \geq \min(\lambda, H_1)$.

The proof shows in particular that almost fully good independence relations can be built in fully tame and short categorical AECs:

Theorem 5.53. *Let $\mathbf{K}$ be a fully $\mathrm{LS}(\mathbf{K})$-tame and -type short AEC with amalgamation and no maximal models. If $\mathbf{K}$ is categorical in a $\lambda \geq \left(2^{\mathrm{LS}(\mathbf{K})}\right)^{+4}$, then:*

1. *$\mathbf{K}_{\geq \min(\lambda, H_1)}$ is almost fully good.*

2. *If $\mathbf{K}$ is fully $< \aleph_0$-tame and -type short, then $\mathbf{K}_{\geq \min(\lambda, H_1)}$ is fully good.*

Proof. As in the proof above. Note that by Corollary 5.46, $\mathbf{K}$ is categorical in H_1. $\square$

Using large cardinals, we can remove all the hypotheses except categoricity:

Corollary 5.54. *Let $\mathbf{K}$ be an AEC. Let $\kappa > \mathrm{LS}(\mathbf{K})$ be a strongly compact cardinal. If $\mathbf{K}$ is categorical in a $\lambda \geq h(\kappa)$, then:*

1. *$\mathbf{K}_{\geq \lambda}$ is fully good.*

2. *If $2^\theta < 2^{\theta^+}$ for every cardinal θ and Claim 2.44 hold, then[53] $\mathbf{K}$ is categorical in all $\lambda' \geq h(\kappa)$.*

Proof sketch. By a result similar to Theorem 3.10, $\mathbf{K}_{\geq \kappa}$ has amalgamation and no maximal models. By Theorem 3.7, $\mathbf{K}$ is fully $< \kappa$-tame and -type short. Now Corollary 5.39 gives the first part. Theorem 5.51 gives the second part. $\qquad\square$

6 Conclusion

The classification theory of tame AECs has progressed rapidly over the last several years. The categoricity transfer results of Grossberg and VanDieren indicated that tameness (along with amalgamation, etc.) is a powerful tool to solve questions that currently seem out of reach for general AECs.

Looking to the future, there are several open questions and lines of research that we believe deserve to be further explored.

1. **Levels of tameness**

   This problem is less grandiose than other concerns, but still concerns a basic unanswered question about tameness: are there nontrivial relationships between the parametrized versions of tameness in Definition 3.2? For example, does κ-tameness for α-types imply κ-tameness for β-types when $\alpha < \beta$? This question reveals a divide in the tameness literature: some results only use tameness for 1-types (such as the categoricity transfer of Grossberg and VanDieren Theorem 4.10 and deriving a frame from superstability Theorem 5.22), while others require full tameness and shortness (such as the development of $< \kappa$-satisfiability Theorem 5.19). Answering this question would help to unify these results.

   Another stark divide is revealed by examining the list of examples of tame AECs in Section 3.2.1: the list begins with general results that give some form of locality at a cardinal λ. However, once the list reaches concrete classes of AECs, every example turns out to be $< \aleph_0$-tame (often this is a result of a syntactic characterization of Galois types, but not always). This suggests the question of what lies between or even if the general results can be strengthened down to $< \aleph_0$-tameness. For the large cardinal results, this seems impossible: the large cardinal κ should give no information about the low properties of $\mathbf{K}$ below it because this cardinal disappears in $\mathbf{K}_\kappa$. The other results also seem unlikely to have this strengthening, but no counter example is known. Indeed, the

---

[53]The same conclusion holds assuming only that κ is a measurable cardinal. Moreover, if $\mathbf{K}$ is axiomatized by an $\mathbb{L}_{\kappa,\omega}$ theory, we can replace $h(\kappa)$ with $h(\kappa + \mathrm{LS}(\mathbf{K}))$ and do not need to assume that $\kappa > \mathrm{LS}(\mathbf{K})$.

following is still open: Is there an AEC **K** that is $\aleph_0$-tame but not $< \aleph_0$-tame?

2. **Dividing lines**

This direction has two prongs. The first prong is to increase the number of dividing lines. So far, the classification of tame AECs (and AECs in general) has focused on the superstable or better case with a few forays into strictly stable [BG, BVa]. Towards the goal of proving Shelah's Categoricity Conjecture, this focus makes sense. However, this development pales in comparison to the rich structure of classification theory in first-order[54]. Exploring the correct generalizations of NIP, NTP_2, etc. may help fill out the AEC version of this map. It might be that stronger locality hypotheses than tameness will have to be used: as we have seen already in the superstable case, it is only known how to prove the existence of a global independence relation assuming full $(< \aleph_0)$-tameness and shortness.

The other prong is to turn classification results into true dividing lines. In the first-order case, dividing lines correspond to nice properties of forking on one side and to chaotic nonstructure results on the other. In AECs, the nonstructure side of dividing lines is often poorly developed and most results either revolve around the order property or use non-ZFC combinatorial principle. While these combinatorial principles seem potentially necessary in arbitrary AECs[55], a reasonable hope is that tame AECs will allow the development of stronger ZFC nonstructure principles. For example, Shelah claims that in AECs with amalgamation, the order property (or equivalently in the tame context stability, see Theorem 5.14) implies many models on a tail of cardinals. However, there is no known analog for superstability: does unsuperstability imply many models?

3. **Interaction with other fields**

Historically, examples have not played a large role in the study of AECs. Examples certainly exist because $\mathbb{L}_{\kappa,\omega}$ sentences provide them, but the investigation of specific classes is rarely carried out[56]. A better understanding of concrete examples would help advance the field in two ways. First, nontrivial applications would help provide more motivation for exploring AECs[57]. Second, interesting applications can help drive the isolation of new AEC properties that might, *a priori*, seem strange.

---

[54] Part of this structure is represented graphically at `http://forkinganddividing.com` by Gabe Conant.

[55] For instance, result the statement "Categoricity in λ and less than the maximum number of models in λ^+ implies λ-AP" holds under weak diamond, but fails under Martin's axiom [She09b, Conclusion I.6.13].

[56] A large exception to this is the study of quasiminimal classes (see Example 3.2.1.(10)) by Zilber and others, which are driven by questions from algebra.

[57] It should be noted that some prominent AEC theorists disagree with this as a motivating principle.

This interaction has the potential to go the other way as well: one can attempt to study a structure or a class of structures by determining where the first-order theory lies amongst the dividing lines and using the properties of forking there. However, if the class is not elementary, then the first-order theory captures new structures that have new definable objects. These definable objects might force the elementary class into a worse dividing line. However, AECs offer the potential to look at a narrower, better behaved class. For instance, an interesting class might only have the order property up to some length λ[58] or only be able to define short and narrow (but infinite) trees. Looking at the first-order theory loses this extra information and looking at the class as an AEC might move it from an unstable elementary class to a stable AEC.

4. **Reverse mathematics of tameness**
The compactness theorem of first-order logic is equivalent to a weak version of the axiom of choice (Tychonoff's theorem for Hausdorff spaces). If we believe that tameness is a natural principle, then maybe the first-order version of "tameness" is also, in the choiceless context, equivalent to some topological principle: what is this principle?

5. **How "natural" is tameness?**
We have seen that all the known counterexamples of tameness are pathological. Is this a general phenomenon? Are there natural mathematical structures that are, in some sense, well behaved and should be amenable to a model-theoretic analysis, but are not tame? Would this example then satisfy a weaker version of tameness?

6. **Categoricity and tameness**
We have seen that tameness helps with Shelah's Categoricity Conjecture, but there are still unanswered questions about eliminating the successor assumption and amalgamation property. For example, does the categoricity conjecture hold in fully $< \aleph_0$-tame and -type short AECs with amalgamation?

Going the other way, what is the impact of categoricity on tameness? Grossberg has conjectured that amalgamation should follow from high enough categoricity. Does something like this hold for tameness?

7. **On stable and superstable tame AECs**
From the work discussed in this survey, several more down-to-earth questions arise:

(a) Can one build a global independence relation in a fully tame and short superstable AEC? See also Question 5.8.

(b) Is there a stability spectrum theorem for tame AECs (i.e., a converse to Theorem 5.14)?

(c) In superstable tame AECs, can one develop the theory of forking

---

[58]Like the example of algebraically closed valued fields of rank one, Example 3.2.1.(15).

further, say by generalizing geometric stability theory to that context?

---

7   Historical remarks

7.1   Section 2

Abstract elementary classes were introduced by Shelah [She87a]; see Grossberg [Gro02] for a history. Shelah [She87a] (an updated version appears in [She09b, Chapter I]) contains most of the basic results in this Section 2, including Theorem 2.9. Notation 2.11 is due to Baldwin and is used in [Bal09, Chapter 14]. Galois types are implicit in [She87b] where Theorem 2.20 also appears. Existence of universal extensions (Lemma 2.25) is also due to Shelah and has a similar proof ([She99, I.2.2.(4)]).

Splitting (Definition 2.22) is introduced by Shelah in [She99, Definition 3.2]. Lemma 2.23 is [She99, Claim 3.3]. The extension and uniqueness properties of splitting (Theorem 2.24) are implicit in [She99] but were first explicitly stated by VanDieren [Van06, I.4.10, I.4.12]. The order property is first defined for AECs in [She99, Definition 4.3].

Definition 2.27 is implicit already in [SV99], but the amalgamation in μ is not assumed (only a weak version: the density of amalgamation bases). It first appears[59] explicitly (and is given the name "superstable"[60]) in [Gro02, Definition 7.12]. Limit models appear in [SK96, Definition 4.1] under the name "(θ, σ)-saturated". The "limit" terminology is used starting in [SV99]. The reader should consult [GVV] for a history of limit models and especially the question of uniqueness. Theorem 2.31 is due to VanDieren [Vanb].

Shelah's eventual categoricity conjecture can be traced back to a question of Łoś [Ło54], which eventually became Morley's categoricity theorem [Mor65]. See the introduction to [Vasf] for a history. Conjecture 2.32 appears as an open problem in [She78]. Theorem 2.33 is due to Shelah [She83a, She83b]. Conjecture 2.34 appears as [She00, Question 6.14]. Theorem 2.35 is the main result of [She99]. Theorem 2.36 appears in [SV99, Theorem 2.2.1] under GCH but without amalgamation. Assuming amalgamation (but in ZFC), the proof is similar. The proof of Shelah and Villaveces omits some details. A clearer version appears in [BGVV]. An easier proof exists if the categoricity cardinal has high-enough cofinality, see [She99, Lemma 6.3]. Question 2.37 is stated

---

[59] With minor variations: joint embedding and existence of a model in μ is not required.

[60] This can be seen as a somewhat unfortunate naming, as there are several potentially nonequivalent definitions of superstability in AECs. Some authors use "no long splitting chains", but this omits the conditions of amalgamation, no maximal models, and joint embedding in μ. Perhaps it is best to think of the definition as a weak version of having a good μ-frame.

explicitly as an open problem in [Bal09, Problem D.1.(2)]. Theorems 2.38 and 2.39 are due to VanDieren and the second author [VV, Section 7].

Good frames are the main concept in Shelah's book on classification theory for abstract elementary classes [She09b]. The definition appears at the beginning of Chapter II there, which was initially circulated as Shelah [She09a]. There are some minor differences with the definition we give here, see the notes for Section 5 for more. Question 2.40 originates in the similar question Baldwin asked for $L(Q)$ [Fri75, Question 21]. For AECs, this is due to Grossberg (see the comments around [She01a, Problem 5]). A version also appears as [She00, Problem 6.13]. Theorem 2.41 is due to Shelah [She09c, Theorem VI.0.2]. A weaker version with the additional hypothesis that the weak diamond ideal is not λ^{++}-saturated appears is proved in Shelah [She01a], see [She09b, Theorem II.3.7]. Corollary 2.42 is the main result of [She01a]. Theorem 2.43 is the main result of [She09b, Chapter II]. Claim 2.44 is implicit in [She09b, Discussion III.12.40] and a proof should appear in [Sheb]. Theorem 2.45 is due to Shelah and appears in [She09b, Section IV.4]. Shelah claims that categoricity in a proper class of cardinals is enough but this is currently unresolved, see [BVb] for a more in-depth discussion. Theorems 2.45, 2.46, and 2.47 appear in [She09b, Section IV.7]. However, we have not fully checked Shelah's proofs. A stronger version of Theorem 2.45 has been shown by VanDieren and the second author in [VV, Section 7], while [Vasc, Section 11] gives a proof of Theorems 2.46 and 2.47 (with alternate proofs replacing the hard parts of Shelah's argument).

7.2 Section 3

The version of Definition 3.1 using types over sets is due to the second author [Vas16b, Definition 2.22]. Type-shortness was isolated by the first author [Bon14c, Definition 3.3]. Locality and compactness appear in [BS08]. Proposition 3.4 is folklore. As for Proposition 3.3, the first part appears as [Bon14c, Theorem 3.5], the second and third first appear in Baldwin and Shelah [BS08], and the third is implicit in [She99] and a short proof appears in [Bal09, Lemma 11.5].

In the framework of AECs, the Galois Morleyization was introduced by the second author [Vas16b] and Theorem 3.5 is proven there. After the work was completed, we were informed that a 1981 result of Rosický [Ros81] also uses a similar device to present concrete categories as syntactic classes. That tameness can be seen as a topological principle (Theorem 3.6) appears in Lieberman [Lie11b].

On Section 3.2.1:

1. Tameness could (but historically was not) also have been extracted from the work of Makkai and Shelah on the model theory of $L_{\kappa,\omega}$, κ a strongly compact [MS90]. There, the authors prove that Galois types are, in some

sense, syntactic [MS90, Proposition 2.10][61]. The first author [Bon14c] generalized these results to AECs and later observations in [BTR, BU] slightly weakened the large cardinal hypotheses.

2. Theorem 3.8 is due to Shelah. The first part appears essentially as [She99, II.2.6] and the second is [She09b, IV.7.2]. The statement given here appears as [Vasc, Theorem 8.4].

3. Theorem 3.10 is essentially [She01b, Conclusion 3.7].

4. This is folklore and appears explicitly on [GK, p. 15].

5. The study of the classification theory of universal classes starts with [She87b] (an updated version appears as [She09c, Chapter V]), where Shelah claims a main gap for this framework (the details have not fully appeared yet). Theorem 3.11 is due to the first author [Bonc]. A full proof appears in [Vasf, Theorem 3.7].

6. Finitary AECs were introduced by Hyttinen and Kesälä [HK06]. That $\aleph_0$-Galois-stable $\aleph_0$-tame finitary AECs are $< \aleph_0$-tame is Theorem 3.12 there. The categoricity conjecture for finitary AECs appears in [HK11]. The beginning of a geometric stability theory for finitary AECs appears in [HK16].

7. Homogeneous model theory was introduced by Shelah in [She70]. See [GL02] for an exposition. The classification theory of this context is well developed, see, for instance [Les00, HS00, HS01, BL03, HLS05]. For connections with continuous logic, see [BB04, SU11].

8. Averageable classes are introduced by the first author in [Bona].

9. A summary of continuous first-order logic in its current form can be found in [BYBHU08]. Metric AECs were introduced in [HH09] and tameness there is first defined in [Zam12].

10. Quasiminimal classes were introduced by Zilber [Zil05]. See [Kir10] for an exposition and [BHHKK14] for a proof of the excellence axiom (and hence of tameness).

11. That the λ-saturated models of a first-order superstable theory forms an AEC is folklore. That it is tame is easy using that the Galois types are the same as in the original first-order class.

12. Superior AECs are introduced in [GK].

13. Hrushovski fusions are studied as AECs in [VZ].

14. This appears in [BET07].

15. This is analyzed in [Bonb].

The Hart-Shelah example appears in [HS90], where the authors show that it is categorical in $\aleph_0, \ldots, \aleph_n$ but not in $\aleph_{n+1}$. The example was later extensively analyzed by Baldwin and Kolesnikov [BK09] and the full version of

---

[61]This was another motivation for developing the Galois Morleyization.

Theorem 3.14 appears there. The Baldwin-Shelah example appears in [BS08]. The Shelah-Boney-Unger example was first introduced by Shelah [Shec] for a measurable cardinal and adapted by Boney and Unger [BU] for other kinds of large cardinals.

7.3 Section 4

The categoricity transfer for universal classes is due to the second author [Vasf] This section presents a proof incorporating ideas from the later paper [Vase]. If not mentioned otherwise, results and definitions there are due to the second author.

Lemma 4.5 is folklore when atomic equivalence is transitive but is [Vasf, Theorem 4.14] in the general case. As for Theorem 4.6, one direction is folklore. The other direction (tameness implies that the good frame can be extended) is due to the authors, see the notes on Theorem 5.26 below. The version with weak amalgamation (Theorem 4.7) is due to the second author.

Theorem 4.10 is due to Grossberg and VanDieren [GV06a]. Definition 4.11 is due to Shelah [She09b, Definition III.3.2]. The account of orthogonality and unidimensionality owes much to Shelah's development in [She09b, Sections III.2 and III.6] but differs in some technical points explained in details in [Vase]. Theorem 4.16 is due to Shelah [She09b, III.2.3]. Theorem 4.17 is due to Shelah with stronger hypotheses [She09b, Claim III.12.39] and to the second author as stated [Vase, Theorem 2.8].

7.4 Section 5

Question 1.3 is implicit in [She99, Remark 4.9]. A more precise statement appears as [BGKV16, Question 7.1].

The presentation of abstract independence given here appears in [Vasa][62]. The definition of a good frame given here (Definition 5.12) also appears in [Vasa, Definition 8.1]. Compared to Shelah's original definition ([She09b, Definition II.2.1]), the definition given here is equivalent [Vasa, Remarks 3.5, 8.2] except for three minor differences:

- The existence of a superlimit model is not required. This has been adopted in most subsequent works on good frames, including, e.g., [JS13].

- Shelah's definition requires forking to be defined for types over models only. However it is possible to close the definition to types over sets (see for example [BGKV16, Lemma 5.4]).

- Shelah defines forking only for a subclass of all types which he calls *basic*. They are required to satisfy a strong density property (if $M < N$, then there is a basic type over M realized in N). If the basic types are all the (nonalgebraic) types, Shelah calls the good frame *type-full*. In the tame

---

[62]There independence relations are not required to satisfy base monotonicity.

context, a type-full good frame can always be built (see [GV, Remark 5.6]). Even in Theorem 2.41, the frame can be taken to be type-full (see [She09b, Claim III.9.6]). The bottom line is that in all cases where a good frame is known to exist, a type-full one also exists.

Question 5.8 appears (in a slightly different form) as [BGKV16, Question 7.1]. Theorem 5.9 is Corollary 5.19 there[63]. As for Proposition 5.10, all are folklore except (2) which appears as [Vasa, Lemma 4.5] and symmetry which in this abstract framework is [BGKV16, Corollary 5.18] (in the first-order case, the result is due to Shelah [She78] and uses the same method of proof: symmetry implies failure of the order property).

Galois stability was defined for the first time in [She99]. The second part of Theorem 5.13 is due to Grossberg and VanDieren [GV06b, Corollary 6.4]. Later the argument was refined by Baldwin, Kueker, and VanDieren [BKV06] to prove the first part. Theorem 5.14 is due to the second author [Vas16b, Theorem 4.13].

Averages in the nonelementary framework were introduced by Shelah (for stability theory inside a model) in [She87b], see [She09c, Chapter V.A]. They are further used in the AEC framework in [She09b, Chapter IV]. The Galois Morleyization is used by the authors to translate Shelah's results from stability theory inside a model to tame AECs in [BVc, Section 4]. They are further used in [GV].

That tameness can be used to obtain a global uniqueness property for splitting (Theorem 5.15) is due to Grossberg and VanDieren [GV06b, Theorem 6.2]. $< \kappa$-satisfiability was introduced as κ-coheir in the AEC framework by Grossberg and the first author [BG]. This was strongly inspired from the work of Makkai and Shelah [MS90] on coheir in $L_{\kappa,\omega}$, κ a strongly compact. A weakening of Theorem 5.19 appears in [BG], assuming that coheir has the extension property. The version stated here is due to the second author and appears as [Vas16b, Theorem 5.15]. Theorem 5.20 is [BG, Theorem 8.2]. The definition of μ-forking (5.16) is due to the second author and appears in [Vas16a]. Theorem 5.17 is proven in [Vasa, Section 7]. Theorem 5.21 is due to the authors [BVc, Theorem 5.16].

Theorem 5.22.(1) is due to the second author [Vasa, Proposition 10.10]. Theorem 5.22.(2) is due to VanDieren and the second author [VV, Corollary 6.10] (an eventual version appears in [BVc], and an improvement of VanDieren [Van16] can be used to obtain the full result). Theorem 5.22.(3-4) are also due to VanDieren and the second author [VV], although (3) and (4) were observed by the second author in [Vas16a] in the categorical case (i.e., when we know that the union of a chain of λ-Galois-saturated models is λ-Galois-saturated).

Theorem 5.23 and Remark 5.25 are due to Grossberg and the second author [GV]. The notion of a superlimit model appears already in Shelah's original

---

[63]Of course the general idea of looking at forking as an abstract independence relation which turns out to be canonical is not new (see for example Lascar's proof that forking is canonical in superstable theories [Las76, Theorem 4.9]).

paper on AECs [She87a] (see [She09b, Chapter I]). Shelah introduces solvability in [She09b, Definition IV.1.4]. Lemma 5.24 appeared in an earlier version of [GV]. When κ is strongly compact, it can be traced back to Makkai-Shelah [MS90, Proposition 4.12].

Theorem 5.26 is due to the authors and appears in full generality in [BVd]. Rami Grossberg told us that he privately conjectured the result in 2006 and told it to Adi Jarden and John Baldwin (see also the account in the introduction to [Jar16]). In [Bon14b, Theorem 8.1], the first author proved the theorem with an additional assumption of tameness for *two* types used to transfer symmetry. Later, [BVd] showed that symmetry transfer holds without this extra assumption. At about the same time as [BVd] was circulated, Adi Jarden gave a proof of symmetry from tameness assuming an extra property called the continuity of independence (he also showed that this property followed from the existence property for uniqueness triples). The argument in [BVd] shows that the continuity of independence holds under tameness and hence also completes Jarden's proof.

Independent sequences were introduced by Shelah in the AEC framework [She09b, Definition III.5.2]. A version of Theorem 5.27 for models of size λ is proven as [She09b, III.5.14] with the assumption that the frame is weakly successful. This is weakened in [JS12], showing that the so-called continuity property of independence is enough. In [BVd], the continuity property is proven from tameness and hence Theorem 5.27 follows, see [BVd, Corollary 6.10].

Definition 5.28 is due to the second author [Vasa, Definition 8.1]. The definition of almost good (Definition 5.29) is implicit there and made explicit in [Vasf, Definition A.2]. Fully good and almost fully good are also defined there. Theorem 5.31 and Theorem 5.37 are due to the second author. A statement with a weaker Hanf number is the main result of [Vasa] (the proof uses ideas from Shelah [She09b, Chapter III] and Adi Jarden [Jar16]). The full result is proven in [Vasf, Appendix A]. What it means for a frame to be successful (Definition 5.32) is due to Shelah [She09b, Definition III.1.1] but we use the equivalent definition from [Vasa, Section 11]. Type locality (Definition 5.34) is introduced by the second author in [Vasa, Definition 14.9]. Corollary 5.39 and Theorem 5.40 is implicit in [Vasa] (with the Hanf number improvement in [Vasf, Appendix A]). Theorem 5.41 is due to the second author [Vasf, Appendix C].

Theorem 5.43 appears in [GV06c]. A version of Theorem 5.44 is already implicit in [Vasa, Section 10]. Shelah's omitting type theorem (Theorem 5.45) appears in its AEC version as [She99, Lemma II.1.6] and has its roots in [MS90, Proposition 3.3], where a full proof already appears. Corollary 5.46 is due to the second author [Vasc, Theorem 9.8]. The categoricity conjecture for tame AECs with primes (Theorem 5.47) is due to the second author (the result as stated here was proven in a series of papers [Vasf, Vase, Vasc], see [Vasc, Corollary 10.9]). The converse from Remark 5.48 is stated in [Vasb]. The categoricity conjecture for homogeneous model theory is more or less implicit in [She70] and is made explicit by Hyttinen in [Hyt98] (when the language is countable, this is due to Lessmann [Les00][64]). More precisely, Hyttinen proves that categoricity in a $\lambda > |T|$ with $\lambda \neq |T|^{+\omega}$ implies categoricity in all $\lambda' \geq \min(\lambda, h(|T|))$. Corollary 5.49 is stronger (as it includes the case $\lambda = |T|^{+\omega}$) and is due to the second author [Vase, Theorem 2]. Theorem 5.50 is due to the second author [Vasc, Corollary 10.6]. Theorem 5.51 is due to the second author (although the main idea is due to Shelah, and the only improvement given by tameness is the Hanf number, see Theorem 2.47). With full tameness and shortness, a weaker version appears in [Vasa, Theorem 1.6], and the full version using only tameness is [Vasc, Corollary 11.9]. The second part of Corollary 5.54 also appears there.

7.5   Section 6

Several of these questions have been in the air and there is some overlap with the list [Bal09, Appendix D]. The question about the length of tameness appears in the first author's Ph.D. thesis [Bon14a]. A question about examples of tameness and nontameness appear in [Bal09, Appendix D.2]. Whether failure of superstability implies many models is conjectured in [She99] (see the remark after Claim 5.5 there) and further discussed at the end of [GV].

The idea of exploring the reverse mathematics of tameness (and the specific question of what tameness corresponds to if compactness is the Tychnoff theorem for Hausdorff spaces) was communicated to the second author by Rami Grossberg. That tameness follows from categoricity was conjectured by Grossberg and VanDieren [GV06a, Conjecture 1.5]. That one can build a global independence relation in a fully tame and short superstable AEC is conjectured by the second author in [Vasa, Section 15].

---

[64] In that case, a stronger statement can be proven: if D is categorical in some uncountable cardinal, then it is categorical in all uncountable cardinals.

Bibliography

[Bal09]   John T. Baldwin, *Categoricity*, University Lecture Series, vol. 50, American Mathematical Society, 2009.

[BB04]   Alexander Berenstein and Steven Buechler, *Simple stable homogeneous expansions of Hilbert spaces*, Annals of Pure and Applied Logic **128** (2004), 75–101.

[BET07]   John T. Baldwin, Paul C. Eklof, and Jan Trlifaj, $^{\perp}N$ *as an abstract elementary class*, Annals of Pure and Applied Logic **149** (2007), 25–39.

[BG]   Will Boney and Rami Grossberg, *Forking in short and tame AECs*, Preprint. URL: http://arxiv.org/abs/1306.6562v9.

[BGKV16]   Will Boney, Rami Grossberg, Alexei Kolesnikov, and Sebastien Vasey, *Canonical forking in AECs*, Annals of Pure and Applied Logic **167** (2016), no. 7, 590–613.

[BGVV]   Will Boney, Rami Grossberg, Monica VanDieren, and Sebastien Vasey, *Superstability from categoricity in abstract elementary classes*, Annals of Pure and Applied Logic **168** (2017), no. 7, 1383–1395.

[BHHKK14]   Martin Bays, Bradd Hart, Tapani Hyttinen, Meeri Kesälä, and Jonathan Kirby, *Quasiminimal structures and excellence*, Bulletin of the London Mathematical Society **46** (2014), no. 1, 155–163.

[BK09]   John T. Baldwin and Alexei Kolesnikov, *Categoricity, amalgamation, and tameness*, Israel Journal of Mathematics **170** (2009), 411–443.

[BKL]   John T. Baldwin, Martin Koerwien, and Michael C. Laskowski, *Disjoint amalgamation in locally finite AEC*, The Journal of Symbolic Logic **82** (2017), no. 1, 98–119.

[BKV06]   John T. Baldwin, David Kueker, and Monica VanDieren, *Upward stability transfer for tame abstract elementary classes*, Notre Dame Journal of Formal Logic **47** (2006), no. 2, 291–298.

[BL03]   Steven Buechler and Olivier Lessmann, *Simple homogeneous models*, Journal of the American Mathematical Society **16** (2003), no. 1, 91–121.

[Bona]   Will Boney, *The Γ-ultraproduct and averageable classes*, Preprint. URL: http://arxiv.org/abs/1511.00982v3.

[Bonb]   Will Boney, *Some model theory of classically valued fields*, In preparation.

[Bonc]   Will Boney, *Tameness in groups and similar AECs*, In preparation.

[Bon14a]   Will Boney, *Advances in classification theory for abstract elementary classes*, Ph.D. thesis, 2014.

[Bon14b]   Will Boney, *Tameness and extending frames*, Journal of Mathematical Logic **14** (2014), no. 2, 1450007.

[Bon14c]   Will Boney, *Tameness from large cardinal axioms*, The Journal of Symbolic Logic **79** (2014), no. 4, 1092–1119.

[BS08]   John T. Baldwin and Saharon Shelah, *Examples of non-locality*, The Journal of Symbolic Logic **73** (2008), 765–782.

[BTR]   Andrew Brooke-Taylor and Jiří Rosický, *Accessible images revisited*, Proceedings of the American Mathematical Society **145** (2017), 1317–1327

[BU]   Will Boney and Spencer Unger, *Large cardinal axioms from tameness*, To appear in Proceedings of the American Mathematical Society.

[BVa]   Will Boney and Monica VanDieren, *Limit models in strictly stable abstract elementary classes*, Preprint. URL: http://arxiv.org/abs/1508.04717v3.

[BVb]   Will Boney and Sebastien Vasey, *Categoricity and infinitary logics*, Preprint. URL: http://arxiv.org/abs/1508.03316v2.

[BVc]   Will Boney and Sebastien Vasey, *Chains of saturated models in AECs*, Archive for Mathematical Logic **56** (2017), no. 3, 187–213.

[BVd]   Will Boney and Sebastien Vasey, *Tameness and frames revisited*, to appear in Journal of Symbolic Logic.

[BYBHU08]   Itay Ben-Yaacov, Alexander Berenstein, C. Ward Henson, and Alexander Usvyatsov, *Model theory for metric structures*, Model theory with applications to algebra and analysis (Zoé Chatzidakis, Dugald Macpherson, Anand Pillay, and Alex Wilkie, eds.), vol. 2, Cambridge University Press, 2008, pp. 315–427.

[EM02]   Paul C. Eklof and Alan Mekler, *Almost free modules: set-theoretic methods*, revised ed., North Holland Mathematical library, vol. 46, North-Holland, 2002.

[Fri75]   Harvey M. Friedman, *One hundred and two problems in mathematical logic*, The Journal of Symbolic Logic **40** (1975), no. 2, 113–129.

[GK]   Rami Grossberg and Alexei Kolesnikov, *Superior abstract elementary classes are tame*, Preprint. URL: http://www.math.cmu.edu/~rami/AtameP.pdf.

[GL02]   Rami Grossberg and Olivier Lessmann, *Shelah's stability spectrum and homogeneity spectrum in finite diagrams*, Archive for Mathematical Logic **41** (2002), no. 1, 1–31.

[Gro]   Rami Grossberg, *A course in model theory I*, A book in preparation.

[Gro02]   Rami Grossberg, *Classification theory for abstract elementary classes*, Contemporary Mathematics **302** (2002), 165–204.

[GV]   Rami Grossberg and Sebastien Vasey, *Equivalent definitions of superstability in tame abstract elementary classes*, to appear in Journal of Mathematical Logic.

[GV06a]   Rami Grossberg and Monica VanDieren, *Categoricity from one successor cardinal in tame abstract elementary classes*, Journal of Mathematical Logic **6** (2006), no. 2, 181–201.

[GV06b]   Rami Grossberg and Monica VanDieren, *Galois-stability for tame abstract elementary classes*, Journal of Mathematical Logic **6** (2006), no. 1, 25–49.

[GV06c]   Rami Grossberg and Monica VanDieren, *Shelah's categoricity conjecture from a successor for tame abstract elementary classes*, The Journal of Symbolic Logic **71** (2006), no. 2, 553–568.

[GVV]   Rami Grossberg, Monica VanDieren, and Andrés Villaveces, *Uniqueness of limit models in classes with amalgamation*, Mathematical Logic Quarterly 62 (2016) 367–382.

[Har75]   Victor Harnik, *On the existence of saturated models of stable theories*, Proceedings of the American Mathematical Society **52** (1975), 361–367.

[HH09]   Åsa Hirvonen and Tapani Hyttinen, *Categoricity in homogeneous complete metric spaces*, Archive for Mathematical Logic **48** (2009), 269–322.

[HK06]   Tapani Hyttinen and Meeri Kesälä, *Independence in finitary abstract elementary classes*, Annals of Pure and Applied Logic **143** (2006), 103–138.

[HK11]   Tapani Hyttinen and Meeri Kesälä, *Categoricity transfer in simple finitary abstract elementary classes*, The Journal of Symbolic Logic **76** (2011), no. 3, 759–806.

[HK16]   Tapani Hyttinen and Kaisa Kangas, *Quasiminimal structures, groups and Zariski-like geometries*, Annals of Pure and Applied Logic **167** (2016), no. 6, 457–505.

424   *Bibliography*

[HL02]   Tapani Hyttinen and Olivier Lessmann, *A rank for the class of elementary submodels of a superstable homogeneous model*, The Journal of Symbolic Logic **67** (2002), no. 4, 1469–1482.

[HLS05]   Tapani Hyttinen, Olivier Lessmann, and Saharon Shelah, *Interpreting groups and fields in some nonelementary classes*, Journal of Mathematical Logic **5** (2005), no. 1, 1–47.

[HS90]   Bradd Hart and Saharon Shelah, *Categoricity over P for first order T or categoricity for $\phi \in \mathbb{L}_{\omega_1,\omega}$ can stop at $\aleph_k$ while holding for $\aleph_0, \ldots, \aleph_{k-1}$*, Israel Journal of Mathematics **70** (1990), 219–235.

[HS00]   Tapani Hyttinen and Saharon Shelah, *Strong splitting in stable homogeneous models*, Annals of Pure and Applied Logic **103** (2000), 201–228.

[HS01]   Tapani Hyttinen and Saharon Shelah, *Main gap for locally saturated elementary submodels of a homogeneous structure*, The Journal of Symbolic Logic **66** (2001), no. 3, 1286–1302.

[Hyt98]   Tapani Hyttinen, *Generalizing Morley's theorem*, Mathematical Logic Quarterly **44** (1998), 176–184.

[Jar16]   Adi Jarden, *Tameness, uniqueness triples, and amalgamation*, Annals of Pure and Applied Logic **167** (2016), no. 2, 155–188.

[Jec03]   Thomas Jech, *Set theory*, 3rd ed., Springer-Verlag, 2003.

[JS12]   Adi Jarden and Alon Sitton, *Independence, dimension and continuity in non-forking frames*, The Journal of Symbolic Logic **78** (2012), no. 2, 602–632.

[JS13]   Adi Jarden and Saharon Shelah, *Non-forking frames in abstract elementary classes*, Annals of Pure and Applied Logic **164** (2013), 135–191.

[Kir10]   Jonathan Kirby, *On quasiminimal excellent classes*, The Journal of Symbolic Logic **75** (2010), no. 2, 551–564.

[Kue08]   David W. Kueker, *Abstract elementary classes and infinitary logics*, Annals of Pure and Applied Logic **156** (2008), 274–286.

[Las76]   Daniel Lascar, *Ranks and definability in superstable theories*, Israel Journal of Mathematics **23** (1976), no. 1, 53–87.

[Les00]   Olivier Lessmann, *Ranks and pregeometries in finite diagrams*, Annals of Pure and Applied Logic **106** (2000), 49–83.

[Lie11a]   Michael J. Lieberman, *Category-theoretic aspects of abstract elementary classes*, Annals of Pure and Applied Logic **162** (2011), no. 11, 903–915.

[Lie11b]   Michael J. Lieberman, *A topology for Galois types in abstract elementary classes*, Mathematical Logic Quarterly **57** (2011), no. 2, 204–216.

[Ło54]   Jerzy Łoś, *On the categoricity in power of elementary deductive systems and some related problems*, Colloquium Mathematicae **3** (1954), no. 1, 58–62.

[Mor65]   Michael Morley, *Categoricity in power*, Transactions of the American Mathematical Society **114** (1965), 514–538.

[MS90]   Michael Makkai and Saharon Shelah, *Categoricity of theories in* $\mathbb{L}_{\kappa,\omega}$, *with* κ *a compact cardinal*, Annals of Pure and Applied Logic **47** (1990), 41–97.

[MS94]   Menachem Magidor and Saharon Shelah, *When does almost free imply free?*, Journal of the American Mathematical Society **7** (1994), no. 4, 769–830.

[Ros81]   Jiří Rosický, *Concrete categories and infinitary languages*, Journal of Pure and Applied Algebra **22** (1981), no. 3, 309–339.

[Shea]   Saharon Shelah, *A.E.C. with not too many models*, Preprint. URL: http://arxiv.org/abs/1302.4841v2.

[Sheb]   Saharon Shelah, *Eventual categoricity spectrum and frames*, Paper number 842 on Shelah's publication list. Preliminary draft from Oct. 3, 2014 (obtained from the author).

[Shec]   Saharon Shelah, *Maximal failure of sequence locality in A.E.C.*, Preprint. URL: http://arxiv.org/abs/0903.3614v3.

[She70]   Saharon Shelah, *Finite diagrams stable in power*, Annals of Mathematical Logic **2** (1970), no. 1, 69–118.

[She78]   Saharon Shelah, *Classification theory and the number of non-isomorphic models*, Studies in logic and the foundations of mathematics, vol. 92, North-Holland, 1978.

[She83a]   Saharon Shelah, *Classification theory for non-elementary classes I: The number of uncountable models of* $\psi \in \mathbb{L}_{\omega_1,\omega}$. *Part A*, Israel Journal of Mathematics **46** (1983), no. 3, 214–240.

[She83b]   Saharon Shelah, *Classification theory for non-elementary classes I: The number of uncountable models of* $\psi \in \mathbb{L}_{\omega_1,\omega}$. *Part B*, Israel Journal of Mathematics **46** (1983), no. 4, 241–273.

[She87a]   Saharon Shelah, *Classification of non elementary classes II. Abstract elementary classes*, Classification Theory (Chicago, IL, 1985) (John T. Baldwin, ed.), Lecture Notes in Mathematics, vol. 1292, Springer-Verlag, 1987, pp. 419–497.

[She87b]   Saharon Shelah, *Universal classes*, Classification theory (Chicago, IL, 1985) (John T. Baldwin, ed.), Lecture Notes in Mathematics, vol. 1292, Springer-Verlag, 1987, pp. 264–418.

[She99]   Saharon Shelah, *Categoricity for abstract classes with amalgamation*, Annals of Pure and Applied Logic **98** (1999), no. 1, 261–294.

[She00] Saharon Shelah, *On what I do not understand (and have something to say), model theory*, Math. Japonica **51** (2000), 329–377.

[She01a] Saharon Shelah, *Categoricity of an abstract elementary class in two successive cardinals*, Israel Journal of Mathematics **126** (2001), 29–128.

[She01b] Saharon Shelah, *Categoricity of theories in* $\mathbb{L}_{\kappa,\omega}$*, when* κ *is a measurable cardinal. Part II*, Fundamenta Mathematica **170** (2001), 165–196.

[She09a] Saharon Shelah, *Categoricity in abstract elementary classes: going up inductively*, Classification theory for abstract elementary classes, College Publications, 2009, Paper number 600 in Shelah's publication list, pp. 224–377.

[She09b] Saharon Shelah, *Classification theory for abstract elementary classes*, Studies in Logic: Mathematical logic and foundations, vol. 18, College Publications, 2009.

[She09c] Saharon Shelah, *Classification theory for abstract elementary classes 2*, Studies in Logic: Mathematical logic and foundations, vol. 20, College Publications, 2009.

[SK96] Saharon Shelah and Oren Kolman, *Categoricity of theories in* $\mathbb{L}_{\kappa,\omega}$*, when* κ *is a measurable cardinal. Part I*, Fundamentae Mathematica **151** (1996), 209–240.

[SU11] Saharon Shelah and Alexander Usvyatsov, *Model theoretic stability and categoricity for complete metric spaces*, Israel Journal of Mathematics **182** (2011), 157–198.

[SV] Saharon Shelah and Andrés Villaveces, *Categoricity may fail late*, Preprint. URL: http://arxiv.org/abs/math/0404258v1.

[SV99] Saharon Shelah and Andrés Villaveces, *Toward categoricity for classes with no maximal models*, Annals of Pure and Applied Logic **97** (1999), 1–25.

[Vanb] Monica VanDieren, *Superstability and symmetry*, Annals of Pure and Applied Logic **168** (2017), no. 3, 651–692.

[Van06] Monica VanDieren, *Categoricity in abstract elementary classes with no maximal models*, Annals of Pure and Applied Logic **141** (2006), 108–147.

[Van16] Monica VanDieren, *Symmetry and the union of saturated models in superstable abstract elementary classes*, Annals of Pure and Applied Logic **167** (2016), no. 4, 395–407.

[Vasa] Sebastien Vasey, *Building independence relations in abstract elementary classes*, to appear in Annals of Pure and Applied Logic.

[Vasb] Sebastien Vasey, *Building prime models in fully good abstract elementary classes*, to appear in Mathematical Logic Quarterly.

[Vasc]   Sebastien Vasey, *Downward categoricity from a successor inside a good frame*, Preprint. URL: http://arxiv.org/abs/1510.03780v4.

[Vasd]   Sebastien Vasey, *Saturation and solvability in abstract elementary classes with amalgamation*, Preprint. URL: http://arxiv.org/abs/1604.07743v3.

[Vase]   Sebastien Vasey, *Shelah's eventual categoricity conjecture in tame AECs with primes*, Preprint. URL: http://arxiv.org/abs/1509.04102v3.

[Vasf]   Sebastien Vasey, *Shelah's eventual categoricity conjecture in universal classes. Part I*, to appear in Mathematical Logic Quarterly.

[Vasg]   Sebastien Vasey, *Shelah's eventual categoricity conjecture in universal classes. Part II*, Selecta Mathematica **23** (2017), no. 2, 1469–1506.

[Vas16a]   Sebastien Vasey, *Forking and superstability in tame AECs*, The Journal of Symbolic Logic **81** (2016), no. 1, 357–383.

[Vas16b]   Sebastien Vasey, *Infinitary stability theory*, Archive for Mathematical Logic **55** (2016), 567–592.

[VV]   Monica VanDieren and Sebastien Vasey, *Symmetry in abstract elementary classes with amalgamation*, Archive for Mathematical Logic **56** (2017), no. 3, 423–452.

[VZ]   Pedro H. Zambrano and Andrés Villaveces, *Hrushovski fusions and tame abstract elementary classes*, Preprint.

[Zam12]   Pedro H. Zambrano, *A stability transfer theorem in d-tame metric abstract elementary classes*, Mathematical Logic Quarterly **58** (2012), 333–341.

[Zil05]   Boris Zilber, *A categoricity theorem for quasi-minimal excellent classes*, Logic and its applications (Andreas Blass and Yi Zhang, eds.), Contemporary Mathematics, American Mathematical Society, 2005, pp. 297–306.

Printed and bound by CPI Group (UK) Ltd, Croydon, CR0 4YY

24/10/2024

01778284-0010